Mammals of the Northern Great Plains

J. Knox Jones, Jr.

David M. Armstrong

Robert S. Hoffmann   *University of Nebraska Press, Lincoln & London*

and Clyde Jones

Mammals of

the Northern

Great Plains

Frontispiece, Black-Tailed Prairie Dog

Copyright 1983 by the University of Nebraska Press

All rights reserved
Manufactured in the United States of America

The paper in this book meets the guidelines for
permanence and durability of the Committee on
Production Guidelines for Book Longevity of the
Council on Library Resources

Library of Congress Cataloging in Publication Data

Main entry under title:

Mammals of the Northern Great Plains

Bibliography: p.
Includes index.
1. Mammals – Great Plains.   I. Jones, Jr., J. Knox.
QL719.G73M35   599.0978   82-2693
ISBN 0-8032-2557-1   AACR2

39, 421

Contents

Figures

Acknowledgments

Work on this book was initiated in 1968 in a graduate seminar under the direction of J. Knox Jones, Jr., at the University of Kansas. In 1969 Robert S. Hoffmann became a co-sponsor of the course. A number of graduate students participated in the seminar and made contributions to this handbook in its formative stages, gathering pertinent literature or drafting preliminary accounts of species. In approximate order of their contribution to the project, those students were: D. M. Armstrong, J. R. Choate, T. H. Kunz, J. W. Koeppl, R. E. Smith, R. K. Rose, R. W. Turner, L. C. Watkins, J. J. Pizzimenti, R. L. Robbins, J. B. Bowles, R. P. Lampe, H. H. Genoways, P. B. Robertson, C. B. Rideout, E. C. Birney, C. J. Phillips, J. P. Farney, R. J. Cinq-Mars, S. M. Kortlucke, J. E. Pefaur, and A. Cadena-G. After several years of dormancy, the project was reinstated in 1977. A contract from the United States Fish and Wildlife Service supported final stages of preparation of the manuscript. All accounts of species have been written or rewritten since 1977, and every attempt has been made to continue to revise accounts on the basis of new literature that came to our attention through 1980. A few publications dated 1981 and 1982 also have been cited.

In recent years, several graduate students have served the project as research assistants. They are James C. Halfpenny, Michael L. Johnson, and Richard L. Stucky (University of Colorado); Gregory E. Glass and Gary McGrath (University of Kansas); and Wm. David Webster and Carole J. Young (Texas Tech University).

For assistance with typing or for other logistical support we thank Mary Lou Eppelheimer, Peggy Johnson, and Deborah Roth (University of Colorado); Jan Elder, Nancy Kreigbaum, Rose Etta Kurtz, Coletta Spencer, and Peggy Wessel (University of Kansas); Sandra Boatman, Betty Johnson, Joyce Langston, Margaret Meter, Shirley Neunaber, and Mary Ann Seaman (Texas Tech University); and Cathy Blount, Robert Fisher, and Mary Layman (Denver Wildlife Research Center).

Line drawings were executed by Debra K. Bennett and Thomas H. Swearingen (University of Kansas). Photographs generously were made available to us by a number of colleagues and agencies, and we thank them collectively here; individual acknowledgments are given in figure legends. Roger W. Barbour of the University of Kentucky, Robert R. Patterson of the University of Kansas, Daniel Q. Thompson of the United States Fish and Wildlife Service, and John L. Tveten of Houston, Texas, were especially helpful in providing photographic assistance.

Personnel of state wildlife agencies and various federal wildlife units in the tristate region have provided unpublished information on species under their care, for which we are grateful. In particular, we should mention Karl Menzel, Nebraska Game and Parks Commission, C. R. Grondahl, North Dakota Game and Fish Department, and

Ron Fowler, South Dakota Department of Game, Fish, and Parks. Philip L. Wright and Chester Loveland (University of Montana) provided unpublished data on *Glaucomys sabrinus*, and Clarence Wales of Lawrence, Kansas, supplied information on current auction prices for pelts of fisher and marten.

Measurements of some species have been taken from major faunal works and systematic revisions. In addition, we have freely used the collections of the Museum of Natural History of the University of Kansas, University of Colorado Museum, the Museum of Texas Tech University, and the National Museum of Natural History. A number of mammalogists have reviewed part or all of the drafts for various sections of the manuscript. For this effort we especially thank Robert W. Seabloom of the University of North Dakota, J. Freeman of the University of Colorado, Richard Curnow of the Denver Wildlife Research Center, and D. L. Pattie of the Northern Alberta Technical Institute (Edmonton). Theodore M. Klein of the Department of Classical and Romance Languages, Texas Tech University, provided valuable assistance in tracing the derivation of scientific names.

Introduction

For the purposes of this book, the Northern Great Plains are defined as the states of Nebraska, North Dakota, and South Dakota. As a physiographic concept, the northern part of the great interior grasslands of North America is, of course, much broader in geographic extent than the Dakotas and Nebraska, but the three states lie in the heart of the region, and thus the title for this work seems appropriate.

Our expectations in writing *Mammals of the Northern Great Plains* were to provide a comprehensive, yet semitechnical, treatment of free-living mammals that would prove useful to specialist and nonspecialist alike. The content and style therefore were developed in a way we hope will interest the inquiring high-school student, on the one hand, and provide a point of reference for the professional mammalogist on the other. Between these extremes, wildlife biologists, conservationists, environmental specialists, college students of vertebrate zoology, and others interested in mammalian natural history should find the present treatment useful for their needs as well.

Historical Background

The earliest published references to mammals on the Northern Great Plains appeared in accounts written by explorers, fur traders, and missionaries who visited the region in the 1700s. It was not until the nineteenth century, however, that systematic exploration of the Northern Plains was begun, sponsored primarily, but not entirely, by the United States government after the purchase of the Louisiana Territory from France in 1803.

The first federally sponsored expedition was that headed by Captain Meriweather Lewis and Captain William Clark, who led a party up the Missouri River on their way to the Pacific coast, beginning in 1804, and back again in 1806. Their journals provide a wealth of information on mammals along the Missouri, and several species occurring on the Northern Plains (*Antilocapra americana, Cynomys ludovicianus,* and *Neotoma cinerea*) were first named and described on the basis of specimens and notes resulting from the trip. Others who traversed the Missouri in the early years and who contributed information about mammalian inhabitants of the region included John Bradbury and Henry M. Brackenridge, both in 1811, Prince Maximilian of Wied-Neuwied in 1833–34, and the well-known naturalist John J. Audubon in 1843.

In 1819–20, Major Stephen H. Long led an exploration party, which included naturalists Thomas Say and Edwin James, across the plains of Nebraska to the Rocky Mountains. In the course of the journey they discovered the mountain peak now bearing the major's name. The party camped over winter on the Missouri River just north of present-day Omaha, Nebraska, at the "Engineer Cantonment." James's later account of the expedition contains a list of mammals that occurred around the winter quarters, including original descriptions by Say of such species as *Blarina brevicauda*, *Canis latrans*, and *Cryptotis parva*, as well as information on mammals encountered elsewhere on the journey. In 1823 Long undertook another expedition, again accompanied by Say, to the United States–Canadian boundary along the Red River, but only a few mammals were seen or taken on this venture. Much of the early history of mammals in the Red River country comes from the accounts of Alexander Henry, who was in charge of the Northwest Company's trading posts there from 1800 to 1808.

After discovery of gold in California in 1848, increased military presence on the Northern Plains resulted in continued exploration and, thus, additional information on the natural history of the region. It was during this period that the first mammalogical exploration of the Black Hills took place, beginning with the Evans and Harney expeditions of 1852 and 1855, respectively. The most important early exploration of that area, however, was by a party headed by Lt. G. K. Warren in 1857, during which scientist F. V. Hayden made observations on 44 species of mammals. These activities culminated in the several decades following the Civil War, principally under the aegis of the federal government's Pacific Railroad surveys. The many specimens of mammals and other animals resulting from these endeavors ultimately found their way to the Smithsonian Institution, and the former served as a partial basis for S. F. Baird's (1858) monumental treatment of mammals (exclusive of bats) of North America, Harrison Allen's (1864, 1894) monographs on the Chiroptera, and the updated compendium on lagomorphs and rodents by Coues and Allen (1877).

In the last part of the nineteenth century and early in the present one, the newly organized operation that was to become the famous Bureau of Biological Survey of the United States Department of Agriculture supported a systematic survey of the distribution of the terrestrial vertebrate fauna of North America. In connection with this survey, various individuals, some of whom became famous names in American mammalogy, visited the Northern Plains in quest of specimens. Among these were H. E. Anthony, Vernon Bailey, Merritt Cary, Remington Kellogg, and E. A. Preble, and many mammals they collected were reported in the *North American Fauna* series published by the agency. Collectors from the American Museum of Natural History also were active in the late 1800s in the tristate region, especially on the Black Hills and in adjacent areas, and a number of specimens thus obtained were recorded in publications by the illustrious ornithologist and mammalogist J. A. Allen.

Vernon Bailey's interest in the Northern Plains culminated in publication in 1927 of his work on the mammals of North Dakota, the only comprehensive statewide treatment of that group. Later, Genoways and Jones (1972) summarized mammalian distribution to the south and west of the Missouri River in the southwestern part of that state, and Wiehe and Cassel (1978) recently have provided an updated checklist of species occurring in North Dakota.

No complete work on South Dakotan mammals has appeared, although Over and Churchill (1945) did issue an annotated checklist (mimeographed), based mostly on the published literature, as Choate and Jones (1982) did more recently. However, several publications dealing with parts of that state are of interest—Andersen and Jones (1971) on Harding County in the northwest, Findley (1956) on Clay County in the southeast, Turner (1974) on the Black Hills, and Wilhelm, Choate, and Jones (1981) on the important ecological area around Lacreek National Wildlife Refuge in Bennett County.

For Nebraska, Swenk (1908) first summarized the whole of the mammalian fauna and, more recently, Jones's *Distribution and Taxonomy of Mammals of Nebraska*

(1964) listed 81 native and five introduced species. Five kinds of mammals have been reported as new to Nebraska and one has been dropped from the list for that state since publication of the latter work (see Jones and Choate 1980).

In relatively recent years, collections of mammals from the three states have begun to accumulate at regional institutions. These mammalian specimens and the field notes accompanying them will provide a rich store of data for those who wish to pursue future studies of mammals occurring on the Northern Great Plains. Such collections may be found at (but are not limited to) the University of North Dakota, North Dakota State University, South Dakota State University, the University of Nebraska, and Kearney (Nebraska) State College, as well as in the Field Museum of Natural History (Chicago) and in museums at the universities of Kansas, Michigan, and Minnesota. We hope this book will stimulate additional work on Northern Plains mammals by pointing up gaps in knowledge of distribution and ecology of included taxa. Furthermore, environmental impact studies, principally in the western part of the region to date (see, for example, Nagel

et al. 1974 and Seabloom, Crawford, and McKenna 1978), doubtless will result in accumulation of additional information and materials for research, as will the work of state and federal agencies charged with management of game and fur-bearing species, protection of endangered animals, and related activities.

Finally, any work of this sort must take into account publications dealing with faunas in nearby regions. Fortunately, relatively recent studies of mammals are available for adjacent states—Hazard (1982) for Minnesota, Bowles (1975) for Iowa, Schwartz and Schwartz (1981) for Missouri, Bee et al. (1981) for Kansas, Armstrong (1972) for Colorado, Long (1965) for Wyoming, Hoffmann and Pattie (1968) for Montana—and these volumes were regularly consulted, as were later, more specific references. No comparable summaries have been published for the Prairie Provinces of Canada, but Beck's (1958) list of Saskatchewan mammals and Soper's (1961) treatment of those in Manitoba proved useful, as did Banfield's (1974) *The Mammals of Canada* and the work by Hall (1981) on North America as a whole.

Organization and Treatment

In the accounts that follow, species of each genus are listed in alphabetical order. Genera, families, and orders are arranged in conventional phylogenetic sequence, and treatment of these higher taxa is deliberately brief. Readers desiring more detail should consult the synopsis by Anderson and Jones (1967) for orders and families and that of Walker et al. (1964 and subsequent editions) for genera. Both scientific and vernacular names of species generally follow Jones, Carter, and Genoways (1979).

Information in the accounts of species is organized under five headings: Name, Distribution, Description, Natural History, and Selected References. In the first section, comment is made on the derivation of the

scientific name of the species; often, alternative vernacular names are provided.

The section on distribution describes the general geographic and ecological range of a species. Subspecies (if recognized) are listed in this section. The geographic ranges of most species are mapped, based on the currently known distribution; the maps, however, are deliberately conservative, and additional fieldwork in various parts of the region certainly will extend the known limits of many mammals. In several cases where too few specimens of a species have been reported to allow the distribution in the tristate region to be shaded with confidence, only the actual localities of record are indicated. At the scale used, range maps are inevitably imprecise, and the reader will need to interpret the distribution as mapped with due attention to the ecological tolerance of each species. For example, the beaver and muskrat are shown as occurring throughout the region, whereas they obviously are restricted to riparian habitats. The former distributions of those species extirpated from the three states since settlement are not mapped, nor are the ranges of introduced mammals.

Descriptions, coupled with information in keys to species, should allow positive identification of most specimens in hand. In some instances, however, it is difficult even for specialists to distinguish between closely related species in the field (species of *Sorex, Myotis, Sylvilagus, Perognathus, Peromyscus,* and *Microtus,* for example). The observer may need to consult particulars of both geographic and ecological distribution to enable tentative identification or, rarely, may need to be content with a generic determination. Descriptions of color are informal. Conventional external measurements are provided in millimeters. For detailed instructions on taking such measurements of mammals, see Hall (1962, 1981) or DeBlase and Martin (1974, 1980). Cranial measurements are not presented in detail; each account includes only some measure of length and breadth of the skull, also in millimeters. Occasionally, additional cranial measurements are presented in keys (all measurements and details of the mammalian skull and lower jaw are defined or figured in the Glossary). In cases in which geographic variation in size is pronounced within the tristate region, measurements of two or more populations may be given; usually, however, measurements of a single sample suffice to characterize a species. We have included a photograph of each species where available. At several points in the text, line drawings of cranial, dental, or external features are provided as an aid to identifying species. For illustrations of skulls of all species, the reader should consult Hall (1981).

The section on natural history includes a summary of what is known of the biology of the species: ecology, behavior, reproduction, development, molt, food habits, predators, parasites, and so on. Wherever possible, data were derived from studies conducted on the Northern Plains or in immediately adjacent states or provinces. When studies farther afield are mentioned, the reader should interpret the information with caution, because details of the biology of mammalian species differ considerably from place to place. That accounts of species vary in length is in part because less is known of the biology of some members of our native fauna than others. Also, we have tended to emphasize somewhat those species that are particularly familiar or especially widespread in the region.

Because this handbook is directed at a broad, general audience, literature citations are not made in text, except in introductory sections and occasionally in ordinal accounts. We have done this to make the text readable, and we realize it may offend some professional mammalogists. Selected references at the end of each account are not meant to provide an exhaustive list of literature consulted and, in fact, have been held to a minimum. They frequently represent general studies and reviews in which additional references will be found. With this in mind, accounts from the series *Mammalian Species,* published by the American Society of Mammalogists and intended as concise but complete accounts of each species treated, have been cited wherever available. Persons interested in subscribing to *Mammalian Species* should write to Dr. Gordon L. Kirkland, Secretary-Treasurer, American Society of Mammalogists, % The Vertebrate Museum, Shippensburg State College, Shippensburg, Pennsylvania 17257.

We have attempted to restrict technical vocabulary to the minimum consistent with accuracy. Technical terms used repeatedly are defined in the Glossary. Also, except for millimeters, in which we record all external and cranial measurements, we have tended to use common English units of measure rather than units of the metric system (some of which are entered in the Glossary).

Study of Mammals

Wild mammals are not easy to study in the field. Many are nocturnal in habits, and many are cryptically colored. Most are secretive, and some are perceptive enough to avoid man much of the time. Still, the study of mammals has its rewards. The native fauna is interesting in its own right, and a better understanding of the adaptations of our mammalian kin can lead to a new appreciation of our own biological heritage. Advances in knowledge of the ecology and behavior of mammals demand both keen observational skills and patience. Therefore a corps of dedicated amateur naturalists could add considerably to the fund of information available in this discipline.

Sometimes mammals can be observed directly. This is especially true of diurnal kinds (such as squirrels), but some crepuscular species also can be watched satisfactorily with equipment no more sophisticated than a pair of binoculars or a spotting scope. Many nocturnal mammals continue to behave normally when illuminated with red light, such as from a flashlight or battery-powered lantern covered with heavy red cellophane.

Frequently, important evidence of mammals is more subtle than that gained by direct observation. Tracks provide reliable information on the presence of some species and also may indicate population densities. With practice, a good tracker can estimate the size of the animal from its tracks and can ascertain the direction and rate of travel. Tracks usually are encountered in native habitats, but some workers prepare tracking pits using fine, moist sand. Commercial trapping scents or carrion baits can be used to attract a variety of furbearers to scent posts, and baits such as peanut butter or grains will attract rodents. Voucher specimens of tracks can be preserved as plaster of paris casts. For identifying tracks of mammals, the handbook by Murie (1954) is the classic in its field, and is recommended highly.

Another common sign of mammals is scat, or fecal droppings. Murie's (1954) book also is an aid to identification of scat, which provides important evidence of the presence of animals. Rough indexes of population size have been developed based on counts of fecal pellets over a period of time. Also, scat can be analyzed as a key to food habits. Microscopic structure of leaf fragments, insect exoskeletons, and mammalian hairs allow close identification of prey items. In this regard, a key by Moore, Spence, and Dugnolle (1974) will aid in identification of dorsal guard hairs of most of the mammals found on the Northern Great Plains.

The distribution of mammalian species is documented by specimens preserved in museums. A standard specimen consists of a museum "study skin" accompanied by a cleaned skull. A complete specimen includes a label with precise notation of locality of capture, sex and reproductive condition, date, and the name and field number of the collector. Standard external measurements are recorded on the label: total length, length of tail vertebrae, length of hind food from heel to tips of claws, and length of ear from notch. Partial specimens (such as odd skulls found in the field), if accompanied by accurate locality data, also may be useful. Complete skeletons and entire specimens preserved in fluid are important additions to museum collections, allowing for study of the hard and the soft anatomy. Field notes are a valuable adjunct to any specimens preserved for future study. For instructions on preparing museum specimens, see Hall (1955, 1962, 1981), Hall and Kelson (1959), or DeBlase and Martin (1974, 1980). Each of these references also includes extensive comments on the preparation of field notes.

A well-prepared museum specimen with complete field data, accompanied by thoughtful field notes and preserved in a well-curated museum collection, represents a wealth of biological information. Not only does it document a locality of occurrence at some point in time, but it provides data on reproduction, anatomy, pelage and molt, geographic and individual variation, microevolution, and even such unlikely subjects as ectoparasitism (lice and ticks are especially prone to cling to their hosts through the preservation process), dental pathology, and environmental chemistry (inasmuch as certain pollutants may accumulate in fur or skin). Papers in the anthology by Banks (1979) suggest the broad scope of museum-based research on wildlife.

Specimens may be obtained from highway casualties or taken deliberately by trapping, shooting, or other means. For details on collecting methods, see DeBlase and Martin (1974, 1980). In any event, collecting of scientific specimens is carefully controlled in most states. Prospective collectors should contact the appropriate state wildlife agency about current regulations.

The regurgitated pellets of owls, and to a lesser extent those of hawks, are a source of specimens that should not be overlooked. Owls feed by swallowing their prey whole. Bones and fur are packed into a tight bolus and regurgitated from the gizzard. These birds are efficient predators, and first records of such small mammals as shrews and pocket mice from a particular locality frequently are taken from their pellets. Keys and descriptions in this book should assist in identification of most mammalian skulls. If in doubt about the identity of remains from owl pellets or other sorts of specimens, do not hesitate to ask at the biology department of the near-

est major college or university. Mammalogists generally are eager to help others learn more about the native fauna, because in so doing they learn more about it themselves.

Research on mammals often is not directly involved with collecting for taxonomic, distributional, or anatomical study. In investigations of ecology and behavior, for example, it may be desirable to interfere as little as possible with natural activities. Most such studies do involve identification of individual animals, however. In some instances individuals can be recognized by color pattern, size, scars, or other peculiar marks. In other cases the animals are captured and marked (with ear tags or dye patterns, for example) for future recognition. Capture involves use of a wide array of live traps, many of which are described and evaluated by DeBlase and Martin (1974, 1980). Techniques for studying populations of game species have been reviewed by Schemnitz (1980).

As a general rule, researchers involved with field or laboratory studies of native mammals should prepare museum specimens of representatives of each species studied. The time invested may seem great at the moment, but in the long run preparation of voucher specimens is inexpensive insurance. As indicated earlier, some kinds of mammals are difficult to identify, and many hours of painstaking field or laboratory work may be questioned, or come to naught, if the identity of organisms studied cannot later be verified with certainty. With voucher specimens, properly prepared and preserved in a museum, such problems as may arise can easily be resolved.

Environment

There is a widespread misconception that the Northern Great Plains are a monotonous and featureless region with little environmental diversity. This unfortunate error arises partly because much of the region now is cropland, the original vegetation having been replaced with the monoculture of grains that makes the Northern Plains an important part of the "breadbasket of the world." Also, most visitors to the region travel on major east-west highways. Because these highways tend to follow river valleys, the ecological contrast of the uplands often is bypassed. In addition, some of the environmental pattern is subtle. One must look beyond the seeming monotony to see the real landscape. Some understanding of ecological pattern is important to those who really wish to understand the mammals of the Northern Great Plains, because ecological pattern, whether regional or local, is reflected in the distribution of mammalian species.

Ecological pattern—interrelated patterns in vegetation, climate, soils, physiography, drainage—may be considered a pattern of resources for the fauna, a pattern of opportunity. We will look at several layers of pattern in turn. There is a danger in doing this, however, for the environment is not made up of independent factors—as we shall see. Rather, one should attempt to visualize the environment as an interacting whole, dynamic in space and time. Thus, the Northern Great Plains can be thought of as a mosaic, a patchwork of interdependent ecosystems—an ecosystem being a volume of environment containing populations of living organisms and nonliving (abiotic) elements. Materials are exchanged between the living and nonliving parts of an ecosystem, powered by a flow of solar energy. Whole ecosystems are unfathomably complex, but such is the structure of life on the macroscopic level.

Vegetation

Plants are essential to mammalian life. Directly or indirectly, they provide food. For many mammals they provide shelter as well. The plant component of an ecosystem—the vegetation—reflects the climate and soil of an area and, in turn, helps mold the abiotic environment. Because vegetation is readily observed, often it is the easiest factor on which to base a quick description of an ecosystem. When we speak, for example, of a deciduous riparian ecosystem, an image comes to mind of a tunnel-like canopy of cottonwoods, willows, and elms arching over a meandering prairie stream. Perhaps we need to be reminded that such an ecosystem also includes a characteristic alluvial soil, other plants, fungi and microorganisms, a wide variety of invertebrates, and several dozen species of native mammals, birds, reptiles, amphibians, and fishes.

To describe the actual vegetation at a given time is no easy task. It may, in fact, be misleading, for the situation at any particular time may be relatively short-lived. Imagine a hypothetical, yet typical, sequence of events in eastern Nebraska. In the 1860s settlers moved into the rich bottomland, felled the trees, and burned slash and stumps. A typical field then was planted to corn and some vegetable crops. The plot remained a basis for subsistence agriculture until about 1890, when it was abandoned after a flood. By 1917 the area had become densely covered with thickets and small trees, but land was in short supply and so the brush was cleared, this time with machinery, and the plot planted to corn again. In the 1930s the area was allowed to go fallow. It was quickly invaded by weeds, followed by some hardy annuals, then perennials, and finally shrubs. It was cleared again and planted to pasture grasses. But cattle could not quite keep ahead of the tree and shrub seedlings, so the field was burned every other year. Eventually it was neglected long enough that trees reestablished themselves, and the field ultimately became valuable forested parkland in a flood-protection zone adjacent to an expanding suburb.

In the sequence described above, succession—partly in response to human intervention, partly in response to natural circumstances—resulted in continual change for a century or more. How would one describe the vegetation of the site? The area supported in turn bottomland forest, crops, herbs, thickets, grassland, and finally another forest. As the vegetation changed, so did the resource base for mammalian populations.

To obviate the problems of succession and human interference, Küchler (1964) developed the concept of potential natural vegetation—that is, the vegetation to be expected in a given area if the influence of man is removed and succession telescoped into a single instant. This concept is not easy to apply rigorously, of course, especially in an area managed thoroughly for human ends. It is a process of informed speculation, which demands thorough acquaintance with successional patterns, physical factors of site quality, and characteristics of relict stands of undisturbed vegetation.

Figure 1 is a simplification of Küchler's map of the potential natural vegetation of Nebraska and the Dakotas. His actual map included 13 major stand types in the tri-state region, but that is more detail than we need at present. A pristine picture of the Northern Plains reveals vast expanses of grassland interrupted by ribbons of deciduous forest along the major rivers and their tributaries. In eastern Nebraska, the riparian woodland is dominated by several species of oaks and hickories. To the north and west, the floodplain forest is characterized by plains cottonwood, willow, and American elm. These forests have a fairly rich understory of shrubs. Probably they are more nearly continuous along watercourses now than when the plains first were settled, because fire and flood are controlled by man to some extent over much of the area. Grazing is common in riparian woodlands, and domestic stock tramples and eats seedlings. Hence many stands are overmature, consisting mostly of old trees. Still, the riparian forest is of profound importance to mammalian distribution, providing narrow corridors of what is essentially eastern woodland habitat that stretch westward into arid areas to which some species would not otherwise be adapted. Man's habit of planting trees in cities and as shelterbelts and ornamentals on farmsteads also has extended the influence of deciduous forest ecosystems on the plains.

Forests of other kinds occur on the few areas of pronounced relief that mark the Northern Great Plains. For example, the Turtle Mountains and the Pembina Hills in northern North Dakota are covered with a woodland of aspen, as well as a savanna of widely scattered bur oak interspersed with tall grasses.

Woodland also occurs on the uplands of

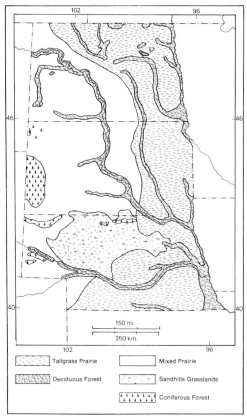

1. Potential natural vegetation of the Northern Great Plains (modified after Küchler 1964).

the western part of the tristate region. On the escarpments of the badlands of the Little Missouri River in North Dakota and the Cheyenne and White rivers of South Dakota, as well as along the crest of the Pine Ridge, Wildcat Ridge, and the breaks of the Niobrara River in Nebraska, are beautiful woodlands of ponderosa pine. They seem to occur on the ridgetops partly because the ridges afford protection from prairie fires and also provide good drainage. The understory in these woodlands includes wheatgrass, blue grama, and needlegrass.

Still better developed is the coniferous forest of the Black Hills. The dominant tree species on the Hills is ponderosa pine, the dark green needles of which appear black from a distance and give this unique landform its name (from the original Sioux words *paha sapa*). Associated with pon-

derosa pine are a variety of shrubs, including big sagebrush, mountain mahogany, rabbit brush, and skunkbush sumac. Aspen also occurs on the Black Hills, and lodgepole and limber pines occur in local stands. On the northern part of the Hills, forests of white spruce are present, especially in cool moist canyons and on north-facing slopes. Canyons of the eastern foothills of the Black Hills support species of eastern deciduous trees, including bur oak, American elm, and hop hornbeam, as well as cottonwood and box elder.

The grasslands of the Northern Great Plains are complex and have been the subject of a vast and contentious literature. *The Phytogeography of Nebraska* (Pound and Clements 1898) is a classic of American plant ecology, a pioneering work in distributional ecology and successional theory. Work on the Northern Plains, especially in Nebraska, influenced the development of F. C. Clements's ideas on plant succession and climax communities. This tradition was continued in the detailed accounts by Weaver (1954) and Weaver and Albertson (1956). This is not the place to review those works and the literature they inspired. For our purposes it probably is enough to realize that the grasslands of the Northern Great Plains are not all alike. The true prairie of the more humid eastern third of the region is an environment quite different from that of the plains grassland to the west, which we might term "mixed-grass prairie," because it includes tall, mid, and short grasses. Furthermore, the grasslands of the Nebraska Sand Hills are sufficiently peculiar to warrant description on their own.

In the eastern third of Nebraska and east of the Missouri Coteau in the Dakotas, the native grassland is the tall-grass, true prairie ecosystem. In the Red River Valley, big and little bluestem, panic grass, and Indian grass predominate. In the James River Valley and farther west and southward on the loess hills of central Nebraska, the dominant grasses are western wheatgrass, big bluestem, and needlegrass. This is a transitional grassland, with elements of both tallgrass and mixed-grass prairies. In south-central Nebraska the grassland also tends toward mixed prairie, with blue grama and sideoats grama becoming dominants and numerous species of forbs becoming prominent.

West of the Missouri Coteau in the Dakotas and west and south of the Nebraska Sand Hills, true mixed prairie predominates. Most of the grasses are shorter than in the more humid prairies to the east, and the grasses tend to be more sparse. Many species are bunchgrasses. Blue grama is the predominant species, mixed with western wheatgrass and needlegrass. In drought years taller grasses are limited to swales, but they migrate upslope when moister periods allow. Overgrazing favors the spread of pricklypear cactus and yucca. Where the thin bunchgrass sod is broken by the plow, or bison, or prairie dogs, blowouts can form, leading eventually to dustbowl conditions.

Despite the fact that they lie in the semi-arid country west of the 99th meridian, the Sand Hills of Nebraska—because of their high water table—support a rich tall-grass flora, with strong eastern affinities, mixed with species peculiar to sandy soils. Dominant grasses are big bluestem, little bluestem, sand bluestem, sand reed, and needlegrass. South of the Platte River, sand sage becomes a codominant on the sand and loess hills, and yucca also is common.

Vegetation of a site provides a handy point of reference for speaking of whole ecosystems. Also, vegetation integrates the physical resources of a site—moisture, soil, physiography—and makes the area habitable for mammals. This is particularly true of smaller species such as rodents and shrews. Given some practice, a naturalist can look at a particular patch of habitat and predict with a high degree of confidence just what small mammals will be found there.

Physiography and Drainage Patterns

Physiography is the branch of the earth sciences concerned with description of landforms. A physiographic view of a region is important to understanding the distribution of life, because landforms reflect the regional history of the earth, providing a sort of summary of geologic materials and processes. Physiography affects the distribution of soils and surface water; it affects, and is affected by, the climate. These factors, in turn, help shape the vegetation upon which mammals depend.

The commonly accepted physiographic scheme for the United States was summarized by Fenneman (1931, 1938). Nebraska and the Dakotas all lie in the Interior Plains Physiographic Division. This division is subdivided as follows:

A. Central Lowland Province
   1. Western Lake Section
   2. Dissected Till Plains Section
B. Great Plains Province
   1. Glaciated Missouri Plateau Section
   2. Unglaciated Missouri Plateau Section
   3. Black Hills Section
   4. High Plains Section
   5. Plains Border Section

These physiographic sections differ in geologic history and general ecology and in the opportunity they afford for mammalian life.

Distinguishing the Great Plains from the Central Lowlands is not easy, except in North Dakota, where the boundary is the eastern edge of the Missouri Plateau. The Great Plains are covered with a mantle of Tertiary sediments eroded from the young Rocky Mountains and deposited primarily by braided streams. In the intervening years these sediments have been carried away as the streams of the Missouri River drainage cut headward. Hence the border is a dynamic one. Furthermore, it is obscured in Nebraska and South Dakota by wind-borne sand and loess.

Despite its physiographic obscurity, the boundary between the Central Lowlands and the Great Plains is an important one; it can be seen in subtle features of the landscape. The boundary corresponds roughly with the 20-inch isopleth of precipitation. East of that line it is more humid. Because of the lower humidity and higher elevation to the west, the sky seems bluer and higher; this is "big sky country." East of the boundary, towns are closer together and farms are smaller; there are more highways and railroads. Counties are smaller. Crops are mainly corn and soybeans in the south and small grains and potatoes northward. West of the line, farms give way to ranches. Some of the land is tilled for wheat, but much of it is used for grazing. In the west, beef cattle and sheep are the predominant livestock, whereas to the east it is feedlot and dairy cattle, and hogs. Moving westward across the tristate region, these changes occur gradually between 99 and 101 degrees west longitude. Similar changes occur in the native ecology of the area; the cultural lines seen in *The National Atlas* are real lines in the landscape, reflected in the native biota.

The Central Lowlands Province includes those parts of the Dakotas drained by the Souris, James, and Red rivers and the eastern one-fourth of Nebraska. The northern part of this region lies in the so-called Western Lake Section. Parts of the section are remarkably flat, for they are basins of Pleistocene lakes: Lake Agassiz in the Red River Valley (of which lakes Winnipeg and Manitoba are remnants), Lake Souris, now the Souris River Valley and the Souris Plain, and Lake Dakota, in what is now the James River Valley. The Turtle Mountains, rising to an elevation of some 2,300 feet above sea level, represent an outlying remnant of a larger Missouri Plateau of Tertiary age. South of these plains, in extreme eastern South Dakota and in eastern Nebraska, is the Dissected Till Plains Section, which is hilly country, the relief cut into glacial drift of Kansan age. Much of the landscape of the Northern Plains is a product of the Pleistocene ice ages. Generally speaking, the features in the eastern

part of the region are remnants of the latest glaciation, the Wisconsin. In eastern Nebraska, however, there are features that date back to Illinoian and Kansan times and to an even earlier glacial advance, the Nebraskan.

The Missouri Plateau constitutes the largest part of the Great Plains Province in the tristate region. The course of the Missouri River follows roughly the southern limit of Wisconsin glaciation. The river thus divides the Missouri Plateau into two sections, glaciated to the north and east, unglaciated to the south and west. The southern limit of the Missouri Plateau is the Pine Ridge of Nebraska. The unglaciated Missouri Plateau Section is characterized by broad river valleys and dissected uplands. Where the uplands are of shale, as along the Cheyenne, White, and Little Missouri rivers, badlands have formed. Badlands are typical of arid areas where soft sediments are positioned well above local base level. The badlands of the Dakotas are important to the distribution of mammalian species.

Before the Pleistocene, the Missouri Plateau seems to have been a coherent physiographic unit. Glaciation changed the character of the northern and eastern sections, cutting down hills, filling valleys, and leaving behind a magnificent moraine, the Missouri Coteau, a line of stone-capped hills running diagonally across North Dakota and southward into South Dakota west of the James River.

South of the Missouri Plateau lies the High Plains Section, the largest subdivision of the Great Plains Province (although not the most extensive in the tristate region). The High Plains are a remnant of a much larger plain that once extended from the Rocky Mountains to the Central Lowlands. Sediments were deposited in Tertiary times by slow-moving braided streams. The general surface of the High Plains is some of the flattest terrain on earth. The Llano Estacado, or Staked Plains, of West Texas and eastern New Mexico—the southern end of the High Plains—was so named because early travelers left lines of stakes behind them so they could find their way back

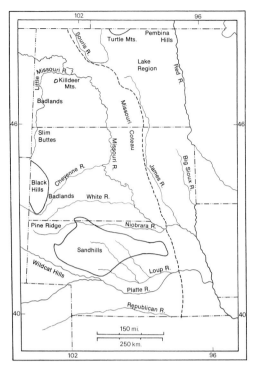

2. *Some important physiographic features of the Northern Great Plains.*

across the featureless landscape. In Nebraska and southern South Dakota, the High Plains do not appear quite so flat because the Tertiary Plain is overlain with Quaternary sand and loess. An exemplary flat landscape remains in southwestern Nebraska as the Cheyenne Table, however. The Nebraska Sand Hills are a magnificent area of some 24,000 square miles. The dunes—which range to hundreds of feet high—are largely stabilized, and in the swales between them are lush grasslands. East of the Sand Hills, the contact between the Great Plains and the Central Lowlands is obscured by loess, deposits of wind-borne silt tens of feet deep. This is another flat landscape, and one with rich soils.

As mentioned above, the High Plains once were more extensive than they are today. Headward cutting by streams has caused the westward retreat of the High Plains and has left a distinctive physiographic section, the Plains Border, between the High Plains and the Central Lowlands. This process is visible in ex-

treme southern Nebraska, in the Republican River Valley. To the north, the boundary between the High Plains and the Plains Border is obscured by loess hills.

Several uplands in the tristate region are referred to as mountains, but only the Black Hills of western South Dakota are worthy of the name from a geologic standpoint. The Black Hills are an elliptical dome about 125 miles long and 60 miles wide, rising some 4,000 feet above the general level of the Missouri Plateau. Harney Peak, at 7,242 feet, is the highest point in the three states. The geologic history of the Black Hills is complex. Several episodes of mountain uplift have occurred in the area, traceable back to Cambrian times. The current dome was uplifted from the late Cretaceous to the Eocene. The ancient Precambrian core pushed up through many tens of thousands of feet of younger sedimentary rocks, standing them on edge and exposing them to weathering. The result is a mountain range with a granitic core surrounded by concentric rings of progressively younger rocks, limestones, sandstones, and shales. The elevation of the Black Hills allows a fairly high effective precipitation, which supports a vegetation of coniferous forest. Hence the Black Hills are an island of mountain habitat surrounded by a sea of grassland. This is of considerable importance to mammalian distribution, and some species of mammals are restricted in the tristate region to the Black Hills. The limestones of these unique mountains have eroded so that they are a major area of caverns, which has influenced the distribution and abundance of several species of bats.

The drainage patterns on the Northern Great Plains are fundamentally important to mammalian distribution. Through geologic time, the rivers (along with glaciers and wind) have carved much of the topography. Rivers provide water to support stands of riparian woodland that thread across the prairies as narrow peninsulae of eastern wooded habitat. This is important to mammalian diversity, especially in the western part of the region. Some mammals—beaver, muskrat, and mink, for example—depend directly on the rivers and other bodies of water for habitat, and many others—such as the fox squirrel, white-footed mouse, and white-tailed deer—depend on the woodland that follows rivers.

The master stream of most of the tristate region is, of course, the Missouri River, which heads in the mountains of western Montana and flows in a sinuous diagonal direction through North and South Dakota and along the eastern border of Nebraska. The Yellowstone joins the Missouri just inside the North Dakota boundary, and the Little Missouri heads near Devil's Tower and courses north around the Killdeer Mountains to join the Missouri. As we have noted, the Missouri River generally follows the southern edge of the more recent continental glaciers. This causes an interesting pattern in the tributary streams. Tributaries from the west are long, and many flow almost due east across the plains. The Heart, Cannonball, Grand, and Moreau rivers drain the unglaciated Missouri Plateau; the Belle Fourche and Cheyenne rivers drain the Black Hills. The White River flows from the nothern slope of the Pine Ridge, and the Niobrara gathers its headwaters from the southern slope, flowing thence along the north edge of the Sand Hills. The North Platte River heads in North Park, Colorado, and flows northward around the Laramie Mountains before turning southeastward to enter Nebraska in the grasslands of Goshen Hole. The South Platte heads in South Park, Colorado, and flows north and east to join the North Platte. The Platte marks the southern boundary of the Nebraska Sand Hills and flows eastward the breadth of Nebraska. The north-central part of that state is drained by the Loup River system, whereas much of the southern part is drained by another eastward-flowing stream, the Re-publican, that eventually turns south to join the Missouri after first merging with the Kansas River.

Western tributaries of the Missouri are mature streams, hundreds of miles long. Eastern tributaries, by contrast, are short streams only a few thousand years old, heading on the moraines of Wisconsin-age glaciers. The James and Big Sioux rivers drain eastern South Dakota; they are the exceptions to the rule about short tributaries on the east, and they are further peculiar because they once flowed northward to Hudson Bay via the Red River and Lake Winnipeg, that is, into glacial Lake Agassiz.

Eastern North Dakota retains its ice-age drainage pattern, the Red River flowing northward to Lake Winnipeg. The so-called Souris Plain also drains northward, via the Souris (or Mouse) River, which joins the Assiniboine and eventually the Red River.

Generally speaking, the Northern Great Plains is a region of moderate relief. Divides between rivers usually rise gradually, to heights of at most a few hundred feet above the valleys. Yet this is enough, especially in the semiarid western part of the three states, to dictate substantial change in vegetation and in other components of ecosystems, including mammals. Sometimes uplands prove impassable barriers to animals. At other times the barriers may be passable, but only barely so. The barriers slow gene flow, and populations on either side are able to differentiate from one another. Patterns of geographic variation, sometimes formalized in subspecific nomenclature, often correspond with hydrologic patterns. Hence rivers have an evolutionary role in addition to their fundamentally important ecological one.

Soils

Soils influence mammalian distribution in several ways. They are the dynamic result of vegetation, climate, physiography, and geologic parent material interacting over time. Soils and vegetation are interdependent, and we already have seen how vegetation influences mammalian distribution. Soils also influence mammals directly. For example, many mammals are burrowers. Some (the eastern mole and the pocket gophers) are highly adapted to a fossorial life. Others, such as the ground squirrels, pocket mice, and kangaroo rats, burrow extensively but forage aboveground. Still other species, the deer mouse and hispid cotton rat, for example, live in subterranean homes but do little excavating of their own. On the featureless prairie, a tunnel in the soil may be the only protection available from hazards such as predators, inclement weather, and fire. Soils affect mammals, and mammals affect soils. The burrowing animals turn over the soil, exposing to weathering materials that had been below the surface. Also, animal wastes and the remains of the mammals themselves decompose and return to the store of soil nutrients.

Because they are dynamic and because they are susceptible to disturbance by man, direct and indirect, soils vary greatly from place to place. This variation is seen in particle size and in physical and organic soil chemistry. Such local variation can affect mammalian populations. A soil too coarse or too wet will be unsuitable to plains pocket gophers. Too much organic matter means poor dust bathing for northern grasshopper mice. If there are too many rocks, pocket mice cannot build proper burrows. Too much clay impedes drainage, pines will not grow, and porcupines cannot make a good living. Some details of the fit between soils and mammals are given in accounts of individual species. Here, however, concern is with broad, regional patterns. Such patterns are fairly obvious in the field, and they correspond well with other environmental features discussed here.

In describing soils of the Northern Great Plains, we risk some confusion. Soil classification in the United States was revolutionized with the publication by the Department of Agriculture of *Soil Classification: A Comprehensive System, Seventh Approximation* (Soil Survey Staff 1960).

The new system differs fundamentally from earlier systems with which many readers may be familiar, for it classifies soils as they occur in the field and does not involve soil-forming processes. There is no one-to-one correspondence between higher categories of soil classification in the old and new systems. Nonetheless, rough equivalence will be suggested to aid communication; although technically incorrect, it may be useful.

Soils of the eastern part of the tristate region are dark and rich-looking, high in organic matter. Formerly such soils were termed chernozem, from the Russian for "black soil." Such black soils now are classified in the order Mollisols. Mollisols of North and South Dakota east of the Missouri Coteau are Borolls, cool soils, with an average annual temperature of less than 47° F. Such soils are suited to small grains. In the Red River Valley, Aquolls predominate; such soils (commonly referred to as humic gley) are saturated seasonally with water. The alluvial soils of the floodplains of the major rivers westward on the plains also are Aquolls. The Mollisols of eastern Nebraska are warmer than those to the north; these "cornbelt" soils are termed Udolls. West of the Missouri Coteau—west of the 99th to 100th meridian—soils are paler and brownish. These Ustolls (formerly chestnut soils) are subject to intense seasonal drought in summer.

The Sand Hills of Nebraska represent another order of soils, the Entisols, soils showing no pedogenic horizons. Soils of the Sand Hills reveal little evidence of weathering beyond the parent material; such sandy soils are termed Psamments (formerly included among the regosols). Soils of the badlands also are Entisols. Here, however, the parent material is shale. These poorly developed soils are typical Entisols, the Orthents. On the Black Hills, Alfisols (suborder Boralfs) prevail. These are typical brownish mountain soils, developed under forest cover, with clay accumulation in the subsurface layers.

Climate

The Northern Great Plains lie squarely in the "temperate zone," midway between the Equator and the North Pole. Yet the climate of the region is hardly "temperate," being one of pronounced extremes—hot in summer and cold in winter. This is because Nebraska and the Dakotas are about midway between the Atlantic and Pacific oceans and thus lie far from the meliorating effects of maritime air masses. The Northern Plains are exemplary of what climatologists term a "continental climate."

Another factor that influences the climate of the tristate region is elevation. Lowest points in all three states are along their eastern boundaries (Missouri River, southeastern corner of Nebraska, 840 feet; Big Stone Lake, northeastern South Dakota, 962 feet; Red River, northeastern North Dakota, 750 feet). The land surface in each of the three states rises gradually toward the west, reaching highest elevations along their western boundaries (Cheyenne Table in Kimball County, Nebraska, 5,424 feet; Harney Peak, South Dakota, 7,242 feet; White Butte, North Dakota, 3,506 feet). The higher the elevation, the thinner the layer of atmosphere. Hence there is progressive diminution from east to west across the plains in each of the components of the atmosphere, such as water vapor. Water vapor absorbs the sun's heat during the day and releases it at night. Thus, it acts as a temperature buffer, one that is increasingly less effective westward across the three states, resulting in ever wider daily fluctuations in temperature.

Another factor in the climate of the Northern Great Plains is the "rain shadow" effect. Pacific air masses tend to enter the region from the north-northwest in winter, passing over the Rocky Mountains. The mountains force the air up, thus cooling it and condensing the moisture it carries, causing precipitation on the western side of the mountains. The plains to the east of the mountains tend to be dry, but this effect is lessened eastward across the tri-

state region (a pattern that is also reflected in vegetation and soil).

The mean frost-free growing season in southeastern Nebraska is more than 160 days; in parts of North Dakota it is hardly more than 100 days. Mean annual temperature at Omaha, Nebraska, is 51.5° F, the January mean is 22.6° F, and the July mean is 77.2° F. For Sioux Falls, South Dakota, and Bismark, North Dakota, those figures are 45.4° F, 14.2° F, 73.3° F, and 41.4° F, 8.2° F, and 70.8° F, respectively. In most summers, temperatures above 100° F are recorded in all three states, and winter lows below −40° F are not infrequent, especially in the Dakotas.

Whereas temperatures are zoned from south to north, precipitation is zoned from east to west, ranging from well over 30 inches per year in eastern Nebraska to less than 12 inches per year in northwestern South Dakota. In all three states, most precipitation comes during the growing season; June is the wettest month, and December and January are the driest months. In contrast to the prevailing winds in winter from the north and west, those in summer are from the south-southeast, laden with moisture from the Gulf of Mexico.

As noted, the climate of the Northern Great Plains is one of extremes. In conjunction with vegetation and soils, it helps over time to shape the fauna. To visualize this, think of the number of burrowing mammals on the plains: eastern mole, black-tailed prairie dog, ground squirrels, pocket gophers, pocket mice, kangaroo rat, grasshopper mouse, swift fox, and badger, for example. These and other mammals escape the severity of the prairie climate by tunneling in the earth (and some species regularly use the abandoned burrow systems of others). Hibernation (or at least periods of inactivity) in winter, aestivation in summer, and annual migrations also are adaptations of some mammals to climatic rigors.

The climate of the Northern Plains is

variable on a scale of hours, days, weeks, and months. It also has been variable over longer periods. Poorly understood cycles of drought, for example, have recurred every 20 to 30 years since settlement, and the climatic fluctuations of the Pleistocene—ice ages alternating with interglacials on a schedule of tens of millennia—have shaped the environment of much of the region. Whatever their scale, climatic fluctuations have strongly influenced the composition of the mammalian fauna. Diel changes dictate patterns of sleep and wakefulness; the march of seasons cues reproduction, molt, and hibernation in a variety of species; and longer climatic cycles contribute to cyclic abundance of organisms, stress the ecological tolerances of species, and lead to shifts in distribution, some of which have been dramatic.

Mammalian Communities

The living components of an ecosystem constitute a biotic community—the organisms occurring together at any particular time and place. Mammals are important to each of the native ecosystems of the Northern Great Plains, and some mammalian species have adapted to living in man-made ecosystems as well.

The mammalian community is structured as a complex web of interactions. Herbivorous mammals eat plants and are themselves eaten by carnivores. Omnivorous mammals eat both plant and animal material, including carrion left as animals die. And, of course, the web of mammalian interactions—predation, competition, commensalism—is but a fraction of the overall network of symbiotic relationships in the biotic community. Owls and weasels compete for voles; ants and pocket mice compete for seeds. Fly larvae and opossums compete for carrion, and the opossum feeds on maggots as the flea and the tapeworm feed on the opossum. A fox squirrel preys on a robin's eggs, only to be preyed upon by a goshawk. Community interactions form an intricate, shifting web across time and space.

The sum of the resources, both biotic and abiotic, appropriated by a species population constitutes the niche of that population. No two co-occurring species have the same niche. If their resource needs were identical, whenever one of the species gained a reproductive advantage it would drive the other to local extinction. However, closely related species, those with a fairly recent ancestor in common, frequently have similar niches. Such species tend not to occur together geographically, however, but to have complementary distributions. An example from our own area is the distribution of the eastern cottontail (*Sylvilagus floridanus*) and Nuttall's cottontail (*S. nuttallii*). These two species have similar requirements but are nowhere found together, being eastern and western in distribution, respectively. Their ranges closely approach each other along the western edge of the Northern Plains from northwestern North Dakota southward to the Black Hills. In other cases, two species with similar niches may live in the same geographic area but occupy different habitats. Examples of this are the distributional patterns of the deer mouse (*Peromyscus maniculatus*) and the white-footed mouse (*P. leucopus*), and the desert cottontail (*Sylvilagus audubonii*) and the eastern cottontail. In each pair of species the former is a mammal of open country whereas the latter occurs in deciduous forest or riparian situations.

A given population of a species occurs where its resource needs are adequately met. Where predators and competitors are few and resources are plentiful, the species likely will prosper. Where competition or predation is strong or resources are marginal, the species population may be marginal as well, perhaps given to local extinction in particularly lean years.

All mammalian species have limits of tolerance. There are biotic and physical bounds beyond which they cannot live. No species—even the most tolerant—can be truly ubiquitous. Because species have limits, because they have peculiar biological needs and challenges, their ecological distributions are fairly predictable. Species with similar limits of tolerance, with similar but compatible niches, tend to occur together in communities. With some experience one can look at a particular piece of landscape, judge its capacity for supporting populations, and predict with a fair measure of accuracy the mammals that will be found there.

The following kinds of mammals are typical of the mixed-grass prairie community: black-tailed jackrabbit, desert cottontail, spotted ground squirrel, thirteen-lined ground squirrel, black-tailed prairie dog, silky pocket mouse, hispid pocket mouse, Ord's kangaroo rat, plains harvest mouse, northern grasshopper mouse, swift fox, black-footed ferret, badger, pronghorn, and bison.

Where sagebrush occurs with mixed-grass prairie, forming a shrub-steppe type of vegetation, such species as Merriam's shrew, Wyoming ground squirrel, northern pocket gopher, and sagebrush vole join with members of the mixed-grass prairie community to form a distinctive shrub-steppe mammalian community.

The tall-grass prairie community includes some species more typical of the drier, mixed prairie to the west but is characterized by several species that occur only locally outside the tall-grass region. Among those are the eastern mole, least shrew, Franklin's ground squirrel, plains pocket gopher, western harvest mouse, hispid cotton rat, and prairie vole. In prairie wetlands, such species as the masked shrew, meadow vole, muskrat, southern bog lemming, meadow jumping mouse, and mink come into prominence.

The distinctive grasslands of the Nebraska Sand Hills have a rich mammalian fauna, sharing species with both the tall-grass prairie to the east and the mixed-grass prairie to the west. Some species, such as

the plains pocket mouse, seem to reach their peak of abundance on the Sand Hills.

The eastern deciduous forest affords opportunities for mammals quite distinct from those of biotic communities elsewhere in the tristate region. Some species are restricted to the oak-hickory forest of the extreme east: the eastern chipmunk, gray squirrel, southern flying squirrel, and woodland vole. Others, however, follow the dendritic fingers of riparian woodland westward past the borders of the region, often to the foot of the Rockies. Among such species are the opossum, eastern cottontail, fox squirrel, white-footed mouse, eastern woodrat, raccoon, red fox, and white-tailed deer.

A number of species of mammals are native to the Northern Great Plains only because of the presence on the Black Hills of a complex of coniferous forests typical of the Rocky Mountains. Among the species of these communities are the Nuttall's cottontail, yellow-bellied marmot, red squirrel, northern flying squirrel, long-tailed vole, and southern red-backed vole. The least chipmunk, bushy-tailed woodrat, and (formerly) bighorn sheep are part of the fauna of the Black Hills but extend eastward and southward beyond the Hills onto pine-clad scarps elsewhere in the western parts of South Dakota and Nebraska.

The deciduous woodlands of the Turtle Mountains and the Pembina Hills support a mammalian community that includes a number of species of the deciduous riparian woodland, supplemented by some distinctive northern species (such as the arctic shrew, marten, snowshoe hare, and moose, and formerly the fisher, wolverine, and caribou) and some species shared with the Black Hills (the least chipmunk, red squirrel, northern flying squirrel, and southern red-backed vole). Some species of these upland islands are relicts of ecosystems more widespread in ice-age environments of the Northern Great Plains. For comments on the history of ecosystems of the region, see the account of zoogeographic patterns below.

Some mammals are sufficiently tolerant that they range through a variety of ecosystems, taking roles in several different mammalian communities. The deer mouse is an example of such a species. So, too, are the coyote, long-tailed weasel, and striped skunk. Most such broadly tolerant species are omnivores or carnivores, which perhaps move readily from one community to another because they are less directly dependent on vegetation than are herbivorous species.

Bats are important components of the mammalian fauna at many localities on the Northern Great Plains, but they usually are not, precisely speaking, part of the earth-bound mammalian community of a particular ecosystem. Although tied energetically to green plants and the sun, they form a community of their own. In terms of the food niche, local bats tend to assort resources among themselves on the basis of hunting strategy or foraging time. They tend to take the "night shift," with swallows and swifts occupying the same foraging space by day and at dusk. Following the cycle of their cold-blooded prey, they are active members of ecosystems only during the warmer months. In terms of roosting space, other groupings of bats can be recognized. For example, the evening bat, silver-haired bat, and red and hoary bats all are tree-roosting species, using primarily deciduous vegetation. The long-eared bat, big

brown bat, small-footed myotis, and fringe-tailed myotis all typically roost in caves, whereas the long-legged myotis and long-eared myotis roost, among other places, in holes or under the bark of conifers.

It is well to recognize that nearly all the mammalian communities of the Northern Great Plains are influenced by man. Ecosystems are cleared, plowed, burned, mined, flooded, or grazed in the service of human economy. Or, if relatively undisturbed, they are subjected to direct effects like hunting and trapping or indirect effects such as acid rain. Shelterbelts have been planted widely on the plains as windbreaks or as a soil-conservation measure, and these have allowed for the expansion of the distribution of the fox squirrel. Raccoons, opossums, and striped skunks thrive in proximity to man, living on garbage or foraging on his crops. Deer respond favorably to the brushy communities that accompany many human activities. Some species, such as the introduced house mouse and the Norway rat, are almost entirely dependent on humans for their survival and might be considered mammalian weeds. They respond well to the largesse of careless civilization, feeding on waste and stored produce, but they seldom are able to invade intact natural ecosystems and compete with the native fauna on its own terms.

Zoogeography

Zoogeography is the study of patterns of animal distribution. One pattern of interest is variation in numbers of species from one place to another. Species occur where they do because they could get there and because, once there, their resource needs were met. Table 1 indicates numbers of species of native Recent mammals in the Dakotas and Nebraska and in adjacent states and provinces. The pattern is complicated. The Northern Plains states each have more species than states to the east, mostly because the badlands of the western part of the region have mammals typical of the Rocky Mountains—kinds that have had no access to Iowa, Minnesota, or Missouri, and to a lesser degree Kansas, and that in any event would not find conditions there suitable for existence. To the north, the number of mammalian species also is fewer, because southern kinds do not occur. North Dakota has fewer species than South Dakota or Nebraska, but more than Saskatchewan or Manitoba (even though the latter province has some true arctic mammals along the shore of Hudson Bay). To the south, Kansas has about the same number of species as do the Northern Plains states, lacking some kinds with northern or western affinities that occur in our region but adding others from the south and southwest.

By far the largest state fauna recorded in the table occurs in Colorado; Wyoming is second. Both are, of course, mountain states, as is Montana, with great topographic and ecological diversity. The disproportionate number of species in Colorado, however, is not due to montane mammals. Species of alpine and subalpine ecosystems are shared with the other Rocky Mountain states, and sometimes with the Northern Great Plains (especially the Black Hills) as well. The disproportionate number of species in Colorado is due instead to a strong complement of taxa with southwestern affinities, which occurs in the western part of the state and also in the foothills of the Eastern Slope. The pattern of species numbers in table 1, then, is due partly to present ecological opportunity and partly to historical circumstances.

In addition to studying variation in numbers of species, zoogeographers seek to understand how they have come to be distributed as they are through the study of the origins and movements of faunas. Understanding something of the origin of the fauna of the Northern Great Plains can provide a context for the dynamic and complex structure of extant mammalian communities. At one end of the scale, zoogeographic patterns merge imperceptibly into local ecological patterns, and the zoogeographer needs to think in terms of ecological tolerances, community succession, and population dynamics. At the other extreme, zoogeographic pattern becomes a function of regional and even continental geologic history, and zoogeographers necessarily must be concerned with the biological implications of such features as mountain building, continental drift, and glaciation.

It is the last of these geologic phenomena that most recently has played a major role

Table 1: Number of Species of Native Recent Mammals Reported
from the Dakotas, Nebraska, and Adjacent States and Provinces

Order	North Dakota	South Dakota	Nebraska	Minnesota	Iowa	Missouri	Kansas	Colorado	Wyoming	Montana	Saskatchewan	Manitoba
	Northern Great Plains			Eastern Boundary			Southern Boundary	Western Boundary			Northern Boundary	
Marsupialia	1	1	1	1	1	1	1	1	1	0	0	0
Insectivora	5	8	6	8	6	6	5	10	8	8	6	7
Chiroptera	9	11	13	7	10	14	15	16	12	13	8	7
Edentata	0	0	1	0	0	1	1	0	0	0	0	0
Lagomorpha	5	5	4	3	2	4	5	7	7	8	4	4
Rodentia	32	33	34	31	24	27	33	56	46	42	28	32
Carnivora	23	21	20	22	19	16	19	24	25	22	22	23[a]
Artiodactyla	8	6	6	7	5	3	5	7	7	9	7	7
Total	83	85	85	79	67	72	84	121	106	102	75	80
Mean		84			73		84		110			77

Sources: North Dakota (Wiehe and Cassel 1978); South Dakota (Choate and Jones 1982); Nebraska (Jones and Choate 1980); Minnesota (Gunderson and Beer 1953; Hazard 1982); Iowa (Bowles 1975); Missouri (Schwartz and Schwartz 1981); Kansas (J. R. Choate, pers. comm.); Colorado (Armstrong 1972); Wyoming (Long 1965; Brown and Metz 1966; Brown 1967b; Jones and Geno-ways 1967a; Thaeler 1972; Hennings and Hoffmann 1977; Thaeler and Hinesley 1979; Stromberg, Biggins, and Bidwell 1981); Montana (Hoffmann and Pattie 1968; Pefaur and Hoffmann 1971; Thaeler 1972; Robinson and Hoffmann 1975; Hennings and Hoffmann 1977; Swenson and Shanks 1979); Manitoba (Soper 1961; Banfield 1974; Nero and Wrigley 1977); Saskatchewan (Beck 1958; Banfield 1974).
[a]Does not include "pinnipeds."

in shaping the Northern Great Plains and its mammalian fauna. The landscapes of much of the region are a product of glacial forces, overlain with a biota derived in Recent times. On more than four occasions in the past several hundred thousand years, glaciers have formed over much of central and eastern Canada and flowed southward over the north-central United States, pushing before them all but the most specialized of arctic animals. As the glaciers retreated, they left behind a virgin land surface, ripe for colonization by a diversity of floral and faunal elements. Whether present times, the Holocene, represent a postglacial era or another inte.glacial episode is a matter for fascinating speculation, but one that cannot be resolved at present.

The point here is that the contemporary faunal and floral assemblages on the Northern Great Plains are products of (1) Wisconsin glaciation, (2) biotic response to glacial recession (in the context of variable Holocene climates), and (3) the effects of aboriginal and European man.

During the Wisconsin glacial stage, extending from more than 70,000 to a mere 10,000 years ago, a continental ice sheet several hundred feet thick overlay a vast area of the Northern Great Plains. The Wisconsin ice sheet obliterated most evidence of previous glaciations in the region. The Missouri Coteau is a glacial moraine, a huge ridge of earth and rock pushed ahead of the flowing ice. The Missouri River skirts the glacial region on the west and south; it carried glacial meltwater to the Mississippi when the climatic pendulum swung toward the warmer, drier conditions we know today. Glacial remnants formed the lake country of eastern North Dakota and the vast lake basins that today are the Souris Plain and the Red and James river valleys.

Nonglacial environments of the Northern Great Plains in Wisconsin times are not known in detail. From various kinds of evidence, however—studies of fossil pollen, mollusks, and occasional vertebrate remains, for example—we can piece together a picture of the regional Wisconsin landscape. Adjacent to the ice sheet was a narrow band of periglacial tundra, home of muskoxen, mammoths, lemmings, and arctic foxes. Beyond the tundra was a boreal forest of spruce, alder, and birch. Shrub steppe, savanna, and woodland ranged southward from the boreal forest onto the Central Great Plains and westward to the Rocky Mountains (which were overtopped with ice caps and valley glaciers of their own). The fauna of these ecosystems was similar to that found today on the Black Hills and at intermediate elevations on the Central and Southern Rocky Mountains. In open country, the fauna was enriched with giant bison, camel-like forms, and horses, species that were to become extinct, owing perhaps to predation pressure from an Old World primate that had arrived on the scene, crossing a Beringian land bridge during the last glacial stage—Paleoindian man.

With the recession of glaciers, vast changes occurred in biotic distributions. The periglacial biota followed the retreating ice northward to where it occurs today, in arctic Canada and on the North Slope of Alaska. Similarly, a formerly widespread coniferous forest biota retreated northward and to refugia on highlands such as the Black Hills and Rocky Mountains. Grassland and savanna biotas moved northward from the southwest and the southeast to fill the void, a void that spelled opportunity for organisms adapted to the rigors of drought and fire in the vast continental interior. Only along watercourses could trees persist over most of the postglacial

plains. Given the high water table of the riparian environment, it was phreatophytic vegetation of willows, cottonwood, and box elder that came to dominate. The coniferous species that remained on the plains were drought-tolerant ponderosa pine and juniper, restricted to rocky scarps like the Pine Ridge or the breaks of the Niobrara where they were protected from prairie fire.

A vision of wholesale movement of faunal elements may strike some as peculiar or remarkable. When thinking of faunal movements, however, we must be cautious not to imagine the changes taking place suddenly. In fact, the vast adjustments of ranges of species that were made during and following Pleistocene glaciation were not much different from those observable today. Glaciers moved erratically, flowing hundreds of miles in thousands of years. Ecosystems were pushed ahead of them, based on rates of movement of at most a few hundred yards annually. With recession, the trend was reversed. These kinds of movements hardly represent frantic flight; rather, they are of the sort that can be seen at present, following from normal expansion and retraction of populations. Indeed, average rates of movement in response to glacial advance and retreat apparently were slower by an order of magnitude or more than documented changes of range of the hispid cotton rat or of the nine-banded armadillo within recent decades.

The climate has not been uniform in Holocene times. On the contrary, it has been marked by pronounced variability, with changes in temperature and moisture regimes on scales of decades, centuries, and millennia. Zoogeographic anomalies sometimes attest to the importance of such climatic shifts to biotic distribution. Although most present distributional patterns can be accounted for by reference to a broad response to postglacial warming and drying, some patterns are more subtle. For example, along the Niobrara River in north-central Nebraska is an isolated population of the eastern woodrat. The contiguous distribution of *Neotoma floridana* lies more than 100 miles to the south. The population along the Niobrara has been interpreted as a relict of the Climatic Optimum, a period some 5,000 years ago when climates were warmer and moister than they are today. Under those conditions, deciduous forest may well have been more nearly continuous than now across eastern and central Nebraska.

Seemingly, climatic amelioration on the plains during historic times has allowed for expansion of ranges of such mammals as the opossum and the hispid cotton rat. The latter species is so near its limits of tolerance on the Northern Plains that a single severe winter storm can alter its distributional limit by tens of miles. Short-term changes in distribution also are seen in northern species. Severe winters in Canada frequently lead to increased numbers of lynx sightings in North Dakota, for example.

A dynamic climate is not the only force for zoogeographic change on the Northern Great Plains. Man also is profoundly influential. Effects of hunting and trapping on the distributions of such species as bison and gray wolf are obvious examples. So, too, are deliberate extralimital introductions (such as those of the mountain goat and the fallow deer) and the possible establishment of feral populations of domesticated mammals (the European rabbit, for example) or the inadvertent introduction of commensals (such as the house mouse). Some human influences, however, are more subtle. Agricultural and grazing practices, for example, seem to have conspired with climatic change in the past several decades to allow the expansion of the range of the black-tailed jackrabbit at the expense of the white-tailed jackrabbit.

The raccoon has become more generally distributed on the Great Plains as a consequence of irrigated farming, the planting of ornamental vegetation and shelterbelts, and the prevalence of refuse dumps. Fox squirrels have expanded from bottomland forests into cities and towns across the plains. The red fox has responded favorably to increased waterfowl production of artificial wetlands following impoundment of rivers. The eastern mole has benefited from the careful cultivation of lawns and golf courses. These are but a few examples. Despite the major and minor effects of man, however, and the effects of short-term climatic change, the greatest influence on the distribution of the biota of the Northern Plains was the last major glaciation.

Because of the Pleistocene history of the Northern Plains, some mammalian communities of the tristate region are young from a geologic standpoint (in contrast with those of the southwestern deserts or the tropical rain forest, for example). Whence came the mammals that now occupy the Northern Great Plains? Taking a broad continental view of the distribution of species, some clues are apparent. Leaving aside for the moment any consideration of ecological distribution, let us think instead of the shapes of ranges of the mammals occurring on the Northern Plains.

If the distributions of the 105 species of mammals native to the tristate region were superimposed on a single map of North America, the figure would appear, at first glance, to be more hodgepodge than pattern; there would be some justice in calling such a figure a "spaghetti diagram." But with patience, important patterns do emerge. The first step in making zoogeographic sense of the picture is to factor out all those species that are so widespread that they can be ascribed to no single areal faunal element. Then the remaining species can be sorted into distinctive faunal elements—groups of species with similar distributional limits—and it becomes apparent that several faunal elements are superimposed on the Northern Great Plains. Given the present-day distribution of these elements and the ecological tolerance of their constituent species, the separate faunal units probably were not sympatric during glacial times. Their superimposition, to form the mammalian communities of the Northern Plains, is fairly recent, a product of the past few thousand years. For example, in a woodlot in eastern Nebraska one might observe an opossum, a fox squirrel, a white-footed mouse, a raccoon, and a woodchuck—a Neotropical species, two eastern taxa, one widespread mammal, and one with boreal affinities. These now form part of the mammalian community of the riparian deciduous forest, an ecosystem type that probably did not exist on the Northern Plains 10,000 years ago. Let us look at the faunal elements that contribute to the mammalian biota of Nebraska and the Dakotas.

Several species (27, or 25.8 percent of the native fauna) of the tristate region are too widespread in North America to identify with any single faunal element. These are:

Small-footed myotis: *Myotis leibii*
Little brown myotis: *Myotis lucifugus*
Silver-haired bat: *Lasionycteris noctivagans*
Big brown bat: *Eptesicus fuscus*
Hoary bat: *Lasiurus cinereus*
Beaver: *Castor canadensis*
Deer mouse: *Peromyscus maniculatus*
Muskrat: *Ondatra zibethicus*
Porcupine: *Erethizon dorsatum*
Coyote: *Canis latrans*
Gray wolf: *Canis lupus*
Red fox: *Vulpes vulpes*
Gray fox: *Urocyon cinereoargenteus*
Black Bear: *Ursus americanus*
Grizzly bear: *Ursus arctos*
Raccoon: *Procyon lotor*
Long-tailed weasel: *Mustela frenata*
Mink: *Mustela vison*
Badger: *Taxidea taxus*
Striped skunk: *Mephitis mephitis*
River otter: *Lutra canadensis*
Mountain lion: *Felis concolor*
Bobcat: *Felis rufus*
Wapiti: *Cervus elaphus*
Mule deer: *Odocoileus hemionus*
White-tailed deer: *Odocoileus virginianus*
Bison: *Bison bison*

Several of these species are broadly distributed ecologically as well as areally. In this connection, note that more than half are carnivores, species at least one step removed from direct dependence on vegetation for food. Others, such as the artiodactyls, are large, mobile species. Still others in this group are specialists in widespread habitats. For example, beaver, mink, muskrat, and (at least formerly) river otter all are potentially as widespread on the plains as are their riparian and wetland habitats. The small-footed myotis inhabits rock shelters and caves, widespread but localized habitats. The silver-haired and hoary bats are broadly distributed as migrants, though fairly limited in their ecological amplitude as residents.

The following 16 species (15.1 percent of the native mammalian fauna) have distributions centered on the Great Plains and hence may be termed the Campestrian Faunal Element:

White-tailed jackrabbit: *Lepus townsendii*
Franklin's ground squirrel: *Spermophilus franklinii*
Richardson's ground squirrel: *Spermophilus richardsonii*
Thirteen-lined ground squirrel: *Spermophilus tridecemlineatus*
Black-tailed prairie dog: *Cynomys ludovicianus*
Plains pocket gopher: *Geomys bursarius*
Olive-backed pocket mouse: *Perognathus fasciatus*
Plains pocket mouse: *Perognathus flavescens*
Hispid pocket mouse: *Perognathus hispidus*
Plains harvest mouse: *Reithrodontomys montanus*
Northern grasshopper mouse: *Onychomys leucogaster*
Prairie vole: *Microtus ochrogaster*
Swift fox: *Vulpes velox*
Black-footed ferret: *Mustela nigripes*
Spotted skunk: *Spilogale putorius*
Pronghorn: *Antilocapra americana*

These species tend to form the core of the mammalian communities of prairie ecosystems. Most are strongly restricted to grasslands, specialized to the rigorous regime of drought, temperature extreme, and fire in the continental interior. More than three-fourths are burrowers, and one is strongly fossorial. Probably during full glacial times these species occurred mostly to the south and west of the tristate region; none is adapted to forested or woodland habitats, although some may occur in savanna.

Another strong component of the mammalian fauna of the Northern Great Plains (some 17 species, 16.2 percent of the native fauna) is the Eastern Faunal Element, the center of distribution of which is the east-central United States. In the tristate region, most of these species are members of the mammalian community of the deciduous forest, although some occur in meadows or prairie wetlands as well. One would suppose that these species were distributed well to the south and east of the glacial border in Wisconsin times, and mostly to the south of the periglacial spruce forest that seems to have clothed the unglaciated parts of the region (particularly in the east). Species of the Eastern Faunal Element are:

Northern short-tailed shrew: *Blarina brevicauda*
Southern short-tailed shrew: *Blarina carolinensis*
Least shrew: *Cryptotis parva*
Eastern mole: *Scalopus aquaticus*
Keen's myotis: *Myotis keenii*
Eastern pipistrelle: *Pipistrellus subflavus*
Red bat: *Lasiurus borealis*
Evening bat: *Nycticeius humeralis*
Eastern cottontail: *Sylvilagus floridanus*
Eastern chipmunk: *Tamias striatus*
Gray squirrel: *Sciurus carolinensis*
Fox squirrel: *Sciurus niger*
Southern flying squirrel: *Glaucomys volans*
White-footed mouse: *Peromyscus leucopus*
Eastern woodrat: *Neotoma floridana*
Woodland vole: *Microtus pinetorum*
Southern bog lemming: *Synaptomys cooperi*

About a quarter of the fauna of the Northern Great Plains is composed of species with continental ranges that center on forested regions to the north or west (or both). One might treat these species as a single faunal element, but we learn more about the mammalian faunal history of the tristate region if instead we think of them

as forming three faunal elements: Boreal, Montane, and Boreomontane.

The Montane Faunal Element has a center of distribution on the Rocky Mountains. It contributes five species (4.8 percent) to the fauna of the three states:

Yellow-bellied marmot: *Marmota flaviventris*
Bushy-tailed woodrat: *Neotoma cinerea*
Long-tailed vole: *Microtus longicaudus*
Western jumping mouse: *Zapus princeps*
Mountain sheep: *Ovis canadensis*

All but one of these species (*Zapus princeps*) are restricted in the region to the Black Hills and adjacent badland country.

The Boreal Faunal Element consists of species with continental ranges centering on central Canada. Some six of the mammals of the Northern Great Plains (5.7 percent) are members of this element. They are:

Arctic shrew: *Sorex arcticus*
Woodchuck: *Marmota monax*
Northern flying squirrel: *Glaucomys sabrinus*
Meadow jumping mouse: *Zapus hudsonius*
Least weasel: *Mustela nivalis*
Caribou: *Rangifer tarandus*

By contrast with the montane mammals mentioned previously, only two Boreal species (*Glaucomys sabrinus* and *Zapus hudsonius*) occur on the Black Hills; rather, Boreal species tend to occur in the eastern parts of the tristate region, mostly in forested ecosystems. Although *Marmota monax* is included in this element, it might equally well be listed as a component of the Eastern Faunal Element.

The Boreomontane Faunal Element includes 14 (13.3 percent) species of the Northern Great Plains:

Masked shrew: *Sorex cinereus*
Water shrew: *Sorex palustris*
Pygmy shrew: *Microsorex hoyi*
Snowshoe hare: *Lepus americanus*

Least chipmunk: *Tamias minimus*
Red squirrel: *Tamiasciurus hudsonicus*
Southern red-backed vole: *Clethrionomys gapperi*
Meadow vole: *Microtus pennsylvanicus*
Marten: *Martes americana*
Fisher: *Martes pennanti*
Ermine: *Mustela erminea*
Wolverine: *Gulo gulo*
Lynx: *Felis lynx*
Moose: *Alces alces*

Boreomontane species co-occur in Canada and southward into the United States along the Rockies and the Appalachians and sometimes along the Sierra Nevada as well. Half of these species have populations on the Black Hills, but none of them on the Northern Great Plains occurs solely there. Although populations on the Black Hills usually are insular (and four of the seven are taxonomically distinct), all Boreomontane species occur elsewhere in the Dakotas and sometimes in Nebraska as well. In terms of ecological distribution, most of these species dwell in forested ecosystems, although some (for example, the masked shrew and the meadow vole) are also mammals of prairie wetlands and others (such as the least chipmunk and the ermine) are broadly tolerant ecologically.

In terms of Pleistocene to Recent distributions, the three faunal elements just described obviously form an interrelated complex. Montane and Boreomontane mammals occur on the Black Hills; Boreal mammals tend not to. The Black Hills evidently were wooded during the Pleistocene. These forests shared with the savannas of eastern Wyoming, western Nebraska, and eastern Colorado a montane mammalian fauna that today occurs not on the High Plains but in the Rocky Mountains. The Boreal fauna was displaced by the glaciers along with its forested habitat to the east and south. With glacial recession, the fauna could move north and west to the once-glaciated Central Lowlands, but not across the semiarid Great Plains as far as the Black Hills. We may envision the Boreomontane fauna as having inhabited

taiga along the edge of the periglacial tundra. This taiga, perhaps discontinuous (especially as interrupted by shrub-steppe in the west), served as a filter-barrier allowing movement of some species populations from west to east and vice versa.

The faunal elements already described make up the bulk of the species of the Northern Great Plains, yet modern ecosystems provide extensive opportunity, and species with distributional centers well removed from Nebraska and the Dakotas contribute to mammalian diversity on the plains. For example, eight Chihauhuan species (7.6 percent of the fauna), mammals with a center of distribution in northern Mexico, occur on the Northern Great Plains. They are:

Fringe-tailed myotis: *Myotis thysanodes*
Townsend's big-eared bat: *Plecotus townsendii*
Desert cottontail: *Sylvilagus audubonii*
Black-tailed jackrabbit: *Lepus californicus*
Spotted ground squirrel: *Spermophilus spilosoma*
Silky pocket mouse: *Perognathus flavus*
Ord's kangaroo rat: *Dipodomys ordii*
Western harvest mouse: *Reithrodontomys megalotis*

Mostly these species are associated with mixed-grass prairie (often on sandy soils) on the semiarid plains. Both Chihauhuan bats are cave dwellers.

The fauna of the tristate region has been enriched further by eight species (7.6 percent) having continental distributions that center on the Great Basin of the western United States. Species of the Great Basin Faunal Element often occur in sagebrush or montane woodland. These are:

Dwarf shrew: *Sorex nanus*
Merriam's shrew: *Sorex merriami*
Long-legged myotis: *Myotis volans*
Long-eared myotis: *Myotis evotis*
Nuttall's cottontail: *Sylvilagus nuttallii*
Wyoming ground squirrel: *Spermophilus elegans*

Northern pocket gopher: *Thomomys talpoides*
Sagebrush vole: *Lagurus curtatus*

Finally, four species (3.9 percent) on the Northern Great Plains obviously are of Neotropical derivation. They are:

Opossum: *Didelphis virginiana*
Mexican free-tailed bat: *Tadarida brasiliensis*
Nine-banded armadillo: *Dasypus novemcinctus*
Hispid cotton rat: *Sigmodon hispidus*

Except for the free-tailed bat (the status of which as a resident of the tristate region is somewhat uncertain), each of these species appears to be expanding its range on the Northern Plains at present, and the two species listed last have come to occur in the region only in the past few years.

There is no reason to believe that the Chihauhuan, Great Basin, and Neotropical faunal elements were represented on the Northern Plains during glacial times. Rather, they have come to occupy the area within the past few millennia, and certainly mammalian communities of Nebraska and the Dakotas still are adjusting to their presence.

Obviously, details of past faunal movements never can be known in detail. Direct evidence of most kinds of mammals in most Pleistocene environments is remarkably poor. Reconstruction of Pleistocene and Holocene events will remain an educated guess based on expanding knowledge of Pleistocene habitats (derived from studies of fossil pollen and ancient soils, for example) coupled with thorough knowledge of the distribution of the native fauna. Despite our imperfect understanding, however, some recognition of the complex history of the native fauna will allow greater appreciation for local ecosystems. Readers interested in more detailed comment on zoogeographic patterns of mammals of the Northern Plains should consult Jones (1964), Hoffmann and Jones (1970), and Turner (1974).

Class Mammalia

About 300 million years ago, in Pennsylvanian times, the synapsid reptiles arose. During the Permian, which followed, synapsids were the most successful of reptiles, represented by such familiar forms as the sail-backed pelycosaurs. The pelycosaurs gave rise to a group of reptiles called therapsids, a diverse lot with both carnivorous and herbivorous representatives. These were active animals with fairly gracile skeletons, the limbs held more nearly beneath the body than in many contemporary reptiles. There was specialization of teeth within the jaw, which was simplified in structure, and a secondary palate had developed. These so-called mammal-like reptiles were, by Triassic times (some 200 million years ago), giving rise to reptile-like, primitive mammals. These newly evolved mammals seemingly arrived at the threshold of a major adaptive radiation and remained there for a hundred million years.

Just about the time mammals arose from their reptilian forebears, another group of reptiles, the archosaurs (or "ruling reptiles"), was undergoing an explosive adaptive radiation, the most familiar group resulting therefrom being the dinosaurs. The archosaurs dominated the ecosystems of the earth throughout Mesozoic times, the "Age of Reptiles." Mammals were present during that whole period, developing the adaptations that allowed them to rise to dominance with the demise of the dinosaurs in the Cretaceous. What are those adaptations that characterize mammals?

There are numerous differences between living mammals and living reptiles—differences in the skeleton, musculature, physiology, and nervous system and behavior, and in reproduction, that are the product of millions of years of evolutionary divergence. As we consider these differences, we must bear in mind that mammals are built on a reptilian foundation, that the traits that typify mammals evolved at different rates, and that mammalian characteristics are interrelated; a simple list of characters does not do the adaptive complex full justice. Speaking quite generally, the hallmarks of mammalian organization are a highly developed reproductive system, a high rate of activity, and a refined nervous system. Try to visualize how these broad themes summarize differences between mammals and reptiles.

Many of the things that separate reptiles from mammals are features of the soft anatomy. These features do not fossilize; thus, their study falls in the realm of comparative anatomy based on examination of living species. The mammalian heart has four chambers. This permits complete separation of oxygen-rich blood being distributed to the general tissues of the body and oxygen-poor blood returning to the lungs. As in other vertebrates, oxygen in the blood is bound to hemoglobin, which is packaged in red blood cells. In mammals these cells are peculiar in that they have no nuclei when mature and are shaped like biconcave disks. This shape—that of an ill-formed bagel—maximizes the exchange surface of the cells and allows rapid unloading of oxygen in the tissues and rapid binding of oxygen in the lungs.

Another distinctive feature of mammalian soft anatomy is the large brain. Not all mammals have the brains of primates, of course, but the mammalian brain in general is distinctive. The essential features of the reptilian brain remain, but in mammals the front part of the brain—the primitive centers that function to interpret smell in lower vertebrates—are greatly elaborated. This is the cerebral cortex, adapted to the task of storing information and integrating past and present experience. This associative cortex, or "neopallium," allows mammals to profit from experience—to learn—to an extent that no other group of organisms can. Such behavioral plasticity certainly has helped make possible the success of mammals in the diversity of environments that characterize Cenozoic time, the "Age of Mammals."

Another characteristic of mammals is hair, a fundamental attribute of the mammalian condition. Hairs are epidermal structures arising from follicles that dip into the dermis. They seem to have evolved as insulative structures between reptilian scales, and their homology to other vertebrate structures is uncertain. A hair is a rod of keratin (the same structural protein that makes up hooves, claws, and nails). The inner part of the rod (medulla) is a hollow, dead air space essential to the insulative value of hair. The outer cortex contains pigment granules. The superficial cuticle is a protective layer, the scaly pattern of which tends to be diagnostic at the level of genus, or even species, when viewed under a microscope. Most mammals (humans are exceptional here as elsewhere) have three types of hairs. The longest are guard hairs, which protect the rest of the pelage from abrasion and often from moisture and usually lend the coat its color pattern. Beneath the guard hairs is a thicker, softer underfur. The vibrissae, or whiskers, are a third kind of hair, usually long, stiff, and functioning in tactile sensation.

Associated with each hair is a minute slip of muscle, the arrector pili. This muscle contracts to raise the hair, thus increas-ing the effective thickness of the insulative layer. Despite the fact that humans are nearly "naked apes," our arrectores pilorum remain, and their contractions are responsible for the "goose bumps" resulting from cold or fear.

Also associated with hair follicles are glands. Sweat glands secrete water (along with dissolved substances such as salt and urea) onto the hairs, and it then evaporates to cool the animal. Sebaceous glands secrete oily substances that lubricate the hair. Despite this protection, the hair is worn by abrasion and by exposure to sunlight and weather. Hence it must be replaced from time to time. The process of replacement is termed molt. Timing and frequency of molt, and the pattern by which molt progresses over the body, vary widely and are described in accounts of individual species.

In some mammals molt changes the color of the animal. This suggests an important secondary function for the pelage; color pattern of mammalian hair is adaptive. Colors may be cryptic, concealing the animal from enemies or prey, or they may be sematic, as in the skunks, warning would-be predators of the presence of other lines of defense.

Specialized skin glands associated with hair may serve a communicative function, characteristic odors serving in recognition of individuals or species, advertisement of territory, or announcement of reproductive condition.

A modification of certain sweat glands gives the class Mammalia its name in that mammary, or milk, glands are characteristic of mammals. In the egg-laying monotremes, milk is secreted onto the fur of the abdomen, but in more advanced mammals the individual milk-secreting glands empty into nipples or teats. Milk is a mixture of proteins, fats, lactose sugar, minerals, and protective antibodies, but its specific composition varies widely with the nutritional needs of particular species. Milk provides young mammals with a developmental head start. Furthermore, it forges a social link between mother and offspring: nursing young are a captive audience, and interaction between the female and her young, and between littermates, provides a basis for mammalian social behavior, a social behavior based on learning.

Another hallmark of the class Mammalia is the reproductive system. As with other mammalian traits, the best way to appreciate mammalian adaptations is to contrast them with the reptilian plan from which they evolved. Reptiles produce shelled eggs. Inside the protective shell are certain important membranes. The chorion is a surface for exchange of gases; the amnion is a fluid-filled sac that protects the developing young from desiccation and from mechanical and thermal shock; the yolk sac contains a store of food; and the allantois is a receptacle for metabolic wastes. The reptilian egg represents a tremendous advance over the amphibian egg from which it arose, because it can be laid on land. Despite this advance, however, most reptilian reproductive strategies remained quantitative—reptiles lay many eggs because they are provided with no direct protection. The two vertebrate classes that arose from reptiles, however, have a qualitative reproductive strategy. Most produce a few zygotes at a time and take good care of them. Birds do this while continuing to employ thoroughly reptilian eggs. Most mammals, however, do not. Only the peculiar monotremes lay eggs. In all other mammals the reptilian eggshell has been relinquished, and there is some degree of embryonic development within the body of the female. The membranes of the reptilian egg remain, however, recruited to new uses. The amniotic sac and its fluid protect the developing fetus; parts of the yolk sac and allantois form the umbilical cord in placental mammals; and the chorion comes into contact with the lining (endometrium) of the uterus. Chorion, allantois, and endometrium all are richly supplied with blood vessels. Through these vessels essential nutrients and oxygen pass from mother to developing young, and waste products pass back in the opposite direction.

The complex of embryonic and maternal membranes and their blood vessels that permits this exchange of materials is termed the placenta. A true placenta characterizes the Eutheria. In marsupials, other than the bandicoots of Australia (family Peramelidae), the "placenta" is derived from the chorion and the yolk sac; in bandicoots and eutherians it is derived from the chorion and the allantois. Eutherian embryos implant in the wall of the uterus to a much greater degree than do marsupial embryos. Complete implantation involves the erosion of the endometrium and the development of fingerlike projections, the chorionic villi. The villi increase the exchange surface between young and mother. The efficiency of exchange also is increased by the progressive elimination of tissue separating mother and offspring. The extreme condition is seen in lagomorphs and some rodents in which only the embryonic capillary wall separates the bloodstreams. In no mammal is there direct mixing of maternal and fetal blood.

The young of marsupials remain in the uterus no more than a few days in most groups and are born far less well developed than the offspring of placental mammals. The gestation period for placentals varies from about two weeks for some kinds of small rodents to 22 months for the African elephant.

Reproduction in wild mammals usually is seasonal. Details appear in accounts of individual species, but the general pattern is for reproduction to be timed so that young are born when resources are plentiful and environmental conditions are optimal. Some mammals reproduce once per season (monestrous), whereas others have two or more reproductive cycles annually (polyestrous). Generally, female mammals are receptive to males during only a certain portion of the reproductive cycle (or estrous cycle), the period of estrus, or "heat." The events of the estrous cycle are coordinated by a fine balance among pituitary and ovarian hormones, and they often seem to be synchronized with environmental cues (day length, for example) via the hypothalamus. Estrus corresponds generally with ovulation, although ovulation may be spontaneous or may be induced by copulation.

Details of structure of the reproductive tract vary among mammalian groups. The uterus, for example, may consist of two distinct horns opening from the oviducts, the horns may fuse to form a single chamber, or the condition may be intermediate. Although the ovaries are paired, only the left ovary is functional in most bats. The mammalian penis also is a highly variable structure, so much so that the form of the penis has been used widely as a clue to taxonomic relationships. Males of many mammalian species have in the genital organ, or penis, a stiffening rod of bone and cartilage, the baculum. Females of those groups often have a homologous structure, the os clitoridis or baubellum. The scrotum is a sac for the testes outside the body wall; it is typical of most mammals. The body temperature of mammals is too high to allow efficient production of healthy spermatozoa, and so the scrotum allows the testes to reside away from the hot core of the body. Usually testes are scrotal only when the male is reproductively active. At other times of the year the testes lie in the abdominal cavity or in the inguinal canal between the abdomen and the scrotum.

We have noted numerous ways modern animals differ from modern reptiles. In reconstructing the evolutionary history of mammals, paleontologists do not have the luxury of a record of soft parts. They have, instead, a record of fossilized teeth and bones. As it happens, however, such fossils can reveal a great deal about the general biology of organisms. Because structure and function are complementary in biological systems, the fossil is not merely a record of structure, but is a record of ecology, behavior, and physiology as well. And what the record shows is another group of traits that add up to a lineage of active, intelligent vertebrates.

Diagnostic features of the mammalian skeleton, like other characters, did not arise all at once in the Triassic. Rather, they evolved at different rates. During the reptilian-mammalian transition, a given line of organisms might have had some features typical of reptiles and others typical of mammals. Because of this, to keep distinctions "clean," paleomammalogists tend to define mammals by a single adaptive complex—the structure of the lower jaw and its articulation to the skull. The lower jaw of reptiles is made up of four bones on either side (dentary, angular, articular, and prearticular). The articular articulates with the quadrate bone at the base of the skull, which articulates in turn with the squamosal. This arrangement has the virtue of being quite flexible. Many reptiles can swallow prey larger around than themselves. But that very flexibility also is a disadvantage, because the jaw articulation lacks sufficient mechanical strength for efficient chewing. Chewing is important to efficient digestion; the smaller the pieces of food that arrive in the stomach, the greater the surface area per volume and thus the more rapid the rate of digestion, and that is important to active animals.

Mammals opted for efficient chewing over width of gape. The jaw was simplified to a single bone, the dentary. The dentary came to articulate directly with the squamosal bone of the skull. This simplification led to a surfeit of bones in the jaw region. The evolving mammals recycled them.

In the reptilian middle ear is a single bone that mechanically increases the amplitude of sound waves reaching the eardrum. Mammals have not one such bone but three—the stapes (derived from the reptilian ear bone) plus the malleus ("hammer") and the incus ("anvil"). The latter bones developed from the reptilian articular and quadrate, respectively. Mesozoic transitional forms now are known in which these elements apparently functioned in both the jaw articulation and the middle ear. The reptilian angular became the tympanic bone of mammals, which helps protect the tiny lever system of the ear ossicles.

Teeth are the hardest structures of the mammalian body and hence the most abundant kind of fossil evidence of mammals. Mammalian teeth are unlike those of reptiles in several respects. First, they are set firmly in sockets (thecodont), unlike teeth of lower vertebrates, which simply perch atop the jawbone (acrodont). Second, mammalian teeth are diphyodont; that is, they come in two (and only two) sets, the permanent dentition and the deciduous (or "milk"] dentition. Permanent teeth, if lost, are not replaced. In reptiles, as in fishes, teeth may be replaced throughout life. Third, mammalian teeth are heterodont; that is, there is specialization in the tooth-row from front to rear. Incisors are for nipping, canines for piercing or grasping prey, cheekteeth (molars, or "cuspids," present only in the permanent dentition, and premolars, or "bicuspids") generally for chopping or grinding. By contrast, reptilian teeth mostly are homodont, all being the same shape.

Numbers of teeth in reptiles are highly variable, even within a species. In mammals, the number of teeth usually is constant within a species and even across more inclusive taxa. Some early mammals (and the primitive marsupials, one of which still lives on the Northern Great Plains) had more teeth, but primitive placental mammals had three incisors, one canine, four premolars, and three molars on each side of the jaw, above and below. This condition may be expressed as a "dental formula," in this case 3/3, 1/1, 4/4, 3/3, total 44.

Variations on this primitive plan have been common. Some of the toothed whales have more numerous teeth that are essentially homodont, but in most mammalian lineages there has been a reduction in total number of teeth. Commonly, herbivorous stocks have lost the canines, for example. Baleen whales and several sorts of ant-eating mammals have lost the dentition altogether.

Elsewhere in the skeleton, mammals also are distinctive. A number of bones in the reptilian skull (postorbital, prefrontal, postfrontal) have been lost, and others (the lacrimal, for example) have been reduced in size. A complex of turbinal bones in the nasal cavity provides support for an expanded olfactory epithelium. Two occipital condyles articulate the skull to the vertebral column. Some ribs have been lost, and the number of vertebrae has become much more nearly constant: usually seven cervicals, 12 to 24 thoracics, five to seven lumbars, two to five sacrals, and a variable number of caudals (but usually fewer than 30). The limb bones have changed in detail, consistent with an increasingly active life-style, and the number of bones in fingers and toes has been reduced from the basic reptilian pattern of 2-3-4-5-3 (read from the thumb or big toe outward) to 2-3-3-3-3. Overall, the mammalian story has been one of simplification. The only new bones "invented" by mammals are the patella, or kneecap, and the genital bone.

One further characteristic of the mammalian skeleton is its high degree of ossification. Limb bones of both reptiles and mammals are formed first of cartilage, which is then replaced by mineral bone. In reptiles the cartilage of bone growth is at the ends of bones. Hence, articulations and muscular attachments are on the soft, actively growing portion of the bone. In mammals, bone growth is different. There are two centers of ossification in mammalian long bones, the central diaphysis (the shaft, roughly) and the terminal epiphyses. The epiphyses provide firm articular surfaces, and the diaphysis provides a firm surface for muscular attachments. Between these two centers of mineralization is a cartilaginous zone where growth occurs. At maturity this zone is obliterated, the epiphyses fuse with the diaphysis, and skeletal growth ceases. Reptiles, by contrast, can continue to grow throughout life.

The basal mammalian stock consisted of small (rat- to cat-sized), agile nocturnal animals that were predators on invertebrates and small vertebrates. From that beginning an adaptive radiation has occurred that is truly astounding. To appreciate better the evolutionary potential of the basic mammalian body plan, think for a moment of the diversity of birds. Even such aberrant kinds as ostriches and kiwis depart relatively little from the avian mode. Given the constraints of flight, early birds "painted themselves into an evolutionary corner," so to speak, closing out many opportunities. In the broad evolutionary history of mammals, there was no such drastic canalization of potential in many groups.

Consider for a moment the adaptive radiation of locomotor modes in mammals. The primitive mammal was quadrupedal and retained the primitive pentadactyl feet seen in salamanders and lizards. Probably early mammals were semiarboreal, and many Recent mammals retain this primitive feature. They may be scansorial, scampering from the ground to trees, or (if larger) they may be ambulatory (walking). Mere walking may be too slow to allow efficient predation or escape, so several groups independently have become cursorial (running). Cursorial mammals tend to have fingers and toes with reduced dexterity. Often they stand on the toes (digitigrade, as in dogs) rather than the soles of the feet (plantigrade, as in bears), or even on the tips of the toenails (unguligrade, as in deer). In several cursorial lines there has been a reduction in number of toes, for greater mechanical strength. In elephants the primitive five toes remain, but they are arranged in a circle, like the flaring base of a column. The locomotion of elephants, and several other lineages of extremely large mammals, is termed graviportal.

Several groups of mammals have sacrificed terrestrial locomotion for efficiency in the water. Natatorial locomotion has reached its zenith in the toothed whales, but baleen whales are not far behind. The seals and their allies are totally committed to aquatic locomotion, as are the manatees, and the sea lions and sea otter are nearly so. The point here is that each of these groups arrived at natatorial locomotion quite independently. Moles and pocket gophers "swim" through the soil (fossorial locomotion). Bats are the only mammals that can truly fly, although several groups (flying squirrels, colugos, marsupial gliders) have developed an efficient gliding locomotion. Bipedal hopping, or saltatorial locomotion, has evolved several times in the rodents as well as in the kangaroos, and gibbons are almost bipedal, but they walk with their hands, swinging gracefully from branch to branch (brachiation). Another line of primates has developed a unique bipedal stride, the human walk.

In short, the primitive body plan has proved immensely plastic over the eons of geologic time. Between living and fossil mammals, the range of adaptation is immense. If the range of fundamentally different life-styles—the breadth of the adaptive radiation—is any criterion of evolutionary success, then mammals are by far the most successful of vertebrates, with only the ruling reptiles coming at all close.

There are about 4,000 living species of mammals, arranged in some 138 families in 20 orders. This diversity is the product of evolution over the past 70 million years, and the Cenozoic, the geologic era following the demise of the dinosaurs, is rightly termed the "Age of Mammals." Mammals range in size from shrews the weight of a copper penny to the vanishing blue whale, at 100 feet long and 150 tons the largest animal that has ever lived. Mammals live on all continents and on most oceanic islands. They inhabit polar seas and tropical deserts, lowland rain forest and alpine tundra. Mammals would be impressive even if humans were not members of the clan.

Living members of the class Mammalia are of two subclasses, the Prototheria (duck-billed platypus and spiny anteaters) and the Theria. The subclass Theria, in turn, is separable into two infraclasses, Metatheria (the marsupials) and Eutheria (placental mammals). Differences between these groups are profound; their most recent common ancestor lived in Mesozoic times. Of the species of mammals occurring on the Northern Great Plains, only one is a marsupial.

Key to Orders of Mammals on the Northern Great Plains

1. First toe on hind foot thumblike, apposable; marsupium present in females; incisors 5/4 —————————— Marsupialia

1.' First toe on hind foot not thumblike or apposable; marsupium absent; incisors never more than 3/3 ——————————2

2. Forelimbs modified for flight _____Chiroptera
2.' Forelimbs not modified for flight ____3
3. Cheekteeth peglike, lacking enamel; dorsum covered by bony carapace _____ Edentata
3.' Cheekteeth not peglike, bearing enamel; dorsum covered with fur _____4
4. Upper incisors absent; feet provided with hooves _____Artiodactyla
4.' Upper incisors present; feet provided with claws _____5
5. Toothrows continuous (no conspicuous diastema); canines present ____6

5.' Toothrows having conspicuous diastema between incisors and cheekteeth; canines absent _____7
6. Canines approximately equal in size to adjacent teeth; size small ____ Insectivora
6.' Canines conspicuously larger than adjacent teeth; size large _____Carnivora
7. Ears of approximately same length as, or longer than, tail; incisors 2/1 _____ Lagomorpha
7.' Ears much shorter than tail; incisors 1/1 _____Rodentia

Checklist of Mammals of the Northern Great Plains

The following checklist of wild mammals occurring on the Northern Great Plains is provided as a ready reference to the 105 native and six introduced species (marked with an asterisk here and in the keys that follow) treated in this work. Page references to species accounts are given. As in the accounts in the text beyond, orders, families, and genera are listed in currently accepted phylogenetic sequence, but species within each genus are listed alphabetically.

34 Class Mammalia

Order Marsupialia

Marsupials are one of the most primitive orders of living mammals and have the longest recorded fossil history (first known from the early Cretaceous of North America, approximately 130 million years ago). In contrast to placental mammals, marsupials are born in a relatively undeveloped or "embryonic" state, and in most species the female carries the young in a fur-lined pouch (marsupium) until they are capable of caring for themselves. The reproductive tract in females is bifid (vagina and uterus divided by a septum); in males the scrotum is anterior to the bifurcate penis.

Marsupials primitively have three premolar and four molar teeth both above and below in the permanent dentition, whereas placental mammals primitively have four premolars and three molars.

Ride (1964) proposed a new classification of the marsupials, elevating the Marsupialia to superordinal rank equal to the Placentalia and including North American marsupials in the order Marsupicarnivora. Kirsch (1977) also recognized Marsupialia as a superorder and revised Ride's classification somewhat, placing the didelphids in the order Polyprotodonta.

Living representatives of the order make up 16 or so families, including approximately 80 genera and more than 200 species. Marsupials occur naturally in two disjunct geographic areas—the Australian region and the temperate and tropical parts of the New World.

Family Didelphidae: Opossums

The family Didelphidae includes the American opossums, which are among the most primitive of marsupials. The family probably originated in North America, but it later was extirpated there, perhaps because of competition with evolving placental mammals. The one species occurring at present in temperate regions of this continent is an immigrant from the American tropics within relatively recent geologic time.

Genus *Didelphis*

The name *Didelphis* is a compound of the Greek *di*, meaning "double," and *delphys*, meaning "womb." The dental formula is 5/4, 1/1, 3/3, 4/4, total 50. This genus contains three nominal species.

Didelphis virginiana: Virginia Opossum

Name. The specific name *virginiana* is derived from the geographic origin, Virginia, of some of the earliest specimens available for scientific study. The vernacular name apparently first was used by Captain John Smith of Virginia, who referred to this animal as "opassum," a name derived from the Algonkian Indian word *apasum.*

Distribution. This opossum is distributed over much of eastern North America, from southern Canada southward to Nicaragua; it has been widely introduced on the Pacific Coast and in several other western states. In the Northern Plains region, the species occurs throughout most of Nebraska northward through central and eastern South Dakota to southern and eastern North Dakota, where it only recently has been recorded. In the western parts of its range in these states, the species frequently is limited to riparian communities associated with the larger river systems. Only one subspecies, *Didelphis virginiana virginiana,* occurs in the region.

The Virginia opossum has not always

occupied an extensive geographic range on the Northern Plains, the species having expanded its distribution both to the north and to the west within historic time. Man has been partially responsible for this dispersal, directly by introducing individuals in places where they did not occur, and indirectly by providing food and shelter and by controlling natural predators.

Description. No other mammal on the Northern Plains is likely to be confused with the opossum. Approximately the size of a house cat, it has a pointed snout, short legs, naked ears, and a long, scaly, scantily haired prehensile tail. In more northerly parts of its range, distal portions of the ears and tail often are missing because they have been frozen. The pelage consists of long, coarse guard hairs interspersed in a thick coat of finer white hairs, most of which have gray or black tips. Guard hairs decrease in frequency on the sides and are absent on the venter. The hair of the venter is essentially the same color as that of the dorsum, but only about two-thirds as long. Although the animals are usually grayish, melanistic opossums are not uncommon, and "cinnamon" and albinistic individuals are known. A claw is present on each of the five toes of the front foot, but only four of the five hind toes are clawed, the hallux (inside toe) being clawless, thumblike, and apposable. Adult females have a fur-lined pouch in which young are carried after birth.

The opossum is the only mammal on the Northern Plains that has five upper and four lower incisors on each side of the jaw and is the only species with a total of 50 permanent teeth. Other characters useful in identification of the skull are premolars 3/3 and molars 4/4, the inflected angular process on the mandible, extremely narrow braincase, and a well-developed sagittal crest in adults.

Mean and extreme external measurements of 11 adults from Kansas and Nebraska are: total length, 722 (643–828); length of tail, 305 (253–378); length of hind foot, 66 (60–75); length of ear, 49 (47–51).

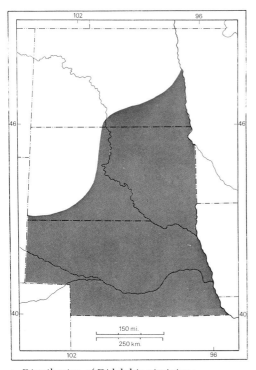

3. *Distribution of* Didelphis virginiana.

Weights of adults vary from 3.5 to 13.5 pounds or more, but individuals weighing 6 to 7 pounds are most common. Means and ranges of the greatest length of skull and zygomatic breadth of eight adults from Kansas and Nebraska are 109.4 (98.3–119.9) and 59.4 (48.3–67.8), respectively.

Natural History. Although found in greatest abundance in wooded situations, especially along streams and rivers, opossums are not uncommon in more open areas, where dens frequently are simple burrows around the roots of bushes or isolated trees. In many cases such dens probably have been excavated by other animals. In wooded habitats, dens are found in such places as brush piles, hollow logs and trees, rock crevices, abandoned squirrel nests, and unoccupied burrow systems. Strawstacks, fencerows, foundations of buildings, and trash piles also make suitable den sites, regardless of surrounding habitat.

It has been claimed that opossums consume almost anything edible because they exercise little choice of foods, but this statement is not entirely valid. Nevertheless, when necessary, they eat an extremely wide variety of plant and animal material. Many opossums feed on carrion, as indicated by the presence of carrion beetles in their stomachs. Corn made up the bulk of the diet of opossums studied one winter in Missouri. Insects, particularly grasshoppers, are among the most important foods in spring and summer. Fruits such as persimmon, mulberry, hackberry, and blackberry have been found in more than 50 percent of fecal droppings collected in autumn and early winter. Other foods include small vertebrates, eggs, numerous kinds of invertebrates, and a variety of vegetable material.

The penis of the male is forked and in copulation penetrates the paired vaginal canals of the female. The forked phallus led to the popular fable that copulation is accomplished through the nose of the female. About 13 days after mating, four to 25 poorly developed young, weighing only a gram or two, are born singly or in groups of up to six. Parturition takes only two to 12 minutes. The forelegs of the young are relatively well developed at birth and the toes possess claws, whereas the hind legs are poorly developed stubs. Hair is absent, and the eyes are incompletely formed and closed. Most young are free of the fetal membranes at birth; those that are not fail to reach the pouch. Young emerge from the vulva swinging their forelegs in such a way as to grasp the abdominal hairs of the mother and thereby make their way to the pouch.

The female does not assist the young directly but assumes a sitting position and relaxes her pouch, which reduces the distance traveled to only about two inches. Newborn opossums make the trip with phenomenal speed; one individual is reported to have reached the pouch in only 16.5 seconds. Although number and arrangement of teats varies individually, they usually total 13, arranged in two lateral rows with one teat between the rows. Sometimes the anterior pair or even the first two pair are small and nonfunctional.

4. *Virginia opossum*, Didelphis virginiana *(courtesy L. C. Watkins).*

A neonate that arrives after all available teats (to which the young firmly attach themselves) are in use, or one that for other reasons fails to secure a teat, does not survive. Pouch young usually number between six and nine (average 8.6 in 23 litters from Nebraska), but extremes of one to 13 have been reported in wild females.

Between the seventieth day (about the time when young first leave the pouch) and weaning, young opossums remain in the den unless they hear "clicking" noises from the female, but they will accompany her if she makes such sounds. Young may accompany their mother by clinging to the fur of either her belly or her back. Later they run along beside her. In captivity, young have been observed responding to the mother as late as 108 days after birth.

Female opossums are sexually mature in the first breeding season after birth. It has been shown that not only does the length of the estrous cycle vary from 22 to 38 days (mean 29.5), but each cycle lasts one to 10 days (mean 3.6) longer than the preceding cycle of that season. On the Northern Plains, females usually have two litters per year, one in late January or February and another in May or June. The second litter usually is slightly larger than the first.

Opossums are preyed on by man, by large raptors, and by a variety of carnivores and are parasitized by ticks, fleas, mites, roundworms, and tapeworms. Diseased individuals are not uncommon. In northern areas, starvation and adverse winter weather kill many individuals, especially young of second litters. Another period of high mortality occurs between three and four months of age, when the young first begin to care for themselves. In the wild, the life span seldom exceeds two years.

Estimates of the size of home range vary from about 10 acres to more than 100 acres. Although some opossums occasionally move to a new den or shift their home range, others are more sedentary. Some, especially young males, may be nomadic. Population densities vary with habitat, season, and year.

No discussion of *D. virginiana* is complete without mention of the famous habit of feigning death or "playing possum." Some individuals feign death merely at the sight of an enemy, whereas others, especially adult males and some adult females, react to danger with harsh, low growls, arched back, and bristling hair.

In the southern United States and in Latin America, opossums are eaten by man. They also are trapped for their fur, which is prime only in winter and not of especially high value. Opossums are of further economic importance because they feed on insects, corn and other grains, fruits, eggs, and even chickens, clean up carrion, and may help to control populations of destructive rodents.

Selected References. General (McManus 1974 and included citations); systematics (Gardner 1973).

Order Insectivora

The insectivores constitute the most primitive order of placental mammals and the stock from which all other kinds arose. Many, however, are highly specialized. The range of diversity from quite primitive to highly derived groups is so great that it is difficult to characterize the order, especially when fossil taxa are considered.

All insectivores have the primitive five-toed condition of the feet, and many retain the well-developed W-shaped pattern of cusps on the molars. Most eat insects, as the ordinal name implies, but many consume other kinds of animals as well as some plant material.

Living representatives of the order are included in at least seven families (although there is some disagreement among specialists, and several additional families, or even separate orders, sometimes are recognized), more than 60 genera, and approximately 400 species. Insectivores occur on all major land masses except Australia, Greenland, Antarctica, and the southern part of South America.

On the Northern Plains, the order is represented by two families, Soricidae and Talpidae.

Key to Families of Insectivores

1. Front feet more than twice as broad as hind feet, possessing stout claws; zygomatic arches and auditory bullae present _____Talpidae

1. Front feet less than twice as broad as hind feet; zygomatic arches absent; tympana ringlike, lacking auditory bullae _____Soricidae

Family Soricidae: Shrews

The family Soricidae includes the smallest of living insectivores, the shrews. Modern shrews generally are classified in approximately 20 genera and 285 species, of which eight, representing four genera, occur on the Northern Plains. Dental formulae of the various genera of shrews are uncertain because of disagreement as to the homologies of certain "unicuspid" teeth. Repenning (1967) proposed a new system of nomenclature for the dentition, but here we retain the more familiar "unicuspid" as a useful descriptive term. Characteristics shared by all shrews include relatively small size, long and pointed snout, diminutive eyes, and relatively short ears that usually are concealed in the pelage.

Key to Shrews

1. Tail less than 40 percent of length of head and body; four or five unicuspids in each upper jaw; if five upper unicuspids (only four usually visible in side view), then condylobasal length more than 20 and cranial breadth more than 11 _____2

1.' Tail more than 40 percent of length of head and body; five unicuspids in each upper jaw (three to five visible in side view); condylobasal length rarely 20 or more (usually less than 19), cranial breadth less than 11 (usually less than 10) _____4

2. Total length 90 or less; dorsal pelage of adults brownish, paler ventrally; four unicuspids in each upper jaw, usually only three visible in side view (fig. 17) _____*Cryptotis parva*

2.' Total length 95 or more; dorsal pelage of adults dark grayish to blackish, not much paler ventrally; five unicuspids in each upper jaw, usually only four visible in side view (fig. 17) _____3

3. Total length rarely less than 125; condylobasal length more than 22; occurring in central and northeastern Nebraska and the eastern Dakotas _____*Blarina brevicauda*

3.' Total length rarely exceeding 120; condylobasal length less than 22; occurring in southern Nebraska _____*Blarina carolinensis*

4. Length of hind foot usually 10 or less; third unicuspid in upper jaw much compressed anteroposteriorly, disklike; usually only three unicuspids in upper jaw visible in side view (fig. 17) _____ *Microsorex hoyi*

4.' Length of hind food usually 10 or more; third unicuspid in upper jaw not disklike; at least four and sometimes five unicuspids in upper jaw visible in side view _____5

5. Pelage dark grayish to blackish dorsally; hind foot longer than 18, fringed with stiff, whitish hairs; condylobasal length more than 19.5 _____*Sorex palustris*

5.' Pelage distinctly brownish to grayish dorsally; hind foot less than 18, not fringed with stiff hairs; condylobasal length 19.5 or less _____6

6. Third unicuspid in upper jaw smaller than fourth (fig. 17); condylobasal length 14.5 or less _____*Sorex nanus*

6.' Third unicuspid in upper jaw as large as, or larger than, fourth (fig. 17); condylobasal length 14.6 or more _____7

7. Pelage distinctly tricolored (dark brown middorsally, decidedly paler on flanks, grayish ventrally); total length greater than 100; condylobasal length greater than 17.5 _____*Sorex arcticus*

7.' Pelage not distinctly tricolored, usually brownish to grayish dorsally (flanks much the same color or only slightly paler), paler ventrally; total length less than 100; condylobasal length less than 17.5 _____8

8. Pelage pale brownish gray to grayish dorsally, whitish ventrally, flank glands usually large and conspicuous; maxillary breadth greater than 4.6; first upper incisors lacking medial accessory cusps _____*Sorex merriami*

8.' Pelage dark brownish to grayish brown dorsally; pale brownish or grayish ventrally; flank glands small and indistinct; maxillary breadth less than 4.6; first upper incisors having medial accessory cusps _____*Sorex cinereus*

Genus *Sorex*

The name *Sorex* is derived from a Latin word meaning "shrew-mouse." The dental formula is: large incisors 1/1, unicuspids 5/1, large premolars 1/1, molars 3/3, total 32. The genus has a mostly boreal, temperate, and montane distribution in Eurasia and North America but ranges southward in Central America to the mountains of Guatemala. There are some 60 nominal species in the genus *Sorex*, five of which occur on the Northern Plains.

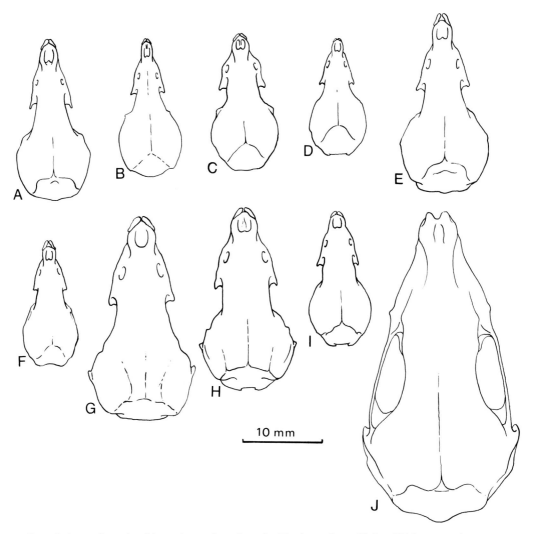

5. *Dorsal views of crania of insectivores found on the Northern Great Plains:* (A) Sorex arcticus; (B) S. cinereus; (C) S. merriami; (D) S. nanus; (E) S. palustris; (F) Microsorex hoyi; (G) Blarina brevicauda; (H) B. carolinensis; (I) Cryptotis parva; (J) Scalopus aquaticus.

Sorex arcticus: Arctic Shrew

Name. The specific name *arcticus* is derived from a Latin word meaning "northern." Some other vernacular names that have been applied to this species are Richardson's shrew, saddle-backed shrew, black-backed shrew, and tricolored shrew.

Distribution. Sorex arcticus occurs from the north-central United States northward into Canada and Alaska and throughout much of Siberia, although one recent study referred Alaskan (and, by inference, Siberian) shrews to another, related species. Two subspecies are known from the Northern Plains—*Sorex arcticus arcticus*, which occurs in northwestern North Dakota, and *Sorex arcticus laricorum*, which is known from eastern North Dakota and adjacent northeastern South Dakota.

Description. The arctic shrew generally re-

sembles *Sorex cinereus* but is larger externally, with longer hind feet and a larger
skull. The tail is of medium length. A long,
narrow flank gland is found in males. In
winter the pelage is distinctly tricolored in
that the rich dark brown of the back is
sharply delimited from the pale brownish
sides and the grayish belly; summer pelage
and the pelage of young animals are paler
and less distinctly tricolored. An autumnal
molt occurs in September or later, resulting
in the acquisition of the distinctive tricolored pattern of the winter pelage. Molt
in spring occurs in May and June.

The two subspecies of *Sorex arcticus* in
the Northern Plains region can be distinguished on the basis of color and size.
Sorex arcticus arcticus differs from *S. a.
laricorum* in having darker pelage dorsally
and laterally, a deeper braincase, a narrower
interorbital region, and a shorter skull.
Also, individuals of *laricorum* average
slightly larger overall than those of
arcticus.

Average and extreme external measurements reported for *arcticus* from Alberta,
followed by those of *laricorum* from Minnesota, are: total length, 112.5 (108–115),
116.3 (115–117); length of tail, 40.0
(38–42), 40.3 (39–42); length of hind foot,
13.7 (13–14), 14.0 (14). Weight ranges from
7 to 11 grams. Representative cranial measurements of *arcticus*, followed by those of
laricorum, are: condylobasal length, 18.9
(18.5–19.4), 19.2 (18.6–19.5); cranial
breadth, 9.4 (9.1–9.6), 9.4 (9.0–9.8). Females
average slightly larger than males. Intergradation between the subspecies
arcticus and *laricorum* probably occurs in
northern North Dakota.

Natural History. In the Northern Plains
region, arctic shrews live in lowland areas
of marsh grasses, mixed forbs, cattails, and
willows, usually in moist valleys and along
lakeshores. Farther north, in the boreal forest, they inhabit damp places in spruce
(*Picea*) and tamarack (*Larix*) forests, but
they also are found in drier, nonforested
habitats. The breeding season extends from
late winter through much of the summer.
Young adults from the previous year make
up most of the spring breeding population.
Juveniles seldom appear before June or later
than September. Early young-of-the-year
sometimes become pregnant by late summer but generally do not breed until the
year after their birth. Some females may
have two or more litters each year, with
four to 10 young per litter. The mean number of fetuses per pregnant female for 24
females from all parts of the range was 6.4.
Field studies of marked individuals indicate
that most live no longer than about 15
months. Two age groups usually can be
recognized when specimens taken at the
same time and place are compared—animals born in that year and adults in their
second year of life.

Home ranges of arctic shrews have a
radius of approximately 150 feet. Population density was 3.5 per acre on a 5.7-acre
plot in Wisconsin. These shrews usually
are nocturnal, but they may be active for
short periods in daytime. During periods of
rest, individuals have been observed to lie

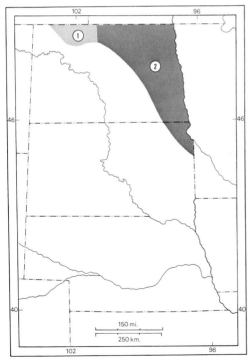

6. *Distribution of* Sorex arcticus:
(1) S. a. arcticus; (2) S. a. laricorum.

7. *Arctic shrew,* Sorex arcticus
(courtesy E. C. Birney).

on their backs and lick the anal region, behavior probably related to coprophagy, which has been reported in other shrews. Grooming is done with the feet, which first are wiped along the mouth.

Arctic shrews are voracious eaters, as evidently are all members of the family. The diet consists of miller moth larvae, grasshoppers, larval and adult click beetles, caterpillars, larval and adult sawflies, fly larvae, aquatic insects, and other small aquatic and terrestrial invertebrates. During periods of unusually high density of prey species, these shrews may kill more animals than they can eat. Curiously, this behavior leads to incomplete digestion and metabolically inefficient utilization of the food. Captive shrews have eaten earthworms, hamburger, mealworms, fly larvae, and grasshoppers. If offered dead rodents, captive animals usually chew flesh, skin, muscle, and bone from the head and eat the brain.

Skeletal remains of arctic shrews have been found in great horned owl pellets. Weasels also prey upon these animals.

Selected References. Natural history (Buckner 1964, 1966; Clough 1963); systematics (Jackson 1928; Youngman 1975).

Sorex cinereus: Masked Shrew

Name. The specific name *cinereus* is of Latin origin and means "ash-colored." Some other vernacular names that have been applied to this species are cinereous shrew, common shrew, and long-tailed shrew.

Distribution. Sorex cinereus is known to occur throughout much of the northern half of North America and in eastern Siberia. The species ranges over most of the Northern Plains, southward to Cass County in eastern Nebraska and Lincoln County in western Nebraska, and to the Kansas border in the central part of that state. A single subspecies, *Sorex cinereus haydeni,* occurs in the region.

Description. The masked shrew is of medium size for the genus *Sorex,* weighing from 3 to 5 grams. However, the subspecies *haydeni* is smaller than many other races, and evidence is accumulating that *cinereus* and *haydeni* may represent distinct species. Summer pelage is dark brownish dorsally, paler brown on the sides, and grayish white ventrally, resulting in a slightly tricolored appearance. Winter pelage is somewhat paler dorsally, generally grayish brown. Seasonal molt occurs between April and June and again in the autumn, usually in October. Extreme color variants are rare, although completely or partially albinistic specimens have been reported. One, a white individual with dark extremities (Himalayan color pattern), was captured near Aberdeen, South Dakota.

External measurements of 10 specimens from Nebraska are: total length, 93.8 (88–97); length of tail, 36.0 (34–38); length of hind foot, 11.8 (11–13). Condylobasal length averages approximately 15.5 in *S. c. haydeni* and cranial breadth averages 7.7. Three adults from Nebraska weighed 3.4, 3.8, and 4.7 grams.

Natural History. No detailed study of the life history of the masked shrew has been made, although it is a common mammal in favored habitats. On the Northern Plains, *Sorex cinereus* seems to pefer moist or damp habitats in deciduous and coniferous forests, as well as marshes, grassy bogs, and other riparian environments. Although they prefer moist conditions, masked shrews ap-

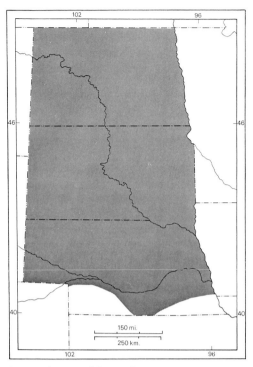

8. *Distribution of* Sorex cinereus.

parently are distributed in a wider variety of habitats than are other members of the genus, but populations frequently are low in marginal areas.

Although masked shrews are in part sympatric with least shrews (*Cryptotis parva*), specimens of these two species seldom are captured in the same trapline, suggesting either that the two seek different microhabitats or that one often excludes the other in competition. In Colorado, populations of *S. cinereus* and those of the montane shrew (*S. monticolus*) have been reported to fluctuate synchronously, indicating that they are affected by the same environmental factors or by each other, or both.

Masked shrews eat a variety of invertebrates such as beetles, flies, moths, ants, bees, wasps, butterflies, sawflies, grasshoppers, crickets, centipedes, worms, spiders, mollusks, and sowbugs. Some small vertebrates such as mice and salamanders also are consumed. Individuals occasionally also eat minute amounts of vegetable matter;

like most other species of shrews, they can be taken in traps baited with rolled oats. A female kept in captivity for 77 days was reported to have eaten more than three times her own weight daily.

The masked shrew breeds from April to October on the Northern Plains. Because individuals can breed at two months of age, shrews born in early spring sometimes reproduce by late summer. Females have from four to 10 young per litter and are polyestrous, bearing two or three litters per year. The gestation period is 19 to 22 days. Young are born in spherical nests four to five inches in diameter, constructed by adults from grass and other vegetable matter. At birth young have no trace of hair, the teeth have not erupted, the eyelids are fused, the feet possess well-defined grooves but are not separated into distinct digits, the pinna of the ear is only a small fold of skin, the body is dark grayish whereas the feet and tail are cream-colored, and the skin is rough. In a recent study in Ohio, eyes were found to open by days 17 and 18 and ears by days 14 to 17. Incisors erupted between days 13 and 14, and weaning occurred after the twentieth day of life. Weight of young averaged 0.28 gram on the day of birth, 2.1 grams on the eighth day, and 3.5 grams on day 20. Young evidently are well developed before leaving the nest. Molt from juvenile pelage, sparser and grayer than that of adults, is directly into summer or winter adult pelage, depending on season.

Nests frequently are hidden in cavities of old stumps or logs, under rocks or in simi-

9. *Masked Shrew,* Sorex cinereus
(courtesy R. S. Hoffmann).

lar situations. These insectivores often use burrows of other small mammals, but occasionally they construct their own, which are about three-fourths of an inch in diameter and about nine inches deep. Temporary runways are formed in loose soil and litter as the shrews forage for food. The network of tunnels may cover 40 square feet; it is used to store food remains and sometimes contains the nest.

Members of the genus *Sorex* do not have submaxillary poison cells or ventral dermal glands, both of which are found in the short-tailed shrews (*Blarina*). However, male masked shrews do have lateral scent glands, similar to those in other shrews and consisting of enlarged sweat and sebaceous glands, that may function in establishing territories.

Ticks, fleas, and hair mites rarely parasitize this species. Frogs, fish, short-tailed shrews, and birds of prey occasionally feed on masked shrews. Carnivores frequently kill shrews but apparently find them unpalatable because of the scent glands.

Masked shrews may be active at any time of the day or night. Like other shrews, *S. cinereus* does not hibernate but remains active throughout the year; individuals have been reported to tunnel in snow in winter. The home range usually varies from half an acre up to an acre and a half.

Although the range of vision is rather restricted, sight plays an important role in the life of the masked shrew. The sense of smell also is important and is an effective means of locating food at short distances. Hearing may be acute for certain sounds. Loud noises of low frequency cause no alarm, whereas sounds that are imperceptible to the human ear often cause great excitement. Three types of vocalization audible to humans have been listed: a rapid series of staccato squeaks when fighting; faint twittering notes when foraging; and a slow grinding of the teeth during the rare periods of inactivity. In addition, these and other shrews probably utter high-frequency sounds above the range of human hearing, which they may employ in echolocation of objects.

Populations of *S. cinereus* usually range in concentration from one to about 10 shrews per acre; in one study in Manitoba, however, 86 individuals were taken on a quadrat of two hectares (about five acres) in white cedar forest with a ground cover of sedge. Individuals can be captured readily in snap traps, but pitfalls serve as a better index to population densities. Recent studies suggest that high densities may overstimulate the adrenal system, thereby causing collapse of such populations. This physiological phenomenon could result in cyclic fluctuations of populations from place to place and from year to year, and possibly also at different seasons in a single year.

Masked shrews are mostly beneficial because of their diet of insects and other small invertebrates that may damage crops. A recent study in Canada, for example, suggested that, by eating larvae, *S. cinereus* provides an effective control on larch sawfly populations.

Selected References. General (Jackson 1961); natural history (Brown 1967b; Buckner 1964, 1966; Forsyth 1976; Wrigley, Dubois, and Copeland 1979); systematics (Jackson 1928; van Zyll de Jong 1976a, 1980).

Sorex merriami: Merriam's Shrew

Name. The specific name of this shrew was proposed in honor of C. Hart Merriam (1855–1942), a noted American naturalist, first chief of the United States Biological Survey and first president of the American Society of Mammalogists.

Distribution. Merriam's shrew is known from only two specimens from the Northern Plains, one from near Medora, North Dakota, and a second from south of Rushville, Nebraska. The North Dakota specimen was referred to the nominate subspecies and the one from Nebraska to the subspecies *leucogenys*, two races recognized for many years. However, a recent analysis of specimens from throughout the range of *S. merriami* revealed it to be a monotypic species.

This shrew occurs from the Northern Plains westward to Oregon and Washington and south to the central parts of Arizona and New Mexico. The specimen from Nebraska is the easternmost known record of the species. There is doubt as to the identity of the specimen reported from North Dakota, which no longer can be found in the collection of the National Museum of Natural History; an earlier investigator noted that only the skin of the posterior part of the body, the hind feet, and the tail of this shrew were available for study.

Description. Sorex merriami is a pale-colored, medium-sized shrew. The upper parts are ash gray or drab, tinged with buff, whereas the sides and underparts are whitish to buffy. The feet and underside of

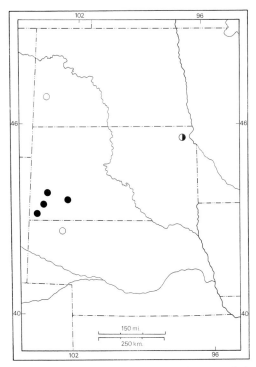

10. Distribution of Sorex merriami *(open circles),* Sorex nanus *(solid circles), and* Sorex palustris *(half solid circle).*

the tail also are white, but the dorsal surface of the tail is tinged with buff. Average and extreme external measurements of specimens from Colorado, Montana, Nebraska, and Wyoming are: total length, 96.6 (90–101); length of tail, 36.9 (32–39); length of hind foot, 11.9 (11–13). Weights from throughout the range of the species average 5.3 (4.2–6.8) grams.

The skull is broad, short, and laterally swollen (fig. 5), and the braincase does not rise above the plane of the rostrum. Condylobasal length of specimens from several western states averaged 16.3 (15.7–17.1), and cranial breadth averaged 8.3 (7.9–8.8). The unicuspid teeth are higher than long, and the unicuspid part of the toothrow appears crowded.

Sorex merriami can be distinguished from other plains-dwelling shrews of the genus *Sorex* by its pale color, relatively short and broad skull, and the well-developed flank glands of the male, which are among the largest found in American shrews.

Natural History. Almost nothing is known of the natural history of Merriam's shrew on the Northern Plains. The specimen from Nebraska was taken along with *Sorex cinereus* in a moist meadow adjacent to a semiarid area in which sagebrush and grasses were the predominant vegetation. The species is known from similar situations in southeastern Wyoming and north-central Colorado. For example, in a study in southern Wyoming, *S. merriami* was the only shrew taken in short-grass prairie habitats, although it also ranged into areas of sagebrush and mountain mahogany inhabited by *S. monticolus* and *S. cinereus.* In Washington and Wyoming, this shrew appears to be most common in areas of sagebrush and bunchgrasses; frequently it is associated with the sagebrush vole, *Lagurus curtatus,* the runways of which reportedly provide Merriam's shrews with protection from predators and with a supply of insect food in an otherwise barren habitat. Analyses of contents of nine stomachs from Washington revealed the presence of caterpillars, larval and adult beetles, cave crickets, and spiders.

Merriam's shrew has distinct summer and winter pelages. Summer pelage reportedly is more brownish and longer than that of winter. Hairs on the feet and tail do not change color seasonally. The spring molt, which occurs in April, first appears in nearly symmetrical patches on the back. Autumnal molt occurs in October or early November and seems to begin on the posterior part of the back and to move anteriorly and laterally.

Judging from autopsy data, the breeding season of this shrew lasts from mid-March to early July in Washington. Records of five, six, and seven fetuses have been reported. Flank glands of males are most evident during the breeding season. The glands probably are responsible for the strong musky odor of these animals, which may attract other individuals.

Owls are the only known predators on Merriam's shrews, which are known to be parasitized by fleas, cestodes, and trematodes. The only probable economic importance of this shrew stems from its insectivorous habits. *S. merriami* has been taken most often in pitfalls such as sunken cans and in unbaited snap traps. Owl pellets also are a potential source of specimens that should not be overlooked.

Selected References. General (Armstrong and Jones 1971 and included citations); systematics (Diersing and Hoffmeister 1977).

Sorex nanus: Dwarf Shrew

Name. The specific name of this shrew is the Latin word meaning "dwarf."

Distribution. This species occurs mainly in subalpine and alpine habitats in much of the intermontane West of the United States. It is known on the Northern Plains only from five localities in South Dakota—three adjacent to the Black Hills in Custer, Fall River, and Pennington counties, and two in the vicinity of Cottonwood, Jackson County (fig. 10).

Description. Sorex nanus is among the smallest of shrews, and its small size distinguishes it from all but a few species in North America. No subspecies are recognized. The pelage is brownish dorsally, sometimes olive brown, and smoke gray ventrally; the tail is indistinctly bicolored. Representative external and cranial measurements of specimens from throughout the range of the species are: total length, 90.2 (85–94); length of tail, 40.2 (36–45); length of hind foot, 10.6 (10–11); condylobasal length, 14.2 (13.8–14.5); cranial breadth, 6.9 (6.6–7.2). Adults weigh an average of about 2.5 grams (three from Jackson County, South Dakota, weighed 1.8, 2.8, and 3.1).

Natural History. Little is known concerning the natural history of the dwarf shrew. Until recently, this species was represented in museum collections by fewer than a score of specimens. Larval and adult stages of various insects and possibly other assorted invertebrates make up the diet of this seemingly rare animal. *S. nanus* usually has been taken in subalpine and alpine rockslides, at elevations of 8,000 to 10,000 feet, and it once was thought to be more or less limited to such situations. However, specimens recently have been reported from north-central Montana and central Colorado at considerably lower elevations, and those known from South Dakota are from short-grass prairie habitats at even lower altitudes. A specimen from near Fair-

11. Dwarf shrew, Sorex nanus (courtesy D. L. Pattie). Note paper match for size comparison.

burn, South Dakota, was found dead in a stubble field, and one from east of Box Elder was taken in a grassy field along with *Sorex cinereus*; the specimen from Fall River County is represented by a left mandible recovered from an owl pellet. Four individuals from the vicinity of Cottonwood, South Dakota, were trapped in pitfalls (number 10 cans) set out in roadside ditches where the vegetative cover consisted of grasses, clover, and clumps of sage.

Ecological distribution of the dwarf shrew seems to be relatively independent of the availability of surface water and moist substrates. These animals seldom are taken in the snap traps and live traps generally used to capture small mammals, but the recent use of pitfalls in South Dakota and elsewhere has proved a successful means of obtaining specimens. With increased use of sunken cans and other such traps, knowledge of the geographic distribution and abundance of dwarf shrews probably will be greatly augmented. Insofar as can be determined from meager data, populations of *S. nanus* are stable throughout the year.

Based on limited data, breeding in alpine populations probably begins after snowmelt, in late spring. First litters are born in late July or early August, and postpartum estrus evidently is commonplace. Of five lactating females reported in one study, four also were pregnant (fetus counts six to eight, average 6.5). Second litters are produced in late summer. There is no evidence that females reproduce in their first year of life.

Selected References. General (Hoffmann and Owen 1980 and included citations).

Sorex palustris: Water Shrew

Name. The specific name *palustris* comes from the Latin word *paluster*, meaning "marshy." Some other vernacular names that have been applied to this shrew include marsh shrew, "beaver mouse," "fish mouse," "muskrat" shrew, and black-and-white shrew.

Distribution. This species occurs in the north-central and northeastern United States, in montane areas in the western part of the United States, and in southern Canada and Alaska. On the Northern Great Plains, the water shrew is known only from Fort Sisseton, South Dakota (fig. 10). Possibly these shrews will be found to occur elsewhere in northeastern South Dakota and northward into North Dakota along the Red River Valley. One subspecies, *Sorex palustris hydrobadistes*, is known from the region.

Description. The water shrew is the largest of the long-tailed shrews occurring in the Northern Plains region. The tail is nearly as long as the head and body, and the hind feet are large and fringed with stiff whitish hairs. The third and fourth toes of the hind feet are joined by a thin web for more than half their length. Males possess a pair of dermal glands on the flanks that become prominent during the breeding season. Short white hairs grow from these glands when they are fully developed, resulting in a white oval patch.

The pelage of water shrews is dense and soft. The dorsal color is velvety black to brownish in winter and sometimes has a greenish to purple iridescence. The ventral surface is paler, silvery white to smoke gray. The tail is distinctly bicolored, being whitish ventrally and darker dorsally. Summer pelage generally is more brownish than winter pelage and also tends to be iridescent. As in other shrews, there is little or no sexual dichromatism, but males tend to be slightly larger and heavier than females. Weight of adults ranges from 12 to 20 grams. Ranges of external measurements of adults from Wisconsin are: total length, 138–164; length of tail, 63–72; length of hind foot, 19–20. Representative cranial measurements of specimens from Minnesota are: condylobasal length, 20.6 (20.2–21.0); cranial breadth, 10.4 (10.0–10.9).

Natural History. Water shrews frequently are found in grasses or shrubs along banks

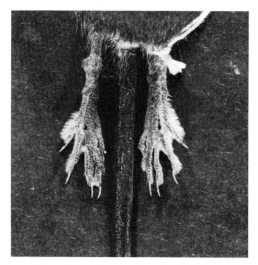

12. Water shrew, Sorex palustris, *hind feet.*

of streams, lakes, and ponds, and around potholes in bogs or forests. In montane areas these shrews occur principally along fast-running, cold mountain streams with banks having rocks, boulders, roots, and overhanging ledges for cover. Sometimes individuals inhabit houses of beavers or muskrats. Water shrews probably breed from late winter through early autumn, and adult females may have two or three litters each year. Females born in early spring may have a litter of their own the first summer, but usually they do not reproduce until a year old. Adult males become sexually active as early as December and remain so throughout the breeding season. Young males, unlike young females, do not mate in their first year. These shrews have five to eight young (most frequently six). They usually live no longer than about two years.

Water shrews are primarily nocturnal—peak periods of activity have been reported as several hours following sunset and the hour before sunrise—but also are active during the day. They are excellent swimmers and divers and also can move along the bottom of streams and pools. Additionally, they reportedly are capable of running on the surface of water for short distances, aided by small air bubbles adhering to the fringes on their feet. The pelage is resistant to wetting; however, if a shrew remains long in water, it dries its fur upon emergence by combing the pelage rapidly with its hind feet. Water shrews generally are secretive, remaining out of sight under overhanging banks and debris. As in most other shrews, sight is poorly developed but is compensated for by well-developed olfactory and tactile senses. Nest-building and tunneling have been documented for animals kept in the laboratory, but little is known of this behavior in the wild.

Nearly half the diet of this shrew consists of aquatic insects. Planaria, small fish, spiders, vegetable matter, annelids, mice, and other shrews also are eaten. Hoarding of food has been reported in captives. Two seasonal molts occur, as in other shrews—one in spring, completed by June, and the other in summer, completed by July or August. The pattern of molt is similar in young and adults. As in other *Sorex*, there is but one maturational molt, from the relatively thin juvenile coat to the appropriate adult pelage of the season.

Water shrews are preyed upon by hawks, owls, weasels, minks, and snakes. They are parasitized externally by mites, fleas, and ticks and internally by roundworms and tapeworms. These shrews have been known to destroy fish eggs and small fish, but because of their insectivorous habits they probably are economically beneficial.

Selected References. Natural history (Conaway 1952; Jackson 1961; Sorenson 1962); systematics (Jackson 1928).

Genus *Microsorex*

The name *Microsorex* is compounded from the Greek word *mikros,* meaning "small," and the Latin word *sorex,* meaning "shrewmouse." The dental formula is the same as in the genus *Sorex*—large incisors 1/1, unicuspids 5/1, large premolars 1/1, molars 3/3, total 32. Two nominal species have been recognized in the past, but recent findings suggest that the genus is monotypic. Pygmy shrews are restricted to boreal and north temperate North America.

Microsorex hoyi: Pygmy Shrew

Name. The specific name *hoyi* is a patronym proposed in honor of Dr. Philo Romayne Hoy (1816–92), American physician and naturalist.

Distribution. On the Northern Plains, this shrew is known only from a few specimens from eastern North and South Dakota. The species has not yet been reported from Nebraska, although it may occur at least in the northeastern part of that state. A single subspecies, *Microsorex hoyi hoyi,* occurs in the region.

Description. In external appearance the pygmy shrew is slightly smaller, with a relatively shorter tail, than the masked shrew (*Sorex cinereus*), with which it occurs sympatrically throughout much of its range on the Northern Great Plains. The length of the tail and that of the hind foot generally are less in *M. hoyi* than in *S. cinereus,* and the tail is reported to be more distinctly bicolored. In spite of these subtle differences, the two species so closely resemble each other externally that one of the two specimens from Racine, Wisconsin, on which S. F. Baird based the original description of the pygmy shrew was later found to represent *S. cinereus.* However, the two are easily distinguished by viewing the teeth in lateral view under weak magnification. Only three upper unicuspids, the first, second, and fourth, usually are visible on each side in *Microsorex,* whereas four or five are visible in all species of the genus *Sorex.* In one recent systematic study, *Microsorex* was treated as a subgenus of *Sorex* rather than as a distinct genus.

Coloration is reddish brown to grayish brown dorsally, grayer in winter than in summer. The sides are paler than the dorsum, and the frequency of gray-tipped hairs is somewhat greater there than on the back in some specimens. The venter is whitish to grayish overall; ventral hairs are relatively shorter than those of the dorsum and are blackish basally with grayish or whitish tips. Some adult pygmy shrews weigh as little as 2 grams, and the species is one of

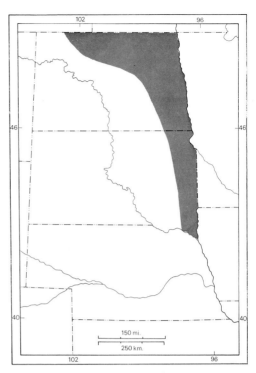

13. Distribution of Microsorex hoyi.

the smallest mammals in North America. Ranges in external and cranial measurements reported from throughout the distribution of the subspecies *M. h. hoyi* are: total length, 75–95; length of tail, 27–34; length of hind foot, 9–11. Cranial measurements recorded for specimens from Minnesota include: condylobasal length, 14.7 (14.3–15.0); cranial breadth, 6.5 (6.4–6.6). The tail of *M. hoyi* generally makes up less of the total length of the animal than in members of the genus *Sorex.*

Natural History. Little is known about the natural history of the pygmy shrew on the Northern Plains. A specimen obtained near Vermillion, South Dakota, was caught in a cattail-rush community on the edge of a slough. In Colorado and Wyoming this species generally is associated with extremely moist sphagnum bogs, but specimens have been taken in spruce forests, meadows, and even in formerly forested areas that had been logged and bulldozed. *M. hoyi* also has been taken in such situations as gardens, rocky slopes, wet and dry forests, piles of driftwood along rivers, and even in areas of human habitation. Some authors have noted that pygmy shrews prefer dry areas, whereas others stress association with moist habitats. There is some evidence that these shrews inhabit drier areas as summer progresses. Requirements of the pygmy shrew may vary either seasonally or geographically; more likely, the animal is less limited in its requirements than available information indicates. A recent study in Manitoba demonstrated that *M. hoyi* has a wide tolerance for variation in moisture, temperature, vegetation type, and cover. This species is active the year around.

Pygmy shrews can be captured in pitfalls but are only occasionally caught in snap traps. In a study in Ontario, 58 of 60 specimens were collected in pitfalls and only two were trapped. Snap traps also have proved ineffective in Montana and Wyoming, where pitfalls have been used with satisfactory results. Several individuals have been taken from within fallen or rotting trees.

Competition with other shrews does not seem to be a limiting factor for pygmy shrews, because they often are found along with masked shrews. In a study in Wyoming, six species of shrews were found living in the same general area, but relative abundance of some species differed in different habitats, and pygmy shrews occurred only in moist sphagnum bogs.

Examination of the stomach contents of wild-taken individuals has yielded butterfly larvae, larval and adult beetles, larval and

14. Pygmy shrew, Microsorex hoyi *(courtesy R. E. Wrigley).*

adult flies, parts of unidentified insects, earthworms, snails, and slugs. One naturalist observed a pygmy shrew that was eating the contents of the burrow of a dung beetle. In captivity these shrews have been noted to feed on mice, other shrews, raw fish, earthworms, grasshoppers, beetles, and several kinds of flies. One captive pygmy shrew was observed to leave fresh meat to eat a biscuit. While eating the biscuit it sat up on its hind legs and held the morsel between its forefeet, much in the manner of a squirrel. In the laboratory it was found that pygmy shrews consumed an average of 7.8 calories per day and metabolized 6.5 calories per day for an overall efficiency of 83 percent. Food also was hoarded when available in abundance.

A pygmy shrew was recovered from the stomach of a garter snake in Ontario, and the species undoubtedly serves as food for owls and also possibly for carnivores and other shrews.

Individuals seem to be active at all times of the day, but slightly more so in hours of darkness. In one laboratory study it was found that pygmy shrews had an average total activity of 239 minutes per day, which was two and a half times that of *Blarina* (94), twice that of *Sorex arcticus* (115), and slightly more than that of *Sorex cinereus* (217). Individual periods of activity were relatively short, averaging only 2.9 minutes, with about 19 activity peaks in a 24-hour period. When shrews are active they wiggle their snouts constantly, suggesting a dependence either on smell or on tactile sensations from the vibrissae. Captives emit short, high-pitched squeaks that may or may not assist in orientation.

Pygmy shrews are adept climbers, make many sudden starts and stops, and can run rapidly when disturbed. They burrow if a proper substrate is provided, making a tunnel not much larger than that made by a large earthworm. In the wild, they make networks of tiny burrows under fallen logs and roots of old stumps, but they also have been trapped in the burrow systems of woodland voles (*Microtus pinetorum*).

Little is known about the intraspecific relationships of pygmy shrews. Although they are generally thought to occur in low densities, this could be a misconception resulting from inefficient census methods. Unlike other soricids, females may bear only a single litter annually; three to eight young per litter have been recorded. Sexually active males have been collected as early as April. Lactating and pregnant females have been taken in June, July, and August. In Colorado, populations of *M. hoyi* attain maximum density annually early in August, two to three weeks earlier than other species of long-tailed shrews.

Two individuals placed in the same cage attacked each other viciously, but this may have been a result of hunger or stress associated with captivity. A yellowish substance with a musky odor is secreted from a lateral skin gland (about 12 mm long) in males. It is not known if musk is emitted for defensive purposes, attraction of mates, or territorial marking, or if it serves some other function.

Spring molt probably occurs typically in late April and early May, and molt in autumn takes place from early October to early November, although molting adults have been recorded from several summer months. It is likely that differences in timing of the molt are related to age or to the reproductive condition of the animal, and they also may be geographically variable within this and other species of shrews occurring on the Northern Plains.

Economically, the pygmy shrew probably is of little importance, although it does feed on insects. Certainly it is not harmful.

Selected References. General (Long 1974 and included citations); systematics (Diersing 1980a; van Zyll de Jong 1976b).

Genus *Blarina*

The generic name *Blarina* was coined arbitrarily and has no Greek or Latin meaning. The dental formula is the same as for *Sorex*—large incisors 1/1, unicuspids 5/1, large premolars 1/1, molars 3/3, total 32. The genus occurs only in eastern North America and at present includes two nomi-

nal species, both of which are found on the Northern Great Plains.

Blarina brevicauda: Northern Short-tailed Shrew

Name. The specific name *brevicauda* is a compound of two Latin words, *brevis,* meaning "short," and *cauda,* meaning "tail." Other vernacular names that have been used for this species include "mole" shrew, shrew "mouse," and blarina.

Distribution. The approximate western geographic limits of this species on the Northern Plains are known reliably only in Nebraska, where it occurs westward to approximately the 101st meridian along the Niobrara River. The species is known definitely in South Dakota only from the eastern part of the state, although it probably ranges at least as far west as the Missouri River, as it does in North Dakota, where it also is well known in the eastern parts.

This short-tailed shrew, a common to abundant inhabitant of the northeastern and north-central United States and southeastern Canada, is represented by a single subspecies in the Northern Plains region, *Blarina brevicauda brevicauda.*

Description. This species can be distinguished from all other shrews that occur on the Northern Plains, excepting *Blarina carolinensis* and *Cryptotis parva,* by its relatively short tail, which makes up less than 40 percent of the total length. From *Cryptotis parva,* this short-tailed shrew is readily distinguished by its much larger size and possession of five, rather than four, unicuspid teeth in each upper jaw. From *Blarina carolinensis,* it differs only in being somewhat larger and in having a different complement of chromosomes, as is explained in the account of *carolinensis.*

Blarina brevicauda has a sensitive, flexible snout that projects slightly beyond the mouth. The eyes are small and probably useful only in the perception of light. The external opening of the ear is large but hidden by the pelage, which is velvety and brushes equally well in any direction,

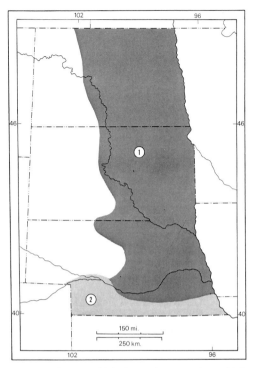

15. Distribution of (1) Blarina brevicauda *and* (2) Blarina carolinensis.

much like that of a mole. The general coloration is grayish to grayish black, slightly darker on the back than on the underparts.

External and cranial measurements (based on specimens from Washington County, Nebraska) are: total length, 132.4 (125–141); length of tail, 26.4 (23–31); length of hind foot, 17.0 (16–18); condylobasal length, 24.0 (22.8–24.9); cranial breadth, 13.5 (12.8–14.0). Weight averaged 25.5 (23.4–29.5) grams.

Natural History. The principal habitat of this short-tailed shrew is deciduous woodland, although in the eastern part of the three-state region it may be found in a variety of habitats including grassy fields, rock ledges, or brush and weeds along fencerows. In some places it is one of the most common small mammals. In the western part of its range, the species tends to be restricted to riparian communities, al-

though it may be as locally abundant there as in the east.

Blarina brevicauda is active both by day and by night, but during diurnal activity it seeks shade or subdued light. Examination of stomachs of shrews trapped by day indicated that they had fed during that time. Most published evidence points to the short-tailed shrew as a solitary animal, but some success in keeping captives together has been reported, and large numbers caught in the same runways over short periods of time suggest reasonable tolerance of other individuals. Estimates of populations have varied between two and 50 per acre. This variation may result partly from local fluctuations in numbers and also the type of habitat sampled. The home range of males may be as large as 4.4 acres but usually is no more than an acre, whereas females evidently have home ranges of less than an acre.

The short-tailed shrew is a short-legged animal that moves with apparent rapidity but little actual speed. It usually runs with the tail elevated after the fashion of the long-tailed soricids. The front feet are relatively wide and strong, adapted to digging in loose soil. In digging, the forepaws are used initially, but the nose and head soon take over the task. Burrows and tunnels of this shrew vary considerably, but most are from one to two inches in diameter. These appear to be made in two zones—a few inches below the surface or in litter directly upon it, and in a deeper zone some 16 to 22 inches below the surface. The two levels are joined at irregular intervals by abrupt connections. Although short-tailed shrews dig burrows and make runways of their own, they frequently use the tunnels and runways of voles.

These shrews occupy two types of nests, a breeding nest and a much smaller resting nest. They construct nests of readily available material such as the leaves of grasses, sedges, forbs, and trees, arranging it into a hollow ball. One breeding nest found in New York was eight inches across the top and almost six inches in depth. The hollow inside was two inches in breadth and slightly less than three inches deep. Resting nests generally are about the size of a large apple.

Sexual activity commences about the first week in February. Testes of males begin to enlarge about this time and reach their maximum size in early April. Although males appear ready to mate at the end of February, females in the wild seldom become pregnant before the middle of March, so that the first litter appears in the first or second week in April. The peak of spring breeding activity ends in the latter part of May. There is little reproductive activity during the summer months. Another peak, however, occurs in late summer and early autumn, with litters being born in September and October. The gestation period is 21 to 22 days, and litter size varies from four to 10 (average between six and seven). A few hours after birth the young are dark pink, wrinkled, and about the size of a large honeybee. By the eighth day they weigh about 6 grams, and hair is visible over most of the body, but teeth are not yet present. The upper incisors appear about the nineteenth day, when the animals weigh about 10 grams. The young are able to move about easily and the teeth are well developed by day 22. Females only a few months old are capable of breeding, but males do not breed in their first summer. In the wild, short-tailed shrews rarely live two full years, and only a few individuals survive two winters. One wild-caught female lived more than 30 months in captivity, and two males born in captivity lived 29 and 33 months.

Based on data from throughout the year,

16. Northern short-tailed shrew, Blarina brevicauda *(courtesy E. C. Birney).*

major food items of the short-tailed shrew (in decreasing order of importance) are: insects, earthworms, vegetable matter, centipedes, arachnids, mollusks, small vertebrates, sow bugs, and the fungus *Endogone*. Almost any species of insect that the shrew happens upon is eaten, but adult beetles and pupal stages of several orders seem to be taken most readily. Where earthworms are abundant, they may be found in fully 50 percent of stomachs examined. Of vegetable matter, small nuts, such as beechnuts, are favored, but it is noteworthy that *B. brevicauda* is easily taken in traps baited with cereals such as rolled oats, and that seeds and pulpy parts of various fruits have been found in the stomachs of short-tailed shrews. Small centipedes frequently are almost whole when they reach the stomach. Small cocoons of spiders containing masses of eggs have been taken from stomachs of shrews, as have several kinds of adult spiders. A short-tailed shrew feeds on snails by attacking them not through the aperture but through the apex and spire, which it neatly bites away to expose the soft parts. Vertebrates eaten are principally salamanders and small mammals. Recent studies indicate that these shrews eat fewer mice than was thought by earlier naturalists. In midwinter, when the temperature is often below 0° F, short-tailed shrews eat a proportionally greater amount of insect food than in summer, mostly dormant beetles and pupae. They also take a large amount of plant material during the winter months. Observations of captive individuals indicate that half the shrew's weight in food is ample over a 24-hour period, and usually only half that much is eaten daily. Short-tailed shrews are known to store food for periods of adversity; the hoarding habit of captive individuals reflects this.

Among the animals known to prey on *Blarina* are water snakes, pilot black snakes, rattlesnakes, copperheads, several species of hawks and owls, ground squirrels, red foxes, long-tailed weasels, striped skunks, and even northern pike and rainbow trout. This shrew is parasitized externally by fleas, mites, ticks, and botfly larvae and internally by acanthocephalans, cestodes, nematodes, and trematodes.

Blarina brevicauda possesses several glands of particular interest. The submaxillary glands produce a venom that is released through a pair of ducts opening near the base of the lower incisors, which project forward, forming a groove along which the venom can flow into a wound. These shrews, which are among the few known venomous mammals, seem to use the venom to disable prey, and possibly also to kill or immobilize invertebrates. No human fatalities resulting from bites are known, but considerable irritation can result. Short-tailed shrews also have a ventral and two lateral scent glands formed from greatly enlarged sweat and sebaceous glands. These glands are well developed in breeding males, less developed in nonbreeding males and females, and poorly developed in females during estrus, pregnancy, and lactation. The scent glands, source of a musky odor, may help in establishing territories and in recognition of individuals.

The first hair to appear on young animals is the juvenile pelage, characterized by a "fuzzy" appearance and retained until the shrew is about adult size. The kind of postjuvenile pelage is determined by the time of year the shrew is born. Shrews born in spring and early summer molt from juvenile to summer pelage, whereas those born late in summer molt directly into winter pelage. Fresh winter pelage of adults is distinguished from fresh summer pelage by its greater length and density. Molt from winter to summer pelage occurs from February to July, but mostly in late spring, and molt from summer to winter pelage can be seen in specimens taken in October and November.

One naturalist described a peculiar call in captives that was often repeated when the animals were disturbed. The call commenced with a shrill grating note, rather high pitched, and ended with a slightly lower pitched chatter, "zeeeee-che-che-che-che." Another note was sometimes heard when animals were feeding and was en-

tirely distinct from the sound of alarm or rage. This was low and continuous, calling to mind the drowsy twitter of sparrows.

Selected References. Natural history (Blair 1940; Dapson 1968; Hamilton 1929, 1931; Pearson 1944, 1945); systematics (Genoways and Choate 1972; Genoways, Patton, and Choate 1977).

Blarina carolinensis: Southern Short-tailed Shrew

Name. The specific epithet *carolinensis* was proposed because this shrew was originally described from a specimen taken in South Carolina.

Distribution. The southern short-tailed shrew occurs throughout the southeastern United States, west to Nebraska and Texas. In the Northern Plains region, it is found only in southern Nebraska (see fig. 15), where it is represented by the subspecies *Blarina carolinensis carolinensis.*

For many years *carolinensis* and its subspecies were regarded as races of *Blarina brevicauda.* However, it has recently been shown that morphologically intermediate individuals are absent where the geographic ranges of the taxa meet across the central part of the eastern United States. The line of demarcation between the two nominal species has been well established in southern Nebraska and adjacent Iowa. An accumulation of evidence argues for specific recognition of *carolinensis,* not only because it evidently meets *brevicauda* geographically without interbreeding, but also because of differences in external and cranial size and in chromosomal complement, and because the fossil record seems to indicate that both species were present in the same deposits in late Pleistocene times. On this basis we have tentatively recognized *B. carolinensis* as a species distinct from *B. brevicauda.* Nonetheless, much remains to be learned concerning the systematics of short-tailed shrews.

Description. Blarina carolinensis closely re-

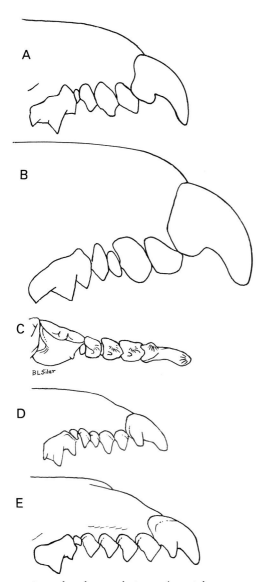

17. Lateral and ventral views of partial upper toothrows of several shrews found on the Northern Great Plains, showing (left to right) first premolar, unicuspid teeth, and first incisor: (A) Cryptotis parva; (B) Blarina carolinensis; (C) Microsorex hoyi; (D) Sorex nanus; (E) S. cinereus.

sembles *B. brevicauda,* differing only in being smaller, in having more pairs of chromosomes (26 as opposed to 25 or fewer), and in possessing a fundamental number of 62 as opposed to 48 chromosomal arms. External and cranial measurements of a

series from southeastern Nebraska are: total length, 112.3 (103–121); length of tail, 22.6 (19–25); length of hind foot, 14.4 (13–16); condylobasal length, 21.1 (20.1–21.9); cranial breadth, 11.5 (11.0–12.2). Weight averaged 14.6 (13.4–15.5) grams.

Natural History. Because *B. carolinensis* long was regarded as the same species as *B. brevicauda,* such differences in their natural history as may exist are unknown. Individuals of both species have been trapped in similar circumstances in Nebraska. See previous account.

Selected References. General (Lowery 1974); also see previous account.

Genus *Cryptotis*

The name *Cryptotis* is a compound of two Greek words—*kryptos,* meaning "hidden," and *otos,* meaning "ear." The dentition differs from that of other North American genera in that there are only four upper unicuspids. The dental formula is: large incisors 1/1, unicuspids 4/1, large premolars 1/1, molars 3/3, total 30. The genus contains about 12 modern species, which are distributed from extreme southern Canada southward into northern South America. Only one species occurs on the Northern Great Plains.

Cryptotis parva: Least Shrew

Name. The specific name of this shrew, *parva,* is the Latin for "small" or "petty." Numerous vernacular names have been used for this animal, among them little short-tailed shrew, small blarina, "bee mole," and cryptotis.

Distribution. The least shrew occurs from southernmost Ontario southward throughout the eastern half of the United States and thence to Costa Rica and Panama. Central South Dakota (Jackson County) represents one of the northernmost latitudes at which the species has been recorded. Nine subspecies of this shrew currently are recognized, only one of which, *Cryptotis parva parva*, occurs on the Northern Plains. In this region the least shrew is found throughout most of eastern and central Nebraska and in southeastern and south-central South Dakota. It is most commonly taken in upland prairies, meadows, and weedy fields, along grassy roadsides, and in similar situations.

Description. The least shrew is an extremely small mammal with a rather robust body and a short tail. The total length is about the same as that of the pygmy shrew, *Microsorex*, but *Cryptotis* has a much shorter tail. In winter the least shrew is dark brown dorsally and dark gray on the belly. In summer the color is paler, nearly sepia on the back and ashy gray on the belly. Color variations such as albinism are rare.

Like other shrews of the Northern Plains,

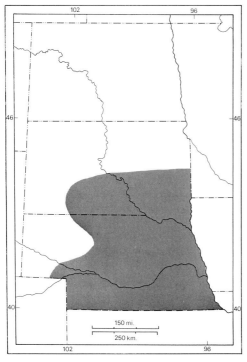

18. Distribution of Cryptotis parva.

Cryptotis parva has a long, pointed snout, tiny ears nearly hidden in soft fur, and small, beady black eyes. The skull is compact and broad. As in other Plains shrews, the teeth are partially pigmented, having a reddish brown coloration on the ridges. The posteriormost of the four upper unicuspids is quite small and cannot be seen in lateral view.

Male and female least shrews are nearly the same size. Average and extreme external and cranial measurements of a series from Washington County, Nebraska, are: total length, 79.1 (73–86); length of tail, 16.4 (15–18); length of hind foot, 10.5 (10–11); condylobasal length, 15.4 (14.9–15.7); cranial breadth, 7.7 (7.6–7.9). Weight of adults generally ranges between 4 and 6 grams.

Natural History. Least shrews are somewhat gregarious or colonial in that several adults have been known to occupy a single nest. One biologist found about 25 least shrews in a nest of leaves under a log in Virginia, and another found at least 31 nesting under a log in Texas; there have been other reports of as many as 12 shrews per nest. Least shrews kept together in captivity fight only when food is scarce. When food is abundant, one captive shrew sometimes will carry items into the nest for another, but this sort of cooperative behavior ceases when food is in short supply.

When moving about above and beneath the ground, least shrews often use runways made by rodents, but they also make their own burrows in soft soil. Two least shrews held in captivity were observed to share the task of excavating a burrow. One shrew did most of the digging, and the other packed loose dirt on the walls of the tunnel. Burrows more than five feet long have been found. The home range apparently is approximately half an acre, based on limited data, and one estimate indicated a density of about 10 animals per acre.

Nests of least shrews usually are constructed of shredded vegetation, especially dried grass and leaves. The nests are small

19. *Least shrew,* Cryptotis parva
(courtesy T. H. Kunz).

(about the size of a man's fist) and roughly spherical, with two entrances. Generally the nest is built in a slight depression under a flat rock, log, or other cover; rarely is it underground.

Least shrews, like other small shrews, are extremely active. They move about in rapid bursts of speed, with constant twitching of their sensitive noses. These shrews are active both by night and by day, but the peak of activity probably is at night. When at rest, least shrews usually lie curled up on their sides, but sometimes they sit hunched up on all four feet. The breathing rate of one resting shrew was timed at about 70 breaths per minute. They sleep soundly and awake slowly and sluggishly, with trembling or convulsive movements.

These shrews devote part of their active time to careful grooming. In captivity they have been observed to wash themselves and to use their feet as combs. Feces are removed from the nest and deposited in one place some distance away.

Least shrews are not noisy animals, but sometimes utter a high-pitched chirping as well as a clicking sound. Some sounds made by these shrews are of high frequency and therefore beyond the range of human hearing.

The breeding season of *Cryptotis parva* on the Northern Plains is undocumented but probably extends from March to October. Shrews born early in the year may

breed before the end of that season, because young females become reproductively active at about three months of age. Recent studies of least shrews in captivity indicate that the gestation period varies between 21 and 23 days. Number of young per litter ranges from two to seven, but the usual litter size is four to six.

Least shrews are naked at birth save for small vibrissae, are translucent pink, and weigh about one-third of a gram. By the sixth or seventh day the young are furred dorsally, and when 10 days old they may weigh more than 2 grams. The ears open on day 10 or 11, and the eyes open by the fourteenth day, at which time the young are nearly as large as adults. At this time the female tenaciously defends them. When they first make short trips from the nest, the young shrews take animal food, but they continue to nurse until 21 or 22 days of age.

The juvenile pelage is thin and "fuzzy," in contrast to adult pelage. As in *Blarina,* molt from juvenile pelage depends on season—shrews born early in the year molt to adult summer pelage, whereas those born in late summer or early autumn molt directly into winter pelage. Seasonal molt in adults takes place in spring and early autumn. Senile animals may molt at any time of year, frequently incompletely, perhaps signaling a breakdown in physiological function. Like other shrews, individual *C. parva* rarely live beyond a second year.

The least shrew feeds on a variety of items. Probably much of its diet consists of small invertebrates, with larger animals, especially dead mice, and vegetative material as a supplement. The most important food items appear to be insects, earthworms, centipedes, snails, and slugs. There are several reports of least shrews nesting in beehives and feeding on larval or pupal bees. Captive shrews sometimes store food in tunnels, and they readily eat hamburger and dead mice; captives also have been reported to drink frequently. Because of their high metabolic rate, individuals eat large quantities of food. In fact, it has been reported that they consume as much as 60 to 100 percent of their body weight every 24 hours.

Owls are probably the most important natural enemy of the least shrew, and remains of specimens often are found in owl pellets. Hawks and snakes also prey on *C. parva*. Carnivorous mammals frequently kill these shrews but rarely eat them. Chiggers, fleas, and mites are known as ectoparasites.

Least shrews are of little economic importance but, like other soricids, probably are useful as a means of controlling populations of insects.

Selected References. General (Whitaker 1974 and included citations).

Family Talpidae: Moles

The family Talpidae, which occurs throughout much of temperate Eurasia and North America, constitutes one of the most highly specialized groups of insectivores. Living moles currently are classified in about 15 genera and approximately 30 species, of which only one, *Scalopus aquaticus,* is known to occur on the Northern Plains. A second species, the star-nosed mole, *Condylura cristata*, has been recorded as occurring in North Dakota, but the report was based on hearsay and never has been substantiated by museum specimens. It is possible that *C. cristata* will be found in the extreme northeastern part of the region, and it therefore is included in the following key.

Key to Moles

1. Tail less than 25 percent of total length; width of front foot equal to or greater than its length; anterior end of snout lacking circular fringe of fleshy processes; premolars 3/3 —————*Scalopus aquaticus*
1! Tail more than 25 percent of total length; width of front foot less than its length; anterior end of snout having circular fringe of fleshy processes; premolars 4/4 —————*Condylura cristata*

Genus *Scalopus*

The name *Scalopus* is based on the Greek word for mole, *skalopos,* which in turn is derived from words meaning "to dig" and "foot." The dental formula is 3/2, 1/0, 3/3, 3/3, total 36. The genus is represented by a single modern species.

Scalopus aquaticus: Eastern Mole

Name. The Latin name *aquaticus,* which means "water-dwelling," is actually a misnomer that was first applied to this species by Linnaeus in 1758, perhaps because of the "webbed" feet.

Distribution. This mole is distributed over much of the eastern and central United States, barely reaching northern Mexico and southernmost Canada in the region of the Great Lakes. Two subspecies are recognized in the Northern Plains region: *Scalopus aquaticus machrinoides* ranges throughout eastern Nebraska to extreme southeastern South Dakota, whereas *Scalopus aquaticus caryi* occurs in central and western Nebraska, northward to southern South Dakota.

The local distribution of moles is limited by moisture and soil conditions. In the western part of the tristate region, these animals are found mostly along rivers and streams and around permanent ponds and lakes. They also colonize irrigated fields and such well-watered areas as gardens, lawns, cemeteries, and golf courses.

Description. The eastern mole is the largest insectivore on the Northern Plains. The distinctly pointed snout is naked anteriorly. The eyes are much reduced and are situated in front of the orbits; the external ears also

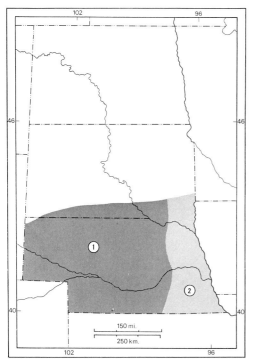

20. *Distribution of* Scalopus aquaticus: (1) S. a. caryi, (2) S. a. machrinoides.

are greatly reduced. The head is coupled closely to the body so that no distinct neck is evident. The palms of the heavily clawed forefeet are wider than long, and the toes of both forefeet and hind feet are joined so as to appear webbed. The tail is rounded and nearly hairless.

Moles are particularly well adapted to their fossorial existence. The pelvic region is greatly reduced relative to the front quarters, allowing the mole to turn around easily in its burrow. The pelvic outlet is so small that the urogenital ducts and large intestine cannot pass through it, and they therefore empty in front of the pubic symphysis. The fur of the eastern mole is dense and short and probably excludes particles of soil. The hairs are constructed in such a way that they bend easily in any direction.

The eyes of moles contain all the elements of the functional mammalian eye, but the structure is so compressed and distorted that probably only light and dark can be distinguished, at best. The sense of touch, on the other hand, is highly devel-

oped. Tactile hairs are situated on the snout and on the margins of the forefeet. The sense of smell also seems to be well developed and no doubt is used in locating food. Hearing, too, is acute and alerts the mole to danger.

The two subspecies of *Scalopus aquaticus* on the Northern Plains can be distinguished on the basis of size and color, *S. a. machrinoides* being larger and darker than *S. a. caryi*. Average and extreme external measurements of eight males and four females, respectively, of *machrinoides* from southeastern Nebraska are: total length, 177.7 (174–185), 171.3 (163–187); length of tail, 33.6 (28.5–38), 33.5 (32–34); length of hind foot, 22.5 (21–24), 21.5 (19–25). Comparable measurements for six males and seven females of *caryi* from northwestern Nebraska are, respectively: 155.4 (145–166), 150.8 (136–166); 25.8 (19–31), 25.4 (22–28); 21.3 (20–22), 20.2 (19–22). Males of both subspecies average larger than females. Recorded weights of adult moles from Nebraska range from 54 to 99 grams.

The color of the eastern mole varies widely, both within and between subspecies. Individuals of *S. a. machrinoides* generally vary between lead gray and dark brownish; those of *S. a. caryi* are paler, silver gray being the usual color. Many moles have a yellowish or orangish patch in the pectoral region, and an occasional individual is uniformly of such a color. Albino moles are extremely rare.

The skull is characterized by its complete zygomatic arches, complete auditory bullae, and lateral inflation (particularly posteriorly).

Mean and extreme cranial measurements for the individuals of *S. a. machrinoides* listed above are: greatest length of skull, 38.0 (36.7–40.2), 37.0 (36.1–38.4); cranial breadth, 19.3 (18.2–20.2), 19.3 (18.6–20.1). Corresponding measurements for eight males and three females of *S. a. caryi* from western Nebraska are: 34.7 (34.4–35.6), 32.3 (31.8–32.9); 17.8 (17.1–18.2), 16.6 (16.4–16.9).

Natural History. Moles spend most of their lives underground. Tunnels are of two

21. *Eastern mole,* Scalopus aquaticus
(courtesy J. L. Tveten).

kinds, those used only for foraging and those of a more permanent sort. Foraging tunnels are constructed an inch or more below the surface, the depth being determined by the amount of moisture in the soil, which in turn dictates the distribution of preferred foods. Such surface tunnels, so-called, form a characteristic ridge of earth, all too familiar to gardeners. These tunnels are formed by the animal as it burrows through the topsoil, in which process the mole employs a lateral motion using only one forefoot at a time. Periodically the tunnels drop abruptly to deeper levels, presumably as a protective measure. Little construction of foraging tunnels takes place when the ground is hard, such as during extended dry periods. When tunnels are damaged, moles repair them by burrowing under the damaged area and heaving up the floor of the old burrow.

More permanent tunnels are constructed six to 10 inches below the surface. These leave no ridges but are marked by frequent mounds of dirt pushed out of the ground. The mounds have been confused with those of pocket gophers, but the two are distinct. Mounds of the eastern mole tend to be nearly circular, because tunnels leading to them are vertical and soil is thrown up equally in all directions. The mounds of pocket gophers, on the other hand, are semicircular, because the tunnels beneath them are oblique and soil is thrown out on only one side. Most permanent, deep tunnels evidently are constructed in spring, but

mounding has been observed at other times of year as well.

Moist humus soils are preferred for burrowing. Sides of hills are used more extensively than are hilltops. Floodplain soils are perhaps the most favored habitat, but such areas are also dangerous to moles, because flooding seems to be the most important natural control on populations. Adults are good swimmers and may survive floods, but in such circumstances young probably perish in large numbers.

In a recent study in Kentucky it was found that males had an average home range of 1.09 hectares (approximately 2.69 acres) and females one of 0.28 hectare (approximately 0.69 acre) over a period extending from 11 months to three years. Individuals were found to be active at all hours of day and night, but activity was concentrated between 0800 and 1600 and again between 2300 and 0400.

Moles do not hibernate but are active throughout the year. In winter they forage in previously built tunnels, mainly below the frostline. There is no evidence of food storage for winter.

A nest four to six inches or more in diameter, lined with grass, leaves, and fine roots, is constructed in a permanent tunnel. Nests typically are situated in well-drained areas. Females give birth to a single litter of naked young annually, in March or April, after a gestation period of five to six weeks. Four is the usual number of young, but two to five have been reported.

Molt takes place twice annually in adults, once in spring and again in late summer or early autumn. The winter pelage is longer and thicker, and usually darker overall, than that of summer.

The eastern mole is alleged to feed on newly planted grain and tubers, but its natural diet seems to consist mostly of invertebrates, preferably adult and larval insects and earthworms. Vegetable matter is not infrequently found in stomachs, but usually in relatively small amounts, and at least some of it may be ingested incidentally. Captive animals have voracious appetites and eat great quantities of insects and earthworms, as well as hamburger and freshly killed mice when offered. They will eat a variety of vegetable foods when deprived of their normal diet.

As noted, floods probably are the greatest single cause of mortality in the eastern mole. Few birds of prey catch moles consistently because they spend so little time on the surface of the ground, but skulls occasionally are found in owl pellets. Foxes, badgers, coyotes, and perhaps skunks dig moles from their burrows; domestic cats and dogs also kill moles but rarely eat them. One investigator reported a mole found in the stomach of a water moccasin, and another recorded an adult from the stomach of a large-mouthed bass.

Moles are of considerable economic importance, at least where they are abundant. On the positive side, they eat great quantities of insects and carry humus to lower levels in the soil, which they also help to aerate. Their fur is of commercial value, although the European mole (*Talpa*) is

more important in this regard than is *S. aquaticus*. Eastern moles also may be detrimental in that the beauty of lawns and gardens is marred by their tunnels, and their predilection for earthworms leads to the destruction of a valuable soil builder. It is uncertain whether moles feed on significant amounts of tubers, seeds, and root crops. However, their burrows provide easy access to underground crops for mice and voles. Several kinds of commercial traps are available for control of moles where they prove to be a nuisance.

The eastern mole is known to be parasitized externally by lice, mites, parasitic beetles, and fleas and internally by nematodes, cestodes, and acanthocephalans.

Selected References. General (Yates and Schmidly 1978 and included citations); distribution (Jones, Choate, and Wilhelm 1978); systematics (Yates and Schmidly 1977).

Order Chiroptera

Bats are the only truly volant mammals, and flight has allowed them wide dispersal. They are absent only from cold latitudes and from certain oceanic islands. Two sub-orders are recognized, the Megachiroptera (a group restricted to tropical and subtropical parts of the Old World) and the Micro-chiroptera. The fossil record of the order extends back to Eocene times, approximately 50 million years ago. Modern bats are representative of 17 families, about 170 genera, and some 850 species.

The flying surface of bats consists of the wing (patagium) and the interfemoral or tail membrane (uropatagium), double-layered extensions of skin of the body that encompass the forelimbs, hind limbs (except the feet), and, in all Northern Plains species, most or all of the tail. The wing is supported internally by elongate bones of the hand and fingers, excepting the first finger (thumb), which is small, clawed, and free of the membrane. The radius is longer than the humerus, and the ulna is rudimentary. The tubercles of the proximal end of the humerus are well developed, and the minor tubercle forms an articulation with the scapula. This condition is particularly advanced in the family Molossidae, species of which are by consequence especially strong and rapid fliers.

Among bats there are kinds that are carnivorous, feeding on small vertebrates, including other bats, a number of species that eat fruit, some that are nectar feeders, a few that catch fish, and one group (vampire bats) that laps blood from mammals and birds. Many bats (including all those treated in this book), however, feed on insects, most of which they catch in flight. When foraging, insectivorous bats continuously emit ultrasonic sounds that allow them not only to echolocate and avoid obstacles in their path of flight, but also to find flying insects. Once detected, an insect is entrapped in the interfemoral membrane or in a wing (and quickly transferred to a cup formed by the interfemoral membrane) and in most instances is eaten in flight. Although the food habits of Northern Plains species vary depending on size and foraging habits, insect orders most frequently represented in analyses of food include Coleoptera (beetles), Diptera (flies), Hemiptera (bugs), Homoptera (aphids, cicadas, and allies), and Lepidoptera (moths).

The teeth of members of the families Vespertilionidae and Molossidae are not

Table 2: Arrangement of Teeth (Upper Followed by Lower) of Bats Occurring on the Northern Great Plains

Genus	Incisors	Canines	Pre-molars	Molars	Total Teeth
Eptesicus	2/3	1/1	1/2	3/3	32
Lasionycteris	2/3	1/1	2/3	3/3	36
Lasiurus	1/3	1/1	2/2	3/3	32
Myotis	2/3	1/1	3/3	3/3	38
Nycticeius	1/3	1/1	1/2	3/3	30
Pipistrellus	2/3	1/1	2/2	3/3	34
Plecotus	2/3	1/1	2/3	3/3	36
Tadarida	1/3	1/1	2/2	3/3	32

much modified from the primitive condition found in the Insectivora, the W-shaped pattern of the molars being especially evident. Even so, the hard parts of insects are mostly discarded in favor of soft tissues, but all ingested food is thoroughly masticated. The maximum number of teeth in bats is 38, and the minimum number 20, although the range in species inhabiting the Northern Great Plains is 30 to 38 (see table 2), variation being found in the number of incisors and premolars.

During the cold months, bats seek suitable hibernacula, sometimes well removed from summer haunts, or migrate long distances to favorable environments south of the Northern Plains. Of the 14 species known from the region, individuals of at least 10 spend the winter in hibernation locally, choosing sites such as caves, abandoned mines, and man-made structures of various sorts where the temperature is relatively constant and, in any event, normally drops no lower than a few degrees above freezing. Before hibernation, which may last up to seven months, bats take on heavy deposits of fat.

In the warm months, daytime retreats are selected to provide seclusion and a suitable microenvironment. The body temperature of a roosting bat approaches that of the immediate surroundings, and so metabolism frequently goes on at a reduced level. Some species congregate in groups, such as maternity colonies. After leaving the roost at dusk or shortly after dark, bats first seek water, which they drink on the fly from ponds, streams, stock tanks, and the like, then forage until the stomach is filled, usually within an hour or two after they first become active. Following this initial period of activity, they may spend several hours resting before another feeding period and subsequent return to a day roost.

Bats molt once a year, in summer, and the molt of pregnant or lactating females frequently is delayed beyond that of other adults. Mating in hibernating species takes place in autumn and winter; sperm is retained in the uterus of the female until spring, when ovulation and fertilization occur. In migratory species, mating evidently takes place during southward migration or on the wintering grounds, because females are gravid when they arrive in the Plains region in spring.

Females of most species occurring on the Northern Great Plains bear but a single offspring; however, those of several species normally have twins, and four is the usual number in the red bat. Young bats, which are left alone by the mother during her nightly foraging flights, mature rapidly and are capable of flight in three to four weeks after birth, by which time they essentially have reached adult size. Incompletely or thinly ossified crania, lack of any wear on the teeth, incomplete fusion of the phalangeal epiphyses, and, frequently, differences in pelage characteristics distinguish volant young-of-the-year from adults. Bats are relatively long-lived, perhaps in part a result of lowered daily and seasonal metabolic rates, and some species of the Northern Plains are known to live 20 years or more.

Natural predators of bats include several kinds of carnivores and snakes, which take animals from daytime or hibernating roosts, and raptors. Man, however, poses the greatest threat to bat populations through harassment or destruction of colonial groups, for a variety of reasons, and by widespread use of chlorinated hydrocarbon insecticides, to which bats are highly sensitive. Disease decimates populations of some species from time to time. Bats are known to carry rabies, and the disease has been reported from all species that occur in the tristate region. The frequency of infection is low in most kinds of bats, however, and they do not readily transmit the disease to other mammals. A few human deaths from rabies have been attributed to bats, which seem to be the only mammals that do not themselves succumb to the disease.

One or more of the species of bats known from the Northern Plains are parasitized internally by protozoans (including trypanosomes), cestodes, nematodes, and trematodes. A variety of ectoparasites has been reported including bat bugs (Cimicidae), both winged and wingless bat flies, chiggers, fleas, mites, and ticks. Individuals of some temperate species, especially those that normally roost singly, are remarkably free of ectoparasites, however.

This general account of the Chiroptera is longer than other accounts of mammalian orders found on the Northern Plains because many of our bats have similar attributes and because little precise information is available on the natural history of some kinds. Some of the species accounts below, therefore, are abbreviated compared with many others in this book. A good general reference for additional information is *Bats of America* by Barbour and Davis (1969), published by the University of Kentucky Press. Czaplewski et al. (1979) have provided a synopsis of bats in Nebraska, and Jones and Genoways (1967b) have published an annotated checklist of those in South Dakota (see also Farney and Jones 1980 on bats of the South Dakota Badlands).

Key to Bats

1. Tail extending conspicuously beyond posterior border of uropatagium (tail membrane); anterior border of ear with six to eight horny excrescences; lower incisors bifid (Molossidae) ____*Tadarida brasiliensis*
1.' Tail not extending conspicuously (5 mm at most), if at all, beyond posterior border of uropatagium; anterior border of ear relatively smooth; lower incisors trifid (Vespertilionidae) _____2
2. Single pair of upper incisors; total number of teeth 30–32 _____3
2.' Two pairs of upper incisors; total number of teeth 32–38 _____5
3. Upper surface of uropatagium essentially naked; one pair of upper premolars (total of 30 teeth) _____ *Nycticeius humeralis*
3.' Upper surface of uropatagium thickly furred throughout; two pairs of upper premolars (total of 32 teeth) _____4
4. Dorsal pelage "frosted" (dark brownish tipped with grayish white); forearm

more than 45; greatest length of skull more than 17.5 _____ *Lasiurus cinereus*
4.' Dorsal pelage reddish orange to yellowish brown; forearm less than 45; greatest length of skull less than 14.5 _____ *Lasiurus borealis*
5. Dorsal pelage blackish frosted with white or, if not, ear tremendously enlarged (30 or more in length from notch); premolars 2/3 (total of 36 teeth) _____6
5.' Dorsal pelage not blackish frosted with white (brown, reddish brown, or yellowish brown); ear not noticeably enlarged (23 or less in length from notch); premolars 1/2, 2/2, or 3/3 (total of 32, 34, or 38 teeth) _____7
6. Upper surface of uropatagium furred proximally from a third to half its length; dorsal pelage blackish frosted with white; ear from notch less than 15 _____*Lasionycteris noctivagans*
6.' Upper surface of uropatagium only thinly furred at base; dorsal pelage pale brownish; ear from notch 30 or more _____*Plecotus townsendii*
7. Dorsal pelage pale reddish brown; upper surface of uropatagium furred proximally for about half its length; two pair of upper premolars (total of 34 teeth) _____*Pipistrellus subflavus*
7.' Dorsal pelage brownish or yellowish brown; upper surface of uropatagium only thinly furred at base; one pair or three pair of upper premolars (total of 32 or 38 teeth) _____8
8. Total length more than 110; greatest length of skull more than 18; premolars 1/2 (total of 32 teeth) _____ *Eptesicus fuscus*
8.' Total length less than 110; greatest length of skull less than 18; premolars 3/3 (total of 38 teeth) _____9
9. Dorsal pelage pale yellowish brown, contrasting noticeably with blackish ears and membranes; forearm usually less than 33; hind foot usually less than 8 from heel; mastoid breadth and interorbital breadth usually no more than 7.5 and 3.5, respectively _____ *Myotis liebii*

9.′ Dorsal pelage brownish, not contrasting noticeably with color of ears and membranes; forearm more than 33; hind foot more than 8 from heel; mastoid breadth and interorbital breadth usually greater than 7.5 and 3.5, respectively _____10

10. Ears long, usually 18 or more in length from notch; greatest length of skull more than 15.7; breadth across upper molars usually 6.0 or more _____11

10.′ Ears short or of moderate length, usually 17 or less in length from notch; greatest length of skull less than 15.7; breadth across upper molars usually less than 6.0 _____12

11. Posterior border of uropatagium with conspicuous fringe of hairs (fig. 30); forearm rarely less than 39.0, usually more than 39.5; ears smaller in direct comparison, usually 18–20 in length from notch _____ *Myotis thysanodes*

11.′ Posterior border of uropatagium not conspicuously fringed with hairs; forearm less than 39.5, usually less than 39.0; ears larger in direct comparison, usually 20–22 in length from notch _____ *Myotis evotis*

12. Ear when laid forward clearly extending beyond tip of nose, length from notch usually 16–18; length of maxillary toothrow more than 5.5; length of mandibular toothrow more than 6.9 _____ *Myotis keenii*

12.′ Ear when laid forward extending little, if at all, beyond tip of nose, length from notch usually 15 or less; length of maxillary toothrow 5.5 or less; length of mandibular toothrow 6.9 or less ____13

13. Braincase rising abruptly from rostrum; ears shorter, usually 11–14 from notch; calcar keeled (fig. 30) ____*Myotis volans*

13.′ Braincase rising gradually from rostrum; ears longer, usually 14–15 from notch; calcar not keeled _____*Myotis lucifugus*

Family Vespertilionidae: Vespertilionid Bats

The family Vespertilionidae is nearly cosmopolitan, having essentially the same distribution as described for the order Chiroptera. It represents the largest family of the order in terms of included nominal species, about 300, arranged in 35 genera. Thirteen species representing seven genera occur on the Northern Great Plains.

Genus *Myotis*

The generic name *Myotis* is derived from two Greek words, *mys*, "mouse," and *otos*, "ear," in reference to the fact that the ears of many common bats resemble those of mice. *Myotis* is the most widely distributed terrestrial mammalian genus except for *Homo* (man). It contains about 80 species, six of which occur on the Northern Plains. The dental formula is given in table 2.

Myotis evotis: Long-eared Myotis

Name. The specific name *evotis*, meaning "good ears," pertains to the conspicuous ears of this bat and is of Greek origin.

Distribution. This bat ranges over much of the western United States—from southern Canada south to the highlands of New Mexico, Arizona, and southern California. *Myotis evotis evotis* is the subspecies recognized on the Northern Plains. Although formerly thought to be more widely distributed in the region, the long-eared myotis is known certainly only from extreme northwestern South Dakota and three localities in the western part of North Dakota.

The local distribution of this species in the Dakotas seems primarily limited to pine-covered scarps and other broken ter-

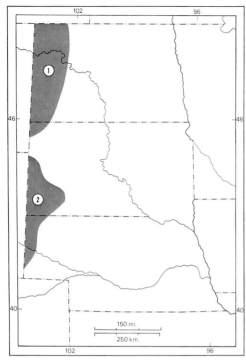

22. *Distribution of* (1) Myotis evotis *and* (2) Myotis thysanodes.

rain. In the intermontane West, it is widely distributed throughout much of the area covered by coniferous forest.

Description. This bat is a relatively large species among those inhabiting the Northern Great Plains and has the largest ears of any American *Myotis*. The basal one-fifth of the uropatagium is haired, but it is otherwise naked except for a few sparse hairs along the free border. The wing joins the foot at the base of the toes. The fur is long and glossy. *M. evotis* is pale yellowish brown dorsally, more so than most Northern Plains species, and approaches the condition found in *M. leibii*, in which dark ears and membranes contrast noticeably with the pale pelage. The hairs are everywhere dusky slate-colored at the base. The ventral pelage is paler than that above.

Among species in this region, *M. evotis* most closely resembles *M. thysanodes*, from which it differs in having a slightly smaller skull, shorter forearm, longer ear,

no conspicuous fringe of hairs along the posterior margin of the uropatagium, less well developed sagittal crest, slightly lower and less inflated braincase, and somewhat less robust molar teeth. Another species with modestly long ears, *M. keenii*, is smaller and somewhat darker and has paler ears that do not contrast much with the dorsal coloration.

Average and extreme external and cranial measurements of 12 specimens from Harding County, South Dakota, are: total length, 93.4 (87–100); length of tail, 40.2 (34–45); length of hind foot, 9.7 (8–11); length of ear, 20.5 (19–22); length of forearm, 38.6 (36.9–39.3); greatest length of skull, 16.3 (15.8–17.0); zygomatic breadth, 9.7 (9.5–10.1). As in other species of *Myotis*, males and females are of approximately the same size, and no differences in pelage are apparent. A series of adults from southeastern Montana weighed between 5.7 and 7.6 grams.

Natural History. The long-eared myotis has been recorded as using sheds, cabins, caves, abandoned mines, and the loose bark of trees for roosting sites. Caves and mine shafts often are used as nocturnal retreats; in fact, one of the most effective ways to capture this species is to net at cave entrances and mine openings at night. Hibernacula are unreported from our region, and, for that matter, little is known of this aspect of the biology. We are aware of

23. *Long-eared myotis,* Myotis evotis *(courtesy T. H. Kunz).*

hibernating individuals having been taken in a mine in eastern Montana, and caves no doubt also serve as favorable winter retreats.

This bat apparently forms small maternity colonies, although not much has been recorded concerning its reproductive habits. Pregnant females have been taken in South Dakota in late May and mid-June, and volant young-of-the-year have been captured the first week of August.

Myotis evotis emerges at late dusk or after nightfall and flies among trees and over water foraging for food. Several species of *Myotis, Eptesicus fuscus, Lasiurus cinereus,* and *Lasionycteris noctivagans* frequent many of the same areas as does the long-eared myotis.

The annual molt takes place in July and August. Five of seven specimens taken in South Dakota in early August were in new pelage, and the others were nearing completion of molt.

Selected References. General (Barbour and Davis 1969); distribution (Jones and Choate 1978).

Myotis keenii: Keen's Myotis

Name. The name *keenii* is a patronym recognizing the Reverend John Henry Keen, who collected the type specimen of this species in 1894.

Distribution. Myotis keenii occurs throughout southeastern Canada and much of the eastern United States, westward to Alberta and eastern British Columbia, eastern Montana, and central Kansas, and southward almost to the Gulf of Mexico. A disjunct population also is known from western British Columbia and adjacent northwestern Washington, but recent investigations suggest that this population may represent a species distinct from the one to the east (in which case the specific name of the bat treated here would become *septentrionalis*).

The distribution of Keen's myotis is not well known in the tristate region. It has been recorded from scattered localities in

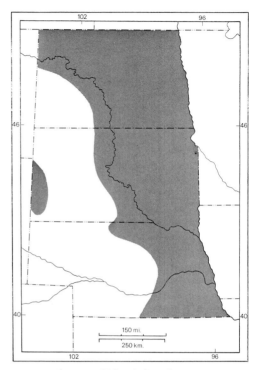

24. *Distribution of* Myotis keenii.

eastern Nebraska, westward at least to Keya Paha and Webster counties, and in the Dakotas, where it possibly occurs in suitable habitats westward to approximately the Missouri River or slightly beyond. It is also a common inhabitant of the Black Hills, but the population there probably is isolated from the main range of the species to the east. The subspecies on the Northern Plains is *Myotis keenii septentrionalis.*

Description. Keen's myotis is similar in appearance to the little brown myotis, *M. lucifugus,* and a combination of characteristics frequently must be used to distinguish between the two species. The ear, which in *M. keenii* extends, when laid forward, well beyond the tip of the nose, may be the single best external character by which the two species can be separated, because in *M. lucifugus* the ear does not extend beyond the tip of the nose (or barely so) when laid forward (length of ear from notch usually 14 or 15 in *lucifugus,* 16 to 18 in *keenii*); the tragus also is longer in *keenii*. The pelage is generally paler and less glossy (appearance may be somewhat grizzled) than in *M. lucifugus*, averaging dull brown dorsally and pale grayish brown ventrally. The ears are pale brownish, the wing membrane extends to the base of the toes, and the calcar is slightly keeled. The skull is longer than in *M. lucifugus* (see measurements) and rises more gradually from the rostrum to the braincase upon direct comparison in lateral view.

Average and extreme external and cranial measurements of 10 specimens from Nebraska are: total length, 93.5 (86–99); length of tail, 39.5 (36–43); length of hind foot, 9.3 (8–10); length of ear, 16.9 (16–18); length of forearm, 35.3 (33.3–36.5); greatest length of skull, 15.3 (14.7–15.6); zygomatic breadth, 9.2 (8.6–9.6). Two males and two females (July-taken) from eastern South Dakota weighed 5.2, 6.7, 7.3, and 8.4 grams, respectively; a December-taken male from Nebraska weighed 6.3 grams.

Natural History. Keen's myotis may be less gregarious, and thus less easily observed and captured, over much of its distribution than many other species of the genus. In the Dakotas and Nebraska it has been collected in summer in wooded habitats—usually along rivers or streams or close to other bodies of water. On the Black Hills this bat has been taken in both deciduous and coniferous habitats. In a recent study in Missouri, *M. keenii* was found to forage mainly over forested hillsides and ridges, frequenting areas just above shrub height under the forest canopy.

Foraging begins about dusk and may continue intermittently during the night, with a second period of marked activity before dawn. Individuals may seek temporary shelter during the night, but they usually return to a more permanent daytime roost—behind bark of trees, under shingles or behind shutters of buildings, or within abandoned or little-used man-made structures, including mines and quarries—spending the day singly or in small groups. Individual *M. keenii* occasionally roost with bats of other species.

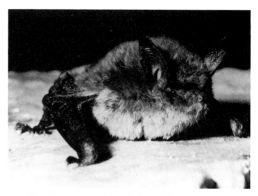

25. *Keen's myotis*, Myotis keenii *(courtesy T. H. Kunz).*

In hibernation, these bats seek caves or mines, and congregations of several hundred have been reported. The period of hibernation is not exactly known, but it probably extends, at least in the southern part of the tristate region, from late September or early October until late March or April. Normally, *M. keenii* hibernates in the same refugia as one or more other vespertilionid species, especially *Myotis leibii, M. lucifugus, Eptesicus fuscus,* and *Pipistrellus subflavus,* and it seems to prefer cool, moist sites where the air is still. Individuals usually overwinter alone or in small groups, hanging in the open or, more often, wedging themselves into cracks or holes. Rarely are winter hibernacula occupied in summer.

Annual molt in *M. keenii* takes place in July and early August on the Northern Plains. Few details are available on growth and development in this bat, but these features likely are similar in all species of *Myotis.*

As in other hibernating bats, copulation takes place in autumn before entry into torpor. A biologist who observed individuals mating in a bat trap reported that males mounted females from the rear and in some instances grasped them by the back of the neck with their teeth. Pregnant females have been reported in late May in Nebraska and from late May to late June in Iowa. Lactating animals are on record from late June to late July in Iowa and as late as mid-August on the Black Hills. Maternity colonies range from a few bats up to 30 or more. Banding studies indicate that individuals may live at least as long as 18.5 years.

Selected References. General (Fitch and Shump 1979 and included citations); systematics (Turner 1974; van Zyll de Jong 1979).

Myotis leibii: Small-footed Myotis

Name. This bat was described in 1842 by John J. Audubon and John Bachman, who named it in honor of Dr. George C. Leib, collector of the type specimen. The technical name has had a somewhat tortuous history. For many years the specific epithet *keenii* was mistakenly applied to the small-footed myotis. The true identity of *keenii* was ascertained in 1928, and the specific name *subulatus* then was employed for the bat we now know as *leibii.* Unfortunately, Thomas Say, who proposed the name *sub-* *ulatus* in 1823, apparently applied it to still another, closely related species, thus necessitating use of the later-proposed name *leibii.*

Distribution. This saxicolous species occurs throughout much of the western United States, from southern Canada to northern Mexico. In the East it is found in much of New England, in adjacent southeastern Canada, south in the Appalachians to Georgia, and in apparently isolated populations in Kentucky, Missouri, and Arkansas. On the Northern Great Plains, one subspecies (*Myotis leibii ciliolabrum*) is known from southwestern North Dakota southward through western South Dakota to the western half of Nebraska. No specimens currently are on record from south of the Platte River in southwestern Nebraska, but *M. leibii* has been reported from adjacent northwestern Kansas. The easternmost record on the Northern Plains is from along the Niobrara River, just west of the 100th meridian, in Keya Paha County, Nebraska.

Description. This handsome little bat is, save possibly for *Pipistrellus subflavus,* the smallest member of the order Chiroptera in the Northern Plains region. The dorsal pelage is pale yellowish brown to golden brown, contrasting markedly with the blackish ears and membranes. There is a blackish mask on the face. The venter varies from pale buff to whitish. The most notable external characteristic other than small size and pale coloration is the strongly keeled calcar.

Average and extreme external and cranial measurements of six specimens (a male and five females) from northwestern Nebraska are: total length, 87.2 (82–89); length of tail, 38.0 (34–44); length of hind foot, 8.2 (7–9); length of ear, 14.3 (13–15); length of forearm, 31.8 (30.8–32.9); greatest length of skull, 14.1 (13.9–14.4); zygomatic breadth, 8.5 (8.2–8.7). Weights of a large series of males and nonpregnant females from the South Dakota Badlands ranged from 3.5 to 5 grams. Pregnant females may weigh as much as 7 grams.

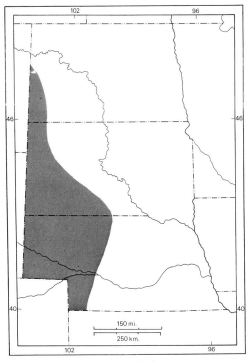

26. *Distribution of* Myotis leibii.

27. *Small-footed myotis*, Myotis leibii
(courtesy M. D. Tuttle).

Natural History. The small-footed myotis is closely associated with rocky habitats throughout much of its distribution. On the Northern Plains it is, for example, a common inhabitant of the Black Hills and Badlands of South Dakota and the Pine Ridge area of northwestern Nebraska, but it occurs in most areas in the western parts of the tristate region where bluffs, dissected breaks and badlands, ridges, cliffs, or major outcroppings prevail.

In winter this bat is known to hibernate in caves and mines. When hibernating, individuals seldom hang in the open, but rather secrete themselves in cracks and crevices, sometimes singly and sometimes in small groups. In summer, day roosts have been found in buildings, behind bark of pine trees, in rocky or eroded crevices and cracks, under rocks on the ground, and even in holes in banks and hillsides and in abandoned swallow nests. Night roosts in summer are typically in caves or mines, beneath rock ledges, and in man-made structures. Individuals forage along cliffs and ledges, among conifers and along deciduous riparian growth, and over streams and ponds.

Females give birth to a single offspring (one case of apparent twins is on record) in July on the Northern Plains, and lactation continues well into August. Eleven pregnant females taken in the Badlands of South Dakota in the last week of June and first week of July carried fetuses measuring

from 13 to 18 mm in crown-rump length. Females and their young usually roost alone by day, although occasional small maternity colonies have been reported. Flying young-of-the-year have been taken in the Badlands as early as 26 July. The known record for longevity in *M. leibii* is 12 years.

Annual molt takes place from late June through July, reproductively active females generally molting later than other adults. Little is known concerning dates of hibernation on the Northern Plains, but in the eastern United States this species is active well into the autumn and emerges from hibernation as early as March.

Selected References. General (Barbour and Davis 1969); natural history (Turner 1974; Tuttle and Heaney 1974).

Myotis lucifugus: Little Brown Myotis

Name. The specific name of this bat is formed from two Latin words that in combination mean "light-fleeing." In the vernacular, the name little brown bat sometimes is used.

Distribution. This well-known species has a broad distribution in North America, from central Alaska southward through most of the United States to central Mexico. Two subspecies occur on the Northern Plains, *Myotis lucifugus lucifugus* in the eastern part of the region and *Myotis lucifugus carissima* in the west. The two subspecies evidently do not meet across at least some of the essentially treeless central parts of the tristate region, but they probably do so in North Dakota and possibly along such east-west drainages as the Niobrara River in northern Nebraska. Accordingly, the accompanying distribution map must be regarded as tentative.

Description. The pelage of the little brown myotis is somewhat more sleek and glossy than that of other species of the genus occurring on the Northern Plains. The dorsal color is dark brown to buffy brown, often with a metallic, coppery sheen; the

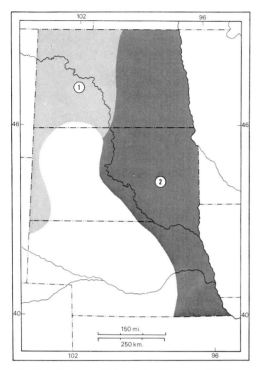

28. Distribution of Myotis lucifugus: (1) M. l. carissima; (2) M. l. lucifugus.

10.6 (10–12), 9.7 (8–11); length of ear, 14.3 (13–15), 14.9 (14–16); length of forearm, 37.4 (35.8–40.2), 36.7 (35.0–38.4); greatest length of skull, 14.7 (14.3–15.1), 14.6 (14.3–15.0); zygomatic breadth, 9.1 (8.9–9.4), 9.0 (8.9–9.2). As in other temperate-zone bats, weight varies both individually and seasonally, but most specimens we have taken in summer in the western part of the tristate region weighed between 5.5 and 8.5 grams. Pregnant females may reach 13 grams or more.

Of the species of *Myotis* known from the Northern Great Plains and adjacent areas, the little brown myotis is likely to be confused with *M. keenii, M. volans,* or *M. sodalis.* For means by which it can be recognized in comparison with *keenii* and *volans,* see accounts of those species. From *sodalis,* which has not been reported from the tristate region but well may be found in southeastern Nebraska, *lucifugus* differs in having glossy (rather than dull) pelage, no keel on the calcar (*sodalis* has a slight keel), and long hairs on the toes, which reach the tips of the claws or beyond (and which are lacking in *sodalis*).

ventral coloration is paler than that on the dorsum, frequently with a slight grayish tinge. The ears, about the same color as the dorsum, are slender, naked, and rounded; the tragus is relatively short and blunt. Albinos have been reported.

The wings and uropatagium are naked except along their proximal margins. The calcar is not keeled. The skull, which is of average size among *Myotis* on the Northern Plains, lacks a distinct sagittal crest. As in many other species of bats, there is a single pair of pectoral mammae.

Of the two subspecies of *M. lucifugus* known from the Dakotas and Nebraska, *M. l. carissima* is paler, both dorsally and ventrally, than *M. l. lucifugus* and slightly larger externally and cranially. Average and extreme measurements of series of adults of the two races from northwestern and eastern Nebraska (*carissima* followed by *lucifugus*) are, respectively: total length, 96.4 (94–99), 91.2 (85–100); length of tail, 39.1 (37–42), 36.8 (33–40); length of hind foot,

Natural History. This species is widely distributed and is one of the most common of North American bats. Thus there is a considerable accumulation of data on its natural history. In summer, maternity colonies usually frequent warm buildings, but they have been found under bridges and occasionally in caves, crevices, hollow trees, and the like; in any event, these colonies usually are near a body of water such as a lake, pond, or stream. Males and barren females may roost in a variety of locations—in stacks of lumber, behind shutters, under loose bark of trees, in rock outcroppings and eaves of buildings, and in similar retreats—singly or in small groups. Large maternity congregations, which may number more than 1,000 bats, sometimes can be located through the noisy chattering of their members. In late summer, swarms of *M. lucifugus* have been reported around potential hibernating sites.

Hibernation on the Northern Plains takes place from late September or early October

29. *Little brown myotis,* Myotis lucifugus
(courtesy T. H. Kunz).

to April or early May, with caves and mines serving as hibernacula; females enter these retreats later and leave them earlier than do males. In torpor, these bats usually hang singly or in small groups, or occasionally wedge themselves into cracks or crevices, but tightly packed clusters have been reported, particularly in the northern part of the range of the species. Temperatures at hibernation sites usually vary from a few degrees above freezing to nearly 60° F in areas of relatively high humidity but occasionally are slightly below freezing. As in the eastern pipistrelle, droplets of water frequently condense on the pelage of torpid individuals. The bats arouse every few weeks and may move within a mine or cave as conditions of the microclimate change. They may take water during periods of arousal, but feeding apparently does not take place. Like most other species, *M. lucifugus* cannot stand prolonged periods of subfreezing temperatures, and this may be a significant cause of winter mortality, especially in young-of-the-year. In eastern Nebraska the little brown myotis occupies the same man-made quarries in winter as *Myotis keenii, Eptesicus fuscus,* and *Pipistrellus subflavus;* in Jewel Cave in the Black Hills, this bat has been taken in hibernation along with *Myotis leibii, M. thysanodes, M. volans,* and *Plecotus townsendii.*

Individuals of this species begin their nightly foraging at late dusk. Apparently many follow a set flight pattern night after night, coursing over water, along the edges of wooded areas and among trees, and over nearby open areas such as pastures or city streets, frequently at heights of 10 to 20 feet above the ground. In a study in Iowa, *M. lucifugus* was found to have a single foraging period, lasting up to five hours after sunset, but no secondary period of activity as was characteristic of *M. keenii,* although clustering at night roosts after an initial feeding period has been reported. When they return to the daytime roost, many bats make several passes by the entrance before finally entering.

The little brown myotis breeds in the autumn and occasionally in hibernacula in winter, the male mounting the female from the rear. As in other hibernating vespertilionids, spermatozoa are stored in the uterus of the female, where they remain viable over the winter months, and fertilization takes place when the bats emerge from hibernation in spring. The gestation period is reportedly 50 to 60 days.

Parturition dates vary with latitude and to some degree individually as well. On the Northern Plains, pregnant females have been taken from late May to late June; volant young-of-the-year have been collected as early as late July. The single young (twins are rare—there were only two cases of twinning in 312 pregnancies in a recent study in Alberta, for example), which weighs 1.5 to 2 grams at birth, clings tightly to a nipple when with its mother in the first days of life. Neonates are flesh-colored and covered with fine, silky hair. The deciduous teeth are almost fully erupted at birth, and the eyes and ears open a few hours afterward. The young bat grows rapidly and can fly in about three weeks, by which time the permanent dentition is almost entirely in place.

In giving birth, the female hangs head up and cups the interfemoral membrane (uropatagium) to catch the newborn young, which emerges from the womb by breech presentation. The young one assists its own delivery by groping with its legs and making various other movements.

As noted, *M. lucifugus* hibernates in the Dakotas and Nebraska and thus is not reckoned as a migratory species. Nonetheless, individuals may disperse 100 miles or more from hibernacula to summer haunts. Flights of up to 50 miles in a single night and speeds of up to 19 miles per hour have been reported. Animals displaced experimentally as far as 270 miles from their home colonies returned to them, indicating a strong "homing" behavior. Males may, on the average, live longer than females. The known record of longevity is slightly more than 30 years, longer than recorded for any other bat.

Individuals spend considerable time grooming themselves in both daytime and nighttime roosts and occasionally upon arousal from hibernation. In doing so they use the claws of the hind feet to comb the pelage and the tongue and teeth to clean the wings.

The annual molt in adults normally takes place in July, the new pelage growing in under the old simultaneously over much of the body; thus, distinct molt "lines" are rarely evident. The pelage of young-of-the-year is sparser, shorter, and somewhat darker and duller than that of adults, and maturational molt takes place in late summer before entry into the first winter of hibernation.

Selected References. General (Fenton and Barclay 1980 and included citations); natural history (Keen and Hitchcock 1980; Schowalter, Gunson, and Harder 1979).

Myotis thysanodes: Fringe-tailed Myotis

Name. The specific name *thysanodes* is of Greek origin and means "fringelike," in reference to the short, stiff hairs along the posterior border of the uropatagium, or tail membrane, in this species.

Distribution. The fringe-tailed myotis is distributed in montane habitats throughout much of western North America, from southwestern Canada to southern Mexico. One subspecies, *Myotis thysanodes pahasapensis,* occurs on the Northern Great Plains, and it is known only from western South Dakota, northwestern Nebraska, and adjacent areas of Wyoming.

As now understood, this subspecies represents a geographically isolated population (see fig. 22), the main range of the species lying to the west.

Description. M. thysanodes is of medium size among the large-eared members of the genus *Myotis* in North America, adults weighing from about 5.5 to 8.5 grams. The dorsal pelage varies from medium brown to rather pale buff, the individual hairs being grayish black basally. Ventrally the pelage is somewhat paler. The ears and membranes, because they are blackish brown, frequently contrast noticeably with the

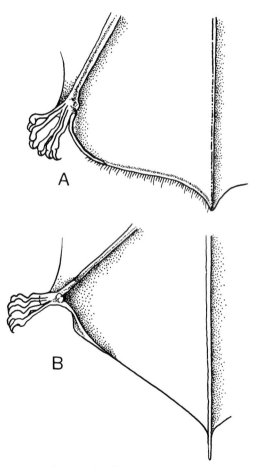

30. *Dorsal view of tail, uropatagium, and hind foot of* (A) Myotis thysanodes *and* (B) M. volans. *Note the conspicuous fringe of hairs on the uropatagium of the former and the distinctly keeled calcar of the latter.*

color of the pelage. A fringe of pale, straw-colored hairs extends posteriorly 1.0 to 1.5 mm beyond the edge of the uropatagium.

Mean and extreme external and cranial measurements of six specimens from the Black Hills are representative: total length, 92.7 (90–96); length of tail, 41.2 (40–44); length of hind foot, 10.7 (10–11); length of ear, 18.8 (17.5–21); length of forearm, 41.2 (39.2–43.3); greatest length of skull, 16.8 (16.4–17.0); zygomatic breadth, 10.2 (10.0–10.5). Superficially, *M. thysanodes* resembles *M. evotis*, but the two can be distinguished as explained in the account of the latter species.

Natural History. This bat seems to prefer montane and upland forests throughout most of its range in the United States, although it also occurs in other wooded habitats, such as desert scrub, and is known from some nonwooded areas as well. On the Northern Great Plains it is relatively common on the Black Hills and is known also from the South Dakota Badlands, the Pine Ridge of northwestern Nebraska, and the Wildcat Hills to the south of the Platte River in the Panhandle of that state. These bats often are observed at dusk foraging along watercourses or over standing water. The species commonly utilizes buildings as day roosts in summer, but it also frequents caves and abandoned mines, especially at night. Usually solitary, *M. thysanodes* occasionally clusters with other species while roosting during the day, and females form maternity colonies that may number several hundred mothers and young. In such colonies there is some evidence of indiscriminate care and nursing of young by females. Individuals of this and other species of *Myotis* can thermoregulate if the energy requirements in so doing are not too great.

The fringe-tailed myotis is resident on the Black Hills throughout the year, individuals hibernating in various caves, including Jewel Cave, and abandoned mines. Elsewhere in the region, these bats may hibernate in man-made structures, small caves and rock fissures, and abandoned mine shafts, although details are lacking.

Local migrations between summer habitats and winter hibernacula have been reported. The period of hibernation extends from September to April or early May but may be interrupted periodically.

Molt, frequently evidenced by new hairs protruding beneath old pelage over both dorsum and venter, has been observed from early July to mid-August. Females give birth to a single young in late June or early July. Three pregnant females captured in mid-June on the Black Hills carried fetuses that measured 6, 8, and 12 mm (crown-rump length). Lactating females have been collected in late July and throughout most of August. The earliest date on which a flying young-of-the-year has been taken in our region is 16 August. The gestation period after implantation is thought to be 50 to 60 days.

Newborn fringe-tailed myotis are hairless and pinkish, with a total length of about 50 and a forearm about 16 mm long. In one population studied, the sex ratio of young was equal. The eyes open and the ears unfold on the first day. Pigmentation develops in the skin by the end of the first week, followed by growth of soft new hairs from the pigmented areas. The young can fly by day 20, and they reach adult size at about the same time. Longevity is not well documented in *M. thysanodes*, but a female banded in New Mexico is known to have lived at least 11 years.

Selected References. General (O'Farrell and Studier 1980 and included citations); distribution (Jones and Choate 1978); systematics (Jones and Genoways 1967a).

Myotis volans: Long-legged Myotis

Name. The specific name *volans* is derived from a Latin word meaning "flying." Hairy-winged myotis is another vernacular name frequently cited.

Distribution. The long-legged myotis ranges over much of western North America, from southwestern Canada south to central Mexico. It principally inhabits open forest lands. Its distribution on the North-

31. *Fringe-tailed myotis.* Myotis thysanodes (*courtesy R. W. Barbour*).

ern Plains evidently is limited to north-western Nebraska and the western parts of the Dakotas. This species appears to be the most common member of the genus on the Black Hills and on the pine-clad buttes of northwestern Nebraska and South Dakota. In North Dakota it reaches the easternmost point of its known geographic distribution in the United States. Only one subspecies, *Myotis volans interior*, occurs in the region.

Description. This bat is of medium size among members of the genus; adults from the Northern Plains have been recorded as weighing from 4.8 to 10 grams, but most fall in the range of 5.5 to 8.5. The pelage is relatively long and soft, extending distally onto the tail membrane and in some speci-mens onto the underside of the wing mem-brane to the level of the elbow. Dorsally, the pelage varies from smoke brown to chocolate brown; tips of individual hairs are slightly burnished. The underparts vary from smoke brown to dull yellowish brown, washed in buff. The ears are low, rounded, and do not reach the nose when laid forward. The calcar is distinctly keeled.

External and cranial measurements of 10 specimens from Harding County, South Da-kota, are: total length, 96.9 (87–104);

length of tail, 43.6 (35–46); length of hind foot, 8.8 (8–9.5); length of ear, 13.8 (13–15); length of forearm, 38.7 (37.0–40.2); greatest

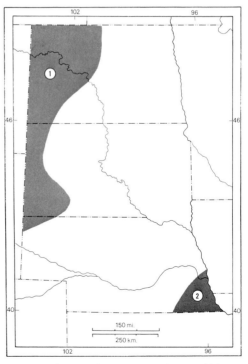

32. *Distribution of* (1) Myotis volans *and* (2) Pipistrellus subflavus.

length of skull, 14.3 (13.6–14.6); zygomatic breadth, 8.9 (8.7–9.2). The skull has a relatively short rostrum, and the braincase is abruptly elevated above the rostrum and appears more inflated than in other species of the genus treated in this book.

Myotis volans somewhat resembles *M. lucifugus*. However, the latter lacks a keel on the calcar, has longer ears, which reach the nostrils when laid forward, has a braincase that rises gradually from the rostrum, and usually is somewhat paler in color.

Natural History. Although the long-legged myotis is common throughout its known range, relatively little is known of the habits of this bat. It seems to prefer open montane forests and often is common in coniferous habitats. Occasionally this species is found in evergreen-deciduous forests, and it evidently tolerates the essentially treeless, barren badlands of northwestern Nebraska and west-central South Dakota. These bats frequently are collected as they forage over woodland meadows or watercourses. Abandoned buildings, caves, old mine shafts, hollow trees, rock fissures, and loose bark are utilized in summer as retreats during the day and for roosts at night after foraging flights. Typically, *M. volans* roosts singly or in small groups, except that it forms large maternity colonies, one consisting of "about 180" females having been

33. Long-legged myotis, Myotis volans (courtesy R. W. Barbour).

found in 1944 in Sioux County, Nebraska. In winter, caves and mines are known to be used as hibernacula, individuals usually hanging singly on walls or ceilings or wedging themselves into cracks and crevices.

The annual molt normally begins in mid- to late June and is completed in July, but, as in many other bats, it may be delayed a month or more in reproductively active females. Curiously, parturition occurs about a month and a half earlier in the western part of the range than in the east. On the Northern Plains, young normally are born from mid-July through mid-August, later than in other bats. Of 19 adult females captured in Sioux County, Nebraska, on 15 August 1975, for example, nine carried large fetuses and an additional four were lactating. Volant young, however, have been taken in South Dakota as early as 20 July, which seems to indicate an unusually prolonged and perhaps variable reproductive season.

Selected References. General (Barbour and Davis 1969); natural history (Turner 1974).

Genus *Lasionycteris*

The name *Lasionycteris* is derived from two Greek words—*lasios*, meaning "hairy," and *nykteris*, meaning "bat." The dental formula is given in table 2. The genus is monotypic and occurs only in North America.

Lasionycteris noctivagans: Silver-haired Bat

Name. The specific name *noctivagans* is a combination of two Latin words, *noctis*, meaning "night," and *vagans*, meaning "wanderer." The vernacular name silver-haired bat is universally used for this species.

Distribution. This handsome and distinctive bat occurs throughout much of the continental United States and is found northward to southeastern Alaska and central Canada and southward into the moun-

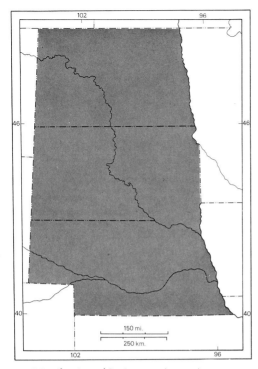

34. Distribution of Lasionycteris noctivagans.

tains of northeastern Mexico; it is also on record from Bermuda. *L. noctivagans* is a migratory species, spending the warm months of the year in forested areas in the northern parts of its range or in montane habitats, where it is locally abundant. Little is known of its precise distribution and occurrence on the Northern Plains, although this species no doubt migrates through much of the region in spring and again in late summer and early autumn. Some individuals spend the summer in the higher elevations of the Black Hills and nearby areas, where young are reared, and this bat also may occur in summer throughout much of North Dakota, elsewhere in South Dakota, and possibly as far south as northern Nebraska. To the east of the tristate region, females are known to be resident in summer as far south as south-central Iowa. Additionally, one November-taken hibernating individual is on record from the Black Hills.

Description. The silver-haired bat can be distinguished from all other species on the Northern Plains by its distinctive coloration. The dorsal pelage is long and blackish brown, "frosted" with silvery white, a situation most conspicuous on the back. There is no sexual dimorphism in coloration, although the pelage in young bats is markedly darker than in adults, and the whitish tips of the hairs are more conspicuous. The membranes are blackish brown, and the ears are short, rounded, and naked; the wings are naked, although the tail membrane is furred on the upper surface. The skull is broad and flat, and the rostrum is strongly concave.

Average and extreme external and cranial measurements of 10 specimens from western South Dakota are: total length, 100.6 (91–108); length of tail, 40.8 (38–45); length of hind foot, 8.7 (7–10); length of ear, 15.2 (12–17); length of forearm, 40.4 (38.7–42.3); greatest length of skull, 16.3 (15.5–16.9); zygomatic breadth, 9.9 (9.7–10.2). Five June- and July-taken males weighed an average of 9.3 (8.7–10.4) grams, and a torpid male captured in a cave in Pennington County, South Dakota, on 19 November 1967 weighed 11.4 grams. Weights of a large series of summer-taken males and nonpregnant females from southeastern Montana varied between 8.5 and 12.5 grams.

Natural History. The silver-haired bat usually roosts singly under loose bark and in hollows of trees, but little is known of its specific roosting behavior owing to its solitary habits. Other roosts have been reported in open buildings, bird nests, and woodpecker holes, and in dense canopy. Migrants frequently are found in outbuildings or in piles of boards, bricks, fenceposts, and the like, especially on the essentially treeless plains. *L. noctivagans* hibernates in hollow trees, buildings, caves, mines, and rock crevices.

Silver-haired bats generally emerge earlier in the evening than other bats and fly in a slow, leisurely pattern not far above the ground. They sometimes can be seen in flight when temperatures are near freezing. On the Black Hills this bat reportedly pre-

35. *Silver-haired bat,* Lasionycteris noctivagans *(courtesy T. H. Kunz).*

fers to forage over grassy valleys that contain a source of standing water and that are lined by forested hillsides. According to findings in a study in Iowa, *L. noctivagans* is one of the species that exhibits a distinct secondary period of foraging in early morning hours; data from southeastern Montana, however, do not reveal a marked bimodality in nocturnal activity. There is some evidence that this species may shift periods of nocturnal activity as a result of competition with other bats.

Mating takes place in the autumn, and the young, usually twins, are born in June or early July. The eyes of neonates are closed, and the body is pink except for the dark membranes. Within several hours after birth the skin becomes darkly pigmented. As in many other bats, young probably can fly at an age of three to four weeks. Most females and their young roost alone, but there is some evidence that maternity colonies occasionally are formed.

As indicated earlier, *Lasionycteris* is a migratory bat, and individuals vacate the northern parts of the range of the species in late summer and early autumn, to return again in late spring. Some silver-haired bats apparently migrate to the southern United States and as far as northeastern Mexico, where they evidently are active during the winter, although details of natural history

on the winter range are scarce. There is increasing evidence, however, that some members of this species hibernate at middle latitudes. There is one record of a November-taken torpid individual from a cave in the Black Hills, and other hibernating bats have been recorded from Minnesota and Illinois eastward through southeasternmost Canada to Long Island, and also in British Columbia. Much remains to be learned of the winter habits and whereabouts of *L. noctivagans,* but it may be a species in which both hibernation and migration to southerly latitudes are options for coping with the severe winters characteristic of most of the summer range of the species. In the warm months, the earliest and latest records from the tristate region are 12 May 1913 in North Dakota and 2 October 1945 in Nebraska.

Selected References. General (Barbour and Davis 1969); distribution (Izor 1979); natural history (Jones et al. 1973; Reith 1980).

Genus *Pipistrellus*

The name *Pipistrellus* is the Latinized form of an Italian word, *pipistrello,* meaning "bat." The dental formula is given in table 2. The genus, which may contain as many as 50 species, is highly diversified in the Old World. Only two species occur in North America, one of which reaches the Northern Great Plains.

Pipistrellus subflavus: Eastern Pipistrelle

Name. The name *subflavus* is a combination of the Latin words *sub* ("below") and *flavus* ("yellowish") in reference to the general color of the underparts of this bat.

Distribution. This species ranges over much of eastern North America, from Quebec to Honduras. One subspecies, *Pipistrellus subflavus subflavus,* is found on the Northern Plains but is known there only from limestone quarries on either side of the Platte River in Cass and Sarpy counties, Nebraska (see fig. 32).

Description. The eastern pipistrelle is one of the smallest bats found in the tristate region. The color of pelage varies from yellowish to brownish gray, somewhat paler on the venter than above. Individual dorsal hairs are tricolored; the base and tip of each hair are dark, separated by a paler middle area. Fur covers the anterior third of the uropatagium. There is no difference in color between the sexes and little or no seasonal variation, although melanism is rather common in the northeastern part of the range of the species. The wing membranes are generally darker than the pelage. The tragus is short, blunt, and curved. The calcar is not keeled. The naked ears are longer than broad, and the tips are rounded; when laid forward they extend slightly beyond the nostrils.

Average and extreme external and cranial measurements of a series of 10 adults from eastern Nebraska are: total length, 83.0 (75–90); length of tail, 37.5 (33–41); length of hind foot, 9.1 (8–10); length of ear, 13.0 (12–14); length of forearm, 33.2 (31.6–34.5); greatest length of skull, 13.1 (12.6–13.4); zygomatic breadth, 8.0 (7.5–8.4). As in other hibernating species, the weight is lowest in spring and greatest before torpor begins in early autumn. In a study in Kansas, males were found to average 7.5 grams in September but only 4.6 in April; averages for females from the same months were 7.9 and 5.8.

Natural History. Although *P. subflavus* is one of the most common bats in the eastern United States, little is known of its habits in summer. Individuals are thought to forage primarily in open wooded areas and along the borders of woodlands, frequently near or over water; rarely do these pipistrelles forage in deep woods or over open fields. This species usually appears early in the evening and can be recognized by its small size and weak, erratic flight. In daytime in summer, eastern pipistrelles are rarely found in buildings (although small maternity colonies have been recorded in such shelters), and they may seek diurnal retreats in the foliage of trees. In the Southeast these bats have been recorded as roosting in Spanish moss during the day. Occasionally individuals are found hanging in open buildings, porches, and the like. Caves, mines, and crevices in rocks are used as hibernacula in winter as well as night roosts in the warm months. Seasonal movements are not well understood, but short migrations of up to about 50 miles are known between summer and winter haunts.

36. *Eastern pipistrelle,* Pipistrellus subflavus (courtesy T. H. Kunz).

In late summer, hundreds of pipistrelles may join other bats swarming about caves that will serve some as hibernacula. Caves and abandoned mines are the best-known hibernation sites. Individuals usually are single in torpor (although small groups have been reported), seeking relatively warm and humid areas within the hibernaculum, and they may arouse and move occasionally during the winter. Because of the humid surroundings, beads of water frequently collect on hibernating pipistrelles. In adjacent Kansas, this bat is known to enter winter quarters in September and to quit them in May and early June.

The reproductive biology of the eastern pipistrelle has not been fully documented. In a study in Missouri, spermatozoa were found in the uteri of females from November to April. Birth of young (one to three, usually twins) evidently takes place in late June or early July at the latitude of Nebraska. Offspring are relatively large at birth and are volant in three to four weeks; their pelage is distinctly darker than that of adults, and molt into the adult coat may not be completed before hibernation. In a population analyzed in West Virginia, survival was poor in the first year, peaked in the third, and gradually declined thereafter. Females were found to live about 10 years, and a few males lived as long as 15.

Selected References. General (Barbour and Davis 1969; Lowery 1974); natural history (Davis 1966; Jones and Pagels 1968).

Genus *Eptesicus*

The generic name *Eptesicus* is of Greek origin, apparently derived from the words *epien* and *oikos,* and means "house-flier." The genus is distributed nearly worldwide in temperate and tropical regions, being represented by about 30 species (somewhat dependent on where generic boundaries are drawn), only one of which occurs in the United States. The dental formula is given in table 2.

Eptesicus fuscus: Big Brown Bat

Name. The specific name *fuscus* is a Latin word meaning "dark."

Distribution. The big brown bat is distributed all across the United States and southern Canada, southward to northern South America. Two subspecies are recognized on the Northern Great Plains, *Eptesicus fuscus fuscus,* which occurs in the eastern third of the tristate region, and *Eptesicus fuscus pallidus,* which ranges throughout the western part.

In summer this bat is abundant in many areas but is more or less restricted locally to places that provide suitable daytime roosts, including lodging for maternity colonies, and a ready source of food. In winter the species hibernates in caves, fissures, or man-made structures such as quarries, mines, buildings, storm sewers, and similar shelters.

Description. Eptesicus fuscus is easily distinguished from most other species on the

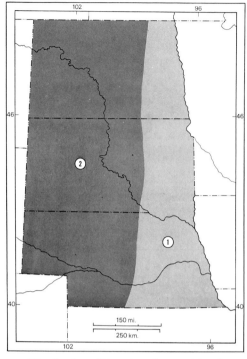

37. *Distribution of* Eptesicus fuscus: (1) E. f. fuscus; (2) E. f. pallidus.

Northern Plains by its large size (only the hoary bat is larger). The pelage is brown in color, paler on the venter than above, and hairs extend only slightly onto the wing and tail membranes. The face is somewhat darker than the rest of the dorsum, and the dark-colored ears are of medium size, thick, and hairless. The tragus is relatively short (less than half the length of the ear) and blunt, and the calcar usually is keeled. The skull is large and heavy, slightly exceeding in overall size that of the hoary bat, and is flattened dorsally. *E. fuscus* has two incisors and one premolar in each upper toothrow, whereas the hoary bat has one and two, respectively, although both species have a total of 32 teeth.

The two subspecies in the Northern Plains region can be recognized by both color and size, *Eptesicus fuscus fuscus* being larger and darker than *E. f. pallidus.* Average and extreme external measurements of 10 specimens of each subspecies (five of each sex of *fuscus* from eastern

Nebraska and 10 females of *pallidus* from the northwestern part of that state) are, respectively: total length, 121.0 (112–130), 117.6 (112–120); length of tail, 46.4 (40–50), 44.2 (41–47); length of hind foot, 11.2 (9–12), 10.9 (10.5–11); length of ear, 18.5 (17–20), 17.0 (16–18); length of forearm, 45.3 (42.7–49.0), 44.1 (43.0–45.5); greatest length of skull, 19.8 (18.9–21.3), 19.0 (18.3–19.4); zygomatic breadth, 13.0 (12.2–13.6), 12.4 (12.0–12.9). Females average approximately five percent larger than males in this species.

As in other hibernating bats, weight varies considerably according to season. From a low in spring, the weight gradually increases over the summer as fat is accumulated in anticipation of winter torpor. Ten summer-taken females of *pallidus* from Sioux County, Nebraska, weighed an average of 19.9 (16.4–28.1) grams, whereas a series of 28 *fuscus* taken in October in south-central Nebraska averaged 28.1 (21.0–33.0); 20 February-taken, hibernating males from the Black Hills of South Dakota had a mean weight of 14.9 (12.1–16.2) grams.

Natural History. Because of its size, abundance, colonial habits, and predilection for using man-made structures as roosts—and thus its tendency to be a relatively conspicuous member of any local fauna—more is known about the ecology of this species than that of any other bat treated in these accounts, save possibly *Myotis lucifugus.* In April, males and barren females disperse from hibernacula, but they usually spend the summer within the general vicinity (some individuals, however, may use the same retreat the year around). Singly or in small groups, these bats seek diurnal roosts in such places as tree hollows, rock fissures, under loose bark, behind shutters of houses, between expansion joints of bridges, and within various man-made structures. Gravid females form maternity colonies, frequently in buildings.

Eptesicus fuscus is a relatively early flier and can be recognized by its large size and normally slow, strong, relatively steady flight, during which individuals may chatter audibly. Like other bats, this species first seeks water after leaving the daytime roost, then begins its crepuscular and nocturnal foraging at heights rarely exceeding 30 feet over waterways, ponds, and clearings in wooded areas, along the edges of riparian woodlands, or around lighted areas in cities and towns. Most feeding activity is confined to the early hours of the evening, although some individuals may forage intermittently throughout the night. There is, however, no marked bimodal feeding pattern as in some other species of bats. Many *E. fuscus* evidently feed over the same areas night after night.

The big brown bat is a hardy animal, and its activity frequently continues well into the autumn. Even on warm days in winter, individuals may awaken from hibernation and leave the roost to fly abroad. During torpor the body temperature approximates that of the surroundings; recorded temperatures of hibernating bats in an abandoned mine in northeastern Kansas varied from 1.8 to 18.3° C. If the ambient temperature drops below freezing, which not infrequently happens because cool, dry, relatively exposed places usually are selected for winter roosts, a big brown bat will arouse and move to a warmer site. For this and other reasons, there is considerable movement of individuals within (and even between) hibernacula, and they may feed, drink, and even copulate during such periods of activity. *E. fuscus* usually hibernates

38. Big brown bat, Eptesicus fuscus *(courtesy T. H. Kunz).*

singly or in groups of two to three, wedged into cracks or crevices or hanging in the open, but it does cluster in places where exposure to the elements is greatest. On the Northern Plains this species likely enters hibernation in late September or early October, males possibly earlier than females.

Copulation occurs before entry into hibernation in autumn and also at times over the winter, but, as in other hibernating species, fertilization and implantation of the embryo are delayed until spring, a reproductive strategy unique to bats. Parturition in this species usually occurs in June and early July in the tristate region, although pregnant females have been taken late in July. At birth, big brown bats average between 3 and 4 grams in weight and measure approximately 54 mm in total length and 16 in length of forearm. The young are nearly hairless when born but are completely furred by the end of the first week of life. Young bats can fly three to four weeks after birth and reach full adult size at about 70 days of age. Volant young have been collected as early as 20 July in western Nebraska and 26 July on the Black Hills of South Dakota and in North Dakota; by contrast, a lactating female accompanied by two large young has been recorded from as late as the first week in September in Nebraska.

The number of young produced by females is geographically variable in temperate North America. Those from the eastern part of the range of the species give birth to two offspring as a rule, whereas those from the intermontane West and Pacific coastal areas generally have only one. Nebraska and the Dakotas lie in the transition zone, and it is therefore not surprising that both single and twin fetuses and nursing young have been reported from the tristate region. Nonetheless, current evidence indicates that bats from the eastern part of the Northern Plains usually bear two young, whereas those from the west usually have one (another means by which the two subspecies occurring in our region generally can be recognized).

Adult males and nonreproductive females usually complete the annual molt by the end of June or in early July (but some do not do so until early August). Lactating females delay the molt until about the time young are weaned. Young-of-the-year molt into adult pelage in mid- to late summer, before entering hibernation. Juvenile pelage is duller, darker, shorter, and somewhat sparser than that of adults. In all cases, molt seems to take place simultaneously over much of the body rather than in a distinct progression.

Big brown bats migrate over short distances—from summer haunts to winter retreats, for example—but long-distance movements are not known under natural conditions. "Homing" ability, however, apparently is strongly developed in this species; some individuals returned to their home roosts in Ohio after having been transported 450 miles. The known record of longevity for *E. fuscus* is 19 years.

Selected References. General (Barbour and Davis 1969); natural history (Kunz 1974; Phillips 1966); systematics (Long and Severson 1969).

Genus *Lasiurus*

The generic name *Lasiurus* is a combination of two Greek words—*lasios*, meaning "shaggy" or "hairy," and *ura*, meaning "tail," in reference to the hair-covered uropatagium. The genus is restricted to the New World, where it is represented by about seven species, two of which occur on the Northern Great Plains. In general the vernacular name "tree bats" is applied to members of this genus. The dental formula is given in table 2; it is of note, however, that the first premolar above is small, peglike, and internal (lingual) to the toothrow. Unlike other North American bats, each species has two pair (instead of a single pair) of mammae.

Lasiurus borealis: Red Bat

Name. The Latin word *borealis* means "northern," which is not particularly descriptive of the distribution of this species. The vernacular name reflects the distinctive coloration of this bat.

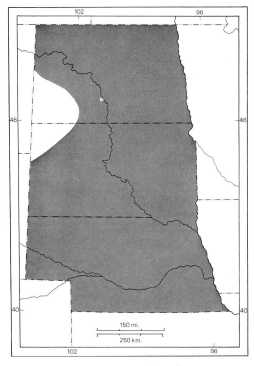

39. Distribution of Lasiurus borealis.

Distribution. The red bat is a partly migratory species that ranges from southern Canada southward through much of the United States and Middle America into South America; migrants have been reported from Atlantic islands such as Bermuda, and related species occur on the Greater Antilles. It is absent from the Rocky Mountain region and from most of the Mexican Plateau. Red bats pass through much, perhaps all, of the tristate region in spring and autumn migrations. In the warm months, *L. borealis* is resident in the Dakotas and Nebraska, albeit restricted in the western part of the three states to urban areas and wooded riparian communities. Little is known of the winter range of the species, although it probably resides at latitudes to the south of the Northern Plains. Only one subspecies, *Lasiurus borealis borealis,* is known from the region.

Description. There is considerable variation in color among red bats, but they easily can be distinguished from all other bats of the Northern Great Plains in that the dorsal pelage ranges from bright reddish orange to chestnut. Adults are of essentially the same size, although sexual dichromatism obtains, males being more reddish than females. This distinction is accentuated because in females many dorsal hairs are tipped with white, imparting a somewhat frosted appearance. The sexual difference in color also is evident in young red bats.

There is a distinctive yellowish white patch of hair on each shoulder. The dorsal and ventral surfaces of the wings are furred outward from the body toward the elbow, and the dorsal surface of the uropatagium is fully furred. The hind feet are well covered with hairs, and the calcar is keeled. The ventral pelage is paler than that of the dorsum. No seasonal variation in color is known.

The wing membranes are brownish black, in contrast to the paler forearms and phalanges. The ears are short, broad, rounded at the tip, and nearly naked inside and on the rims. Hair is dense on the basal two-thirds of the ear, however, and on its dorsal surface. The tragus is triangular and broad at the base. The skull is short, broad, and deep, the rostrum being nearly as broad as the braincase.

Eight specimens from Nebraska and South Dakota (one male, seven females) had the following average and extreme external dimensions: total length, 113.1 (106–121); length of tail, 48.0 (43–54); length of hind foot, 9.1 (8–10); length of ear, 12.2 (10–14); length of forearm, 39.4 (37.8–40.9). In six females from the same series, the greatest length of skull averaged 13.6 (13.2–14.1), and zygomatic breadth averaged 10.1 (9.8–10.5). Based on a large series from Kansas, adult red bats generally weigh between 8 and 14 grams, although males have been recorded weighing as little as 6 grams and pregnant females may weigh 20 or more. Weight is seasonally variable in that autumn migrants normally have taken on substantial fat reserves.

The red bat differs from its larger cousin, the hoary bat (*Lasiurus cinereus*), in color and in its noticeably smaller size.

Natural History. Red bats roost in trees and occasionally in other vegetation and are among the most conspicuous bats in the eastern part of the Dakotas and Nebraska, where their preferred wooded habitat prevails. They are relatively rare in the western part of the three states, being found only on the Black Hills and the Pine Ridge, in the few suitable wooded areas along major watercourses, and occasionally in cities and towns; except during migrations, the species almost certainly is absent from vast areas of the essentially treeless plains.

As noted, these bats spend the daylight hours hanging in the foliage of trees, singly or in family groups comprising a female and her young. Elms seem to be preferred, but box elder, fruit trees, and a variety of other woody plants also have been recorded as harboring red bats. The bat hangs by one foot from a leaf petiole or from a twig or branch and thus resembles a dead leaf from a distance. The usual individual roost site is densely shaded above and to the sides, but open below, and may be as low as four feet above the ground or as high as 40 feet or more in tall trees. The highest roosts seem to be occupied by family clusters or individual young-of-the-year after they have left the female. Typically, trees in "edge"

40. *Red bat,* Lasiurus borealis, *family group (courtesy T. H. Kunz).*

situations are selected for daytime roosts— trees at or near the borders of wooded areas, along streams, or lining city streets, golf course fairways, and the like—not far distant, in any event, from a source of free water.

The red bat emerges earlier in the evening than most other species, and individuals frequently can be seen flying over open areas and along wooded borders before the sun has set. Most red bats evidently forage over the same area night after night, ranging up to 1,000 yards or so from the daytime roost, which also may be used again and again (or the same roost may be used by different individuals). The early flight pattern is rather high and relatively slow and erratic, but as darkness falls the bats drop to treetop level and below to feed, at speeds that have been timed at up to 40 miles an hour in straight flight. In a study in Missouri utilizing ultrasonic detection equipment, members of the genus *Lasiurus* could be distinguished in flight from other bats because they emitted a musical "chirp" instead of a nonmusical "beep." Research in Iowa revealed a marked period of activity in the first two hours after sunset, followed by reduced activity during the remainder of the night.

Lasiurus borealis quits the Northern Plains in late summer and early autumn, to return again in spring. It is not known whether individuals return year after year to the same summer or winter haunts. The earliest and latest known dates of capture in the tristate region are 15 April and 19 September. For years it was assumed that autumn migration took red bats to the southern United States and adjacent warm regions, where they overwintered. However, no apparent increases in population size have been reported in southern states during the cold months of the year; also, it now is known that these bats are common in winter in the Ohio River Valley, southern Missouri, and probably elsewhere at middle latitudes in the eastern half of this country, where they have been observed, sometimes in substantial numbers, flying about in the afternoon or early evening on warm winter days. In such circumstances, individuals almost certainly are torpid during cold weather, seeking retreats inside hollow trees, but details are lacking.

L. borealis is known to swarm with other bats in late summer and autumn around the entrances of caves and mines and occasionally is found in torpor in such places (although it probably cannot survive over winter in them because of the constant low temperatures). In the Gulf states, however, red bats evidently are active during most or all of the colder months. Obviously, much remains to be learned about the biology of this species in winter. On the basis of scanty present knowledge, it appears that some populations migrate long distances, whereas others do not, and that red bats regularly hibernate in some areas but not in others. It apparently is necessary for hibernating individuals to take up winter residence at latitudes at least far enough south that some warm winter days (about 60° F) will occur—days on which the bats awaken to feed and drink.

Mating evidently takes place during migration in late summer and early autumn, and possibly also on the wintering grounds. In any event, females are gravid when they return to the Northern Plains in spring. Copulation between mating pairs in flight has been observed. The known number of fetuses or young ranges from one to five, but the average litter size is between three and four, more than in any other species of bat, and the mode is four for pregnancies and litters reported from the Northern Plains. The young, which are hairless at birth and have eyes closed, are born mostly in June in the tristate region and probably are in flight in about a month, although they may nurse for five to six weeks; the latest record of a lactating female from the Dakotas and Nebraska is one taken on the Black Hills on 20 August.

Females with nursing young occasionally are dislodged from trees by strong winds, heavy rain, or hail, or for other reasons, and are unable to escape from the ground because of the weight of their offspring; such a family group is especially susceptible to predation and also accounts for the relatively large number of mothers with young

in museum collections. Females have been collected in flight with young attached, but, as in other bats, this probably suggests nothing more than that the young are being moved from one location to another; it is certainly known that foraging females leave their young in the tree roost, returning unerringly to them after feeding.

Several authors have noticed the paucity of adult males in Plains states in summer and have suggested that most spend the warm months elsewhere. This phenomenon is strongly developed in *Lasiurus cinereus* and deserves additional study in the red bat. Both sexes are found in late summer and in migrating congregations in autumn, but data are unclear as to when the adult males join such groups.

Tree bats have a different set of predators than do colonial species, blue jays being an important predator on red bats, especially the young. As noted in the ordinal account, bats with solitary roosting habits tend to be relatively free of ectoparasites. The timing and progression of the annual molt have not been reported.

Selected References. General (Barbour and Davis 1969); natural history (Jones, Fleharty, and Dunnigan 1967; Kunz 1973; LaVal and LaVal 1979).

Lasiurus cinereus: Hoary Bat

Name. The specific name *cinereus* is a Latin word meaning "ash-colored." This and the vernacular name both refer to the grayish white dorsal color of this bat.

Distribution. The hoary bat has an interesting and disjunct distribution, limited to the Western Hemisphere and some nearby islands. In North America one subspecies, *Lasiurus cinereus cinereus*, occurs from Canada south to the highlands of Central America. Another race is known from South America and a third from the Hawaiian Islands, where the hoary bat is the only native terrestrial mammal. On the Northern Plains the species occurs throughout the region as a migrant, although when resident in the warm months it is re-

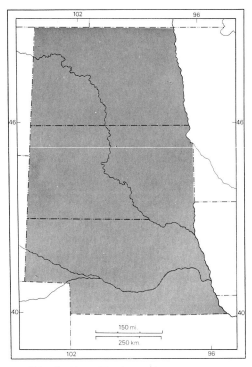

41. Distribution of Lasiurus cinereus.

stricted, as a "tree bat," to suitable wooded habitats.

This is a strongly migratory bat and, just as happens in migratory birds, individuals occasionally deviate from customary routes. This accounts for the fact that apparent migrants have been taken as far north as Southampton Island in northern Canada, on Iceland, Bermuda, and the Caribbean island of Hispaniola, and even from the Orkney Islands off Scotland.

Description. Lasiurus cinereus generally resembles the red bat but can be distinguished from it and all other Northern Plains species by a combination of its distinctive coloration and large size. The ears are short, broad, and rounded, lightly furred and accentuated by a dark, naked border, and the tragus is short and blunt. The wings are relatively long and narrow, and this species is a strong flier. The dorsal pelage varies from yellowish brown to mahogany, frosted with silver to impart a "hoary" appearance. The hairs on the neck are longer than those on the back, forming a slight "ruff." The wing is furred outward from the body to the level of the elbow, and the dorsal surface of the uropatagium is covered with hair. There is a yellowish white patch on each shoulder, sometimes more or less continuous across the chest, and a cream-colored spot near the wrist. The skin of the wings and uropatagium is brownish black. The feet are furred above, and the calcar is moderately keeled. The ventral pelage is paler than that above, whitish to yellowish brown.

Males and females are similarly colored, although the latter average somewhat the larger. Average and extreme external and cranial measurements of 10 females from South Dakota are as follows: total length, 139.0 (133–147); length of tail, 55.0 (46–65); length of hind foot, 12.8 (11–14); length of ear, 18.7 (17–20); length of forearm, 54.5 (53.0–56.2); greatest length of skull, 17.8 (17.0–18.7); zygomatic breadth, 12.6 (12.2–13.2). Weight of the same females (nonpregnant) averaged 25.0 (19.6–34.5) grams. The rather robust skull has a broad, short rostrum that angles upward to a broad and heavy braincase. Only *Eptesicus fuscus* among Northern Plains bats exceeds *L. cinereus* in cranial size.

Natural History. Like the red bat, *L. cinereus* seeks retreats in trees during the day, and the habits of the two species thus are similar in that regard. The usual roost is well covered above with vegetation, open below, and situated 10 to 15 feet above the ground. Apparently both deciduous and coniferous trees are used. This species emerges later in the evening than do most smaller bats and is swift in flight. In a study in Iowa, activity was greatest three to seven hours after sunset, followed by a marked lull and a second surge of activity before return to the daytime roost. These bats regularly emit a chattering during flight that is audible to the human ear.

The hoary bat is a strongly migratory species, and at least some migrants apparently fly in groups, which may number several hundred bats. Individuals depart the Northern Plains in late summer, to return

42. Hoary bat, Lasiurus cinereus
(courtesy J. L. Tveten).

Newborn young are blind, lightly covered with fur, and have a forearm length of about 19. The young cling tightly to the female during the day, but she leaves them alone when she forages at night. A family group may remain at the same roost until the young are weaned, but a female will move her family if disturbed. As in the red bat, females with young attached occasionally fall to the ground, where they are essentially helpless. Young are volant approximately four to five weeks after birth. Postjuvenile molt apparently is completed before southward migration; the annual molt of adults takes place in July.

Because these bats are essentially solitary and because they tend to fly later than other species, they seldom are observed by man. There is some indication that migrants are active earlier in the evening (and perhaps later in the early morning) than are residents, and many migrating groups have been observed or captured. Even in summer, individuals occasionally can be seen over water at dusk with other bats, but the best collecting techniques require the use of mist nets or specially built bat traps after dark. For some unknown reason, there seems to be a greater tendency for hoary bats and red bats to become impaled on barbed-wire fences than is the case in other species.

Selected References. General (Barbour and Davis 1969); distribution (Findley and Jones 1964); natural history (Kunz 1973; Turner 1974).

Genus *Nycticeius*

The generic name *Nycticeius* is derived from the Greek word *nyktos,* meaning "night," and the Latin *eius,* "belonging to." The dental formula is given in table 2. The genus contains about seven species, but only two occur in the New World, one on the mainland of North America and a closely related species on the island of Cuba. Other members of the genus occur in Africa and nearby southwestern Asia, and in the Australian region.

again in April and May. The wintering grounds are not certainly known but may be in the southern states and in the mountains of Mexico southward to northern Central America. A few individuals have been reported from middle latitudes in the eastern United States in winter, however, and there is at least one report of an apparent hibernating hoary bat. In summer, males and females are mostly segregated, the former occurring at higher elevations and latitudes than the latter, which are earlier migrants. In the tristate region, adults of both sexes are known to occur together only on the Black Hills and adjacent pine-clad buttes of western Nebraska and northwestern South Dakota. Only females and young-of-the-year have been reported from elsewhere except during migrations.

Females normally bear twins, which are born from late May to early July. The earliest record of a litter from the Northern Plains is of a female with two young taken on 6 June in Nebraska; lactating females have been recorded as late as the last week of July on the Black Hills. Volant young have been reported from the last week of June in Nebraska but usually are not taken until mid-July in most areas.

Nycticeius humeralis: Evening Bat

Name. The specific name *humeralis* is of Latin origin and comes from *humerus* ("upper arm") and *alis* ("pertaining to").

Distribution. The evening bat is distributed throughout much of the eastern United States, from Nebraska, southern Michigan, and central Pennsylvania south to Florida and northeastern Mexico. The subspecies on the Northern Great Plains, *Nycticeius humeralis humeralis,* is known there only from deciduous woodlands and riparian communities in the southeastern part of Nebraska, which represents the northwestern limit of the known distribution of the species.

Description. In appearance the evening bat somewhat resembles the big brown bat, *Eptesicus fuscus,* but it is notably smaller and has only one pair of upper incisors instead of two. It also resembles the little brown myotis, *Myotis lucifugus,* in having a short, blunt tragus and dark brownish coloration above (somewhat paler below), but it can be distinguished readily from any member of the genus *Myotis* by the dental formula (see table 2). The wings and tail membrane lack fur, and the calcar normally is not keeled. Until approximately six weeks old, young-of-the-year are noticeably darker than adults. No albino individuals have been recorded, but varying amounts of white pelage are known to occur in both juveniles and adults, as happens in several other species of Northern Plains bats.

External measurements of eight adult females from Nebraska are: total length, 90.7 (83–96); length of tail, 37.1 (35–39); length of hind foot, 8.5 (8–9); length of ear, 13.1 (12.5–14). Length of forearm in four of these specimens averaged 35.9 (35.1–36.3). No adult males have been taken on the Northern Plains, but females average slightly larger than males elsewhere in the known distribution of the species. The greatest length of skull and zygomatic breadth, respectively, of four Nebraska-taken females averaged 14.8 (14.2–15.1) and 10.1 (10.0–10.3).

The skull of the evening bat is broad and flat, with only 30 teeth in total, the fewest of any bat in the region, and the upper incisors are separated from the canines. Weight of adults, somewhat dependent on season, ranges between 7 and 14 grams.

Natural History. In southern Iowa and Missouri, where the evening bat is locally abundant, individuals roost during the day in hollows in trees and in a variety of man-made structures. Small colonies also have been found behind the loose bark of trees. *Nycticeius* is a slow, deliberate flier and begins to forage at late twilight. In flight this bat usually courses over clearings, farm ponds, and openings in trees bordering creeks and rivers. The evening bat is migratory, and individuals leave northern parts of the range for unknown destinations to the south in late September and October. Females return in late April and early May. Adult males are not known from Nebraska or neighboring states to the east and south

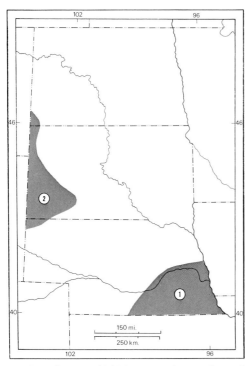

43. *Distribution of* (1) Nycticeius humeralis *and* (2) Plecotus townsendii.

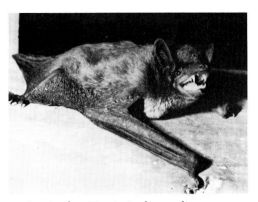

44. *Evening bat,* Nycticeius humeralis
(courtesy T. H. Kunz).

and evidently do not occur there in summer. In the southeastern United States this bat is active in all months of the year, but torpid individuals have been reported from southern Arkansas. The extent to which hibernation may be widespread in *N. humeralis* is, however, unknown.

Maternity colonies, frequently numbering hundreds of individuals, are established in old buildings or hollow trees, where females give birth to one to three (usually two) young (one report of four fetuses is on record). Parturition probably occurs from early June to early July in the northern part of the range of the species, somewhat earlier in the south. Breeding occurs in autumn or winter, with implantation of embryos delayed until early in the spring.

At birth young are pink and hairless and weigh slightly less than 2 grams; the eyes are closed. The eyes open within 12 to 30 hours, and the body is covered with soft, dark fur by the ninth day. Deciduous dentition (2/3, 1/1, 2/2, 0/0, total 22), present at birth, is fully replaced by the adult dentition by the time the bat is 30 days old. The growth rate of young *N. humeralis* is somewhat variable, depending on roost type and, to some extent geographic locality. Females normally do not carry young with them in flight, but they can do so if necessary.

Young evening bats can fly at 20 days of age or shortly thereafter, but they may not become completely independent of maternal care for several more weeks. Volant young have been taken in Nebraska as ear-

ly as the last day of June. A study in Indiana demonstrated that this species has considerable "homing" ability; many adult females and some young returned to their home colony after being released nearly 100 miles away.

Selected References. General (Watkins 1972 and included citations).

Genus *Plecotus*

The name *Plecotus* is derived from the Greek words *plekos,* meaning "braid" or "twist," and *otos,* "ear." The dental formula is given in table 2. The genus occurs in both the Palaearctic and Nearctic regions and is represented by five Recent species, one of which is found on the Northern Great Plains.

Plecotus townsendii:
Townsend's Big-eared Bat

Name. The specific name *townsendii* was proposed in honor of an early American naturalist, J. K. Townsend (1809–51). The vernacular name of this bat makes reference to the unusually large ears. Another name sometimes used, western lump-nosed bat, refers both to the principal distribution of the species in North America and to the enlarged dorsolateral glandular areas on the muzzle.

Distribution. This big-eared bat ranges from central Appalachia to the Ozark Highlands in the eastern United States; in western North America it is widely distributed from southern Canada to central Mexico. One subspecies, *Plecotus townsendii pallescens,* occurs on the Northern Plains but is known there only from western South Dakota (in the Black Hills, the Badlands, and the northwestern corner) and from a single locality on the Pine Ridge in northwestern Nebraska. Probably it also will be found in badland areas in extreme southwestern North Dakota, because it has been recorded from adjacent counties in Montana and South Dakota (see fig. 43).

P. townsendii is a cavernicolous species throughout most of its range, and local distribution probably is restricted by the availability of caves and analogous man-made structures, such as abandoned mines.

Description. This bat can be distinguished from all others on the Northern Plains by its conspicuously large ears and prominent facial glands. The ears are thin, narrowly rounded at the tips, and sparsely furred along the margins. Blood vessels appear as a prominent network on the ears when they are fully extended. The facial glands are paired masses between the eyes and the nostrils; their function is unknown.

This big-eared bat is of medium size; the wings and tail membrane are hairless. The pelage is overall pale brownish, and the individual hairs are distinctly bicolored, being gray basally and cinnamon to brownish at the tips. The venter is somewhat paler than the dorsum. Immature bats are noticeably darker than adults. The skull is characterized by a narrow and depressed rostrum in contrast with a highly arched braincase.

External and cranial measurements of 10 males from the Black Hills of South Dakota average as follows: total length, 101.5 (95–105); length of tail, 46.5 (46–47); length of hind foot, 10.9 (9–12); length of ear, 35.6 (30–38); length of forearm, 43.0 (40.8–44.3); greatest length of skull, 16.5 (16.2–16.9); zygomatic breadth, 8.8 (8.5–9.1). The weight of bats in this series averaged 9.8 (7.9–11.3) grams. There is no sexual dimorphism in size in this species.

Natural History. This long-eared species remains in most of its restricted range on the Northern Plains throughout the year. In winter these bats hibernate in caves and mine shafts, where it is not uncommon to find them at sites exposed to wind currents and fluctuating temperatures, frequently near the opening of the hibernaculum, in total darkness or in the twilight zone.

Individuals may roost singly in hibernation or form clusters of up to about 40 bats, including both sexes. In general, the larger the hibernaculum the more big-eared bats

45. *Townsend's big-eared bat,* Plecotus townsendii *(courtesy J. P. Farney).*

it contains and the larger the clusters. Restricted local movement does occur in winter, usually within a hibernaculum (but also between them, even in extremely cold weather), apparently to secure favored roost sites having relatively low temperatures. This behavior evidently assures the lowered metabolism necessary to conserve fat reserves during hibernation. In any event, this long-eared species readily awakens from hibernation and may change roosting sites frequently during winter, apparently in response to changes in temperature. Males tend to select warmer hibernating sites than do females and generally seem to be more active in winter.

In spring, females leave their place of hibernation and select sites that later will serve as nursery colonies. Such sites tend to be relatively warm, a situation that, along with clustering, maintains temperatures that may aid in the digestion and assimilation of food and also facilitates gestation, lactation, and development of young. Males and nonreproductive females are usually solitary in summer, although occasionally they are found in the same retreats as are used for nurseries, albeit separated from females with young. In August, about the time young are weaned, maternity colonies begin to break up, and the bats subsequently return to winter quarters. Summer retreats, like hibernacula, usually are in caves or mines, but use of buildings also has been reported, and the latter sometimes are utilized as night roosts. Local movements are frequent in the warm months, when males and non-reproductive females may move from roost to roost within an area. Females with young also may move to a new maternity site if disturbed. There is little direct evidence of significant migration in this species; "homing" tendencies have been demonstrated, but generally over far shorter distances than recorded for other colonial vespertilionids.

When at rest, the big-eared bat normally hangs free, suspended only by its feet. In winter the ears typically are curled along the neck like the horns of a ram. The fingers are spread so that the wings cover the ventral body surface, and the tail membrane is curled downward to cover the posterior part of the venter and the distal parts of the wings; the long hairs are erected perpendicularly. Thus a layer of air is trapped that serves as insulation. In the warm months the ears are usually fully extended and the wings folded against the body when at rest.

The breeding season extends from early October to late February. Sperm can be found in most females by the end of October, although repeated copulation occurs throughout much of the winter, sometimes while the female is torpid. Females breed in their first autumn, but males probably do not breed until the year after their birth. As in other hibernating bats on the Northern Plains, sperm is stored in the uterus until ovulation occurs in late winter or early spring. The duration of gestation may depend to some degree on the body temperature of the female, which possibly accounts for the wide range in gestation period (56 to 100 days) reported for California populations.

On the Northern Plains, females give birth to a single young in late June or July (the latest date for which we have a record of pregnancy is 20 July). During parturition the female assumes an inverted position (head up, that is); birth occurs by breech presentation. The mother grooms the baby by vigorously licking it, and within minutes the young bat becomes attached to one of her teats. Newborn *P. townsendii* have been described as grotesque in appearance, owing to the large, floppy ears and disproportionately large feet and thumbs. As in most, but not all, other species that rear young in colonies, the mother recognizes her own offspring; upon returning to a colony from nightly foraging flights, females search through the clustered young until they find their own.

The growth of young bats, which weigh about 2.5 grams at birth, is rapid. At two and a half to three weeks of age they are capable of short flights, although they probably do not leave the maternity roost to forage with adults until they are approximately six weeks old. Females continue to nurse their young until the latter are nearly two months old. The average life span for this species probably is about five years among individuals that survive their first winter; the known record for longevity is 16.5 years.

Long-eared bats are swift and agile fliers, although they also may assume a slow, butterflylike flight, and they are extremely adept at avoiding obstacles: they are not often caught in mist nets, for example. At one cave in the Black Hills, the entrance is barred by a steel gate from September through May each year, but these bats maneuver through the bars with little difficulty. In flight the ears are usually held erect and pointed forward almost parallel to the body. This species does not leave daytime roosts until well after dark, and thus individuals rarely are observed in flight.

An annual molt takes place in August, usually occurring earlier in males than in females. There is no particular pattern of loss or replacement of hair. Young bats attain adult pelage at approximately six to eight weeks of age.

Selected References. General (Barbour and Davis 1969); natural history (Humphrey and Kunz 1976; Martin and Hawks 1972; Pearson, Koford, and Pearson 1952; Turner 1974); systematics (Handley 1959).

Family Molossidae: Free-tailed Bats

The family Molossidae includes some 85 species arranged in 11 genera. The distribution of the family includes most temperate and tropical areas of the world, except the deserts of North Africa and the interior of Asia. Molossids have narrow wings and are fast, strong fliers. The tail projects noticeably beyond the posterior border of the uropatagium.

Genus *Tadarida*

The genus *Tadarida* includes about 15 nominal species and occurs in tropical, subtropical, and some temperate regions in both the Old World and the New World. Three species are known from the United States, but only one ranges as far north as the Northern Plains. The dental formula is given in table 2. The generic name was coined in 1814 by pioneer naturalist Constantine Rafinesque (1783–1840), who did not divulge its derivation; possibly it comes from the Greek *ta darida*, meaning "long ones."

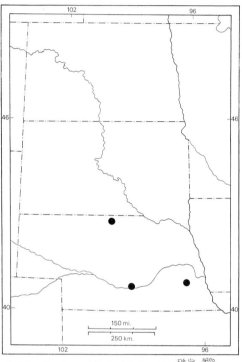

46. *Distribution of* Tadarida brasiliensis.

Tadarida brasiliensis:
Brazilian Free-tailed Bat

Name. The name *brasiliensis* denotes that the original description of the species was based on specimens from Brazil. The vernacular name guano bat is frequently used.

Distribution. Tadarida brasiliensis has one of the broadest distributions of any New World bat, ranging from Chile and Argentina northward to the southern and western United States; it also occurs throughout the West Indies. Some populations in this country are migratory, moving southward many hundreds of miles in late summer and autumn to mostly unknown locations in Mexico and returning again in late winter and spring. In the Northern Plains region the species is known only by six specimens from three localities in

Nebraska, one along the Niobrara River in Keya Paha County being the northernmost record of occurrence in North America. These reports evidently represent young-of-the-year that wander northward in late summer from the colonies in which they are born (the nearest currently known being in southern Kansas and adjacent parts of Oklahoma), with the exception of a pregnant female taken in Lincoln, Nebraska, on 27 June 1931. The latter gave birth the day following her capture and likely "overshot" in migration an intended destination somewhere to the south. The subspecies found on the Northern Plains is *Tadarida brasiliensis mexicana*.

Description. The Brazilian free-tailed bat is readily distinguished from other bats of the Northern Plains by a tail that extends well beyond the posterior border of the uropatagium. The ear is rounded and relatively short and bears horny excrescences on its anterior border. Unlike the larger *Tadarida macrotis* (which has been captured in Colorado, Kansas, and Iowa but is unknown from Nebraska or the Dakotas), its ears are not joined at the base, and there are three lower incisors on each side instead of two.

The dorsal pelage is dark grayish brown, but individual hairs are whitish at the base. The venter is slightly paler than the dorsum. The range in external measurements of seven specimens from Colorado and Nebraska is as follows: total length, 97–108; length of tail, 30–38; length of hind foot, 9–11; length of ear, 16–18; length of forearm, 42.5–44.0. The weight of adults generally varies between 8 and 14 grams. The greatest length of the skull and zygomatic breadth, respectively, of three Nebraska-taken specimens are 16.5, 16.7, 16.8, and 9.5, 9.7, 9.7.

Natural History. Like other molossid bats, *T. brasiliensis* is a strong, fast flier, attaining speeds up to 60 miles an hour. After the young are weaned in summer, populations tend to disperse from maternity colonies, and individuals may appear well to the north of their usual haunts, as is reflected in most of the records from Nebraska. Aside from the pregnant female mentioned earlier, the other five specimens, all apparently young-of-the-year, are males taken in Lincoln in August of 1913 and 1956, a male netted in Keya Paha County in August 1972, and a male and female collected in a roost occupied by big brown bats (*Eptesicus fuscus*) in the Buffalo County Courthouse in 1973—one picked up dead in September and the other found mummified in October.

Populations of this free-tailed bat have been studied extensively at various locations in the southern United States. Geographic variation is apparent in preferred roosting sites and in migratory habits. Populations in the Southeast and in California are mostly sedentary and live primarily in buildings. In Texas and the adjacent Southwest, however, the species is chiefly, but

47. *Mexican free-tailed bat,* Tadarida brasiliensis *(courtesy J. L. Tveten).*

by no means exclusively, a cave inhabitant and is migratory.

Colonies of females and young tend to be highly localized. Perhaps the most famous of these is the one at Carlsbad Caverns, New Mexico, but a number of caves in Texas and Arizona contain much greater concentrations of bats. An estimated 20 million inhabit Bracken Cave in Texas, for example. Summer habits of males are less well known, but evidently they are widely dispersed and congregate in relatively small groups, sometimes near maternity colonies. In Oklahoma and Texas, the cave bat (*Myotis velifer*) commonly is found in association with *T. brasiliensis*. Annual molt takes place in adult males mostly in July and in females mostly in August.

Breeding occurs on the winter range in February and March, and young are born in June after a gestation period estimated to be from 77 to 100 days. Young are naked and blind and weigh about 2.7 grams. As in other bats, females do not carry the young in flight, except rarely to move them to a new roost area. There appears to be little mutual recognition of mother and offspring among the thousands or millions of bats in a maternity colony. A given female will suckle the first one or two young to locate her mammae as she enters the nursery area.

Brazilian free-tailed bats are significant to man because of the relatively high incidence of rabies reported in some populations. Also, their voracity for insects is of obvious benefit, at least locally, and the guano produced by millions of roosting bats is sufficient to allow its limited commercial exploitation in some places as a nitrogen-rich fertilizer. Bracken Cave, for example, produces some 60 tons of guano annually. At some caves the crepuscular exodus of *T. brasiliensis* (about 15 minutes after sunset) is an attraction to tourists—millions of bats depart the cave mouth with a mighty roar and appear as a dark cloud visible for some distance. Individual bats are known to fly as high as 10,000 feet and to forage up to 50 miles away from the home cave.

Selected References. General (Barbour and Davis 1969); distribution (Czaplewski et al. 1979).

Order Edentata

The edentates are confined to the New World, occurring from the central United States southward through tropical America to Argentina. Modern members of the order, which has a long fossil history, are assigned to four families—Bradypodidae and Choloepidae (sloths), Dasypodidae (armadillos), and Myrmecophagidae (anteaters)—only one of which occurs outside tropical regions. They represent 13 genera and approximately 30 species.

The anteaters are toothless, as the ordinal name implies, but both armadillos and sloths possess premolars and molars. These are, however, peglike, single-rooted, and lacking in enamel. Deciduous teeth are present only in armadillos. At least some digits (number varies among groups) bear long recurved claws. The testes are abdominal.

Family Dasypodidae: Armadillos

This family includes eight genera and about 20 species and has the broadest distribution (the same as for the order) of any of the edentate groups. Only one species, however, occurs north of Guatemala, the main concentration of living armadillos being in South America.

Genus *Dasypus*

The generic name is thought by some to be derived from a combination of two Greek words that generally are regarded as meaning "hairy-footed" or "rough-footed." There is some evidence, however, that it may come from a rough translation into Greek of the Aztec word *azotochtli,* meaning "tortoise-rabbit," used in the original nontechnical description of the animal by a Spanish explorer. Armadillos have no incisors or canines but do possess peglike cheekteeth (premolars and molars, possibly including some that are "deciduous"). In this genus there are seven to nine such teeth in each upper and lower jaw, a total of 28 to 36. *Dasypus* contains six nominal species.

Dasypus novemcinctus: Nine-banded Armadillo

Name. The specific name *novemcinctus* is a combination of two Latin words, *novem,* "nine," and *cinctus,* "banded," and describes the nine prominent bands on the distinctive bony carapace.

Distribution. This armadillo ranges from South America northward to the southeastern and south-central United States, barely reaching the Northern Plains in extreme southern Nebraska. So many individuals have been released in places where the species previously did not occur that the precise natural distribution cannot be defined. There also is evidence that the nine-banded armadillo is dispersing northward on the Great Plains. Records from Kansas are fairly commonplace, and several specimens have been taken in southern Nebraska in recent years. An account for the species is therefore included here, albeit in somewhat abbreviated fashion. The distribution is not mapped. The subspecies is *Dasypus novemcinctus mexicanus.*

Description. The nine-banded armadillo is such a distinctive mammal that it cannot be confused with any other on the Northern Great Plains or, for that matter, in temperate North America. Most of the animal is covered by a bony carapace, composed of an anterior scapular shield followed by nine flexible bands and a terminal pelvic shield. The tail is encased in a series of bony rings, and the top of the head is covered by a bony plate. The face, chin, neck, belly, and upper parts of the legs are not "armored" but are covered by tough skin that is thinly haired. Hairs also arise from beneath the posterior scutes of the flexible dorsal bands. The legs and feet are covered with scutes. The forefoot has four digits and the hind foot five, all heavily clawed. The prominent ears are thick-skinned and leathery. The anterior parts of both upper and lower jaws are toothless, and the rostrum is long and narrow. The general dorsal color is grayish brown.

48. *Nine-banded armadillo,* Dasypus novemcinctus *(courtesy J. L. Tveten).*

Two females and a male from Louisiana had the following measurements: total length, 730, 763, 730; length of tail, 350, 373, 310; length of hind foot, 65, 70, 65; length of ear, 32, 35, 38. Adults generally weigh between 7 and 10 pounds. Greatest length of skull and zygomatic breadth, respectively, in four adults from Louisiana averaged 96.5 (93.7–100.0) and 37.4 (35.6–38.5). The sexes do not differ in size.

Natural History. This armadillo occurs principally from woodlands to open savanna and scrub; those few records from Nebraska and northern Kansas generally are associated with river valleys. It is predominantly, but not exclusively, nocturnal. Food consists mostly of insects and other invertebrates, such as snails and earthworms, but it also eats fruit, berries, seeds, mushrooms, and eggs of ground-nesting birds and perhaps reptiles.

Dasypus novemcinctus digs its own burrows and also "roots" and digs in the ground in search of food. In places where it is common, such activities are at best a nuisance in gardens, lawns, and agricultural lands, but the animal generally is beneficial because of the large number of noxious insects it eats and because its burrows serve as retreats for a variety of small mammals and other forms of wildlife. Active burrow systems have two or more tunnel openings, but only one is regularly used; the nest of leaves and grasses may amount to as much as half a bushel in volume.

Breeding is said to occur in middle to late summer, but implantation of the fertilized ovum is delayed (about 14 weeks) until later in the autumn. Normally four young, identical quadruplets from a single fertilized egg, are born in late winter after a gestation period of 120 days. They are precocial and can walk about a few hours after birth, and they reach sexual maturity in a year.

Armadillos have few natural enemies, although large carnivores are known to catch and eat them. If disturbed, they can run with surprising speed, generally seeking refuge in a burrow or in dense underbrush. When caught in the open, individuals can curl up in such a way that the bony carapace provides substantial protection from a predator. Many armadillos are killed on highways, partly because the animals seem unaware of approaching vehicles until the last minute when, because of a nervous response to danger, they jump directly upward and are struck by an automobile or truck that might otherwise have passed over them. In some places in Latin America they are hunted for food; the flesh when fried has a delicious taste described as "somewhere between chicken and shrimp." Because the species neither hibernates nor stores food, climate certainly is the most important factor limiting northward dispersal of the nine-banded armadillo. It is difficult to comprehend, for example, free-living individuals in our region surviving a cold winter with deep snow like that of 1978–79.

Vocalizations known to be made by armadillos include "wheezy" and piglike grunts, piglike squeals, a buzzing noise made by frightened individuals, and a "chuck" uttered by pairs during the mating season. Hearing and sight are not especially well developed, but these animals do have an excellent sense of smell. Armadillos can swim but also are reported to cross streams by walking on the bottom. In a study in Florida, the maximum home range calculated for eight individuals varied from 1.1 to 13.8 hectares (about 2.7 to 34 acres).

Selected References. General (Wetzel and Mondolfi 1979 and included citations; McBee and Baker 1982 and included citations).

Order Lagomorpha

The order Lagomorpha comprises two living families, the Leporidae (rabbits and hares) and the Ochotonidae (pikas). There are some 12 genera of lagomorphs and about 62 species; thus the order has not undergone extensive adaptive radiation. Lagomorphs are nearly cosmopolitan, however, being present throughout Eurasia, Africa, North America, and northern South America, and they frequently are important members of the ecosystems they occupy. In Australia, the European rabbit, *Oryctolagus*, was introduced by man, with disastrous environmental consequences.

For many years lagomorphs were considered rodents. Like rodents, they have ever-growing incisors, a long diastema, and high-crowned cheekteeth adapted to a vegetable diet. Unlike rodents, however, they have two upper incisors on each side, the second being a peglike tooth situated immediately behind the first. The fossil record of lagomorphs is traceable back to the Paleocene and suggests an ancestry independent of the rodents. Despite their superficial resemblance to rodents, rabbits and hares may be more closely related to ungulates. Lagomorphs of the Northern Great Plains are all members of the family Leporidae.

Family Leporidae: Rabbits and Hares

The family Leporidae includes some of our most familiar native mammals. There are 10 genera and about 43 species of living leporids, which are native to all continents except Australia and Antarctica. They occur in habitats ranging from tropical forests to arctic and alpine tundra. Leporids may be divided conveniently into rabbits, which bear altricial young, and hares, which bear precocial young. On the Northern Plains, the "cottontails" are true rabbits, whereas the snowshoe hare and jackrabbits are hares.

Key to Leporids

1. Hind foot 110 or less (usually less than 105); interparietal bone distinct (fig. 55) _____2

1'. Hind foot 105 or more (usually more than 110); interparietal bone indistinct, fused to parietal (fig. 55) _____5

2. Total length greater than 460; length of ear greater than 71; greatest length of skull more than 80 _____*Oryctolagus cuniculus**

2'. Total length less than 460; length of ear less than 71; greatest length of skull less than 80 _____3

3. Auditory bullae relatively large and inflated (fig. 54); ear longer than 61 _____ *Sylvilagus audubonii*

3'. Auditory bullae relatively small and not inflated (fig. 54); ear shorter than 61 ___4

4. Total length usually greater than 390; ear relatively sparsely furred within; diameter of external auditory meatus less than crown lengths of upper molars (fig. 54) _____*Sylvilagus floridanus*

4.' Total length usually less than 390; ear densely furred within; diameter of external auditory meatus greater than crown lengths of upper molars (fig. 54) _____ *Sylvilagus nuttallii*

5. Ear of adult less than 90; no anterior projection on supraorbital process _____*Lepus americanus*

5.' Ear of adult greater than 90; pronounced anterior projection on supraorbital process _____6

6. Dorsum of tail white or, if darker, with middorsal line that does not extend onto back; groove in first upper incisor simple, resulting in simple fold on occlusal surface of tooth; mesopterygoid fossa relatively broad, usually 11.5 or greater (see fig. 58) _____*Lepus townsendii*

6.' Dorsum of tail black, with black patch extending onto back; bi- or trifurcate groove in first upper incisor, resulting in complex fold on occlusal surface of tooth; mesopterygoid fossa relatively narrow, usually 11.0 or less (see fig. 58) _____*Lepus californicus*

Genus *Sylvilagus*

The name *Sylvilagus* is derived from the Latin word *silva*, which means "forest," and the Greek word *lagos*, "hare." The genus includes 14 species, which range from southern Canada to Argentina and Paraguay; three species occur on the Northern Plains. The dental formula is 2/1, 0/0, 3/2, 3/3, total 28.

Sylvilagus audubonii: Desert Cottontail

Name. The name *audubonii* is a patronym for the well-known artist and naturalist John James Audubon (1785–1851). Sometimes this species is referred to in the vernacular as Audubon's cottontail, but the name desert cottontail is an apt description of its habitat in the semiarid West.

Distribution. This cottontail ranges from northern California northeastward to Montana and the Dakotas and southward through Texas into south-central Mexico. Only one subspecies, *Sylvilagus audubonii*

baileyi, is found on the Northern Plains, where it occurs in the western portions of Nebraska and the Dakotas.

Locally the desert cottontail may be found in a variety of habitats. Most commonly it occupies dry uplands, but occasionally it occurs in mesic situations where one is more likely to encounter the eastern cottontail.

Description. S. audubonii is a small rabbit with long hind legs and long, sparsely furred ears. The upper parts are pale grayish washed with yellow, with a few black hairs in the middorsal region. The sides are paler than the back, and the underparts are whitish except for an orangish brown throat patch that extends to the chest. The hairs on the feet are short. The pelage of this species is similar in color to that of the eastern cottontail and Nuttall's cottontail, but *S. audubonii* is paler on the head, nape, and dorsum. Females have four pair of mammae, one pectoral, two abdominal, and one inguinal. Average and extreme external measurements of eight specimens from

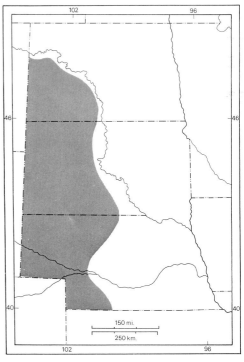

49. *Distribution of* Sylvilagus audubonii.

northwestern Nebraska are: total length, 413.7 (390–435); length of tail, 45.0 (39–58); length of hind foot, 96.4 (90–101); length of ear, 65.6 (62–69). Three males and three females from the same area weighed 940, 1,040, 1,090, 1,090, 1,230, and 1,375 grams, respectively.

Cranially, *S. audubonii* can be distinguished from *S. floridanus* and *S. nuttallii* by its large, inflated auditory bullae (fig. 54), larger external auditory meatus, shorter maxillary toothrow (average about 13 mm as opposed to 15), larger cheekteeth, and more prominent, upturned supraorbital process. Representative cranial measurements for seven *S. audubonii* (three males, four females) from northwestern Nebraska are: greatest length of skull, 71.0 (69.2–72.8); zygomatic breadth, 35.6 (34.8–37.2).

Natural History. Natural history of the desert cottontail has been studied extensively in the western portion of its range in California and Nevada. These studies indicate habits similar to those of *S. floridanus.* In winter the greatest activity occurs between dawn and dusk, whereas in the milder months it is between 3 P.M. and 9 A.M. Patterns of activity often are influenced by conditions of the local habitat. Rabbits inhabiting open areas will limit foraging to times of low light intensity. In areas where thick stands of vegetation provide adequate cover, activity may begin earlier in the afternoon and last longer into the morning. Feeding becomes irregular from an hour after sunrise until late afternoon. The home range of females is about one acre, whereas males may occupy as much as 15 acres. Densities as high as 16 per hectare (about seven per acre) have been reported. Females show no agonistic behavior, and as many as four may occupy the same area. Resident males, however, will not tolerate territorial intrusion by other males. The intruder may simply be chased out, or combat may ensue.

S. audubonii makes use of "forms" situated in tall grass for resting, but it also frequents burrows during the daily periods of inactivity. There is evidence that *S. au-*

50. Desert cottontail, Sylvilagus audubonii (courtesy T. W. Clark).

dubonii occasionally constructs its own burrows, but more often it uses dens abandoned by other mammals such as badgers, prairie dogs, foxes, and skunks. In some parts of the range, thickets of brambles are preferred cover. Desert cottontails are more abundant on moderately grazed rangeland than in either overgrazed or ungrazed areas.

The desert cottontail forages in grassy hollows, dry washes, and gullies and at the edges of cultivated fields. Physiography and vegetation are important factors for concealment. During the late morning and early evening—times of greatest light intensity—feeding takes place in brushy cover or in shaded areas. Open grassy areas are exploited only during times of low light intensity. Wind tends to restrict foraging activity to sheltered areas in thickets, perhaps because the resultant commotion limits ability to perceive danger. Rabbits living in open situations rely heavily on grasses for nourishment. In other areas they eat a wide variety of vegetation, including species of sedge, rush, willow, oak, blackberry, and wild rose. Sagebrush supplements the diet in winter. Stems are clipped at a 45° angle; only the more succulent terminal shoots are eaten.

Water is obtained from succulent vegetation and from dew. Habitat probably prevents the desert cottontail from drinking, except perhaps rarely. The kidneys have a greater relative medullary thickness than those of *S. floridanus*, which is correlated with the ability to concentrate urine (thereby reducing water loss)—an important physiological factor in arid climates. This adaptation undoubtedly has played an important role in the exploitation of more arid regions by *S. audubonii*.

As in other species of *Sylvilagus*, the main line of defense is escape. This behavior may take several forms, depending on the situation. If danger is perceived at a far distance, the animal may "freeze" to avoid detection. If in an exposed situation, it may run to brushy cover, where it will momentarily pause at the edge. If danger persists, the rabbit hops into the brush, where it hides at the edge of the thicket. If further disturbed, it penetrates deeper and seems to disappear. In cases of sudden surprise by a predator at close range, the cottontail springs into the air to clear low vegetation and other obstacles. Using open ground for escape, it can reach a speed of 15 miles an hour on a zigzag course over 100 yards or more. When a rabbit is running directly away from a pursuer, its white-tufted tail is easy to follow; but should the animal suddenly change direction or stop, it seems to vanish instantly because the rest of the pelage blends well with the surroundings. This behavior is adaptive in confusing predators. Rabbits are known to thump the ground with the hind feet, which may be a signal of danger. This action has been observed most often at night, but it is not unknown during the day.

The breeding season varies geographically, beginning earlier in areas with mild winters and later in places with harsh winters. In some parts of the range breeding may occur throughout the year. In California, reproduction takes place between December and June, whereas in Colorado onset of reproduction may be as late as April, lasting until late July or early August. Females may have two or more litters per year. One to five young per litter have been recorded, with the average about three. The latter figure is somewhat less than that found in the eastern cottontail or

Nuttall's cottontail, but a longer breeding season offsets the effect of smaller litter size. Nests of the desert cottontail may consist of shallow, pear-shaped depressions, or they may be in shallow burrows where vegetative cover is sparse. In either case a soft mat of grasses and fur is used to protect the offspring. At birth the young are about 80 to 90 mm in total length and are altricial. Growth is rapid, however, and the eyes are open by day 10. The young remain in the nest until about two weeks old. They can reproduce at about 100 days. Males are capable of breeding throughout the year, although few are in reproductive condition in autumn. Females may produce two litters during the first year of life. Full maturity is reached at about six to nine months of age in both sexes.

Molt of desert cottontails has not been studied in detail, although distinct nestling, juvenile, and subadult pelages seem to occur, as in *S. floridanus*. Adults molt twice each year, in late spring and early summer and again in the autumn.

Mortality is high, as in other species of *Sylvilagus*, and few individuals live longer than two and a half years in the wild; average longevity is less than a year. Natural enemies include foxes, coyotes, bobcats, hawks, owls, eagles, and rattlesnakes. Parasites include cestodes, nematodes, protozoans, botfly larvae, ticks, and fleas.

Selected References. General (Chapman and Willner 1978 and included citations).

Sylvilagus floridanus: Eastern Cottontail

Name. The name *floridanus* is the latinized name of the state of Florida, the locality from which the type specimen of this species was first described.

Distribution. The eastern cottontail ranges as far north as southern Manitoba and southward to Central America. Within the United States, it has a continuous distribution from the eastern seaboard westward in suitable habitat to the foothills of the Rockies, but it does not occur in northern New England. The species was introduced into Washington and has subsequently spread into adjacent parts of British Columbia and Oregon.

Of the two subspecies that occur on the Northern Plains, *Sylvilagus floridanus similis* is the more widespread, being found throughout most of the Dakotas and in the western three-fourths of Nebraska. *Sylvilagus floridanus mearnsii* occupies the eastern fourth of Nebraska and probably extreme eastern South Dakota.

Locally, this species is somewhat restricted ecologically, to mesic situations in riparian communities or adjacent agricultural situations where dense plant growth provides good cover. With settlement, *S. floridanus* probably expanded its range westward at the expense of *S. nuttallii*. Where *S. floridanus* occurs sympatrically with *S. audubonii*, the two generally are separated ecologically, with *audubonii* occupying the drier and more sparsely covered uplands. However, in

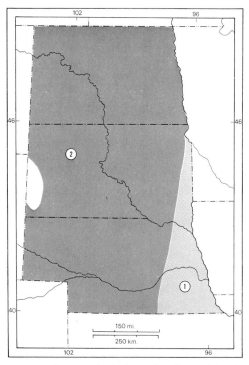

51. *Distribution of* Sylvilagus floridanus: (1) S. f. mearnsii; (2) S. f. similis.

some transitional areas the two species have been found together.

Description. S. floridanus is the largest cottontail found on the Northern Plains. It is similar in appearance to both the desert cottontail and Nuttall's cottontail but can be distinguished by its larger size, darker color, and relatively shorter ears. The feet, underpart of the tail, and underparts are white, whereas the long, dense dorsal pelage varies from grayish to brownish. There is a distinct rusty patch on the nape of the neck. There are several reports of melanistic and albino individuals in the literature.

The two subspecies that occur on the Northern Plains show no cranial differences but can be distinguished by color of pelage. *S. f. mearnsii* is darker on the upper parts than *S. f. similis.* The differences are more easily discerned in winter pelage. Mean and extreme external measurements of 10 *similis* from Cherry County, Nebraska, followed by those of eight *mearnsii* from Richardson County, Nebraska, are: total length, 420.3 (400–452), 419.2 (408–432); length of tail, 50.5 (41–71), 46.2 (36–65); length of hind foot, 97.5 (83–104), 96.6 (90–103); length of ear, 55.9 (52–61), 58.1 (55–61). Weights of adults range from 2 to 3 pounds; females average slightly heavier than males.

Cranial details of the eastern cottontail show wide geographic variation and cannot be described adequately so as to distinguish the species from all other members of the genus. Locally, however, it is possible to distinguish skulls of *S. floridanus* from sympatric species of *Sylvilagus. S. floridanus* is distinguished from *S. audubonii* and *S. nuttallii* by its smaller, uninflated auditory bullae (fig. 54), by the posterior extension of the supraorbital process, and by the external auditory meatus being less than the crown length of the first upper molar. Representative average cranial measurements of six adult *S. f. similis* from north-central Nebraska, followed by those of seven *S. f. mearnsii* from the southeastern part of the state, are: greatest length of skull, 72.6 (70.8–75.3), 73.0

(70.8–74.8); zygomatic breadth, 36.0 (35.6–36.4), 36.0 (34.3–37.6).

Natural History. Eastern cottontails are solitary animals. They may be active at any time of the day or night but are most active during twilight hours. The home range of females usually covers one to five acres but may be 15 acres or more where food and cover are sparse. Adult males usually have larger home ranges than do females and juveniles—up to 100 acres. Home ranges of most individuals may be superimposed broadly, but there is some evidence that breeding females are territorial. Foraging movements are 50 to 100 meters per day, over a haphazard course. Daylight hours often are spent resting in a "form" or concealed in brushy areas and thickets. A typical form is a well-concealed shallow, oval depression, which the animal constructs by digging and trampling the ground. It may be lined with grass or leaves. Cottontails do not migrate or hibernate but are active

52. *Eastern cottontail,* Sylvilagus floridanus *(courtesy J. K. Jones, Jr.). Note molt lines.*

throughout the year. They occasionally "hole up" in their forms or other shelter for short periods in unfavorable weather. During times of particularly cold weather and heavy snows they may temporarily occupy underground dens of other mammals. In the warmer months daylight hours may be spent sunbathing or resting in a cool, shady spot.

There are two daily feeding periods. One begins just before sunrise and lasts several hours; the other occurs in the late afternoon, lasting until after sunset. In summer the diet of eastern cottontails consists mostly of herbaceous plants, of which they consume a wide variety. The choice of foods depends on regional and seasonal availability as well as individual preference. Cottontails seem to be especially fond of legumes and grasses. They also eat stems and shoots of shrubs and woody vegetation, but mostly in winter when preferred foods are dormant or covered by snow. Twigs, buds, and bark of trees are consumed when green vegetation is scarce. As winter snows accumulate, the animals may be able to feed on the lower branches of trees. In times of great food shortage these cottontails have been known to eat snails, pupae of moths, and even carcasses of their own kind. Water is obtained from succulent vegetation and dew, and sometimes by drinking.

As is true of most lagomorphs, members of the genus *Sylvilagus* regularly reingest their own fecal pellets. This practice has been termed "coprophagy" or "reingestion" and also is known to occur among certain rodents and insectivores. Leporids produce two types of fecal pellets; soft pellets are taken from the anus and re-eaten, but hard pellets are not. Evidently the digestive system of these animals is incapable of extracting the total nutritional value from plant material that is ingested only once. By reingestion the digestive system makes more efficient use of the available food resources. Also, reingestion allows the animals to obtain vitamins synthesized by bacteria in the cecum.

Cottontails move slowly by short hops except when attempting to escape from

danger. When escaping a predator they can attain speeds of 18 miles an hour for short distances. Escape behavior is characterized by long leaps followed by a series of evasive zigzag hops. They not only evade predation by flight but often "freeze" into a motionless squatting position to avoid detection. If cornered, they can strike hard blows with their hind feet, which may momentarily stun an attacker and permit escape. They can swim but evidently rarely do so. The eyesight of cottontails is excellent. With their protruding eyes set at the sides of the face, they can detect movement at almost any angle. While foraging they often sit up on their hind legs to gain a better view of the surroundings. Vocalizations are rare, but *S. floridanus* can make a high-pitched cry when attacked or greatly disturbed.

The breeding season varies geographically but is confined to the milder months, extending at our latitude from about March through September. Ovulation is induced by copulation. Females bear from one to nine offspring per litter, with four or five most common. Litter size varies geographically, with larger litters in northern parts of the range. Fecundity in Missouri has been shown to vary with soil fertility. Postpartum estrus occurs, and a female may have seven or more litters per year. The gestation period is 28 to 30 days.

As the time of parturition approaches, the female digs an elliptical nest about 10 inches long and six inches deep. This is lined with vegetation and wads of soft fur pulled from her shoulders, flanks, and legs. The young are altricial at birth—blind, naked, and helpless. The female nurses them at dawn and dusk and otherwise keeps them covered with a soft mat of grass and fur. She spends much of her time in a form near the nest and may defend the nest from intruders. At birth the young are 150 to 200 mm long and weigh about 40 grams. Within two weeks the eyes are open, the young are furred, and they can move about and nibble vegetation. By this time they look like miniature adults. At four to five weeks of age the young are completely independent of the female, and they reach adult weight in about four months. Females of early litters may breed in their first season (males do so rarely).

S. floridanus has a complex series of pelages. Nestlings have a dense, wavy, gray coat. At about five to seven weeks of age a thin "salt-and-pepper" juvenile pelage grows through the nestling fur. Subadult pelage becomes apparent on the rump and sides at 11 to 15 weeks of age; it is paler, thinner, and more buffy than adult pelage, which comes in at about 30 weeks of age. Adults molt twice annually. In spring and early summer, the worn winter pelage is shed in patches and replaced by the summer coat. Winter pelage is attained in September and October.

Cottontails have numerous enemies, and most of the young perish before becoming reproductively active. They are an important source of food for coyotes, foxes, weasels, hawks and owls, rattlesnakes, and numerous other carnivorous vertebrates.

Although eastern cottontails may live 10 years in captivity, in the wild only 10 percent reach one year of age, and only one percent reach much over two years. About 75 percent die before the age of five months.

This cottontail is an important game animal, and it is not uncommon for the annual kill to exceed one million animals in some states. The average cottontail yields about one and a half pounds of dressed meat, and the species thus is of considerable economic importance to man. The fur sometimes is used for trim on clothing. Cottontails are quite vulnerable to traffic and are one of the mammals most commonly killed on roads. Alteration of the environment by fire, clearing of land, and agriculture leads to destruction of many nests and a short-term decrease in this cottontail, although the long-term effect may be to expand suitable habitat. Adverse weather also takes a heavy toll, especially among the young. Fleas, ticks, mites, botflies, flukes, and tapeworms all parasitize cottontails, and several diseases infect them, including tularemia, or "rabbit fever," which is transmissible to man.

At high densities, eastern cottontails can do extensive damage to cultivated crops such as alfalfa and clover. Their bark-eating habit in winter is injurious, if not fatal, to many trees. Commercial repellents are available to prevent crop damage. However, in places where it is legal, shooting or trapping may be the most effective—and environmentally the least detrimental— means of control.

Selected References. General (Chapman, Hockman, and Ojeda C. 1980 and included citations).

Sylvilagus nuttallii: Nuttall's Cottontail

Name. The name *nuttallii* is a patronym for the ornithologist and naturalist Thomas Nuttall (1786–1859).

Distribution. Nuttall's cottontail occurs in all western states, ranging from the eastern slopes of the Sierra Nevada eastward to western North Dakota and the Black Hills. *Sylvilagus nuttallii grangeri* is the subspecies reported from the Northern Plains.

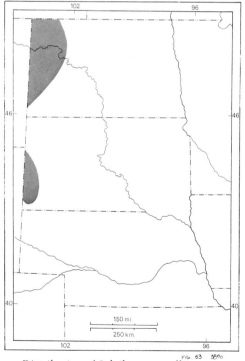

53. *Distribution of* Sylvilagus nuttallii.

Locally this cottontail is most commonly found in timbered areas, at higher elevations than the eastern or the desert cottontail. Its distribution in the Dakotas is limited to the vicinity of the Black Hills and to the badlands country of western North Dakota. In the latter area it is being replaced by *S. floridanus*. Nuttall's cottontail and the desert cottontail are sympatric in the western part of the Northern Plains. The two species are separated ecologically, however, with *nuttallii* occupying dense vegetation along streams and bottomlands, whereas *audubonii* is found in the rougher, more open uplands.

Description. Nuttall's cottontail is a medium-sized rabbit similar in color to both the desert and eastern cottontails. It is somewhat darker than *audubonii* with more black hairs dorsally. The ears of *nuttallii*, smaller than those of *audubonii*, have a thin line of black on the margins, and the nape has a bright rusty color. The hind legs also are relatively smaller than those of the desert cottontail. Compared with *S. floridanus*, *S. nuttallii* has paler dorsal pelage, a duller brownish throat patch, and more densely furred ears. The hind feet of *nuttallii* are covered with long, dense hairs, distinguishing it from both *S. floridanus* and *S. audubonii*. Average and extreme measurements of seven specimens from the Black Hills of South Dakota are: total length, 372.8 (353–401); length of tail, 48.3 (32–64); length of hind foot, 92.0 (83–100); length of ear, 65.3 (60–70). Six males from South Dakota averaged 809.5 (666–984) grams in weight. Females generally are heavier than males.

Cranially, *S. nuttallii* is distinct from *S. floridanus* in having a larger external auditory meatus and a more slender rostrum. The auditory bullae are of medium size, smaller than in *S. audubonii* and larger and more inflated than in *S. floridanus* (fig. 54). Additional characters distinguishing *S. nuttallii* from *S. audubonii* and *S. floridanus* are found in the key to the species of *Sylvilagus* and in accounts of *S. floridanus* and *S. audubonii*. Cranial measurements of five males from South Dakota are: greatest length of skull, 66.6 (63.2–69.0); zygomatic breadth, 35.5 (33.3–35.9).

Natural History. Nuttall's cottontail inhabits rocky canyons and ridges where thick stands of sagebrush provide cover, food, and shelter. It often is associated with stands of coniferous trees. This species makes use of "forms" for resting, but where vegetation is sparse it will use burrows or rocky crevices. *S. nuttallii*, like other cottontails, is mostly crepuscular—foraging early in the morning and late in the afternoon and evening, but it may be active at any time of the day or night. Heavy rains diminish foraging activity, but cold temperatures alone probably have little effect. Tracks of this species often are found in snow after a storm. Adverse weather undoubtedly reduces visu-

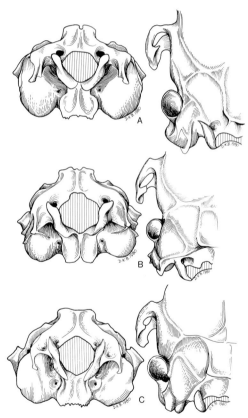

54. *Rear (left) and ventral (right) views of auditory region of skull of three species of* Sylvilagus *occurring on the Northern Great Plains:* (A) S. audubonii; (B) S. floridanus; (C) S. nuttallii

al and auditory acuity in rabbits, which would severely limit their detection of predators. Consequently, restricting activity during inclement weather increases survival.

Studies in the western portion of its range show that sagebrush is the principal food of *S. nuttallii.* Juniper also is consumed, but grasses, which are available in spring and summer, are the preferred food. This cottontail appears to be more solitary than other species of *Sylvilagus*, but groups may be found foraging together in select areas such as moist grassy clearings bordered by brush.

If it senses danger, the animal usually will run a short distance, stop, face away from the source of danger, and hold its ears erect. Low scrub helps to obscure its location, and the dorsal pelage, which blends with the surroundings, further aids concealment. If danger is imminent, the rabbit may flee in a semicircular path, thereby disguising the intended direction of flight.

The breeding season is in the spring and summer. Litter size averages four to five (range, one to eight). The young are cared for in a nest similar to those of other cottontails and reach full maturity within one year, although young of early litters may breed in their first summer. Females can bear four or five litters per year. *S. nuttallii* has the greatest mean litter size of the cottontails of the Northern Great Plains. Often species that inhabit more northern latitudes have larger mean litter sizes than those to the south, perhaps an adaptation to shorter breeding seasons that limit females to fewer litters per year.

Natural enemies of this cottontail include the coyote, bobcat, red fox, red-tailed hawk, marsh hawk, Swainson's hawk, long-eared owl, and possibly rattlesnakes.

Selected References. General (Chapman 1975 and included citations).

Genus *Lepus*

Lepus is an ancient Latin word meaning "hare." The dental formula is 2/1, 0/0, 3/2, 3/3, total 28. The genus is distributed widely in Eurasia, Africa, and North America and includes about 20 species, three of which are native to the Northern Great Plains.

Lepus americanus: Snowshoe Hare

Name. The specific name was proposed with reference to the American origin of this hare.

Distribution. Lepus americanus occurs in suitable habitat across the width of northern North America, from New England and the Maritime Provinces to the Pacific Coast, ranging northward to above the Arctic Circle and southward along the principal mountain ranges to Tennessee, New Mexico, and central California. The nominate subspecies, *Lepus americanus americanus*, occurs across Canada from the Atlantic seaboard to western Alberta, and southward to central North Dakota (see fig. 57).

Description. The snowshoe hare is the smallest member of the genus *Lepus* in the Northern Plains states, about midway in size between the cottontails and the jackrabbits. Its smaller size, shorter ears and tail, and relatively longer hind feet distinguish it readily from *Lepus townsendii*, even when both species are in white winter pelage. Average and extreme external measurements of seven adults from Manitoba and Saskatchewan are: total length, 429.3 (370–463); length of tail, 32.2 (25–38); length of hind foot, 117.8 (105–125). The length of the ear generally ranges from 62 to 70. Adults weigh 1.5 to 4 pounds.

The summer adult pelage of *Lepus americanus* is generally a rusty brown, the back being somewhat less reddish than the crown. There is an indistinct line of darker brown hairs on the dorsal midline. The chin is white, and the nape is grayish brown. The upper surface of the tail is black, but the underside of the tail and the belly are grayish white. The winter pelage is pure white throughout most of the range of the species, although the ears are black

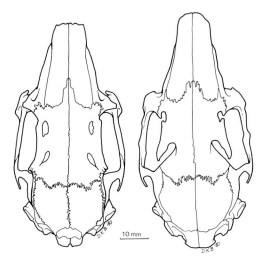

55. *Dorsal view of skulls of* Sylvilagus audubonii *(left) and* Lepus americanus *(right) showing condition of the interparietal bone in rabbits as opposed to hares.*

at the tips and may be tinged with buff.

The long, wavy pelage of neonates is a dull grizzled gray above, somewhat darker on the tail. The underparts are grayish white, and there are small black spots at the tips of the ears. By the age of two months, young hares are more nearly the brownish color of adults.

The skull of *Lepus americanus* is readily distinguished from that of other native North American hares because there is no anterior projection on the supraorbital process. Also, it is smaller than that of other hares. Greatest length of skull averages about 75; zygomatic breadth averages 37.

Natural History. The local distribution of *Lepus americanus* on the Northern Plains is determined mostly by the availability of suitable habitat. Areas with forests and shrubbery afford cover and food. Snowshoe hares rarely leave wooded areas except at the edges of brush patches, where they are close enough to retreat quickly if necessary. They occupy both hardwood forests and coniferous woodlands if an understory of shrubs is present. Alder swamps and burned areas with heavy regrowth of woody vegetation are especially favored.

These hares spend the daylight hours resting in simple, unlined "forms" beneath shrubbery, unless disturbed. They occasionally occupy hollow logs or abandoned burrows of other mammals. They forage at night, ranging from the form over well-worn runways. These runways may be quite obvious if snowshoe hares are abundant. Another sign of these mammals is a dust bath, an open spot where the hares, usually intolerant of each other, may congregate in small numbers. The peak of activity is about 11 P.M., and more than half the activity in summer takes place between 8:30 P.M. and 1:30 A.M.

Intraspecific aggression frequently has been noted among snowshoe hares. Pregnant females are actively hostile toward males, and males may fight among themselves during the breeding season, biting each other on the back and sometimes removing patches of fur. Captive males have been reported to be hostile toward young. The home range of females is three to four acres. For males the home range is much larger, being the sum of the home ranges of the females they court.

Lepus americanus appears to breed from early March to late August. The period of gestation is about 36 days. Litters of one to seven young have been reported, although there are only three pair of mammae, one pectoral and two abdominal. Litters most often contain three young. An individual female may produce one to three litters each year, although three probably is uncommon. The young are precocial and readily move about soon after birth if disturbed, but normally they remain huddled quietly together. No nest is constructed, and the young are simply placed in a well-concealed form.

Young snowshoe hares grow rapidly. The birth weight is doubled in eight days and trebled in 12 days. Milk is quickly supplemented by solid foods, and the young are weaned by the age of one month, when the weight has increased some nine times over that at birth. Adult weight is reached at about five months, but young-of-the-year usually do not breed.

The molt cycle of *Lepus americanus* has received the attention of several investigators. Autumnal molt from brown to white pelage occurs over a period of 70 to 90 days

56. Snowshoe hare, Lepus americanus (courtesy V. B. Scheffer).

from September to December. The onset of molt is controlled at least in part by day length. Molt begins on the ears, wrists, and feet and extends to the lower legs, lower rump, and tail. The molt then extends progressively toward the midline, occurring last on the shoulder region and crown. The underfur is not shed in the autumn molt. The spring molt sequence is essentially the reverse of the autumn pattern and takes place over about 70 days from March to May. The color changes produced by the molting cycle do not necessarily correlate with the weather of a given season. Internal control mechanisms have evolved that relate color to the average conditions of a region. In years of late snowfall, white hares are highly conspicuous against the landscape, but during the usual winter their coloration serves well to help protect them from predation.

The food habits of L. americanus are generally similar to those of their relatives of the open plains, the jackrabbits. In summer individuals forage on herbaceous woodland vegetation. During winter they browse on buds, twigs, and bark and on the young shoots of evergreens. Unlike most other lagomorphs, these hares eat some carrion. They seldom drink free water. Snowshoe hares frequently girdle young trees, interrupting the transport of nutrients in the phloem and thus killing saplings. The effect of this habit on forests has been the subject of considerable debate, but the current consensus seems to be that the long-term damage is slight. The hares seldom kill trees in concentrated patches, but girdle saplings at random over wide areas. The net effect serves the best interests of the forester by thinning stands of young trees and allowing the survivors better conditions for growth. The value of this species as food and game probably counterbalances any negative effect it may have on man's interests. Inasmuch as they are largely confined to woodlands, these hares have little effect on crops and are perhaps the least detrimental of the three species of Lepus on the Northern Plains.

Snowshoe hares are an important link in the food chain throughout their range. They are staple items in the diet of many carnivores including red fox, coyote, lynx, and bobcat. Smaller carnivores such as weasels feed on the young. The larger birds of prey, including hawks, eagles, and owls, depend heavily on snowshoe hares for food. The dependence of these predators on L. americanus is illustrated by the way cyclic abundance of the predators correlates with population cycles of the hares. The snowshoe hare is subject to severe internal and external parasitism.

Population cycles of L. americanus have been the subject of research and debate for decades. Populations differ by a factor of 25 or more from high to low. In high years, densities of five or six per acre are common. Under such conditions it is hard ever to get out of view of at least one snowshoe hare. The cycles of abundance are of interest because there is a coincidence of highs and lows over wide areas, and these trends are closely followed by a population cycle of the lynx, a valuable fur-bearer. There are several hypotheses about the underlying cause of this cycle. It has been thought that increased density led to increased incidence of parasitic disease. "Shock disease," a general stress syndrome, has been associated with high population densities. Recent studies indicate shortages of winter food as a significant factor in population decline, along with decreased survival of adults and especially young.

Selected References. Natural history (Aldous 1937; Grange 1932a, 1932b; Keith and Windberg 1978).

Lepus californicus: Black-tailed Jackrabbit

Name. Lepus californicus is most commonly called the black-tailed jackrabbit but is also known as Great Plains jackrabbit and gray-sided jackrabbit. The scientific name was applied because the specimen upon which the original description was based was from California.

Distribution. This jackrabbit ranges throughout much of western North America, from western Arkansas and Missouri westward to the Pacific Coast, and from Washington and Idaho south to the Mexican states of Hidalgo and Querétaro. *Lepus californicus melanotis* is the subspecies of the Northern Great Plains, ranging from southern South Dakota and southeastern Wyoming, south through eastern Colorado and New Mexico, and east through northern Texas to Arkansas and Missouri. *L. californicus* appears to have extended its range in recent years, displacing the white-tailed jackrabbit, *L. townsendii,* in wide areas. Possible explanations for this displacement are noted in the account of the latter species.

Description. The black-tailed jackrabbit is generally dark in color. The dorsum is gray to blackish tinged with buff. The tail has a black middorsal stripe that extends onto the back. The sides are gray, and the venter and hind feet are white. The ears are white posteriorly and black on the tips. The sexes are colored alike, and adults molt only once each year. Thus there is no marked seasonal color change like that seen in the other hares of the Northern Plains.

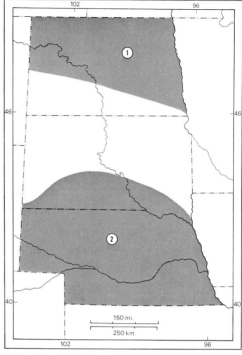

57. *Distribution of* (1) Lepus americanus *and* (2) Lepus californicus.

At birth the young are darker than adults, and paler areas of the pelage tend to be yellowish. The forehead of young black-tailed jackrabbits is marked with a white blaze, most conspicuous at two to three weeks of age, which disappears at about three months as subadult pelage is assumed. Full adult pelage is apparent the first winter by six to nine months after birth.

Lepus californicus is a large lagomorph, although somewhat smaller on the average than *L. townsendii.* External measurements of 10 adults (four males, six females) from central Nebraska are: total length, 568.4 (535−585); length of tail, 76.3 (71−87); length of hind foot, 130.5 (128−134); length of ear, 118.5 (112−125). Ears of museum specimens may shrink appreciably in drying; the measurements above were recorded from fresh specimens. Adults weigh from 5 to 7 pounds, females averaging slightly more than males.

The skull is characterized by its pro-

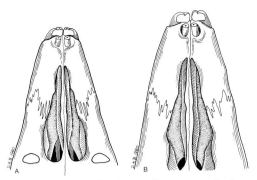

58. Ventral view of palate and incisor teeth of jackrabbits found on the Northern Great Plains: (A) Lepus californicus; (B) L. townsendii.

nounced supraorbital processes and fenestrate maxillary bone. Representative cranial measurements of adults from Nebraska are: greatest length of skull, 98.4 (95.3–101.3); zygomatic breadth, 44.4 (41.4–45.8).

Natural History. Black-tailed jackrabbits are animals of the open plains. An ideal habitat includes short grasses and herbs for food and ease of locomotion, with scattered brush for cover. They often become abundant on overgrazed land because grazing encourages this type of vegetation. They do not readily move into areas of tall grass or forest where visibility is obscured. During the day the jackrabbit rests in a "form," a simple, unlined depression in the ground, usually beneath protective vegetation. The animal ventures forth to forage in the late afternoon or evening, using well-marked routes.

During the warmer months black-tailed jackrabbits feed on grasses, herbs, and crops, primarily in open country. They are especially likely to move into alfalfa fields, where the usually solitary animals may congregate when other forage is meager. The winter diet consists of dried grasses and forbs, buds and bark of woody vegetation, pricklypear, small fruits, roots, and young winter wheat. When other food supplies are inadequate, numbers of jackrabbits may congregate at haystacks or straw piles. An adult eats about 250 grams of forage per day, about one-sixth the amount taken by one sheep. Jackrabbits seldom drink, be-

cause they derive sufficient moisture from the vegetation in their diet.

Jackrabbits, like other leporids, produce two kinds of feces. The characteristic hard, dark droppings seen in the field are the final waste product, voided during the feeding period. During the day, when the animal is resting, soft feces are formed. These pellets are taken from the anus and swallowed whole. The significance of this process of reingestion (also called coprophagy or cecotrophy) is not completely understood. It no doubt functions in part to enhance the utilization of nutrients in the high-cellulose diet of the hare, but perhaps it developed as a mechanism for saving vitamins synthesized by the intestinal and cecal flora. The young ingest the soft feces of their mothers and thereby establish their own colony of cellulose-digesting, vitamin-producing bacteria. Waste droppings of jackrabbits may contain substantial quantities of seeds, and germination of some seeds is enhanced by passing through the digestive system of a hare. A study in Kansas showed that seeds of sand dropseed were "sown" in droppings at higher rates than range managers recommended for artificial restoration of abandoned fields.

Black-tailed jackrabbits bear one to four litters of one to eight young (usually two to four) each year after a gestation period of 41 to 47 days. Larger litters correlate with better nutritional conditions. Ovulation is induced by copulation, which occurs repeatedly after long, vigorous chases. Pregnant females have been taken from late January to late August in Kansas. The young are precocial: they are fully haired, have open eyes, and can move about at birth. At two to three days they are able to hop well. At about 10 days milk is supplemented with solid food. The young are reared in an open hollow in the soil, usually in an area of dense grass. Young jackrabbits grow rapidly, achieving 90 percent of their adult size and 65 percent of adult weight in 10 weeks. Normally, young-of-the-year do not breed, but all young of a given year probably breed in the following season.

The single annual molt of the black-

59. Black-tailed jackrabbit, Lepus californicus *(courtesy K. Steibben, Kansas Fish and Game Commission).*

tailed jackrabbit begins, depending on latitude, between late August and early October. The first evidence of new pelage is seen on the top and sides of the head, as a strip extending from the nose backward between the eyes to the base of the ears and as lateral bands from the sides of the nose to the cheeks. The strips of new pelage meet under the chin in the midventral line. At about the same time new hairs appear on the back, advancing toward the nape and rump. The venter is the last region to attain new pelage, when molt lines advancing from either side meet irregularly along the midventral line.

The senses of sight and smell are keen, warning jackrabbits of potential danger. The animals rely on protective coloration and speed to escape predation. They can travel at 35 to 40 miles an hour when pursued, progressing by leaps of two to three yards or more. At intervals they make a somewhat higher "observation leap" or "spy hop."

Despite these protective adaptations, predators such as coyotes, foxes, snakes, and almost all the larger birds of prey take their toll of black-tailed jackrabbits. Further, they are parasitized by botfly larvae, fleas, lice, ticks, mites, tapeworms, and nematodes. They are highly susceptible to various microbial diseases, some of which, such as tick fever, spotted fever, and tularemia, also infect man. A wise rule of thumb is to beware of any rabbit or hare that obviously is diseased or emaciated or is behaving strangely.

Lepus californicus is of value to man as small game, for food and "sport." However, owing in large part to human alteration of the environment, black-tailed jackrabbits have become a nuisance in many areas. Numbers have increased as a result of predator-control programs, new areas have been invaded because of agriculture and deforestation, and overgrazing by domestic animals has led to further damage to the range by jackrabbits, which tend to crop vegetation closer than cattle do. Jackrabbits may feed on cultivated crops, nursery and orchard stock, and hay. Where they have become detrimental to man's interests, they should be controlled by appropriate fencing or shooting. The use of poisoned hay and other such indiscriminate measures in jackrabbit control is to be discouraged.

Selected References. Natural history (Hansen and Flinders 1969; Lechleitner 1958; Tiemeier et al. 1965; Vorhies and Taylor 1933).

Lepus townsendii: White-tailed Jackrabbit

Name. The specific name of this species was proposed in honor of J. K. Townsend (1809–51), a pioneer American ornithologist, who collected the type specimen.

Distribution. The white-tailed jackrabbit originally occurred from southern Kansas northward to central Saskatchewan and Alberta. The Mississippi River marks the approximate eastern limit of its range. To the west of the Rocky Mountains it occurs across the Great Basin to the Sierra Nevada and northward in suitable situations to southern British Columbia. The subspecies on the Northern Great Plains is *Lepus townsendii campanius,* which ranges over the region east of the Rockies.

Although *Lepus townsendii* formerly occurred over all the Northern Plains except

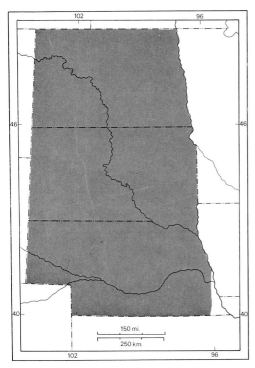

60. *Distribution of* Lepus townsendii.

perhaps in extreme southeastern Nebraska, it now appears to be limited to the region north of the Platte River. The attenuation of its range northward and westward corresponds with an expansion of the range of *Lepus californicus*. The black-tailed jackrabbit seems more adaptable to cultivated lands than is the white-tailed jackrabbit. The extension of the range of *L. californicus* is probably due in part to the increasing cultivation of the plains. It also has been suggested that the gradual warming trend of the past century has allowed for the northward expansion of the black-tailed jackrabbit, essentially a southwestern species.

Description. The summer pelage of *Lepus townsendii* is pale buffy gray on the dorsum and neck, whereas the ventral region is pale gray or white. The tail is white, although there may be a median dorsal line of black hairs, which does not extend onto the back. The posterior aspect of the ears is white, and the ears are tipped with black. The winter pelage is white over its range

on the Northern Plains, but in other regions the color change is not complete. The tips of the ears remain black, and there may be an obscure buffy tinge to the pelage of the ears, feet, face, and back.

The juvenile pelage is similar to the summer pelage of adults, but guard hairs are sparser, exposing considerable amounts of the underfur. There are relatively more guard hairs on the tail. The postjuvenile pelage is intermediate between that of juveniles and adults in summer.

Lepus townsendii is a large, heavy-bodied hare, although its ears and hind limbs are proportionately smaller than those of *L. californicus*. Average and extreme measurements of 17 specimens (12 males and five females) from the Northern Great Plains are: total length, 609.6 (540–640); length of tail, 85.1 (70–112); length of hind foot, 144.6 (126–165); length of ear, 108.3 (95–114). Females tend to be larger than males.

The skull of this species is similar to that of the black-tailed jackrabbit. No method is known for distinguishing the skulls of the two species throughout their ranges, but the nature of the folds of the enamel on the occlusal surface of the first upper incisors as described in the key to leporids should distinguish the skulls of specimens from the Northern Plains. Average and extreme cranial measurements of 10 adults from Nebraska and the Dakotas are: greatest length of skull, 96.5 (94.1–99.8); zygomatic breadth, 47.4 (45.9–49.6). Large adults may weigh 10 pounds or more, but the average is about 7 pounds.

Natural History. Lepus townsendii is typically a mammal of the open plains, where its keen hearing and swiftness afford it protection from predation. In the Rocky Mountains it occurs to timberline and above but still holds strictly to open areas. Where the species is sympatric with *Lepus californicus*, it is found in more open vegetation (the black-tailed jackrabbit prefers areas with scattered shrubbery). White-tailed jackrabbits retreat to the shelter of the edges of woodlots and riparian communities only in the most severe winters.

They do not burrow in the soil but may tunnel beneath snow in search of food. In their usual habitat on the plains, they crouch low in their "forms" during the day. Depending on their coloration for concealment, they are more easily approached than black-tailed jackrabbits. When danger is imminent they escape by running, and they may attain a speed of 40 miles an hour when pursued.

L. townsendii breeds from late February to mid-July. Late-lying snow cover may delay the onset of breeding. Ovulation is induced by copulation. Females bear two to four litters of one to nine precocial young. Litter size declines in successive litters each year, averaging three to four overall. The young weigh about 100 grams at birth and are concealed in burrows, cavities, or shallow forms. The female's milk is supplemented at an early age with green plants, and the diet of half-grown young consists entirely of vegetable matter. The young gain little weight during their first week but gain about 9 grams per day from day 20 to day 80, reaching adult weight at 95 to 125 days of age.

White-tailed jackrabbits feed in the early morning and from late afternoon well into the evening. Their diet in summer consists of native grasses and herbs and, in agricultural areas, of clover, alfalfa, and garden

61. *White-tailed jackrabbit,* Lepus townsendii, *in winter pelage (courtesy U.S. Fish and Wildlife Service).*

greens. In winter these animals eat dried vegetation, but they obtain most of their food by browsing buds, young shoots, and bark of woody plants. Although not typically gregarious, these hares may gather at haystacks in areas where other food supplies are inadequate.

There are two molts annually. In spring, molt begins around the eyes and on the lower back, proceeding to the head and cheeks and then the rump and sides. The ears molt last. The autumn molt reverses the foregoing sequence and changes the pelage to white. There are two maturational molts, which progress much as does the spring molt.

A wide variety of mammalian and avian predators prey on young white-tailed jackrabbits, but the most serious enemies of adults probably are diseases and parasites (including protozoans, tapeworms, roundworms, ticks, and fleas) and the activities of man, who affects the numbers of these animals directly, by hunting, and indirectly, by breaking up the native grasslands for agriculture.

Lepus townsendii is of value to man as a source of food and fur. The species generally is considered superior to the black-tailed jackrabbit as food, because whitetails are somewhat heavier and the meat has a more pleasant taste. At times, white-tailed jackrabbits have been harvested for food in commerical quantities. In the winter of 1894–95, a single commission house in Saint Paul, Minnesota, handled some 12,000 white-tailed jackrabbits, most of which were taken in the Dakotas. The fur of these mammals has been used in the manufacture of hatter's felt.

Although the food habits of the white-tailed jackrabbit may attract it to hay crops and young tree plantations, the animals do not cause appreciable damage to cultivated crops. It is unlikely that the species is ever sufficiently abundant on the Northern Plains to be significantly detrimental to man's interests.

Selected References. Natural history (Bear and Hansen 1966; James and Seabloom 1969*a*, 1969*b*).

Order Rodentia

The order Rodentia represents the largest and most diverse of Recent mammalian orders, comprising more than 30 modern families, containing about 490 genera and 1,620 species (some 40 percent of all those among living mammals), and rodents are the most abundant mammals in many regions of the world. The order first is known in geologic time from the late Paleocene of North America, about 60 million years ago. Representatives of this group are nearly cosmopolitan in distribution, being native to all land areas except some arctic and oceanic islands, New Zealand, and Antarctica; furthermore, rodents have been introduced by man in most of the few places in which they did not occur naturally. Seven families are native to the Northern Great Plains (and representatives of two others have been introduced into the tristate region). The 42 species indigenous to the Dakotas and Nebraska range in size from the smallest harvest mice and pocket mice, some adults of which weigh less than 10 grams, up to the porcupine and beaver.

The typical rodent is strictly terrestrial, but some are scansorial or semiarboreal, dwelling mostly in trees (including species that glide), others are fossorial, living most of their lives underground, and still others are semiaquatic. Members of the order are characterized by a single pair of relatively large and ever-growing incisors, separated from the cheekteeth by a distinct diastema, in both the upper and lower dental arcades, and a dental formula that never exceeds 1/1, 0/0, 2/1, 3/3, total 22. The upper and lower toothrows are capable of partial or complete apposition at the same time. Many rodents are obligate herbivores, but some also eat insects and other small animals.

Key to Families of Rodents

The following key includes not only rodent families native to the Northern Plains, but also two (Myocastoridae and Muridae) that have been introduced from outside North America. Representatives of these two families are treated in a special section of this book.

1. Modified for semiaquatic life; hind feet webbed; lower incisor more than 6.0 in width at alveolus _____2
1.' Not especially modified for semiaquatic life (except *Ondatra*); hind feet not webbed; lower incisor less than 5.5 (less than 4.0 in all except Erethizontidae) in width at alveolus _____3
2. Tail flattened dorsoventrally, its breadth approximately 25 percent of its length; infraorbital canal smaller than foramen magnum _____Castoridae
2.' Tail not flattened dorsoventrally, its breadth less than 10 percent of its

length; infraorbital canal larger than foramen magnum _____Myocastoridae*

3. Sharp quills on dorsum and tail; infraorbital canal larger than foramen magnum _____Erethizontidae

3.' No quills on any part of body; infraorbital canal never larger than foramen magnum _____4

4. Hairs on tail usually distichous; skull with distinct postorbital processes _____Sciuridae

4.' Hairs on tail not distichous (except in *Neotoma cinerea*); skull lacking distinct postorbital processes _____5

5. External fur-lined cheek pouches present; cheekteeth 4/4 _____6

5.' No external fur-lined cheek pouches; cheekteeth 4/3 or 3/3 _____7

6. Tail more than three-fourths length of head and body; hind feet larger than forefeet; tympanic bullae exposed on posterodorsal part of skull _____ Heteromyidae

6.' Tail much less than three-fourths length of head and body; hind feet smaller than forefeet; tympanic bullae not exposed on posterodorsal part of skull ___Geomyidae

7. Tail much longer than head and body; hind feet noticeably elongate; cheekteeth 4/3 _____Zapodidae

7.' Tail about equal to, or shorter than, head and body; hind feet not noticeably elongate; cheekteeth 3/3 _____8

8. Annulations of scales on tail nearly or completely concealed by pelage (except in *Ondatra*, in which the tail is laterally flattened); cheekteeth with two longitudinal rows of cusps or prismatic _____Cricetidae

8.' Annulations of scales easily visible on sparsely haired tail; cheekteeth with three longitudinal rows of cusps _____ Muridae*

Family Sciuridae: Squirrels

This family is nearly worldwide in occurrence, the Australasian region, southern South America, and Antarctica being the only continental areas in which squirrels are absent. Sciurids are considered relatively primitive rodents, partly because of the structure of their teeth and jaws and the associated muscles. There are three basic "kinds" of squirrels—the nocturnal "flying" squirrels, the diurnal tree squirrels, and the ground squirrels. The first two sorts spend much time in trees and build their nests above the ground, whereas the latter are, as their name implies, terrestrial and usually nest underground in burrows. Because of their dependence upon trees, flying squirrels and tree squirrels are mostly marginal in distribution on the Northern Plains, and of patchy occurrence within the region, in contrast to the ground squirrels, which are widespread and often abundant in grassland habitats.

About 50 genera and 250 species of squirrels currently are recognized, of which seven genera and 15 species inhabit the Northern Great Plains. All share the typical rodent incisors and complete set of molar teeth, but the upper premolars are variable in size and number. The geologic record dates from the Oligocene of Eurasia and North America.

Key to Squirrels

1. Skin between forelimbs and hind limbs noticeably loose, forming gliding membrane; narrow interorbital region V-shaped _____2

1.' Skin between forelimbs and hind limbs not noticeably loose; interorbital region not V-shaped _____3

2. Hair of venter white to base; total length less than 260; greatest length of skull less than 36 ___*Glaucomys volans*

2.' Hair of venter gray, white only at tips; total length more than 260; greatest length of skull more than 36 _____*Glaucomys sabrinus*

3. Length of hind foot usually more than 75; postorbital processes at right angles (or nearly so) to longitudinal axis of skull ————————————————4

3.' Length of hind foot usually less than 75; postorbital processes not at right angles to longitudinal axis of skull ——5

4. General color tawny; face black, with white mark on nose; postorbital processes at slightly less than right angles to longitudinal axis of skull ——————————— *Marmota flaviventris*

4.' General color brown; face brown to black but lacking white mark on nose; postorbital processes at right angles to longitudinal axis of skull —————— *Marmota monax*

5. Crown of second upper molar as long as wide; tail usually more than 40 percent of total length ——————————————6

5.' Crown of second upper molar noticeably wider than long; tail usually less than 40 percent of total length ————8

6. Anterior border of orbit opposite fourth premolar ventrally; greatest length of skull less than 55; upper parts olive brown to reddish orange; venter white ————————————*Tamiasciurus hudsonicus*

6.' Anterior border of orbit opposite first molar ventrally; greatest length of skull more than 55; pelage color as described below ——————————————————7

7. Pelage usually predominately gray above and white below; tips of hairs of tail white; two upper premolars usually present but the first (P3) much reduced and occasionally absent ——————————————*Sciurus carolinensis*

7.' Pelage orange brown above, paler below; tips of hairs of tail orange; only one upper premolar (P4) present ——————————————— *Sciurus niger*

8. Pelage always striped above; stripe(s) present above eye; infraorbital foramen pierces zygomatic plate ————————9

8.' Pelage striped, spotted, mottled, or plain; no stripe above eye; infraorbital canal separates zygomatic plate and rostrum ——————————————————10

9. Only one upper premolar present; dorsal stripes not continuous to base of tail, indistinct on face —*Tamias striatus*

9.' Two upper premolars present, the first (P3) reduced; dorsal stripes continuous to base of tail, distinct on face ————————————*Tamias minimus*

10. Tail less than one-fourth length of body; tip of tail black; upper toothrows strongly convergent posteriorly and teeth laterally expanded ————————————*Cynomys ludovicianus*

10.' Tail more than one-fourth length of body; upper toothrows nearly parallel and teeth not laterally expanded ——11

11. Dorsal pelage marked with large stripes or spots; length of maxillary toothrow less than 8.5 ————————————————12

11.' Dorsal pelage uniform or grizzled in color; length of maxillary toothrow more than 8.5 ————————————————13

12. Dorsal pelage marked with a series of alternating dark brown, pale brown, and broken pale and dark brown longitudinal stripes; postorbital constriction less than 12 ————————*Spermophilus tridecemlineatus*

12.' Upper parts beige to cinnamon, with scattered white spots of moderate size; postorbital constriction more than 12 ————————————*Spermophilus spilosoma*

13. Upper parts buffy cinnamon; center pinkish or yellowish; tail less than 120 ————————————*Spermophilus richardsonii* and *S. elegans* (see text)

13.' Upper parts with a "salt and pepper" appearance and with olive to tawny buff overwash; venter white; tail more than 120 ——————*Spermophilus franklinii*

Genus *Tamias*

Tamias is a Greek word meaning "steward" or "treasurer," in reference to the food-storing behavior of these animals. The dental formula is 1/1, 0/0, 1–2/1, 3/3, total 20 to 22. The genus contains about 24 species, of which two occur on the Northern Great Plains. All chipmunks are found in North America except for one species that is widespread in Eurasia.

These animals formerly were placed in two genera, *Eutamias* for the chipmunks of western North America and Eurasia, and

Tamias for the chipmunk of eastern North America, and some authorities still retain both. However, recent evidence indicates that all chipmunks are so closely related as to make their separation into two genera unwarranted. The vernacular name is of uncertain origin. While it usually is attributed to the "chip" call note of members of the genus, it more plausibly was derived from an American Indian word, *achitamon* or *chetamon*, which means "headfirst," in reference to the way chipmunks (and tree squirrels) descend tree trunks.

Tamias minimus: **Least Chipmunk**

Name. The Latin word *minimus* means "least." This species includes the smallest of the chipmunks, although there is considerable geographic variation in size.

Distribution. The least chipmunk ranges over most of western North America, ex-

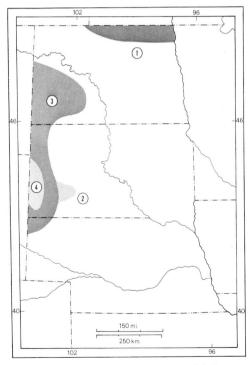

62. *Distribution of* Tamias minimus: (1) T. m. borealis; (2) T. m. cacodemus; (3) T. m. pallidus; (4) T. m. silvaticus.

tending eastward into central Canada and the northern Great Lakes states. It occupies the widest geographic and altitudinal range of any chipmunk. Local distribution is determined by the presence of an adequate combination of food and cover. It is found in a great variety of biotic communities, including alpine tundra, sagebrush deserts, and coniferous and broad-leafed forests. Four subspecies are found on the Northern Plains: the extremely pale *Tamias minimus cacodemus* and the dark-colored *Tamias minimus silvaticus* are restricted to the Badlands and Black Hills, respectively, of South Dakota, whereas *Tamias minimus borealis*, also relatively dark, is found in northeastern North Dakota, and the pale *Tamias minimus pallidus* frequents the western part of the tristate region.

Description. This is the smallest and most variable of all chipmunks. Typically the upper parts are mixed tawny and gray, grading into clear tawny on the sides. A blackish median stripe and four dark brown to blackish side stripes mixed with rufous run along the dorsum, separated by grizzled gray stripes. The outermost pair of side stripes is white. The tail is carried straight up when the animal runs and is ochraceous fringed with buff; the underparts are white. There are two molts each year. The major molt, from faded winter to fresh summer pelage, occurs in June or July, except in breeding females, in which it may be delayed until August. The molt into winter pelage is less pronounced and occurs in late September or October.

Average and extreme external measurements for a series of four males and four females, respectively, of *T. m. borealis* from Fergus County, Montana, are: total length, 197.5 (190–203), 196.0 (186–213); length of tail, 89.2 (88–91), 94.8 (90–99); length of hind foot, 30.8 (30–31), 30.8 (30–31); length of ear, 15.2 (15–16), 15.2 (15–16). Other populations are slightly larger. Comparable measurements for seven males and ten females, respectively, of *T. m. cacodemus* from the South Dakota Badlands are: 199.6 (191–212), 205.4 (191–218); 88.6 (81–95), 91.8 (78–108); 32.1 (31–34),

33.1 (30–35); 14.6 (12–17), 16.1 (15–18). Average weights of these specimens were 36.9 (29.1–46.2) and 38.4 (31.2–53.2) grams. The same measurements for 10 males and 10 females of *T. m. silvaticus* from Lawrence County, South Dakota, are: 192.4 (189–203), 203.1 (191–214); 80.2 (69–86), 90.2 (81–115); 30.8 (30–32), 30.7 (29–32); 15.8 (14–19), 16.2 (15–17); weights of these animals were 45.1 (37.6–50.1) and 48.6 (37.3–55.3) grams for males and females, respectively. The largest subspecies thus is *silvaticus* of the Black Hills, followed closely by *pallidus; cacodemus* of the Badlands is somewhat smaller, and *borealis* of the northern forests is smallest.

This relationship also is borne out by cranial measurements. Average and extreme measurements of the specimens of *borealis* are: greatest length of skull, 30.7 (29.9–31.2), 30.7 (29.8–32.0); zygomatic breadth, 17.6 (16.8–18.3), 17.7 (17.2–18.7). The same measurements for *cacodemus* are: 31.7 (31.2–32.4), 31.7 (30.9–33.3); 17.9 (17.6–18.2), 18.0 (17.5–18.4); and for *silvaticus*: 32.1 (31.3–33.5), 32.0 (30.8–32.6); 17.9 (17.2–19.1), 18.3 (18.0–18.7).

The skull has a high, narrow braincase with weak postorbital processes, and the third upper premolar is present but small. The sexes are only weakly dimorphic in size, but chipmunks are unusual among rodents in that females frequently are larger than males.

Natural History. The species of chipmunks in western North America vary considerably in habits and habitats, but in general they are more agile and sprightly than the eastern chipmunk and are more often seen in trees. They also appear to be less shy and thus are more often observed.

On the Northern Plains, least chipmunks occupy arid badlands, brushy areas along watercourses, and isolated stands of mixed deciduous or coniferous forests. *Tamias minimus borealis* is a characteristic inhabitant of the northern coniferous forest and is found in such habitat only in extreme northeastern North Dakota. This northern subspecies is characterized by a karyotype differing slightly from that of the more southerly races. The Black Hills are inhabited by *Tamias minimus silvaticus,* found in a variety of forested and brushy habitats. The western edge of the Northern Plains and the badlands of the Dakotas are occupied by pale-colored populations—*Tamias minimus pallidus* and *Tamias minimus cacodemus,* the latter restricted to the White River Badlands of southwestern South Dakota. *T. minimus* digs burrows in the ground or in old logs, but little is known of the details of their construction. Generally a tunnel two or three feet long terminates in a chamber about a foot below the surface. This chamber contains a nest as well as a store of seeds for winter food. Occasionally, least chipmunks inhabit tree holes.

In the northern part of its distribution this chipmunk disappears during the colder months, and hibernation probably is complete where snow lies all winter. However, in more southerly areas individuals are active throughout the winter. The species is hardy and does not retire to winter quarters until late in the autumn, frequently after the onset of snow and freezing weather. In spring these animals may burrow out through late-lying snow to forage, usually appearing in March. Breeding begins soon, peaking in April but continuing until mid-May.

The young are born after a gestation of 28 to 30 days in May or early June. Litter size ranges from three to nine, but the average is five or six. There is only one annual litter in our area; although second litters have been reported elsewhere, this is suspect.

The altricial young are hairless at birth, with eyes and ears closed, and weigh 2 to 2.5 grams. The eyes and ears are open after about four weeks, and full juvenile pelage is attained in about five weeks. Weaning begins about this time (females have four pair of mammae), and young chipmunks are seen outside the natal burrow by about mid-June. The period of lactation is variable but may be 40 to 50 days, and even for late litters it is terminated by early to mid-August. By the time they are two months of age, young chipmunks are more than

63. *Least chipmunk,* Tamias minimus *(courtesy E. P. Haddon).*

although there is considerable overlap. Population densities also vary with different habitats, but typical late summer densities are between five and 15 chipmunks per acre.

The food of the least chipmunk includes a variety of nuts, fruits, grains, and vegetable matter. Where nuts are not available, the animals eat large amounts of wild berries and seeds. Piñon nuts (not found in our area) are a favorite food, as are the seeds of most conifers (pines, spruces, firs, tamarack, and hemlock). Cactus fruit is eaten when available. Insects are frequently eaten, especially beetles, grasshoppers, and caterpillars.

These animals are preyed upon principally by marsh hawks, red-tailed hawks, weasels, mink, red foxes, and feral domestic cats. Parasites of the least chipmunk include several species of nematodes and an acanthocephalan internally and fleas, ticks, and lice externally. Tularemia has been found in some populations.

Selected References. General (Jackson 1961); natural history (Forbes 1966; Sheppard 1972; Skryja 1974; Turner 1974); systematics (Nadler et al. 1977).

Tamias striatus: Eastern Chipmunk

Name. The Latin word *striatus* means "striped," a reference to the conspicuous pattern of stripes on the dorsal pelage.

Distribution. This species occurs in eastern North America, from southern Manitoba, Ontario, and Quebec south to about latitude 34°, and eastward from the eastern parts of North and South Dakota, Nebraska, Kansas, and Oklahoma to the Atlantic Ocean. Its preferred habitat is brushy areas in broad-leafed deciduous forest, often on wooded bluffs bordering streams and rivers. It is frequently rare to absent locally along the western margin of its range, probably because of forest thinning and the clearing of brush for pasturage or agriculture. Local occurrence may depend upon the availability of cover and

two-thirds grown and independent of the female.

Vocalizations are similar to those of the eastern chipmunk. The common alarm note is a sharp "chipping," which varies in pitch and intensity. Another common note is a low "clucking" sound. These vocalizations are undoubtedly related to communication between individuals and to the social organization of these chipmunks, but remarkably little has been published on this aspect of the biology. Least chipmunks appear to be considerably more social than eastern chipmunks. Home ranges overlap broadly, and several individuals often may be seen foraging close together in the same area. Home range size varies from less than an acre to several acres, depending on habitat quality. Home ranges of adult males average twice as large as those of females,

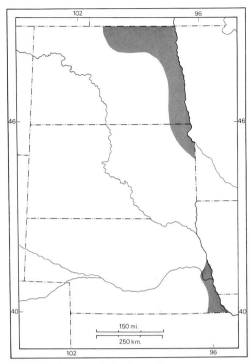

64. Distribution of Tamias striatus.

food, such as nuts, acorns, and berries. Only one subspecies, *Tamias striatus griseus*, occurs in the tristate region.

Description. The eastern chipmunk is a small, mostly terrestrial squirrel with a flattened, bushy tail and indistinct facial stripes. The ground color of the body is a grizzled rusty red to chestnut; there are five black stripes on the sides and back, separated by pale stripes; the stripes fade into the reddish flanks and rump. The underparts are white to buffy, and the feet are tan. The head is characterized by prominent short ears edged with russet. Many observers have reported only a single annual molt in this species, between late June and early September, in contrast to the two molts of the least chipmunk. Juvenile animals certainly seem to molt only once, assuming adult pelage in July and August (second litters may not molt until October or November). However, there is some evidence in support of a second molt in adults,

occurring in late autumn or early winter. The sexes are alike in color, and internal cheek pouches are well developed. Average and extreme external measurements for a series of 11 males and females from Hennepin County, Minnesota, are as follows: total length, 268.4 (253–299); length of tail, 101.3 (93–110); length of hind foot, 36.6 (35–38); length of ear, 13.7 (12–16.5). Weight of three males and one female from Saint Louis County, Minnesota, averaged 97.7 (88–110) grams.

The skull is lightly built and narrow, and the postorbital processes are small and weak. In contrast to *T. minimus*, there is only one premolar in the upper jaw. Average and extreme cranial measurements of two males and a female from Johnson County, Kansas, are: greatest length of skull, 40.4 (38.5–43.0); zygomatic breadth, 21.4 (19.6–23.5).

Natural History. Tamias striatus is rarely seen at any distance from wooded or brushy areas. These chipmunks seem to require protective cover and are not at home in the open. They are mostly terrestrial and rarely climb trees, in contrast to *T. minimus*. Favored habitats are wooded hillsides, rocky ravines, and dry, upland timber. Social behavior is not strongly developed, but they may associate in family groups of females with offspring, and in winter several individuals may be found occupying the same den and subsisting on a common store of provisions.

The eastern chipmunk digs its own burrow, often beneath rocks, stone walls, the roots of trees, or buildings. The holes are small and inconspicuous, and rarely does any dirt accumulate at the entrance. A burrow may have several branches and openings, and there may be four or five storage and nest cavities from one to three feet below the surface. Some storage cavities may hold up to a bushel of food.

All through the summer and especially in early autumn the eastern chipmunk gathers food, which it carries to the den in its well-developed internal cheek pouches. Foods include a considerable variety of fruits, grains, vegetable matter, and such nuts as acorns, hazelnuts, beechnuts, hickory nuts, and chestnuts. Corn, wheat, and oats also are consumed. Wild fruits and berries furnish a considerable part of the diet, and mushrooms may be eaten. The animals also prey opportunistically on invertebrates and small vertebrates.

Most eastern chipmunks become sexually mature in spring of the year after they are born, although some females born in the spring may breed later that summer. Most individuals that mature live only a year or two, but some live longer, and a lucky few may live as long as eight to 10 years. *T. striatus* is therefore one of the longer-lived small mammals in our area.

Mating occurs the last of March or first of April; the gestation period is 31 days. A single annual litter is usual in the northern part of the range, but second litters are the rule farther south. Two to seven (usually four to five) young are born hairless and with closed eyes, weighing about 3 grams. After a month in the nest, during which they are nursed (females have four pair of mammae), the young are well furred, their incisors, deciduous premolars, and first molars have erupted, and their eyes are open. They weigh 30 to 35 grams at this age and are ready for weaning; they begin to venture out from the nest when about five weeks old. The young appear aboveground in June in Minnesota. Partly grown young may be seen with their mothers for several weeks, but by six to seven weeks of age they begin to disperse. While some individuals may move long distances, most seek vacant burrows near the natal burrow and establish home ranges adjacent to, or even overlapping, that of their mother.

In the southern part of their geographic range, eastern chipmunks are more or less active throughout the winter, although they remain in the nest during periods of severe cold and snow, subsisting on stored food. However, farther north they hibernate for varying lengths of time, with reduced body temperature and lowered metabolic functions. Eastern chipmunks are "light" hibernators, in contrast to most ground squirrels, and thus do not lay on much body fat during autumn. The extent of hibernation appears to vary not only geographically, but also by age and sex and even from individual to individual. Variation in fat accumulation accounts for some of this, because obese individuals are more prone to enter torpor.

The common call note of the eastern chipmunk is a low-pitched "cluck" or "chuck," which at times is repeated rapidly and gives the suggestion of a song. When frightened and about to take refuge in a burrow, the chipmunk utters a rapidly trilled whistle while twitching its tail. The home range usually covers one-half acre to one and a half acres. Chipmunks are strongly territorial and aggressively defend the area immediately around the burrow. Home-range size may increase in autumn, as foraging for food intensifies, and long-distance movements to particular food sources may occur at any season.

Males have larger home ranges than females, and in the breeding season they may range well beyond the area they usually occupy, taking part in mating chases of estrous females with other males. Home ranges of individuals in a population show extensive overlap, and social structure is maintained by a dominance hierarchy enforced by threats and chases. Population densities vary with habitat; they may be higher than 20 per acre in prime range within the eastern deciduous forest, but in

65. *Eastern chipmunk,* Tamias striatus *(courtesy E. C. Birney).*

the tristate area, at the margin of the range of the species, they are generally much less, on the order of two to five per acre.

Tamias striatus is preyed upon by raptors such as marsh, Cooper's, and red-tailed hawks, and a variety of native carnivores including weasels, mink, foxes, and raccoons. Fleas, lice, and mites infest the skin and fur of this species, and it is host to a considerable number of internal parasites, including trematodes, cestodes, nematodes, and acanthocephalans. Botflies also attack eastern chipmunks, and individuals with the conspicuous subcutaneous "warbles" that contain the larvae of this fly frequently are taken.

Selected References. General (Forbes 1966); natural history (Elliott 1978; Mares, Watson, and Lacher, 1976; Svendsen and Yahner 1979; Tryon and Snyder 1973; Yahner 1978a, 1978b, 1978c); systematics (Nadler et al. 1977).

Genus *Marmota*

The name *Marmota* is the Latin for marmot and was originally applied to the marmot of the European Alps, *M. marmota.* The dental formula is 1/1, 0/0, 2/1, 3/3, total 22. There are 14 species in the genus, of which two occur on the Northern Great Plains. The remainder are scattered throughout the mountains of western North America and the mountains and steppes of Eurasia. Species of the genus *Marmota* are among the largest of rodents, but they may be considered "giant" ground squirrels, because the two groups are closely related.

Marmota flaviventris:
Yellow-bellied Marmot

Name. The specific name *flaviventris* is derived from two Latin words—*flavus,* meaning "yellow," and *venter,* referring to the underside of the body, or belly. Other vernacular names sometimes used include "rockchuck" and "whistle-pig," referring to the habitat and to the voice.

Distribution. The yellow-bellied marmot occurs in North America as far north as south-central British Columbia and extreme southern Alberta. It reaches the southern limits of its range in the southern Sierra Nevada and White Mountains of California, Toquima and Pine Valley mountains in Nevada, and Sangre de Cristo Mountains in New Mexico, where it is found only at higher elevations, usually above 6,000 feet. Yellow-bellied marmots penetrate only the extreme western edge of the Northern Great Plains, where an isolated population (the subspecies *Marmota flaviventris dacota*) is found on the Black Hills (see fig. 67).

Description. This mammal is of medium size, but large for a rodent; as noted, it may be considered a "giant" ground squirrel. Weight of adult males averages 3.9 kilograms (2.95–5.22), about 6.5 to 11.5 pounds, that of adult females averages 2.8 kilograms (1.59–3.97), about 3.5 to 8.7 pounds. The body is thickset and the head is short and broad. The underfur is soft, dense, and somewhat woolly, chiefly on the back and sides; longer and coarse outer guard hairs cover the entire body. The overall color is yellow brown to tawny, often with a "frosted" appearance owing to pale tips on many of the dorsal guard hairs. The belly is yellowish to tawny orange. Considerable variation in color occurs, even within the Black Hills, and partial to complete melanism is not uncommon in some areas of the Southern Rocky Mountains. There is one annual molt, usually from mid-June through July, its timing depending on the age and reproductive status of the individual. The ears are small and well furred, and the tail is relatively long. The feet bear stout, slightly curved claws on all digits except the thumb, which is rudimentary. Females have five pair of mammae. The cheekteeth are high crowned, and the anterior face of the incisors is yellowish white.

Average and extreme external measurements of eight males and nine females, respectively, from the Black Hills are: total length, 574.1 (472–680), 569.1 (490–640); length of tail, 171.6 (138–225), 167.3

(144–193); length of hind foot, 79.8 (72–86), 76.3 (72–80); length of ear, 31.0 (29–34), 30.0 (25–36). Average and extreme cranial measurements of six males and eight females, respectively, from the same area are: greatest length of skull, 84.0 (76.9–93.5), 81.4 (73.1–91.0); zygomatic breadth, 53.1 (44.8–58.3), 51.8 (44.0–58.6).

Natural History. Yellow-bellied marmots are semifossorial and typically inhabit vegetated talus slopes or rocky outcroppings in meadows. The rocks serve as sunning and observation posts and as support for the burrows, which usually are found on well-drained slopes. Burrows provide nest sites, refuges from predators, and hibernacula. The main entrance of a marmot burrow (there may be up to three) extends to a depth of about two feet, and the main passageway extends another 12 to 15 feet horizontally into the hillside. Several short, blind tunnels branch from the main passageway and from the nest chamber, the

66. *Yellow-bellied marmot,* Marmota flaviventris *(courtesy J. M. Ward, Jr.).*

latter situated at the end of the burrow beneath a large rock. Well-defined trails connect adjacent burrow systems.

Yellow-bellied marmots usually live as members of a colony, typically consisting of a territorial adult male, several adult females, and their offspring. However, there is probably a habitat continuum ranging from sites harboring one animal, through one-harem colonies, to contiguous multiharem sites. In a study in Colorado, 75 percent of the marmots were colony residents, 16 percent resided outside colonies, and nine percent were transients.

Marmots are herbivores, consuming a wide variety of plant species including grasses, flowers, and forbs. In late summer they eat large numbers of seeds. It was recently demonstrated experimentally that marmots feed selectively and exhibit preferences; the plants or plant parts that were rejected contained toxic compounds. Heavy grazing by livestock may remove a large amount of the vegetation in their meadow habitat in a short time, and such reduction in potential food supply during the period of fat accumulation, before the onset of hibernation, may have a serious adverse effect on survival.

Hibernation lasts about eight months on the Black Hills, from early September to May. Young marmots lose about 50 percent of their body weight during torpor, and it thus is crucial that they obtain a threshold weight before entering hibernation. The earlier young are born, the more they will weigh at the time of hibernation. Thus, the earlier the arrival of spring in the natal year, the greater the number of young that survive to become yearlings. Failure to survive hibernation is probably a major source of mortality for all marmots. Females living where the growing season is short may not produce litters in consecutive years; the shortness of the postweaning period when weight gain is possible may preclude sufficient fat deposition to provide energy for both overwintering and subsequent reproduction.

There is only one litter per year. Reproductive behavior is concentrated in the first two weeks after emergence from hibernation. Testes of adult males enlarge several weeks before that time and regress afterward. In some populations yearling females do not produce litters, but in others a significant proportion of yearling females are pregnant. Gestation in marmots requires about 30 days; newborn young average about 110 mm in length and 34 grams in weight. The young remain in the burrow for 20 to 30 days, by which time they are nearly weaned. Most litters are weaned in the first half of July, but emergence has been observed from mid-June to early August. Litter size ranges from three to eight; most commonly it is five. It is likely, however, that litter size varies both geographically and from year to year. Body weights of young marmots average about 590 grams for males and 500 for females in the first week after emergence. Among all age groups, males are significantly heavier and longer than females as a result of higher growth rates. In one study, growth rates of young marmots of both sexes were greater in subalpine habitats (about 11,200 feet) than in montane habitats (about 9,500 feet).

Predation on marmots has seldom been directly observed, but gray wolves, coyotes, and badgers are known to prey on them. Bobcats, hawks, owls, and golden eagles are additional predators. Adult marmots have been observed to chase long-tailed weasels and martens, which may prey on pre-emergent young. Predation often is a lesser source of mortality for colonial animals than for isolated individuals, losses of the former being attributed primarily to mortality during emigration and hibernation.

Marmots harbor fleas, lice, mites, and ticks as well as various internal parasites. The fleas may be important vectors of sylvatic plague.

Marmots typically emerge from their burrows soon after sunrise, defecate, and spend a short time sunning and grooming. Foraging activity peaks in midmorning, followed by sunning, sometimes more grooming, and long intervals in the burrow. A second peak of feeding occurs in late afternoon. The general activity pattern shows seasonal variation, perhaps coinciding with changes in temperature and day length—sunning may have a thermoregulatory function.

Communication is chiefly by auditory and visual cues and by scent marking. There are three basic calls—the whistle, the undulating scream, and the tooth chatter. Variations of the whistle may have several functions, including alert, alarm, or threat. Screams are a response to either fear or excitement, and the tooth chatter is a threat.

Play occurs frequently between young, between young and adults, and between yearlings, but rarely between adults. Aggressive behavior includes some grooming, social mounting, alerting, chases, and flight.

Colonial, territorial males direct most of their aggression toward male yearlings and adults. Some males patrol their territories with conspicuous tail flagging, in which the extended tail is waved in an arc above and behind the body. Anal glands may be used in this context to further advertise the male's presence. The mean size of 24 typical territories was 0.67 hectare (about an acre and a half). Home range size of females in a study in Colorado was 0.13 to 1.02 hectares (about a third of an acre to two and a half acres). The aggressive behavior of females with litters may be partly responsible for dispersal from the colony by both sexes, and aggressive behavior of a territorial male may encourage the dispersal of male yearlings. Adult males live, on the average, about two and a half years in the colony; adult females live three or four years. Maximum longevity is eight to 10 years.

All male offspring disperse, as do many females, usually as yearlings. However, there is a tendency toward delayed dispersal of both sexes in colonies at high altitudes. Also, dispersal among male yearlings may be delayed when there are many present in a colony, when they are underweight, and when the level of amicable behavior between yearling males and adults is high. Low level of adult aggression and high amicability delays female yearling dispersal. Interestingly, dispersal of female yearlings evidently is independent of the number of females in the colony. Dispersing individuals, initially at least, tend to inhabit marginal areas, in spruce forest rather than in meadows occupied by colony residents. Dispersal may take individual marmots several miles from suitable habitat.

Selected References. General (Frase and Hoffmann 1980 and included citations).

Marmota monax: Woodchuck

Name. The Greek word *monas* means "alone" or "solitary" and here refers to the behavior of the species. A variety of vernacular names have been applied to this animal. The origin of "woodchuck" is obscure, but it probably comes from American Indian words (*wejack*, Algonkian; *wenusk*, Cree). "Groundhog" is commonly used, especially in the southern part of the distribution of the species, and the names "whistler" and "siffleur" are occasionally heard.

Distribution. The distribution of *Marmota monax* covers the eastern half of the United States (except Florida and Louisiana), almost all of Canada, and east-central Alaska, in both deciduous and coniferous forest regions. The species reaches the western extent of its range in the United States along the eastern edge of the Northern Plains, in the ecotone between eastern deciduous forest and grassland. Four nominal subspecies occur in the tristate region: *Marmota monax bunkeri* in eastern Nebraska, *Marmota monax monax* in southeastern South Dakota, *Marmota monax rufescens* in northeastern South Dakota and adjacent North Dakota, and *Marmota monax canadensis* along the northern edge of the region.

Description. The woodchuck is heavy-bodied, short-legged, and yellowish brown to brown in color (the venter is paler than the back, but never yellow to orange). The animal has a grizzled appearance because of an intermixture of whitish or buffy tips on

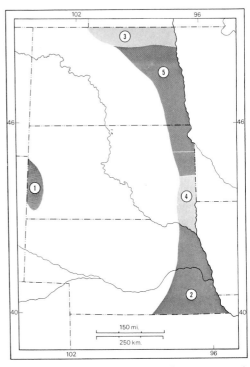

67. *Distribution of* (1) Marmota flaviventris *and* (2–5) Marmota monax. *Subspecies of the latter are:* (2) M. m. bunkeri; (3) M. m. canadensis; (4) M. m. monax; (5) M. m. rufescens.

the cinnamon-to-brown guard hairs. The underfur is dark at the base but tipped with ochraceous or cinnamon. The feet are black or dark brown. The tail is relatively shorter than in the yellow-bellied marmot. The claws of the feet are strong and used effectively for digging. The head is broad and short, with small, rounded, rather inconspicuous ears. Females have only four pair of mammae, one pair fewer than in the yellow-bellied marmot.

The skull has a broad interorbital region, and the postorbital processes project at right angles to the long axis of the skull or even slightly forward, in contrast to the more acute angle found in the yellow-bellied marmot. Average and extreme external measurements of four males and four females, respectively, of *Marmota monax bunkeri* from southeastern Nebraska are as follows: total length, 597.2 (579–654), 581.5 (543–624); length of tail, 142.0 (125–155),

136.8 (124–142); length of hind foot, 93.5 (93–94), 89.3 (86–97); length of ear, 24 (one specimen), 31.0 (28–34). Weight varies seasonally owing to fat deposition in late summer and autumn, ranging from 5 to 12 pounds. Average and extreme cranial measurements of six males and six females, respectively, of *M. m. bunkeri* from eastern Kansas are: greatest length of skull, 99.1 (87.5–104.7), 103.2 (98.5–107.4); zygomatic breadth, 64.5 (58.8–66.9), 66.6 (63.2–70.3). *M. m. monax* closely resembles *M. m. bunkeri,* and there is some question whether the latter should be recognized as distinct from *monax. M. m. rufescens* is somewhat smaller and much more reddish than *bunkeri; M. m. canadensis* is even smaller, though not so reddish.

Natural History. Woodchucks live for the most part as single adults or in family groups. Adults of both sexes live in burrows, which they dig for themselves; they prefer rocky bluffs or stone walls as den sites but often live in meadows devoid of rocks, where burrows are surrounded by an abundant growth of grass or clover. However, woodchucks are primarily inhabitants of forest openings or edges and never are found too far from cover; they are not characteristically found on the open plains. Clearing for agriculture along the forest margin of the Northern Plains has greatly improved the habitat for woodchucks. *M. monax* usually is much less social than the yellow-bellied marmot; adult individuals maintain separate burrow systems, and family groups of females and young usually share the same burrow only during the natal summer.

The most conspicuous feature of a woodchuck burrow is the large accumulation of soil at the mouth. At an active site, the soil is usually fresh and shows many tracks. This elevated mound of soil has been removed from the subterranean chambers and serves to deflect rain as well as providing a place to sun or watch for predators.

Burrows vary greatly with the nature of the terrain. Usually there is one main entrance to each burrow along with a secondary entrance. The main gallery may be as much as 20 to 30 feet long and leads to a nesting chamber 16 to 18 inches in diameter and 10 inches high. The hibernating chamber is similar to the nesting chamber but always is plugged with soil when the animal is inside so as to isolate it from the rest of the burrow, which may be occupied by other mammals such as striped skunks, cottontails, opossums, or even foxes during the winter months. Leading off from the main burrow are tunnels to the surface, the openings of which are generally hidden among shrubs or grasses and lack the telltale accumulation of soil found at the main entrance hole. These holes are used as escape routes whenever the conspicuous front entrance is entered by a predator. Denning sites are generally in soft, well-drained soils on slopes, under roots of dead trees, or in rocks.

The woodchuck is primarily diurnal but may be active on brightly moonlit nights. During cloudy and stormy weather individuals are less active and spend more time in their burrows. The home range is restricted to within 200 to 300 feet of the burrow.

The principal foods of *M. monax* are native grasses and forbs, and cultivated clover and alfalfa. The animals do considerable damage to these crops, mostly by trampling much that they do not eat. While foraging around the burrow, a woodchuck moves slowly, with a waddling gait, feet widespread and belly close to the ground. The animal may climb small trees to search for food or simply to rest and sun. Woodchucks can produce several calls, one a loud shrill whistle, but calls are given rarely. They drink water from streams and pools as well as obtaining it from plants and dew.

Hibernation and preparation for it occupy much of the woodchuck's time. In autumn, after accumulating a thick store of fat, the animal moves into a winter den and, after plugging all entrances, curls up in a tight ball. In a short time, respiration, body temperature, and heartbeat decrease to a fraction of normal summer rates. Woodchucks hibernate for periods varying from four to six months, from about October to March,

68. Woodchuck, Marmota monax *(courtesy K. H. Maslowski and W. W. Goodpaster).*

depending on latitude and local conditions. In spring, adult males emerge first, having lost from a third to half of their autumn body weight. They continue to subsist on the remaining fat reserves while awaiting the emergence of females. Mating then occurs immediately, in late March or early April, and both adult females and many yearling females take part. After a gestation period of 30 to 32 days, two to nine young (generally four or five) are born hairless, helpless, wrinkled, and with eyes closed. In a month a short juvenile pelage is present and the eyes open. About a month and a half after birth the young are weaned and begin foraging outside the den; in two months they are capable of independent existence and disperse from the natal burrow. They attain full size in two years, and longevity may be as much as six years, although many die in their first winter or as yearlings.

The principal natural enemies are red foxes, coyotes, bobcats, eagles, and large raptors. Weasels and mink may prey on juveniles. Human persecution is often heavy; woodchucks are shot for sport or as garden pests, and in some regions they are regularly eaten.

Where woodchucks are common, home ranges around burrows may overlap, and spacing depends upon aggressive interactions between individuals when they meet. Dominance hierarchies are well developed among adjacent animals, and subordinate woodchucks generally avoid dominant neighbors. Aggression directed against juveniles by adult females may be important in initiating dispersal of the young.

Selected References. General (de Vos and Gillespie 1960; Grizzell 1955; Hamilton 1934); systematics (Howell 1915).

Genus *Spermophilus*

The generic name *Spermophilus* is derived from two Greek words, *sperma,* meaning "seed," and *philos,* meaning "loving," an appropriate name based on the observation that seeds constitute a significant part of the diet of many ground squirrels. Another generic name often seen in older literature is *Citellus,* but this name, applied first by Oken in 1816, has been rejected by the International Commission on Zoological Nomenclature. In most publications the two names are synonymous insofar as their systematic (as opposed to nomenclatural) significance is concerned.

The dental formula is 1/1, 0/0, 2/1, 3/3, total 22. The first upper premolar is relatively better developed in most species of *Spermophilus* than in either *Tamias* or *Marmota.* These "true" ground squirrels are of a medium size for rodents—larger than the "pygmy" ground squirrels, or

chipmunks, and smaller than the "giant" ground squirrels, or marmots. There are about 35 species in the genus, of which five occur on the Northern Plains. Ground squirrels are primarily animals of open landscapes throughout North America and Eurasia, some species living in desert, steppe, and even tundra habitats. Because of their semifossorial existence, ground squirrels exhibit a number of morphological characteristics that distinguish them from their arboreal relatives—for example, relatively short limbs, a moderately developed tail, relatively short pelage, and small ears. Internal cheek pouches are well developed, as in *Tamias.* These characteristics also are common to many other fossorial or semifossorial animals. Although ground squirrels often are referred to as "gophers," they should not be confused with the truly fossorial pocket gophers of the family Geomyidae.

Spermophilus elegans:
Wyoming Ground Squirrel

Name. The vernacular name of this ground squirrel reflects the fact that the type specimen was collected in Wyoming, although the species also is found in several nearby states. Another name used mostly by ranchers is "picket-pin," an allusion to the bolt upright posture this squirrel adopts when alarmed, which suggests a stake to which a horse is tethered, or "picketed." The specific epithet *elegans* is Latin and conveys the sense of "trim" or "neat."

Distribution. This species occurs in three areas that are probably isolated from one another—in northern Nevada and adjacent southeastern Oregon and southwestern Idaho, in northeastern Idaho and southwestern Montana, and in southern Wyoming, northern Colorado, and extreme western Nebraska. This third population thus barely enters the Northern Plains in the southern Nebraska Panhandle and now may be extirpated there. The slopes of the Bighorn and Wildcat ridges are likely places to seek surviving colonies of this ground squirrel.

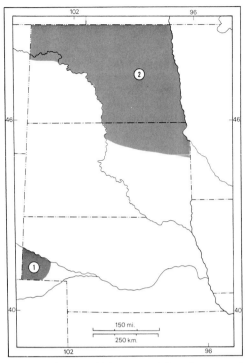

69. *Distribution of* (1) Spermophilus elegans *and* (2) Spermophilus richardsonii.

The subspecies in our area is *Spermophilus elegans elegans.*

Description. This species and the similar Richardson's ground squirrel can be distinguished from other species of *Spermophilus* on the Northern Plains in that they lack dorsal stripes or spots and have rather short tails. In this respect both species resemble the black-tailed prairie dog, which, however, is larger and has a considerably shorter, black-tipped tail. The dorsal fur generally is buffy brown, with grayer shades on the shoulders, neck, and head; the nose is more pinkish or cinnamon. The sides and venter are somewhat yellowish buff, especially in summer pelage, but become more grayish as the pelage wears. There may be only one annual molt in adults, which occurs in spring or summer and varies with sex and reproductive condition. The ears are short and rounded; the eye has a conspicuous white ring surrounding it. The tail is colored dorsally like the

back, but the lower surface is pale orange, with a fringe of outer hairs that are black, edged by conspicuously buffy or whitish tips. The feet are large, with long, stout claws, as befits a mammal that excavates extensive burrows.

Average and extreme external measurements of two males and seven females of *S. e. elegans* from southwestern Wyoming are: total length, 282.9 (275–295); length of tail, 72.8 (60–85); length of hind foot, 43.0 (40–46). Recorded measurements for length of ear vary between 11 and 18 (average 13 to 14). Cranial measurements of the same specimens are: greatest length of skull, 44.5 (43.0–47.2); zygomatic breadth, 24.9 (24.2–25.8). Average weights of 11 male and six female *S. e. elegans* from Wyoming and Colorado were, respectively, 270.7 (230–350) and 314.7 (272.7–366.5) grams.

Natural History. The favored habitat of this species is well-drained upland slopes covered by dry grassland or shrub-steppe, especially sagebrush. Heavy clay or shale-derived "gumbo" soils are avoided, because they are difficult to excavate, but rocky soils, if otherwise friable, are acceptable. Waterlogged soils or loose sand also are unsuitable. The characteristic sign of these ground squirrels is a bare dirt mound, at the side of which a burrow 70 to 90 mm in diameter descends on a steep angle into the ground. These squirrels are highly colonial, and mounds may be found in close proximity. Large colonies may cover many acres, but the density of mounds usually is lower toward the periphery of a colony. The burrow system itself appears not to have been described but is undoubtedly like that of its close relative, the Richardson's ground squirrel. Burrow systems have several secondary entrances in addition to the one at the main mound, and the several burrows interconnect, with a nest chamber situated centrally. Adult females and their young, individual adult males, or groups of sibling subadults occupy separate burrow systems in the colony. In a large colony, several subunits may be found.

In early spring, usually sometime in March (although this varies with elevation and exposure), adult and yearling males emerge from hibernation. They range rather widely in the colony at first but gradually establish a behavioral dominance hierarchy and territorial boundaries that restrict individual movements. The largest, oldest males occupy the central area of the colony, whereas younger, less dominant males tend to be found toward the periphery or in less densely settled parts of the colony. Yearling males are sexually mature but probably do not do much breeding.

Females emerge from hibernation a week or so later than males; they come into estrus within a few days, and mating takes place. Adult females may appear first, followed by yearlings, which are now sexually mature. Dominant males in the central part of the colony have several females within their territories; subdominant males have fewer, perhaps only one or none. After copulation there is a gestation period that is reported to be about a month long (according to one study only 18 days, but the correct gestation period probably is 22 to 23 days), and the one to 11 young are born in a grass-lined nest. Average litter size varies geographically and from year to year but usually is between six and seven; only one litter is produced each year. Females have five pair of mammae. The young weigh about 6 grams at birth and are naked, with eyes and ears closed. By about two weeks of age hair covers most of the body and the lower incisors have erupted; the upper incisors appear after three weeks of age, at which time the eyes and ears open. The first month of life is spent in the underground nest, but by then the offspring weigh about 80 to 100 grams and have molted from their natal pelage to a juvenile coat that is much like that worn by adult squirrels. The young first may appear above ground at this time, although they stay close to the natal burrow. Weaning also commences about this time. In midsummer the postjuvenile molt occurs and adult pelage is acquired.

After all the females within a male's territory have come into estrus and been bred, the males in the central area of a colony usually abandon their territories and move to the periphery. Thus from late April on through the active season most of the colony area is occupied by females and their young. Agonistic behavior among males is much reduced after the breeding season, and adult females are then dominant over males. Occasional aggression is seen between females, although they do not defend definable territories, except for their burrows. As the summer proceeds and the young feed more and more independently, agonistic behavior within the colony becomes rare, and by late summer the young, now having achieved adult size, have dispersed throughout the colony and often beyond its borders. Many more juvenile males than females leave the colony. Adults have been accumulating fat during the summer—males since their reproductive efforts ceased and females since their young were weaned. Weight of adults increases at a rate of about 6 grams per day, until males achieve a total increase of 150 to 200 grams and females about 100 grams. All have then

70. *Wyoming ground squirrel*, Spermophilus elegans *(courtesy J. L. Tveten).*

achieved the requisite physiological state for torpor earlier than the growing young.

Young ground squirrels accumulate fat at about the same rate as do adults, but, because they are also growing to full body size, the total weight increase in their first summer is about 300 grams. Adult males may remain below ground in their nests beginning as early as late July, followed by adult females and then by juveniles in late August or early September. The onset of torpor, with its inactivity and greatly reduced metabolism, often is referred to as "aestivation" when it commences during the warm summer, to distinguish it from winter "hibernation," but physiologically they are identical, and the Wyoming ground squirrel passes insensibly from aestivation in August into hibernation through the autumn and winter without any prolonged interruption of torpor. During this period of inactivity, the body temperature drops to about 4°C, and the animal lives on its fat reserves, which it burns at a rate of about 0.7 gram per day.

The food that fuels fat accumulation is primarily vegetable. Seeds, flowers, stems, leaves, roots, and inner bark of many species of forbs, grasses, and shrubs are eaten. In one Colorado study sagebrush (*Artemesia frigida*) and vetches (*Astragalus, Oxytropis*) were important forbs, whereas brome (*Bromus*) and bluegrass (*Poa*) were the major graminoids in the diet. Seasonal changes also were apparent; for example, cinquefoils (*Potentilla*), sandwort (*Arenaria*), and rushes (*Juncus*) were insignificant in the spring diet but important in late summer. There is no indication of sex or age differences in the diet of Wyoming ground squirrels, and they are generalized, nonselective feeders. Although usually herbivorous, they may feed opportunistically on insects and other arthropods, particularly in late summer when such food is abundant. These ground squirrels also scavenge to some extent and are even cannibalistic (they sometimes may be seen along roadsides feeding on the flesh of an unfortunate animal that has been killed by a passing vehicle).

There are no observations of these ground squirrels attacking and killing another member of a colony as has been described in other species of *Spermophilus,* even though they may feed on carcasses. However, the species does have many predatory enemies, including golden eagles, large hawks, prairie falcons, coyotes, red foxes, and badgers. Long-tailed weasels and several species of snakes may prey on nestling young.

These ground squirrels, like other colonial species, have a well-developed system of calls, some of which are given in response to the appearance of a potential predator near the colony. The most usual alarm call is a "chirp," a series of short, high-pitched notes. Another call, the "churr," is of longer duration and may be employed as a predator alarm call that carries over a longer distance than the chirp. Both also may be used in agonistic behavior when coupled with other visual signs such as tail-flicking.

Epizootics of sylvatic plague may greatly reduce or even wipe out colonies; the plague bacillus is carried by fleas, of which several species infest the Wyoming ground squirrel. Other external parasites include ticks, mites, and lice; internal parasites include stomach and intestinal roundworms, and tapeworms.

Predators, diseases, and the hazards of an extended period of hibernation all take their toll, and mortality among Wyoming ground squirrels in their first year of life is high; often fewer than one in four survive to become reproductively active yearlings the next spring. Adult squirrels have a lower mortality rate and may live three or four years.

In good habitat and in favorable years, these ground squirrels may become abundant. Population densities in colonies may be as high as 10 to 20 per acre. Home-range size in dense colonies is correspondingly restricted and may be only 25 to 50 yards in diameter. Breeding males defend even larger territories in spring, but territorial defense in both sexes usually is restricted to the vicinity of the inhabited burrow system later in the summer.

These squirrels may become serious crop

pests when colonies are adjacent to grain-fields; barley, oats, and rye seem especially susceptible, although wheat also may be damaged. Locally, truck and kitchen gardens also are vulnerable. Damage to range grasses often has been cited, but such claims seem to be exaggerated; much of the diet of the Wyoming ground squirrel consists of shrubs and forbs that are not particularly palatable to cattle. Rather than the indiscriminate poisoning that has been so common in the past, a better economic strategy would be local control of colonies only where they are adjacent to croplands.

Selected References. General (Clark 1970*a*); natural history (Clark 1970*b*; Hansen and Johnson 1976; Zegers and Williams 1977); systematics (Koeppl and Hoffmann 1981).

Spermophilus franklinii: Franklin's Ground Squirrel

Name. The specific name of this ground squirrel honors Sir John Franklin (1786–1847), an English explorer who searched for a "Northwest Passage" across the arctic between the Atlantic and Pacific and eventually lost his life doing so. Other vernacular names include gray ground squirrel and gray "gopher."

Distribution. Franklin's ground squirrel is one of the mammalian species of which the distribution is almost completely restricted to the Northern Great Plains and adjacent Central Lowlands. The species inhabits tall- and mid-grass prairies from southern Kansas and central Illinois northward through central Saskatchewan into east-central Alberta. Throughout this region it coexists with the more common and widespread thirteen-lined ground squirrel (*Spermophilus tridecemlineatus*).

Description. The most distinctive feature of Franklin's ground squirrel is its long, fairly bushy tail, which causes it to resemble superficially the gray squirrel (*Sciurus carolinensis*). However, the color of the back and sides, while gray, has both pale and darker speckles or bars, and the whole dor-

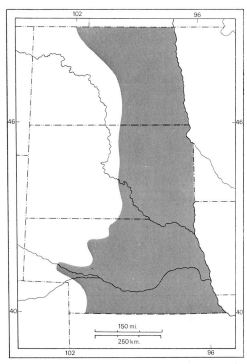

71. *Distribution of* Spermophilus franklinii.

sal surface has an olive buff to tawny buff cast, except for the gray head. The underparts vary from yellowish white to olive gray and generally are less white than the venter of the gray squirrel; finally, the ears are much shorter and rounded, and the tail is somewhat shorter and less bushy than in the tree squirrel. Like most of its relatives, it probably undergoes two annual molts each year as an adult, the first in spring after the breeding season and the second in late summer. There seem, however, to be no published descriptions, and at least one author has stated that there is only one prolonged molt in summer. No subspecific variation has been described for this mammal. Like other ground squirrels, this species has internal cheek pouches and well-developed anal glands. Females usually have five, but sometimes six, pair of mammae. Males are only slightly larger than females.

Average and extreme external measurements of four males and two to four females, respectively, from eastern

Nebraska and South Dakota are as follows: total length, 390.2 (364–412), 367.0 (363–371); length of tail, 137.2 (132–145), 127.0 (124–130); length of hind foot, 55.8 (54–58), 48.0 (42–54); length of ear, 16.1 (13–18), 16.0 (15–17). Weight of one male was 540 grams, and the average of four females was 390.8 (369–408). The skull is characteristically long and narrow, with a flattened outline. Average and extreme cranial measurements for the specimens mentioned above are: greatest length of skull, 56.2 (53.8–58.8), 52.9 (52.2–53.9); zygomatic breadth, 31.6 (29.8–32.4), 29.6 (29.0–30.0).

72. *Franklin's ground squirrel,* Spermophilus franklinii *(courtesy R. W. Barbour).*

Natural History. The favored habitat of this ground squirrel is tall-grass prairie, with tall, dense cover of coarse grasses, weeds, shrubs, or even small trees. It regularly inhabits forest edges and openings and has benefited from clearing of forests for agriculture.

Franklin's ground squirrel is considered uncommon throughout much of its range, but this is due in part to its wary nature, its ability to conceal itself in dense grass and weeds, and the fact that it retreats inconspicuously to its burrow when disturbed. Because of this, little is known about the biology of the animal.

These ground squirrels are not as social as some other species. Single individuals are sometimes found, although *S. franklinii* normally lives in small colonies of 10 or more. Like other ground squirrels, it is strictly diurnal. Even though it is typically terrestrial, it may climb low shrubs and occasionally trees. The burrow system has one or two entrance holes, usually on sloping ground, leading down three to eight feet into a rather extensive set of subterranean tunnels that contain the nest and storage cavities. The entrance holes usually are bordered by conspicuous mounds of dirt. The burrow is dug in well-drained soil where the grass is tall, typically near the edges of wooded areas, thickets, marshes, or cultivated fields. Railroad rights-of-way often provide suitable habitat.

Franklin's ground squirrels, like thirteen-lined ground squirrels, tend to be more omnivorous than most members of the genus. As much as 30 percent of the diet may consist of animal matter, including insects and other invertebrates, nesting birds and eggs, and occasionally other small mammals. Seasonal changes in diet are evident during the relatively short period of activity each year. In spring they eat much vegetation, including new leafy growth as well as roots and other subterranean parts. Consumption of animal food peaks in mid-summer, and by late summer seeds and other fruits have assumed greater importance. Franklin's ground squirrels may at times be significant predators on ground-nesting birds and small mammals, although usually they are too scarce to have a major impact on populations of potential prey species.

This ground squirrel breeds in April or May. Adult males emerge from hibernation first, in breeding condition. Several weeks later, females emerge and mating takes place. A pair may jointly occupy a burrow system for this brief period, but after mating the male departs and becomes solitary. In May or June, after a gestation period of about 28 days, the female gives birth to two to 13 (usually six to nine) blind, naked, pink young. In 16 days the young are furred. After 20 to 27 days their eyes open, and several days later they begin to forage. (Young reared in the laboratory grow significantly more rapidly than field-reared young.) Independence follows about two weeks later, and both sexes mature in the following spring. There is only one litter per year, but litter size appears to vary geographically. Mortality rates also differ from place to place and year to year, but average longevity has been reported as less than that of some other ground squirrels, which is puzzling in view of the large size and long period of hibernation of this species.

Hibernation lasts longer than in many other ground squirrels. *Spermophilus franklinii* emerges later in spring than either the thirteen-lined or Richardson's ground squirrels (which appear from March to early April) inhabiting the same areas. Similarly, Franklin's ground squirrels quickly lay on fat and enter torpor as early as July, adult males first, followed by adult females. Juveniles remain active until September. In contrast, the other two species remain active into October or even early November. There is some evidence that small groups of these ground squirrels hibernate together.

Predators known to kill Franklin's ground squirrels include coyotes, red foxes, long-tailed weasels, and mink. Badgers are perhaps the most serious predators, but skunks, hawks, and snakes are occasional enemies. External parasites include fleas and lice, and a cestode, several nematodes, and two protozoans have been found internally.

Home ranges are difficult to determine for Franklin's ground squirrels. Observations suggest that they may forage several hundred feet from their home burrows, implying a small home range. However, they may move long distances to visit another area and then return to the home burrow. Moreover, they have a tendency to move from one burrow system to another, perhaps three or four times in one or several seasons. Such "nomadism" is unusual in ground squirrels. Population density may exceed six per acre in favorable habitat.

Selected References. General (Iverson and Turner 1972; Murie 1973; Sowls 1948; Turner, Iverson, and Severson 1976).

Spermophilus richardsonii: Richardson's Ground Squirrel

Name. The name of this species was coined in honor of Sir John Richardson (1787–1865), who first discovered the squirrel in 1820. Richardson served as physician on two exploring expeditions led by Sir John Franklin, during which he collected a number of new mammals that subsequently were named by Sabine. Other vernacular names in local use are yellow "gopher" and "flickertail," the latter a reference to the familiar tail-jerking agonistic display.

Distribution. The range of *S. richardsonii* includes the plains of southern Alberta and Saskatchewan, southwestern Manitoba, northern and central Montana, most of North Dakota, northeastern South Dakota, and the western edge of Minnesota (see fig. 69). The species, which is monotypic, formerly included those populations now referred to *S. elegans*, which has a more western distribution. The two species come into contact only in one small area in southwestern Montana, and there is no evidence of a significant amount of genetic exchange between them. Hybridization occurs rarely, but the offspring of such crosses have reduced viability.

The Missouri River appears to have served as a barrier to Richardson's ground squirrel until rather recently. The species is found only to the north and east of the river in the Dakotas, with the exception of one recent sight record from McKenzie County, North Dakota.

Description. This medium-sized squirrel somewhat resembles Franklin's ground squirrel but has a shorter tail and more buffy yellowish coloration. In fresh pelage, the dorsum is nearly uniform pinkish buff to cinnamon buff; the underparts are pinkish to yellowish. The tail is brownish above, peppered with black, and buffy to yellowish below; it is fringed with black hairs that are pale buff to white at the tips. The pelage becomes more smoke gray with wear. As in *S. elegans* but not our other ground squirrels, there may be only one molt annually in adults, which takes place in spring or summer, varying in time with sex and reproductive condition.

Richardson's ground squirrel is so similar in external appearance to the Wyoming ground squirrel as to be indistinguishable in the field. Fortunately, the vocalizations of the two species are distinctive. *S. elegans* has a weak, high-pitched, short "chirp" call that can best be described as "cricketlike," whereas *S. richardsonii* has a somewhat louder, longer "chirp" that drops in frequency. The "churr" calls also differ in an analogous manner. Overall, *S. richardsonii* is somewhat larger, with slightly darker (less grayish brown) dorsal pelage, and the underside of the tail is less intensely orange than in *S. elegans*. Only slight dimorphism is apparent between the sexes. Average and extreme external measurements of four males and three females from North Dakota, Manitoba, and Alberta are: total length, 308.3 (289–324); length of tail, 79.3 (75–83); length of hind foot, 46.9 (45–48). The ear usually measures between 10 and 15 in length. A male and female from South Dakota weighed, respectively, 574 and 386 grams. A series of 20 of each sex from Gallatin County, Montana, averaged 339.2 (260.2–633.2) and 370.1 (265.6–469.0) grams.

The dorsal outline of the skull is convex, with the greatest height between the postorbital processes; the braincase is narrow and deeply constricted anteriorly, and the postorbital processes are slender. Cranial measurements for the same specimens listed above are: greatest length of skull,

73. *Richardson's ground squirrel,* Spermophilus richardsonii *(courtesy D. L. Pattie).*

48.8 (48.0–50.0); zygomatic breadth, 26.4 (25.7–27.6).

Natural History. This squirrel is a creature of open, well-drained land such as is found on the prairie and in pastures, cultivated fields, and river terraces. Its distribution is limited by a combination of favorable soils and vegetation. Originally this species had a more or less continuous distribution over the Northern Plains, and early explorers reported that it was present in great numbers. Elliott Coues, in 1875, stated that it occurred "by hundreds of thousands over as many square miles of territory. . . . Millions of acres of ground are honeycombed with its burrows. . . ." Like the Wyoming ground squirrel, this species forms large colonies in areas of suitable habitat. Greatest numbers are often in areas of short grasses and rough terrain, especially on heavily grazed slopes. *S. richardsonii* is less well adapted to arid shrub-steppe habitat than is *S. elegans.*

With the spread of grainfields over much of the Northern Plains, this species quickly colonized this new habitat with its abundant food. The squirrels did not confine themselves to edges of fields but occupied the middle of large cultivated areas and made their burrows in plowed ground, in the midst of growing grain. However, more efficient mechanized tillage has now eliminated such burrows in most places, except along the edges of fields.

This species seemingly prefers to dig its burrows in sandy loam or gravelly soils. The burrows are approximately three to three and a half inches in diameter, and each has a mound of excavated dirt at the entrance. They rarely go straight down; they range from 12 to 50 feet in length and are from four to five feet below the surface. Spherical nesting chambers, about nine inches in diameter, are lined with dry or green grass, and straw or oat hulls may be found. Similar chambers for food storage also have been reported, but little is known about the significance of food hoarding in this squirrel.

Although cohesive behavior is not strongly developed in this species, burrow systems often are close together. Densities of 10 squirrels per acre are considered average, but higher densities are common in favorable habitat. Large, solitary males living in isolation sometimes are reported, but this is not common.

Like other ground squirrels, these animals appear aboveground in spring throughout the warmer daylight hours. As heat increases at midday, activity becomes bimodal, with peaks in the morning and afternoon as long as the weather is not inclement. By the time the hot days of mid- and late summer are reached, most activity is restricted to the cooler morning hours.

The seasonal cycle of activity also follows that of other ground squirrels. Adult and yearling males emerge from hibernation in mid- to late March, followed a week later by adult and yearling females. Studies in southern Canada indicate no differences in time of emergence related to age, previous reproductive experience, or time of entry into torpor the previous year. Spring emergence did vary by about three weeks from year to year, however, and was correlated with warming of the surface layer of the soil.

Social organization in *S. richardsonii* resembles that already described for *S. elegans*. In spring, males mostly occupy burrows peripheral to the colony, but upon emergence they move into the colony and establish territories that include female burrows, breeding with the residents when they emerge. The single annual litter is born 26 days, on the average, after the female appears aboveground; 22.5 to 23 days of this period are devoted to gestation (earlier records of 28 to 32 days of gestation appear to be in error—see also account of *S. elegans*).

The number of young in litters ranges from three to 11 (usually six to eight). Litters are born in underground nests in late April or early May. By late May to early June, the young come out of the burrows and begin to forage for their own food while still being suckled by the female (there are five pair of mammae).

After breeding, males cease to defend ter-ritories within the colony and usually retire to the periphery or beyond, where they soon commence the annual molt. The few females who fail to conceive, or whose pregnancies terminate prematurely, also may molt early, but most delay molt until after parturition. This difference in timing allows males to accumulate sufficient body fat to enter torpor in July, followed by nonbreeding females. These are followed by females that have bred, then juvenile females, and finally juvenile males in September or even October. There is much variation in time of entry into hibernation (immergence), this being strongly affected by time of previous spring emergence, which in turn is influenced by spring temperature (see above), under the control of both latitude and local weather. Thus, emergence tends to be later in more northerly populations, which in turn causes immergence to be later in summer or early autumn. Progressive shortening of the growing season toward the north on the Plains thus may be partly responsible for setting the northern limits of distribution of this species. Summer weather conditions also influence the ability of *S. richardsonii* to accumulate fat, the result being much year-to-year variation in the timing of onset of torpor. However, the annual cycle is resynchronized the following spring, because males and females tend to emerge about the same time (males before females) in response to increasing spring warmth.

The naked young weigh about 6 grams at birth; growth and development are similar to that in *S. elegans*, particularly during the first month, until weaning commences. The young begin to appear aboveground in late May or early June in southern Canada and weigh about 80 grams at that time. They forage within the mother's territory for their first month of aboveground activity (second month of life), but then the females, having completed lactation and the annual molt and having accumulated fat, begin to immerge into torpor in their burrows, leaving the weaned juveniles on their own. The young undergo a post-juvenile molt in midsummer. Juvenile females have a strong tendency to remain within the mother's territory and they eventually hibernate there, but juvenile males are more likely to wander in late July and August; some may leave the colony entirely, whereas others may eventually hibernate in the colony, but away from their natal territories.

This tendency of young males to disperse leads to a pronounced shift in sex ratio between age classes, owing to the high mortality of juvenile males that accompanies dispersal; the 1:1 juvenile sex ratio becomes about 0.3 male to one female among adults. It is not only juvenile males that disperse; in spring following their emergence, the now-mature yearling males disperse to an even greater extent, with concomitant high mortality. Overall, only about 25 percent of the juvenile females and 12 percent of juvenile males born in one year survive to start the next; survival of adults is about 50 percent for females and 30 percent for males. Adults rarely live more than three or four years.

Much of this mortality is due to vulnerability to predators during dispersal. However, even when established in a colony, the Richardson's ground squirrel has many enemies. Badgers regularly dig them out and feast upon them, from early spring until the ground becomes well frozen and the badgers enter winter quarters. Long-tailed weasels enter their burrows and readily kill them, especially juveniles (adults have been observed to defend themselves successfully). Striped skunks probably dig out a few young, and foxes, coyotes, and bobcats also help reduce numbers. Hawks, prairie falcons, and golden eagles feed extensively on these squirrels, and even Cooper's hawks and goshawks may occasionally take one; burrowing, short-eared, and long-eared owls may capture some during early evenings or on cloudy days. Among the reptiles, bull snakes and prairie rattlesnakes feed upon juveniles to a considerable extent.

Other interspecific relationships of *S. richardsonii* also deserve brief comment. Both this species and *S. elegans* are (or were) often found living within black-tailed prairie dog towns, occupying burrow sys-

tems that appeared to have been first excavated by the prairie dog. Competition for food plants seems likely in such a situation but has not been investigated. *S. richardsonii* also has been observed to chase the smaller *S. tridecemlineatus.*

Richardson's ground squirrel has food habits similar to those of the Wyoming ground squirrel, including its propensity for carrion, but it probably eats more graminoids and fewer forbs and shrubs. The leaves and stems of grasses, and a great variety of forbs, are consumed in spring and early summer. Later in summer and in autumn, when the seeds of the prairie plants and grasses begin to ripen, they constitute the principal foods of these squirrels. Their cheek pouches are so capacious that when well filled they often make the head appear more than double its natural size. Seeds and fruits gathered are rapidly carried to the burrow and deposited in caches. These food stores may be important in early spring, before much new plant growth has occurred, but nothing is known about possible geographic or sexual differences in caching behavior. When grasshoppers or other arthropods are abundant in late summer, the squirrels sometimes feed extensively on them, but this varies with season and locality.

Foraging is restricted to the home range or territory. Breeding males defend territories of up to about three-fourths of an acre in size (average about an eighth of an acre), but thereafter they defend only a small area around the home burrow, while foraging beyond it for a variable distance. Such home ranges overlap, but individual distances between males are maintained by agonistic behavior that includes vocalization, posture, tail-flicking, and chases. Females defend smaller territories (average about one-twentieth of an acre) within the territories of males during the breeding season and maintain them throughout the summer. Adult females tolerate their own young within the territory, even in subsequent years, and often a female offspring will inherit her mother's territory when the mother disappears. Female siblings also remain tolerant of each other after they mature and may share territories and sometimes burrow systems.

Close colonial living undoubtedly contributes to the transmission of parasites and diseases. *S. richardsonii* harbors several fleas, ticks, mites, and lice, as well as intestinal roundworms and tapeworms. The most common internal parasite is a stomach nematode; in one study, 65 percent of the stomachs contained these roundworms. The protozoan *Eimeria* also is a frequent parasite. *Spermophilus richardsonii* is susceptible to sylvatic plague, and its ticks carry spotted fever and the virus of equine encephalomyelitis.

Because they are disease vectors, as well as because of their tendency to multiply rapidly on the edges of cultivated fields, these animals often are pests locally. The annual loss of hay and grain crops they occasion is high. Bounties always have proved ineffective, but carefully controlled poisoning may reduce damage, although broadcast application of poisons is to be strongly discouraged.

Selected References. General (Sheppard 1979); natural history (Michener and Michener 1977; Michener 1979); systematics (Michener 1980).

Spermophilus spilosoma: Spotted Ground Squirrel

Name. The species name *spilosoma* is made up of two Greek words, *spilos,* meaning "spot," and *soma,* meaning "body."

Distribution. Spermophilus spilosoma ranges from South Dakota to the states of San Luis Potosí and Aguascalientes in central Mexico. Twelve subspecies are recognized, but only one, *Spermophilus spilosoma obsoletus,* is found on the Northern Great Plains, where it is limited to approximately the western two-thirds of Nebraska and extreme south-central South Dakota. Its preferred habitat is dry, sandy soils, such as the Sand Hills of Nebraska. This squirrel is sympatric with *S. tridecemlineatus* on the Northern Plains, but it favors drier habitats and is nowhere as

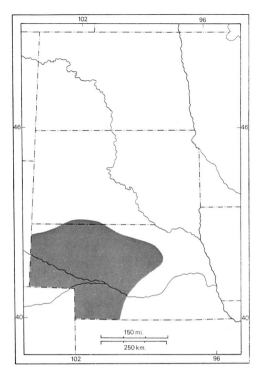

74. *Distribution of* Spermophilus spilosoma.

abundant. The events of the life cycle lag behind those in *S. tridecemlineatus* by two to four weeks, which also reduces competition between the two species.

Description. This species is similar to the thirteen-lined ground squirrel in size but has a shorter tail. The upper parts are grayish beige to cinnamon; the white dorsal spots are nearly square, of moderate size, and randomly scattered rather than arranged in lines. The sides of the body are paler than the back, and the venter is white. There are two molts in adults each year, the first in late May and the second, less conspicuous, in late summer and early autumn. Juveniles molt once, into adult pelage, in autumn.

Average and extreme external measurements of two males and five females from Cherry County, Nebraska, are as follows: total length, 216.6 (201–226); length of tail, 64.3 (61–69); length of hind foot, 31.4 (30–33). Length of ear in three females from Nebraska measured from 8.5 to 12. No size difference is apparent between the sexes; a male and a female weighed, respectively, 194.5 and 152.5 grams. The skull is similar in appearance to that of *S. tridecemlineatus* but has relatively larger auditory bullae and a broader, more inflated braincase as well as broader, shorter nasals. Average and extreme cranial measurements of six males and four females from the same locality as above are: greatest length of skull, 39.2 (38.0–40.2); zygomatic breadth, 23.1 (22.3–23.8).

Natural History. Like other ground squirrels, this species is semifossorial. The burrows are situated in dry, sandy soils, usually beneath vegetation or rocks, about 18 inches below the surface of the ground. Burrows up to 12 feet in length have been recorded. A spherical chamber lined with grass usually is found at the end of the burrow. A single burrow system may have several entrances.

S. spilosoma is strictly diurnal, being seen aboveground mostly during late morning and early afternoon. The animals seldom travel far from the burrow and, when they do, proceed with short runs, stopping frequently to watch and listen. Their "whistle" is similar to that of other ground squirrels.

Hibernation apparently is incomplete, particularly in the southern part of the range of the species, and young seem to be more active than adults in winter. Aestivation may occur intermittently during hot, dry weather, and hibernation usually begins in September in the tristate region. This may account, in part, for the apparent scarcity of the species in some areas. On the Northern Plains, emergence usually takes place in early April.

Unlike most species of ground squirrels, two litters per year are common but occur most frequently in the southern parts of the range. Seven young constitute the usual litter, but as many as 12 and as few as four have been recorded. Parturition takes place in spring and again in late summer after a gestation period of about 28 days. However, only one litter per year may be the rule in the tristate region.

75. *Spotted ground squirrel,* Spermophilus spilosoma *(courtesy D. P. Streubel).*

The young at birth weigh about 4 grams and are naked, with eyes and ears closed. After about four weeks, the juvenile pelage is complete and the eyes are open. In two more weeks, the young weigh about 40 to 50 grams and are mature enough to begin the weaning process; they emerge from the burrow about that time and begin to forage on their own. The female has five pair of mammae.

The mean home range size in a study carried out in Colorado was about three and a half acres. In the breeding season, males had larger home ranges (up to about seven and a half acres) than females, and these broadly overlapped those of the latter. Several males have been known to congregate around an estrous female. Females have overlapping home ranges from about an acre to three acres or more in size. Territorial behavior appears to be restricted to the vicinity of the burrow.

The food of this species consists mainly of seeds or succulent vegetation. Pulp and seeds of cacti frequently are taken. Grasshoppers and beetles also are eaten to a lesser extent, but this species appears to be less carnivorous than *S. tridecemlineatus.* Like other diurnal rodents, the spotted ground squirrel is preyed upon by a variety of raptors, carnivorous mammals, and snakes. Fleas, ticks, and lice infest it externally, and nematodes have been found internally. Because of its relative scarcity and preference for rather barren habitat, this ground squirrel probably is of negligible economic importance.

Selected References. General (Streubel and Fitzgerald 1978*a* and included citations).

Spermophilus tridecemlineatus: Thirteen-lined Ground Squirrel

Name. The Latin adjective *tridecem-lineatus* describes the conspicuous longitudinal stripes on the back of this squirrel—a combination of *tridecem,* meaning "thirteen," and *lineatus,* meaning "lined." This species has also been termed the striped ground squirrel.

Distribution. The thirteen-lined ground squirrel ranges over all of the Northern Plains, inhabiting such areas as roadsides,

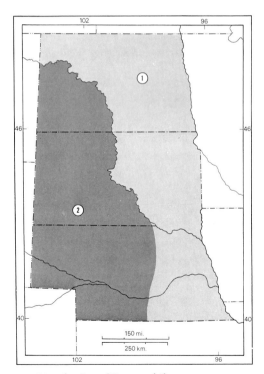

76. *Distribution of* Spermophilus tridecemlineatus: (1) S. t. tridecemlineatus; (2) S. t. pallidus.

golf courses, and pastureland where the soil is well drained. Two subspecies occur in the region: *Spermophilus tridecemlineatus tridecemlineatus* is found generally east of the Missouri River in North and South Dakota and in the eastern quarter of Nebraska, whereas *Spermophilus tridecemlineatus pallidus* occurs west of the Missouri River in the Dakotas and in western Nebraska.

Description. This ground squirrel is of moderate to small size among members of the genus and has a slender body. As in many diurnal squirrels, the eyes appear large, especially in the young, and the external ear is small. The limbs are short, bearing claws of moderate size. The tail is approximately a third the length of the body and is well haired but only moderately bushy. The upper parts are marked with a series of alternating dark (brownish or blackish) and pale longitudinal stripes, with nearly square white spots regularly spaced along each of the dark stripes. Externally, the subspecies found in the Northern Great Plains differ in color and size. *S. t. tridecemlineatus* is darker dorsally and slightly larger than *S. t. pallidus.* The sides, feet, and venter are yellowish orange in *tridecemlineatus* and nearly white in *pallidus.* Average and extreme external measurements of five males and six females of *S. t. tridecemlineatus* from southeastern Nebraska, and of 14 males and four females of *S. t. pallidus* from western Nebraska are, respectively: total length, 250.6 (222–287), 245.2 (230–267); length of tail, 90.5 (82–103), 84.8 (78–91); length of hind foot, 34.9 (34–37), 34.2 (32–36); length of ear, 10.4 (8–12), 8 (one specimen). In autumn the weight of an adult may be almost double what it was the previous spring. Weights of two male and two female *S. t. tridecemlineatus* from eastern South Dakota averaged 172.4 (156.0–185.3) grams; those of nine male and two female *S. t. pallidus* from western North Dakota averaged 110.1 (83.0–145.0).

Because of the extensive distribution of this species and the lack of major zoogeographic barriers on the Northern Great Plains, clinal variation is apparent. In general, paler pelage is found in consistently xeric areas toward the west. Albinos are known but rare. As in *S. spilosoma*, adults exhibit two molts each year, the first in May and the second, less distinct, in autumn. Juveniles molt only once, when they assume adult pelage in late summer to early autumn.

The skull of *S. t. tridecemlineatus* is relatively long, narrow, and weakly built. The zygomatic arches are stout but not expanded. The molariform toothrows are slightly convergent posteriorly. Average and extreme cranial measurements for four male and six female *S. t. tridecemlineatus* and for six male and three female *S. t. pallidus* from the same localities mentioned above are, respectively: greatest length of skull, 39.5 (37.9–41.7), 39.5 (37.2–40.9); zygomatic breadth, 22.4 (20.3–24.4), 23.7 (22.2–24.7). Females usually have six pair of mammae.

Natural History. Although they often occur in loose colonies, thirteen-lined ground squirrels are solitary in behavior for the most part, the exception being early in the spring breeding season. A ground squirrel may dig several burrows, some for temporary shelter and others for more permanent use. Permanent burrows are in well-drained soil where the grass is short; there may be several entrances. The burrow descends steeply at first, then levels off at a depth of four inches to four feet and may be 20 feet in length. In spring, mounds of excavated earth may be seen at the entrances of burrows, but the dirt is gradually dispersed and none marks the entrance in the summer and autumn. During hibernation and at other times, an earth plug may block the entrance to the burrow. Permanent burrows contain spherical chambers used for nesting or storage.

Aboveground, a runway system reminiscent of those constructed by microtine rodents sometimes may be seen; it is used for feeding or traveling from one burrow to another. It is probable that the runways serve as a means of orientation in open, featureless habitat. If one tries to catch these ground squirrels when they have

77. *Thirteen-lined ground squirrel,* Spermophilus tridecemlineatus *(courtesy U.S. Fish and Wildlife Service).*

strayed some distance from a burrow, they invariably seek a runway to follow back to the burrow; without this system of runways the animals probably would be more susceptible to predation.

S. tridecemlineatus is strictly diurnal and almost never is seen in the early morning or when the sky is heavily overcast. On sunny days the animals may be seen sitting erect near burrow entrances, alert to intruders. Tall grass inhibits their view of the surroundings, and consequently these ground squirrels are not found in such situations. In contrast, Franklin's ground squirrel, a species found throughout much of the range of *S. tridecemlineatus*, prefers such habitats. When frightened the animals may utter a rapid birdlike trill or trembling whistle.

The date at which this ground squirrel enters hibernation varies with locality and temperature, but usually it is in late September or early October. Upon emerging in late March or April, the animals are ready to breed. Usually, males emerge a week or so earlier than females. This species is known to ovulate only after coitus, a condition known as induced ovulation. The gestation period is 27 to 28 days, and the young, which may number from five to 13 (usually eight or nine), are born in May or early June. One litter per year is the rule in the northern part of the range, but two litters per year have been reported in Texas.

By July the gonads of males have regressed.

The newborn young, naked and with eyes and ears closed, weigh about 4 grams. They develop juvenile pelage in the third and fourth weeks of life, and the eyes are open by day 28. Weaning may be complete by the time the young are a month old; they first emerge from the natal burrow at that time but continue to use it for a week or two longer. Young males disperse farther from the burrow than do young females (average about 260 and 175 feet, respectively, in a Wisconsin study) and establish their own home ranges in late summer.

The thirteen-lined ground squirrel is omnivorous, feeding on a variety of wild plants and seeds as well as considerable animal matter. In spring the diet consists mostly of grass, leaves, and roots; in summer and autumn the animals feed increasingly on insects and seeds as such food becomes more plentiful. Occasionally they feed on mice and carrion, including animals killed along roads.

Although this is the most common ground squirrel on the Northern Plains, the high birthrate indicates a high rate of mortality. Major limiting factors on numbers may be floods and a variety of predators, such as diurnal birds of prey, feral cats, badgers, weasels, coyotes, and snakes. External parasites include fleas, ticks, mites, and lice, and internally this species harbors trematodes and nematodes.

Population density estimates range from two and a half animals per hectare up to 25 per hectare (about 10 per acre), being several times higher after the young emerge. In one study the home range of males averaged about 11 and a half acres and was largest during the breeding season; that of females averaged about three and a half acres and was largest during pregnancy and lactation. Home ranges overlap, but home burrows are defended.

Where abundant, thirteen-lined ground squirrels may destroy crops such as peas, beans, cucumbers, squash, strawberries, wheat, and oats. Their burrowing activities are not appreciated on golf courses and in cemeteries. In contrast to these destructive tendencies, the role of these squirrels as insectivores and soil builders must be considered.

Selected References. General (Streubel and Fitzgerald 1978b and included citations); systematics (Armstrong 1971).

Genus *Cynomys*

The name *Cynomys* is derived from the Greek words *kynos*, meaning "dog," and *mys*, "mouse." The dental formula is 1/1, 0/0, 2/1, 3/3, total 22. The genus is divided into five nominal species, of which only one occurs on the Northern Great Plains. Prairie dogs are found only in North America. The vernacular name "dog" is presumed to be in reference to their barklike alarm call.

Cynomys ludovicianus: Black-tailed Prairie Dog

Name. The name *ludovicianus* is the Latinized form of Ludwig (Louis). The animal was first collected for science by members of the Lewis and Clark expedition of 1804–06 to the Louisiana Territory and was described by George Ord in 1815 as the "Louisiana marmot."

Distribution. The black-tailed prairie dog is found throughout the Great Plains from southern Saskatchewan south to just beyond the Mexican boundary. The western edge of its range runs along the base of the Rocky Mountains, and the eastern edge formerly followed the ecotone between tallgrass and mid-grass prairie. Typically the species is uncommon east of the 98th meridian, where its favored short-grass habitat gradually gives way to associations with denser ground cover of the sort favored by Franklin's ground squirrel.

Local distribution is favored by the presence of overgrazed or denuded pastureland. Before fencing of the range, prairie dogs depended upon the trails and wallows of *Bison* for such disturbed habitat.

Description. The prairie dog is a specialized robust, burrowing ground squirrel. The

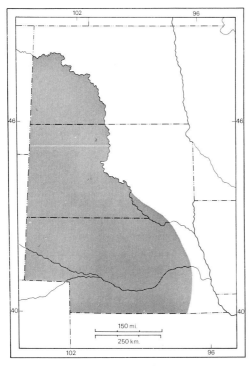

78. *Distribution of* Cynomys ludovicianus.

short tail (a fourth of the total length) is well haired but flat. The ears are short, not extending beyond the pelage. The dorsal surface is clothed with coarse hair of pinkish cinnamon grizzled with buff and black. The tail is tipped with black hairs, and its ventral surface is paler than that above; the underparts are white to buffy white. The sexes are similar in color, and young can be distinguished from adults by the ochraceous cinnamon color of the upper parts and fewer white and black hairs. There are two molts per year—one into a harsh, thin pelage in early summer, succeeded by an autumn molt into winter pelage, which is longer, softer, and more buff and gray than summer pelage. The head is broad and rounded, with moderately large eyes; the face is paler than the dorsum. Cheek pouches are less well developed than in true ground squirrels. The legs are short, with the wrist and heel well furred and a tuft of hair in the center of the palm. The feet are large and five-toed; the toes bear well-developed claws, especially

on the forefeet. Females, which are slightly smaller than males, have eight mammae.

Average and extreme external measurements of 12 adult specimens (eight males and four females) from Hamilton County, Kansas, are: total length, 363.5 (312–410); length of tail, 83.2 (72–99); length of hind foot, 59.6 (50–67); length of ear, 10.3 (8–14). Weights of these individuals averaged 884 (680–1,134) grams. The skull is broad and massive, and the cheekteeth are large and laterally expanded. Average and extreme cranial measurements of 17 adult specimens, 10 males and seven females, from the same locality are: greatest length of skull, 60.7 (55.8–63.4); zygomatic breadth, 43.2 (39.7–45.6).

Natural History. Prairie dogs are diurnal, herbivorous, highly gregarious, and remarkably curious. They inhabit dry upland pastures. Grasses and forbs are their principal foods, but about a fourth of the species used are not eaten by livestock. They rarely eat insects. These animals thus may exert a beneficial influence on range vegetation by eating plant species typical of early successional stages, or overgrazing. However, in times of drought and intense overgrazing by cattle, prairie dogs may help destroy the range completely. In feeding, prairie dogs cut vegetation at its base and eat from the basal portion distally. They also clip, but do not eat, much vegetation within the colony. This produces open ground and thus an unobstructed view of approaching predators.

In summer prairie dogs spend about a third to half of the daylight hours feeding, a third in social interactions with other colony members, working on mounds and burrows, and responding to alarm calls, and the rest in the burrow. Large colonies, or "towns," may be divided by topography into smaller units, termed "wards," five to 10 acres in size, within which may exist one to many social groups or "coteries." The number of prairie dogs in a coterie was two to 35 in one study, averaging eight and a half. Members of a coterie defend a territory from invasion by members of other coteries. However, all animals within a co-

79. Black-tailed prairie dog, Cynomys ludovicianus *(courtesy U.S. Fish and Wildlife Service, photograph by D. E. Biggins).*

teric move freely within the territory except during the reproductive period. Females nursing young defend their burrows from other members of the coterie until pups have been weaned and appear aboveground. Once this occurs, more tolerant social behavior is resumed. Communication within the dog town consists of the alarm bark, the predator warning cry, the "all clear" call, snarls, growls, fear screams, and tooth-chattering. Intraspecific contacts among prairie dogs are generally friendly and include mouth contacts, anal recognition rites, huddling, grooming, playing, and feeding.

Changes in humidity or barometric pressure do not influence activity appreciably. However, strong winds, precipitation, or temperature extremes, both high and low, restrict activity relative to severity. Light intensity influences activity, and without sufficient light these animals seem unable to see well—they bump into objects and fail to show fright or alarm. Nonpredatory animals are generally tolerated or ignored. Thirteen-lined ground squirrels, which sometimes live in prairie dog colonies, provoke growls and chasing.

This prairie dog mates from early March to early April, and, after a gestation period of 28 to 33 days, the single annual litter of two to 10 (generally four to six) helpless young are born hairless and with eyes closed. In 13 days after birth fine hair is evident, and by 26 days the young are well furred. In 33 to 37 days the eyes are open and the pups can bark. In five to six weeks they come aboveground and forage on green vegetation, but they are not completely weaned until a week or so later. In June the young are approximately half the size of adults, and in July or early August they undergo a postjuvenile molt; by the end of the summer they have nearly reached adult size and take part in the regular autumn molt in October. However, full adult weight is not reached until July of the second year. Yearling females (in the second year of life) produce fewer young than do older adults, if they breed at all. Yearling males usually are still immature sexually. Captive prairie dogs have lived for more than eight years, but in natural populations three to four years is the usual maximum life span.

Aside from man, the badger now is the principal predator on prairie dogs. The black-footed ferret, which is believed to prey almost exclusively on this animal, was at one time a most formidable predator. However, control programs of the past 50 years or so have greatly reduced the former abundance of prairie dog towns and have resulted in the near extinction of this mustelid. Other predators include coyotes, foxes, bobcats, hawks, and eagles. Bull snakes and rattlesnakes, which sometimes inhabit prairie dog burrow systems, will take young, but predation attributed to snakes probably has been overemphasized in the past.

Permanent burrows are dug in spring or autumn. The openings tend to be in the direction of downward slope. Particular permanent burrows provide spatial stability to the colony, and from these burrows the animals make repeated forays to peripheral areas where burrows are occupied only temporarily. Depth and length of the underground tunnels vary according to type of soil and terrain.

The cone-shaped mounds around burrow entrances vary in height from a foot to three feet, and in width from three to 10 feet. Mounds are compacted from loose earth thrown out of tunnels as the prairie dog forcefully drives its nose into the soil. These rims of soil around entrances provide vantage points from which to watch for approaching enemies and barriers around the hole in case the colony should be flooded. If water does enter the burrow, air pockets in the tunnels usually permit survival until floodwaters subside. The subterranean tunnels are extensive; after descending three to 10 feet perpendicular to the surface, they angle horizontally for another 10 to 15 feet. Tunnels and entrances are arranged so that winds induce air movement within the burrow system, thus providing ventilation. Several blind side tunnels are excavated lateral to the main tunnel, one terminating in a nest chamber lined with dry grasses and other vegetation. During the night and in the hotter part of summer days, prairie dogs spend their time resting in these subterranean tunnels and chambers.

Much time also is spent in the nest in cold, snowy, or otherwise inclement weather, but black-tailed prairie dogs seem not to become torpid and hibernate, as do most other species of prairie dogs and their closest relatives, the true ground squirrels. Instead they remain dormant in the nest, but maintain normal resting metabolism, for several days, and when the weather improves they appear aboveground again. However, little is known about winter activities on the Northern Plains.

Prairie dog colonies vary greatly in size. Historical records suggest that, before settlement of the Great Plains, some "towns" extended for miles. Most colonies today are small, ranging from about 20 to several hundred individuals; the average colony has fewer than five prairie dogs per acre.

Because of their gregarious nature, prairie dogs are easily poisoned and whole towns are thus completely destroyed. The farmer-rancher should think of control of these animals instead of their total destruction.

Proper management of cattle will ensure a good cover of grass that is of monetary value to the farmer and will at the same time control numbers of prairie dogs without annihilating the species.

Selected References. General (Hoogland 1979; King 1955; Koford 1958; Smith 1967); systematics (Pizzimenti 1975).

Genus *Sciurus*

The name for this genus of common tree squirrels is a compound word from the Greek—*skiouros*, derived from *skia*, meaning "shade" or "shadow," and *oura*, meaning "tail." A tree squirrel sitting upright with the tail curled up its back and over its head is thus "shaded" by the tail. The dental formula for the genus is variable, owing to the presence or absence of the small anterior upper premolar (P3): 1/1, 0/0, 1–2/1, 3/3, total 20 to 22. The genus is widespread in the Northern Hemisphere. Three species occur in Eurasia, and about 15 inhabit North America. In Central and South America there are an undetermined number of species of tree squirrels, and some frequently are placed in the genus *Sciurus*, but such allocation should be regarded as tentative. Only two species inhabit parts of the Northern Great Plains.

Tree squirrels differ from ground squirrels in possessing long, bushy tails, longer limbs that are capable of great flexibility at the joints, and prominent ears and in lacking internal cheek pouches. They display great agility in trees, and neither dig underground dens nor hibernate.

Sciurus carolinensis: Gray Squirrel

Name. The specific name *carolinensis* was given because the original describer, Gmelin, received the specimen on which the species was based from "Carolina." The vernacular name is based on the usual dorsal coloration of most individuals, although black, silver, and even albino color phases occur in some localities.

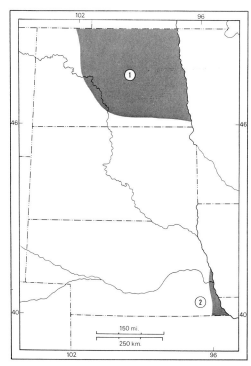

80. *Distribution of* Sciurus carolinensis: (1) S. c. hypophaeus; (2) S. c. pennsylvanicus.

Distribution. The gray squirrel inhabits the eastern broad-leafed deciduous forests of the United States and the Canadian border, from extreme southern Saskatchewan, Manitoba, Ontario, and Quebec south to the Texas Gulf Coast and eastward to the tip of Florida. On the Northern Plains it is naturally restricted to stands of oak-hickory and oak-maple-basswood forest along the zone of contact between grassland and deciduous forest. Gray squirrels have been widely introduced into city parks in the midwestern United States, and populations now are established at a number of points outside the native range, including the Killdeer Mountains of North Dakota. Two subspecies occur in our region—*Sciurus carolinensis pennsylvanicus* in the south, and *Sciurus carolinensis hypophaeus* in the north. The latter subspecies is somewhat larger and darker than the former.

Description. This graceful squirrel is characterized by its long, bushy tail, which it uses for balance, as a "parachute," and perhaps as a distraction display against predators. Underparts are whitish, sometimes lightly suffused with brown. The head, back, and sides are gray, often with a tawny to brown cast. Tail hairs are buffy at the base, black in the middle, and tipped with white. The pelage is soft and dense, the ears are prominent and taper toward the tips, and the eyes are circled by a white ring.

Seasonal differences in color are due in part to fading, and gray squirrels in faded winter pelage may appear quite reddish. However, hair replacement also is involved; there is a single annual molt, the timing of which depends upon the age and reproductive status of the animal. Adult males molt earlier than females, beginning in April, whereas breeding females may not commence until July. The population as a whole usually has acquired the new pelage by September or October. There is no noticeable difference in size or color of the sexes. The young, however, are paler and more grayish than adults. Females possess four pair of mammae.

Average and extreme measurements of six males and five females, respectively, from Richardson County, Nebraska, are: total length, 469.0 (462–492), 476.6 (447–498); length of tail, 206.0 (189–222), 222.2 (201–232); length of hind foot, 67.7 (65–70), 65.8 (62–69); length of ear, 33.0 (32–35), 31.8 (29–32). Weights of 15 males and six females from southeastern Minnesota averaged 581 (383–713) and 565 (373–630) grams, respectively.

The skull is more lightly built than that of the fox squirrel, with a narrower rostrum. A small third upper premolar usually is present, in contrast to its invariable absence in *S. niger*. Respective average and extreme cranial measurements of five males and five females from Richardson County, Nebraska, are: greatest length of skull, 62.3 (62.0–62.6), 62.1 (60.2–63.5); zygomatic breadth, 35.0 (33.7–36.2), 35.3 (34.4–36.5).

Natural History. The preferred habitat of the gray squirrel is mature, broad-leafed, deciduous forest, in particular stands having an abundance of nut- and other mast-bearing trees such as beech, oak, and hickory. Chestnuts probably were important food trees before their near extirpation by the introduced chestnut blight. Gray squirrels apparently reach their greatest densities in undisturbed climax forest, in contrast to many other deciduous forest mammals, in particular the fox squirrel (*Sciurus niger*), which prefers open forest and woodland. Gray squirrels also range northward into the mixed deciduous-coniferous forest of the United States–Canada border area, where pines and hemlock concur with maple, beech, and basswood. In the Southeast this squirrel also occupies upland, evergreen, broad-leafed forest or mixed forests of pine and broad-leafed trees.

Within suitable forests, gray squirrels utilize den trees—either living trees with natural cavities or standing dead trees in which woodpecker holes can be enlarged. Within the den cavity is a nest of leaves and plant fibers. If no tree hole is available, the gray squirrel will build a leaf nest—a roughly spherical mass of interwoven twigs and leaves. In the center of this water-shedding ball, which is lodged in a fork of the tree or twined into the terminal branches, is the fibrous nest. In the extreme north, gray squirrels sometimes resort to ground-level nests under tree roots or in hollow stumps or logs. These are covered by the blanket of winter snow and thus insulated. Tree squirrels, unlike ground squirrels, do not show social tendencies, and each individual occupies its own nest except for a female and her young.

The gray squirrel is diurnal, and its activity extends from early morning to late evening in both summer and winter. Peaks of activity occur around dawn and again at dusk, however, and *S. carolinensis* tends to be more crepuscular than either the fox squirrel or ground squirrels. It is more at home in trees than on the ground and often is observed feeding on buds at the ends of branches, with the body stretched out to its full extent and the tail hanging straight down. At other times it may be seen resting on a horizontal limb with the tail arched over the back, possibly screening itself from the sun or a potential predator. Sometimes an individual lies on its belly with its legs dangling down on each side of a branch. The gray squirrel is shy and will attempt to place a tree trunk or limb between itself and an intruder. It is more arboreal than its larger relative, *S. niger*, and it can negotiate any tree, limb, or telephone wire. If a gap must be crossed, it will jump from limb to limb or tree to tree, sometimes as far as eight feet, with its tail fully spread. On the ground, the animal searches for food by hopping or moving with an ambling walk. It rests on its hind quarters while feeding, adeptly manipulating food with the front feet. When disturbed it moves by leaps and bounds, with great speed, either on the ground or in trees, and from an elevated refuge it calls back with a characteristic scolding "bark," its tail flipping with each call. Gray squirrels are quite vocal and, in addition to the often-heard "bark," employ a variety of "rattles," "grunts," and even a "song" in communication, all usually accompanied by tail movements.

The food of the gray squirrel is principally nuts, which are harvested beginning in summer before they have ripened and gathered at an even greater pace in autumn. It eats seeds, fruits, buds, leaves, flowers, inner bark, and occasional insects in season, and, because of this squirrel's ability to travel through trees, it sometimes takes eggs and young birds from nests. Whereas most foods are dictated by seasonal availability and are consumed immediately, large, hulled fruits (nuts) also are cached and consumed later. Both gray and fox squirrels "scatter hoard" nuts, burying one to several an inch or so below ground before moving on to a new site. Rarely, large hoards are cached in hollows in trees or logs (or attics). Buried nuts are relocated by the keen sense of smell and are eaten at various times during the winter or following warm season. Nuts not recovered are potential seedlings.

Gray squirrels have an extended breeding season, from late December through the

81. Gray squirrel, Sciurus carolinensis *(courtesy W. E. Clark).*

following September. Males are promiscuous breeders, and during the two reproductive peaks, in January–February and again in May–June, "mating chases" ensue in which several males may aggressively pursue an estrous female. The gestation period is approximately 40 to 45 days. There are thus two litters per year, one born between February and April and another between late June and August. Females producing spring litters are both yearlings born the previous spring and adults; adult females and yearlings born in the previous summer may produce a summer litter. However, not all adult females produce two litters each year, and some may skip a breeding season entirely. One to

eight (usually two to four) naked, pink young are born in a helpless stage. The head and especially the feet are proportionately larger than the rest of the body. Both eyes and ears are closed, the ears opening at about 28 days of age and the eyes four to eight days later. The body is covered with hair in six to nine weeks. Weaning takes place at about two months of age, but young born in spring remain with their mother until she bears her next litter. They usually disperse when the summer litter is born, and those born in summer remain with the female through the early winter. Sometimes the young occupy the natal nest and the female disperses instead.

Adults have a high survival rate, as rodents go, and may regularly live five or six years. Old squirrels may reach 12 years of age in the wild, and captives have exceeded 20 years. This longevity may seem surprising in view of the number of potential predators on gray squirrels—red and gray foxes, coyotes, bobcats, mink, long-tailed weasels, hawks, and owls. Raccoons, opossums, and snakes may attempt to take nestlings, although females often defend them successfully. Because of its alertness and agility, this species seems to be captured by predators only rarely.

S. carolinensis is host to many parasites: three species of lice, three mites (including the mange mite, Sarcoptes), at least six species of fleas, several ticks, and botfly larvae. Internal parasites are also numerous and include in aggregate at least 15 species of nematodes, cestodes, acanthocephalans, and protozoans.

Gray squirrels occupy overlapping home ranges, ranging from a fifth of an acre to several acres in size. Although they defend the territory in the immediate vicinity of their nests, they are usually tolerant of other squirrels when foraging, and several individuals, both gray and fox squirrels, may feed side by side in the same tree or at a feeder. Adult males range more widely than do females, and immature squirrels are even more restricted in their movements. Home range appears also to be influenced by population density, which may range from an average of less than one squirrel per acre to more than 10 per acre; at lower densities, home ranges seem to be larger. Density fluctuations are probably controlled by changes in the "mast" (acorns and other nuts) produced by forest trees from year to year, because availability of such food strongly affects both reproduction and mortality in S. carolinensis.

Before their deciduous forest habitat was so disturbed and fragmented by human occupation, gray squirrels periodically reached such high densities that mass migrations were triggered. Such events are much less common now and on a much smaller scale, and their significance to gray squirrel biology is poorly understood.

This species, together with the fox squirrel, remains a popular game animal in those parts of its range where it is fairly abundant. In the tristate region, however, it is of minor importance. Season and bag limit regulations should be carefully controlled to avoid overhunting the gray squirrel on the margins of its range.

Selected References. General (Packard 1956; Uhlig 1955); natural history (Horwich 1972); systematics (Jones and Cortner 1960).

Sciurus niger: Fox Squirrel

Name. The Latin adjective niger means "dark" or "black." The species was named by Linnaeus based on "Catesby's Black Fox Squirrel," a melanistic individual. The vernacular name stems from the reddish dorsal pelage of the typical color phase.

Distribution. Like its relative the gray squirrel, the fox squirrel inhabits the eastern broad-leafed, deciduous forest in North America. Overall, the distribution of Sciurus niger does not extend quite as far north

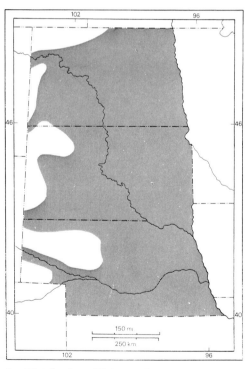

82. Distribution of Sciurus niger.

as that of *S. carolinensis,* and the species is absent in New England and much of the northern Great Lakes region. The fox squirrel, however, ranges farther west, following riparian forest and woodland, as well as shelterbelts and other tree plantations, far out onto the Northern Plains. The species has extended its range actively in recent years, both to the north and to the west. It also has been widely introduced in many places west of its native distribution.

Description. As in other tree squirrels, the most striking feature of the fox squirrel's appearance is its long, bushy tail. The species is somewhat larger and more reddish orange than the gray squirrel, with a buffy to orange, rather than whitish, fringe of hair on the tail. The dorsal fur is tawny brown, grizzled with gray, and the underparts are pale to bright rufous or yellowish brown, sometimes even whitish. The ears are prominent, ochraceous inside, and patches behind the ears are fulvous rather than white as in the gray squirrel. The feet are tawny and the eye-ring is buffy. The black color phase is rare in the tristate area. Young are somewhat paler than adults, the single annual molt of which occurs in April and May, except that it may be delayed in breeding females until August or September. One of the curious differences between fox and gray squirrels is that the former exhibits a reddish to pinkish staining of the bones in varying intensity, the significance of which is unknown.

The skull of the fox squirrel is more robust than that of the gray squirrel, with a broader rostrum and a wider, flatter cranium. A small third upper premolar, which usually is present in the gray squirrel, is lacking in *S. niger.* Average and extreme external measurements of eight males and two females from southeastern Nebraska are: total length, 511.1 (490–528); length of tail, 233.8 (219–245); length of hind foot, 71.2 (67–76); length of ear, 30.6 (27–32). Weights of a large series of males and females from southeastern Minnesota averaged 746 (507–930) and 751 (620–1,000) grams, respectively. Average and extreme cranial measurements of 10 adults from Nebraska are: greatest length of skull, 64.0

(63.0–65.4); zygomatic breadth, 35.7 (33.9–37.3). Only one subspecies, *Sciurus niger rufiventer,* occurs on the Northern Great Plains.

Natural History. The fox squirrel is a common resident of suitable forest and woodland habitats in most of the tristate region, especially in eastern Nebraska and South Dakota, and of the riparian communities along watercourses traversing the middle and western parts of all three states. In contrast to the gray squirrel, *S. niger* seems to prefer open forest and woodland to mature forest stands, but both species may be found together in many places. Only the fox squirrel is likely to be found, however, in the narrow strips of bottomland forest of cottonwood, bur oak, ash, elm, and box elder that border the streams and rivers in most of Nebraska, South Dakota, and western North Dakota. On the Black Hills, the fox squirrel is common in deciduous forests at low elevations but gives way to the red squirrel in higher coniferous stands.

Like the gray squirrel, this species feeds heavily on acorns and large nuts such as hickory nuts and walnuts when they are available. It is not so dependent on these foods as the gray squirrel, however, and feeds on a wider variety of items, both in summer and in winter, including osage orange, or "hedge apple," and corn, which is particularly important in the western part of the Northern Plains where the mast-bearing trees of the eastern forest mostly do not occur. In spring, as the buds of trees and shrubs swell, fox squirrels supplement their diet with buds and flowers, as well as with tubers, bulbs, roots, arthropods, and birds' eggs.

In all seasons this squirrel spends more time than the gray in foraging on the ground, where it moves about with an awkward walk or hop, searching for food with its keen nose. Summer foods show a shift from buds to fruits, berries, and seeds (including grain), but a wide variety of plant and animal material is eaten until the ripening of the fall mast crop once again preoccupies the fox squirrel's attention. It "scatter hoards" nuts in late summer, autumn, and early winter in shallow holes in

83. *Fox squirrel,* Sciurus niger
(courtesy L. Miller).

the ground, but if approached during its harvesting it will retreat at full speed to the nearest tree. Once sheltered by a tree it will hide behind the trunk or a limb, often head downward, until the intruder passes by. This escape behavior differs from that of the more arboreal gray squirrel, which flees into the higher branches.

However, the fox squirrel does forage and carry on other activities in trees. It commonly may be seen resting on a horizontal limb with its tail arched over its back, or lying on its belly with its tail hanging down and its legs dangling down on each side of the limb. The nest is almost always in a tree as well, within a home range that contains one or more leaf nests or dens in hollow trees. It is often held that tree dens are preferred to leaf nests (sometimes termed "dreys"), but the only real evidence for this is that tree dens often are occupied year after year, whereas leaf nests are much more transitory. Construction of leaf nests apparently is identical in the two species of tree squirrels on the Northern Plains and reaches a peak in late summer. Nests vary greatly in size and in whether they are roofed. Fox squirrels fare well in cold winters in leaf nests that are large and well insulated, whereas gray squirrels do better in tree dens; consequently, in areas where both occur, most available tree dens may be occupied in winter by gray squirrels, whereas most of the fox squirrels reside in leaf nests. This ability is also important for fox squirrel survival in the western part of the Northern Plains, where tree hollows are scarce.

Fox squirrels are more strictly diurnal in activity than grays; they are active only between sunrise and sunset, not during dawn or evening twilight. Season has some effect on the diel cycle, but overall there are three peaks of activity—several hours after sunrise in the morning, around noon, and sometimes a minor peak just before sunset. In cold seasons of the year, late afternoon activity is suppressed, whereas midday activity is minimal on hot summer days. Autumn and early winter are times of greatest seasonal activity, because of nest-building and food-hoarding.

The breeding cycle follows the same pattern as in the gray squirrel (although fox squirrels tend to breed earlier) and may be less rigidly defined seasonally. There are two breeding peaks, the first in December and January, the second from April to June. However, males are in breeding condition from November into July, and enough variation in estrus occurs so that some young may be born in every month from February through September. After a gestation period of about 45 days, the female produces a litter of one to six young, averaging three (in a Minnesota study, covering four years, the average was 3.11). Newborns weigh about 15 grams and have proportionately large heads and feet. The young are born without hair and with eyes and ears closed. The ears open in 25 days, but the eyes do not open until 32 to 35 days, at which time the juvenile pelage is completely developed. At 10 to 12 weeks of age the young are out of the nest and can climb; weaning begins at about week 10. The young remain dependent on the female until they are between three and four months old, gradually acquiring foraging skills both in trees and on the ground.

Young born in late winter to a female that then gives birth to a second, spring litter separate from their mother when the second litter is born, but young from second litters may remain with the female for five or six months, finally dispersing in autumn. Sexual maturity in both sexes is achieved in approximately 10 months. Breeding males can be identified easily in the field by their large, hairless, black-pigmented scrota. In nonbreeding periods the testes are abdominal. Reproductive females have eight mammae that are conspicuous during lactation. Adult longevity is four to seven years, although life expectancy at birth is less than four months; in captivity these squirrels may live for 15 years or more. Mating chases, with several males pursuing an estrous female, are as conspicuous in fox squirrels as in gray squirrels. In one case a male fox squirrel in breeding condition was seen to pursue a female gray squirrel, but the chase ended when the female eluded both the fox squirrel and a male gray that was also participating. Inasmuch as hybridization between these two species never has been confirmed, even though they are broadly sympatric, it seems likely that such interspecific chases are rare.

A characteristic vocalization of the mating chase is a high-pitched whine, but most humans are more familiar with the "bark" of the fox squirrel, which is somewhat lower in pitch and more "guttural" than the corresponding call of the gray squirrel. "Barks," "chatter barks," and "breathy barks" all are accompanied by tail-jerking and indicate different degrees of excitement resulting from the presence of an intruder in the home range, either another squirrel or a potential predator. Tree squirrels often chase one another, and this has been interpreted in the past as territorial defense. However, such chases and other agonistic behavior are not restricted to discrete boundary areas but may occur anywhere within the squirrel's home range. Fox squirrels are less tolerant of close approach by other tree squirrels of other species than are gray squirrels, and they are more likely to be solitary.

Several recent studies have shown that the home ranges of tree squirrels tend to be elliptical rather than circular; they are also, of course, three-dimensional rather than two-dimensional like those of strictly terrestrial mammals. A recent study in eastern Nebraska woodlots revealed that the average home range size in spring and summer for adult males was 7.56 hectares (about 18.5 acres); for adult females it was 3.55 hectares (about 8.75 acres), and for juveniles, 3.07 hectares (about 7.5 acres).

Yearling squirrels had the largest home ranges, averaging 15.2 hectares (about 37.5 acres). These estimates were based on radio-telemetry and reveal larger home ranges than earlier observational studies, but they confirm the relative distances covered by the different age and sex classes. Fox squirrels nevertheless may have larger home ranges than gray squirrels, this being associated with the greater openness of their habitat and corresponding lower density of food-producing trees.

As usual, home-range size and population density go hand in hand, and both are determined mainly by habitat quality and the size of the seasonal food crop. Maximum autumn population densities may be two or more fox squirrels per acre in excellent habitat, but in marginal habitat, such as is found in much of the tristate region, they are much lower—on the order of one squirrel to 10 or more acres. In such marginal habitats, where woodlots or stands of mature trees may be discontinuous, fox squirrels may move long distances across uninhabitable fields to reach foraging areas. Such disjunct subareas of a home range may be a half mile or more apart and clearly expose the individual to extreme danger from predators as it travels.

Known predators of the fox squirrel are coyotes, red and gray foxes, bobcats, raccoons, long-tailed weasels, several species of hawks and owls, and the timber rattlesnake. However, none exacts a heavy toll. Fox squirrels are more vulnerable to terrestrial predators and raptors than are gray squirrels because they spend more time on the ground, and overall they may suffer heavier mortality. Parasites and diseases also may be important on occasion. Fox squirrels harbor several species of fleas, ticks (including the spotted fever tick *Dermacentor*), lice, and mites. Particularly serious is the scabies mite, *Sarcoptes scabiei*, the cause of sarcoptic mange. This may become epidemic at times and seriously weakens afflicted squirrels, in some cases leading to death. Internal parasites include gastrointestinal nematodes, cestodes, an acanthocephalan, and protozoans. Tularemia is occasionally encountered in fox squirrel populations, as well as in those of other rodents and rabbits, and squirrel hunters should be alert to the most obvious sign—small white spots on the surface of the liver in an infected animal.

The fox squirrel is more intensively hunted than is its gray cousin, because of its larger size and because it lives in more open habitats where hunting is easier. Of greater importance in most parts of the Northern Plains, however, is the pleasure the sight of a fox squirrel affords people, hunters or not.

Selected References. General (Moore 1957; Packard 1956); natural history (Adams 1976; Longley 1963).

Genus *Tamiasciurus*

The name *Tamiasciurus* is a compound word derived from the same roots that resulted in the generic names *Tamias* and *Sciurus*, the former meaning "steward," and the latter, "shade-tail." The dental formula is 1/1, 0/0, 1/1, 3/3, total 20 (sometimes vestigial third upper premolars also are present). The genus is restricted to North America and contains only three species; one is restricted to the Pacific Coastal region, one to Baja California whereas the range of the third includes the margins of the Northern Great Plains. The systematic relationships of *Tamiasciurus* to the larger tree squirrels of the genus *Sciurus* are at present controversial.

Tamiasciurus hudsonicus: Red Squirrel

Name. The specific epithet *hudsonicus* refers to the fact that the type specimen was obtained on the coast of Hudson Bay, in northern Ontario. The vernacular name is only partly appropriate, inasmuch as the summer pelage is brownish, and some populations are not reddish even in winter. Another vernacular name is "chickaree," of unknown derivation, but possibly of American Indian origin or else an allusion to the call notes of this species.

Distribution. Just as gray and fox squirrels are associated with broad-leafed deciduous

forests, so is the red squirrel characteristic of boreal and montane coniferous forests. The species ranges from the tree line in the north, from Alaska to Labrador, southward to the montane and subalpine forests at high elevations in the Rocky Mountains of Arizona and New Mexico, and in the Appalachians as far south as Georgia. Red squirrels also are found in the mixed coniferous and deciduous forests of the upper midwestern and eastern states, and farther south in the "northern hardwood" forests of maple-beech and maple-basswood. On the Northern Plains, red squirrels inhabit the mixed forests of the Turtle Mountains and Souris River Valley in North Dakota (*Tamiasciurus hudsonicus pallescens*), the maple-basswood forest in eastern North Dakota (*Tamiasciurus hudsonicus minnesota*), and the complex forest associations on the Black Hills (*Tamiasciurus hudsonicus dakotensis*). The last subspecies probably is isolated from neighboring red squirrel populations on adjacent upland outliers of the Rocky Mountains, and also

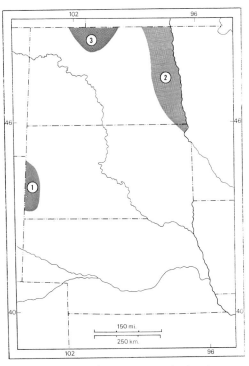

84. Distribution of Tamiasciurus hudsonicus: (1) T. h. dakotensis; (2) T. h. minnesota; (3) T. h. pallescens.

from the northern and eastern populations, by the grasslands of the Northern Plains. Red squirrels also have been reported from the pine-clad buttes of northwestern Nebraska and northwestern South Dakota, but we know of no authenticated populations in these areas.

Description. The red squirrel is much smaller than either the gray squirrel or fox squirrel and spends more time on the ground, but in most respects it is a typical tree squirrel. Seasonal changes in pelage also are more pronounced than in *Sciurus.* In contrast to that genus, it molts twice a year, first in May or early June into summer pelage, and again when it dons the winter coat in late August or September. In the summer coat, the back and sides are brownish red and the venter is white. In North Dakota populations, a narrow black line separates the belly from the sides, but this is indistinct or absent in red squirrels from the Black Hills. The tail is reddish dorsally, grayish to yellowish below. The ears are prominent but lack tufts, and there is a distinctive white eye-ring. Winter pelage is longer and denser than that of summer. A bright red stripe extends from the neck to the tail, and the dorsal fur on the tail also is more reddish. The sides are yellowish gray, and the venter is white to grayish white; the black side stripe is absent in winter in members of all populations in the tristate region, and the ears bear prominent reddish brown tufts. Red squirrels on the Black Hills are generally somewhat paler in both summer and winter pelage than are those from North Dakota, and they average larger.

Average and extreme external and cranial measurements of 10 specimens of *T. h. pallescens* from the Souris River area in North Dakota are: total length, 333.4 (324–360); length of tail, 132.9 (120–143); length of hind foot, 48.1 (46–49); greatest length of skull, 47.6 (46.9–49.0); zygomatic breadth, 26.6 (25.0–27.7). Average and extreme external measurements of three males and two females of *T. h. minnesota* from Walsh County, North Dakota, are: total length, 348.3 (316–430); length of tail, 141.6 (120–184); length of hind foot, 49.7

(47–52). Cranial measurements of two males and two females from the same locality are: greatest length of skull, 46.0 (45.3–46.7); zygomatic breadth, 25.6 (25.3–26.0). Average and extreme external and cranial measurements of 10 male and nine female specimens of *T. h. dakotensis* from the Black Hills are: total length, 332.7 (300–400); length of tail, 133.6 (122–152); length of hind foot, 52.5 (49–57); length of ear, 27.7 (24–31); greatest length of skull, 48.6 (45.3–52.4); zygomatic breadth, 28.0 (25.7–31.5). Weights of two adult *T. h. minnesota* from North Dakota were 203.4 and 238.8 grams; those of 15 adult *T. h. dakotensis* from the Black Hills averaged 238.4 (157.8–337.5).

Natural History. The habitat preferred by the red squirrel is the spruce-fir coniferous forest that exists as a transcontinental belt of boreal "taiga," part of which forms the northern boundary of the Great Plains. In our region spruce forests are found at high elevations on the Black Hills, where red squirrels are common. At lower elevations on the Black Hills spruce is absent, but the squirrels also occur there at lower densities in montane ponderosa pine forests. Elsewhere along the northern and eastern margins of the Northern Plains, low-density populations of red squirrels inhabit mixed forests of hemlock and juniper, maple and basswood. They are scarce to absent in the oak-hickory forests favored by gray squirrels and prefer mature, continuous forest to the isolated stands and open woodlands favored by fox squirrels, but all three species may at times be found together.

This species nests in tree cavities when they are available (especially in ponderosa pine or in broad-leafed trees), but cavities are scarce in spruce forest, and red squirrels are adept at constructing outside nests of grasses, bark, and twigs. These are placed on horizontal branches, usually close to the trunk and are more than a foot in diameter, with an inner nest of fine plant material. More rarely, red squirrels nest in hollow logs or even beneath the roots of trees or in some other space below the surface.

This squirrel usually is highly territorial, defending not only the nest but also its

85. *Red squirrel*, Tamiasciurus hudsonicus
(courtesy R. B. Fischer).

entire home range, unlike *Sciurus*. Adjacent home ranges thus tend not to overlap; boundaries are defended by a territorial advertising call, or "cherr." If this is ignored, vocal and visual display involving tail-flicking and foot-stamping becomes more vigorous. Finally the intruder is chased, gray and fox squirrels as well as other red squirrels. Intensity of territorial behavior varies both seasonally and with different habitats. It is most strongly expressed during the period when seeds are cached and in coniferous forest habitats, and is least expressed during the late winter and spring and in deciduous forest habitats.

This difference in behavior between *Tamiasciurus* and *Sciurus* is related to the different nature of their principal foods and the modes of handling them. As noted above, gray and fox squirrels utilize large nuts that they "scatter hoard" individually. The staple foods of *T. hudsonicus* are conifer seeds (spruce, pine, hemlock) as well as seeds of maple, basswood, and elm, which are small. The most efficient way to harvest these seeds is to cut off the cone or inflorescence before the seeds have ripened and scattered; cones and seed clusters are then placed together in a "midden" or seed cache rather than buried singly. Such middens are often on stumps or logs, in tree hollows, or similar storage places and may amount to several bushels of cones and seeds. One or more middens are included within the territory of each red squirrel and

thus defended from exploitation by other squirrels. The midden also serves as a feeding site, and over the years a large heap of cone scales and cores accumulates as the squirrel tears apart cones to extract the seeds.

The seed cache is gathered during summer and autumn and serves the squirrel as a larder through the winter and early spring. It is at this time that territorial defense is strongest. However, the red squirrel is omnivorous, especially in warmer weather, and consumes buds, flowers, fruits (including berries and nuts), inner bark, blister rust cankers, mushrooms and other fungi, various insects and other invertebrates, and occasionally small mammals and eggs and nestlings of small birds. Such foods are not cached. Territorial defense is weakest during this period, especially in deciduous forest habitats.

Red squirrels are strictly diurnal; in warm weather they are most active in early morning and late afternoon, but as winter cold approaches they gradually shift their activity to midday to take maximum advantage of light and warmth. As in other tree squirrels, the breeding season is extended—from February through September. The female is receptive to breeding only for one day during estrus, and at this time she may move to or beyond the boundary of her territory. Breeding males are attracted from considerable distances, probably by scent, and a temporary breakdown of the usual spatial organization occurs. One or more (up to 10) male red squirrels move into the territory of an estrous female. One male establishes dominance over the others and eventually succeeds in mating with the female; a special "appeasement" vocalization is associated with his approach to the female.

After a gestation period of 35 to 38 days the blind, hairless young are born. Litter size ranges from one to 10 but usually is three to five; the female has four pair of mammae. As in other tree squirrels, development is fairly slow. The juvenile pelage is complete at about 28 days of age, the ears open at 18 to 22 days, and the eyes open at 26 to 36 days. After the eyes open, young may leave the nest, and when about five to six weeks old they begin to eat solid food. Weaning is, however, prolonged until the young are nine to 11 weeks old, by which time the mother has established them in a nest of their own toward the periphery of her territory and begun to defend the bulk of the latter against their intrusions.

The seasonal reproductive pattern of *Tamiasciurus* is like that of *Sciurus*, in that breeding peaks may occur both in late winter–early spring and in late spring–early summer. In our region some adult females may produce two litters per year, and yearlings one, in spring or summer depending on their birth date. However, farther north only one litter per year may be the rule, with adults giving birth to early litters and yearlings to those born later in the year. There is much variation in timing of breeding and litter size as well, probably determined by weather conditions and food supply.

It is a common belief that the most serious natural enemy of the red squirrel is the marten (*Martes americana*), a large semiarboreal member of the weasel family. It is also asserted that other predators are not usually important and that natural predation is not a significant factor in red squirrel mortality. Known predators, aside from the marten, include coyotes, red foxes, lynx, and bobcats and several species of weasels, hawks, and owls. The marten is rare (if not extinct) in the tristate region; in any event the influence of predation by martens on red squirrel populations needs more study, but they may account for only about 20 percent of the total mortality in places where the two occur together. Food shortages and severe weather also are factors, but parasites and diseases are not a major cause of death. Red squirrels are known to harbor at least two species of fleas, three lice, three mites, and two ticks. Internal parasites include a trematode, a cestode, and six species of nematodes.

Territory and home-range size vary seasonally, annually, and in different habitats, in response to the breeding cycle and the nature of the food resource. In deciduous forest home ranges are large, sometimes exceeding 10 acres, although defended territories within them may be smaller. In coniferous forest the home range is roughly coterminous with the defended territory. In one study the home range comprised from 0.23 to 2.91 acres, and in another it ranged from 0.19 to 1.11 hectares (0.47 to 2.74 acres), depending in each case upon the availability of food (seeds). Population densities are known to vary from much less than one per acre in deciduous forest to about 0.36 to 2.75 per acre in coniferous forest. Annual fluctuations are great and are correlated with the size of seed and cone crops.

Although rarely hunted as game because of their small size, red squirrels are trapped for their fur where populations are sufficiently dense to make their harvest profitable.

Selected References. General (Layne 1954); natural history (Rusch and Reeder 1978; Smith 1968, 1970); systematics (Howell 1942; Turner 1974).

Genus *Glaucomys*

The name *Glaucomys* is derived from a combination of two Greek words, *glaukos*, meaning "silver" or "blue gray," and *mys*, meaning "mouse." The dental formula is 1/1, 0/0, 2/1, 3/3, total 22. *Glaucomys* is restricted to North America and is represented on the Northern Great Plains by both known species in the genus. Other small flying squirrels, which may be related to the North American kinds, are *Pteromys* in Eurasia and *Hylopetes* in Southeast Asia.

Flying squirrels, though they glide rather than "fly," differ in certain fundamental ways from other tree squirrels. All have a loose fold of skin along each side that can be stretched between wrist and ankle of the front and hind limbs to form a "gliding membrane." They are nocturnal rather than diurnal and rarely spend much time on the ground. Whether flying squirrels form a biologically natural group is currently under debate; different genera may be more closely related to certain tree squirrels than to each other. Other unrelated rodents (anomalurids in Africa) and even marsupial possums (in Australia) independently have evolved gliding membranes.

Glaucomys sabrinus: Northern Flying Squirrel

Name. The species name *sabrinus* is derived from the Latin name for the Severn River in England. The type specimen was collected at the mouth of a different Severn River, in Ontario, along the southwestern shore of Hudson Bay. The vernacular name contrasts the distribution of this species with that of the closely related *Glaucomys volans.*

Distribution. Glaucomys sabrinus is the northern representative of the American flying squirrels, occurring in the transcontinental coniferous forest of northern North America as well as southward in suitable habitat along the Sierra Nevada, Rocky Mountains, and Appalachians. Some two dozen subspecies are recognized, and two are found on the Northern Plains. *Glaucomys sabrinus canescens* occurs in the northeastern quarter of North Dakota in the Turtle Mountains, Pembina Hills, and Red River Valley. An isolated population, *Glaucomys sabrinus bangsi,* is restricted to the Black Hills of South Dakota and adjacent Bear Lodge Mountains of Wyoming.

Description. The northern flying squirrel closely resembles its southern relative, the main distinction being subtle differences of size and pelage color (see account of *G. volans* beyond). Adults of *G. sabrinus* are larger than those of *volans* (hind foot never less than 34, greatest length of the skull always 36 or more). The hairs of the underparts are dark slate gray at the base, although the tips are white (the ventral fur of *volans* is white to the roots). There is a single annual molt in autumn, as in the southern flying squirrel, and there are four pair of mammae.

The two subspecies on the Northern Plains differ slightly in color and size. *G. s. canescens* is pale pinkish cinnamon above; the head and face tend to be pale smoke gray; the eye-ring and upper patagium are fuscous; the feet are marked with grayish white; and the tail is dark cinnamon shaded with brown above and pale pinkish cinnamon edged with brown beneath. The underparts and soles are creamy white. The upper parts of *G. s. bangsi* are pale wood brown with shades of cinnamon or fuscous, whereas the underparts are whitish with a wash of cinnamon.

Average and extreme external measurements for six adults (a male and five females) of *G. s. bangsi* from the Black Hills are: total length, 300.8 (290–315); length of tail, 133.7 (129–142); length of hind foot, 40.0 (36–46); length of ear, 23.3 (16–29). Males and females are alike in size; the weight of five adults from the Black Hills averaged 141.5 (108.5–163.8) grams. One adult female *G. s. canescens* from Walsh County, North Dakota, mea-

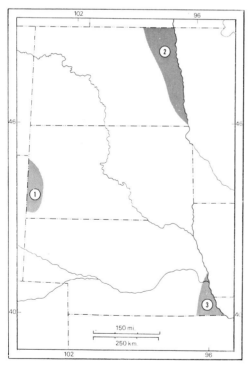

86. *Distribution of* (1–2) Glaucomys sabrinus *and* (3) Glaucomys volans. *Subspecies of* G. sabrinus: (1) G. s. bangsi; (2) G. s. canescens.

sured 311, 140, 39, and 21 externally; no weight was recorded. The skull is medium to large with a rather long rostrum; the auditory bullae are moderately to considerably inflated. Representative average cranial measurements of nine adult *G. s. canescens* from Walsh County, North Dakota, followed by those of 12 adult *G. s. bangsi* from the Black Hills, are: greatest length of skull, 39.3 (38.3–40.6), 40.2 (38.4–41.5); zygomatic breadth, 24.1 (23.4–24.8), 24.0 (22.2–25.1).

Natural History. The habitat of this squirrel is dense coniferous and mixed forests. On the Black Hills it is restricted to spruce forest at higher elevations. It has, however, been known to occupy attics and to take over bird boxes. In the northeastern United States, northern and southern flying squirrels are geographically sympatric in southern New England, the Great Lakes region, and the Appalachian Mountains. However, they actually occur together only rarely. *G. sabrinus* occurs at higher elevations in the mountains, and (where *volans* is in the area) seems restricted to coniferous forests.

Considerably less is known about the biology of *G. sabrinus* than about that of *G. volans.* Throughout much of the range these squirrels nest in tree cavities, often taking over old woodpecker holes. On the Black Hills, the average height of five nest trees was 18.6 feet, ranging from 15 to 25; nest openings were 12.8 feet (10 to 18) above the ground. The opening of one nest hole was oval (57 by 67 mm), and the nest cavity itself was 100 mm wide and 145 mm deep. The spherical nest within was composed of dry grasses and lichens. Leaf nests also are constructed, especially where *G. volans* also occurs and preempts nesting cavities, or in coniferous forests where natural cavities are scarce. Such leaf nests are five to 35 feet above the ground in a conifer and composed of twigs and strips of bark, coarse on the outside, but becoming progessively more finely shredded internally, culminating in the grass and lichen nest proper. Old red squirrel or crow nests sometimes are remodeled. Leaf nests often are used in the warm seasons, but if lack of

tree holes requires their use in winter, the flying squirrel adds insulation until the nest may be more than a foot in diameter.

Like its southern counterpart, this species spends a great deal of time in trees and displays the same specializations of locomotion. It may, however, be somewhat more active in foraging on the ground than is *G. volans,* although the only evidence for this speculation is that *G. sabrinus* seems to be taken more readily in traps set on the ground.

Ground foraging may, however, be related to the rather unusual food habits of this species. In contrast to *G. volans,* which is heavily dependent upon hickory nuts and the like, the staple foods of *G. sabrinus* are lichens and fungi, various kinds of which can be harvested both in trees and on the ground. Nuts and seeds are eaten only sparingly, and an attempt to maintain a northern flying squirrel on spruce seeds was unsuccessful. Other foods, such as staminate cones of conifers, catkins, fruits, and buds, also are eaten to some extent, and the species is reported to eat arthropods as well as the eggs and nestlings of birds. However, *G. sabrinus* may be less carnivorous than *G. volans;* reports of food storage also are lacking, as might be expected in a species that does not rely on nuts or seeds.

87. *Northern flying squirrel,* Glaucomys sabrinus *(courtesy N. M. Gosling).*

At the southern edge of its range, *G. sabrinus* follows the same reproductive cycle as *G. volans*—two litters per year, one born in spring and the other in late summer. For example, lactating females in New York have been found in late May and again in mid-September. However, throughout much of the northern range, females may have only one litter per year. Breeding takes place in spring, from late March through May, with young being born from May to July after a gestation period of 37 to 42 days. Litter size ranges from three to six (usually four). The young, which are blind and hairless at birth, weigh 5 or 6 grams. At about a month old, their eyes are open, they are fully furred, and their coordination is well developed. The young are able to leave the nest for short periods by day 40. During the third month after birth they are weaned gradually and make their first "flight," and when four months old they are almost indistinguishable from adults. They do not breed until their second year.

Fully grown young probably remain with their mother all winter, and they may be joined by other individuals to form wintering aggregations. However, little is known about social structure, territoriality, or home range of this species. Some observations suggest that an adult male associates with the female during the breeding season, occupying a separate but nearby nest. Preliminary radio-tracking studies of movement patterns in Pennsylvania and North Carolina indicate that the home range may have a radius of 100 to 200 meters, thus a size of about seven and a half acres (assuming circular shape). This is considerably larger than the home range of *G. volans* but not unexpected in a species dependent on widely distributed, low-quality foods such as lichens and fungi. An unpublished live-trapping study in Montana gave much smaller values for home-range size.

The northern flying squirrel is less vocal than the southern, and its chirp is lower-pitched and less birdlike. It also utters a sharp squeak, and one observer has described a prolonged musical whistle not unlike that of the red-eyed vireo.

Owls, foxes, weasels, and martens are known predators of northern flying squirrels, and there is one report of an individual found in the stomach of a trout. External parasites include fleas, mites, and lice; nematodes and cestodes have been found internally.

The low reproductive rate of *G. sabrinus* relative to *G. volans* implies a lower mortality rate and greater adult longevity. Individuals lived six years in one unpublished Montana study, and this probably is close to maximum natural longevity. Survival rate of juveniles was less than 30 percent but improved to nearly 50 percent for adults. Little has been published on the population dynamics of this species.

The presence of northern flying squirrels has no detrimental effect on human economic activities, and it confers a great aesthetic benefit on those fortunate enough to catch a glimpse of one.

Selected References. General (Jackson 1961); natural history (Weigl 1978); systematics (King 1951; Turner 1974).

Glaucomys volans: Southern Flying Squirrel

Name. The specific name *volans* is the Latin word for "flying." The name first was applied to this species by Linnaeus in 1758. The vernacular name "southern" refers to the general geographic distribution of the species.

Distribution. The southern flying squirrel occupies the deciduous forest region of eastern North America, from Minnesota, southern Ontario, and Maine south to the Gulf Coast from Texas to Florida. Disjunct and isolated populations also are found in suitable habitat at higher elevations in mountains from northwestern Mexico to central Honduras. The northern flying squirrel, *Glaucomys sabrinus*, occurs mostly north of this range, but both species occur in the Great Lakes region, southern New England, and the Appalachian Mountains. The one subspecies in the tristate region, *Glaucomys volans volans*, is known only from extreme southeastern Nebraska (see fig. 86).

Description. Except for the gliding membrane, or patagium, the spur of cartilage extending out from the wrist to support it, the flattened tail, and the large eyes, flying squirrels look like small tree squirrels.

The soft, silky pelage of *G. volans* is of moderate length. The upperparts vary in color from drab brown through pinkish cinnamon. The sides of the face are smoky gray with a brownish eye-ring; the sparsely haired ears are brown or drab. The upper part of the patagium is dark brown to blackish brown, and the undersurface is edged with pinkish cinnamon. The underparts are creamy white, the hairs being white to the base except on the hind legs. The tail is broad, dorsoventrally flattened, and rounded at the tip, brown above and pale pinkish cinnamon to pinkish buff below. The feet are buffy white to brown and bear well-developed claws. The single annual molt is completed in autumn. New pelage appears first on the sides, then progresses dorsally and anteriorly, until the animal is clothed in long, unworn fur as winter commences. By spring wear is becoming evident on the pelage, and the worn summer fur has a slightly darker, redder cast. The large, dark eyes are surrounded by black rings.

Average and extreme external measurements of two male and three female specimens from Douglas County, Kansas, are: total length, 227 (218–236); length of tail, 93 (81–103); length of hind foot, 30 (28–34); length of ear, 18 (15–20). Adult male and female flying squirrels do not differ in size and usually weigh from 50 to 85 grams, but one female from Leavenworth County, Kansas, weighed 102.8 grams. There are four pair of mammae— one pectoral, two abdominal, and one inguinal.

The skull is small, characterized by a short rostrum, auditory bullae that are moderately inflated, and a narrow V-shaped interorbital region. Average and extreme

cranial measurements of two males and one female from Douglas County, Kansas, are: greatest length of skull, 33.9 (33.4–34.9); zygomatic breadth, 20.5 (20.2–20.9).

Natural History. Southern flying squirrels, like gray squirrels, prefer mature broadleafed forests, which provide an abundance of mast, particularly acorns and hickory nuts. Although the species occupies mixed forest in the northern part of its distribution, oaks are almost always a component of the habitat, and *G. volans* does not range as far north as the gray squirrel. On the Northern Plains this flying squirrel is

88. Southern flying squirrel, Glaucomys volans *(courtesy New York Zoological Society).*

known to occur only in oak-hickory forests along the Missouri River and its tributaries in eastern Nebraska, but it may possibly be found also in bur oak stands in southeastern South Dakota.

These are the most arboreal of our squirrels and, being nocturnal, are best located by their calls—soft "chuck" and high-pitched "tseet" sounds, birdlike chirps, and an occasional sneezelike call or squeal. The best method for observing this squirrel is to search with a flashlight at night. It can be spotted by its large eyes, which reflect a brilliant reddish orange. Sight and sounds of its feeding and gliding activity also will identify this rodent. From its position in a tree it leaps from a limb with gliding membranes outstretched and soars either directly or circuitously to the base of a nearby tree, where it alights facing upward, having flicked its tail up at the last moment to reduce speed and to bring the head and forepart of the body up. Rate of descent is controlled by varying the tension on the patagium, enabling the animal to travel different distances depending on the height from which the glide is initiated. In some cases glides may be as long as 50 yards and at an angle of 30 degrees from horizontal, but usually they are considerably shorter. Steering is accomplished by the long, flat tail, which may be shifted in various planes, enabling the squirrel to avoid obstacles by veering nearly at right angles almost instantly. Immediately after landing the squirrel usually scampers to the far side of the tree and climbs upward a short distance, perhaps to confound any winged predator that may have been in pursuit. It then climbs higher in the tree, where it either forages or prepares for another gliding maneuver, moving in this way throughout its home range. It also moves through the upper branches of trees by running or jumping from limb to limb or, less frequently, moves on the ground in its search for food. A way to find this elusive squirrel in daytime is to locate trees with woodpecker holes or natural cavities. Shaking the tree or striking its base with a stick may cause the squirrel inhabitant to come to the entrance of the hole to satisfy its

curiosity or even to leave the nest. However, many times this trick fails, either because the tree is uninhabited or because the squirrel will not be disturbed.

Southern flying squirrels are much more dependent on tree cavities (or nest boxes) for denning in the northern part of their range than are either their larger northern cousins (*Glaucomys sabrinus*) or their tree squirrel neighbors. This probably is due to their small body size, which necessitates greater protection from winter weather. Like tree squirrels, flying squirrels do not hibernate, and they maintain a high body temperature and metabolic rate all winter long. The nest cavity may be in any sort of tree, and it is lined with plant fibers such as shredded inner bark, leaves, grasses, and moss. Farther south the species is less restricted to cavities and inhabits outside nests, sometimes using old leaf nests of tree squirrels.

Related to their habit of nesting in cavities as protection from cold is the tendency of southern flying squirrels to aggregate in winter nests, thus sharing body heat. As many as 19 have been found sharing the same nest, and the average group numbers five or six in winter, but only two or three in warmer months. Farther south the tendency to aggregate disappears.

Nocturnal foraging and other activities show two distinct peaks, an hour or two after sunset and again before dawn. In cold weather southern flying squirrels forage more, and a third peak of activity may occur around midnight. The animals frequently do not appear on extremely windy nights.

Southern flying squirrels resemble gray squirrels in diet, except that they are considerably more omnivorous. These squirrels both "scatter hoard" individual nuts on the ground and make caches in the nest cavity or other unoccupied holes in trees. Nut-storing peaks in November, and in winter these animals rely mainly on stored nuts, primarily acorns, hickory nuts, beechnuts, and pecans. At other times of the year they eat pine nuts and a variety of other seeds, berries, buds, blossoms, and bark. They actively pursue insects and other invertebrates. Eggs and nestlings of birds are hunted, as are nestling mammals, and flying squirrels occasionally capture adult shrews and mice. They eat carrion, and individuals often are caught in meat-baited traps. Examination of nutshells opened by flying squirrels reveals a small oval or circular hole, with the edge even and smooth rather than jagged and cracked as it often is in nuts eaten by tree squirrels.

Glaucomys volans usually breeds twice annually, once in late winter and again in midsummer. The young are born in April or May and August or September, after a gestation period of 40 days. The number of young in a litter may vary from two to seven but most commonly is three or four. Newborn young are hairless and pink, with eyes and ears closed. At three weeks of age the ears open and the body is furred, and in another week the eyes open. The young are weaned when about six to eight weeks old and soon are capable of short glides. They remain with the mother until shortly before she gives birth to another litter. Sexual maturity is reached at one year. It is not known whether a single adult female will give birth to two litters in a season, but some evidence suggests that females breed only once a year and that yearling females breed in late winter or summer depending on whether they were born in a spring or summer litter the previous year.

Owls are important predators on flying squirrels, and other night-hunting predators such as bobcats and foxes occasionally surprise them on the ground. Weasels, raccoons, and some snakes prey on flying squirrels in their nests. However, predation is probably no heavier a source of mortality than in tree squirrels. Natural maximum longevity may be about five years, but individuals have lived as long as 13 years in captivity. Like other nest-dwelling, gregarious species, *G. volans* has more than its share of parasites, including fleas, mites, and lice externally and nematodes and protozoans internally.

Adult flying squirrels remain in the same general locality throughout their lifetime, usually never straying more than several hundred yards. Home ranges have been reported to vary from 0.4 to two hectares (about one to five acres). The sexes nest separately and defend the territory around the nest cavity unless aggregating in winter. Territorial defense in some circumstances may be extended to the boundaries of the home range by adult females, but males seem always to tolerate overlapping home ranges. Population densities range from about two up to 12 per hectare (about 0.8 to 4.9 per acre), but in our region they are likely to be low.

Flying squirrels are relatively gentle and tame, are easy to raise, and make good pets. They are of little economic importance, even in light of their occasional predation on insects, small mammals, and birds and their eggs.

Selected References. General (Dolan and Carter 1977 and included citations); natural history (Weigl 1978); systematics (Diersing 1980b).

Family Geomyidae: Pocket Gophers

Members of the family Geomyidae, or pocket gophers, are highly adapted for a fossorial (underground) existence. They are closely allied to members of the family Heteromyidae—the pocket mice and kangaroo rats. Species of both families possess fur-lined external cheek pouches and share certain cranial and dental characteristics.

The evolution of pocket gophers is strictly North American, but the present distribution of the family encompasses an area extending from southwestern Canada to northern Colombia in South America, and from coast to coast with the exception of part of the area east of the Mississippi Valley and north of the Savannah River.

Extant pocket gophers are classified into six genera and perhaps as many as 40 species. A great deal of recent work on their systematics has resulted in a number of species being recognized as distinct, but the taxonomy of many groups still is not well known. Two genera are found on the Northern Great Plains, each represented there by a single species. The geologic record dates back to early Miocene times.

Key to Pocket Gophers

1. Upper incisors smooth, occasionally bearing indistinct groove near the anteromedial margin (fig. 90); hind foot less than 30 _____*Thomomys talpoides*
1'. Upper incisors with two distinct grooves (fig. 90); hind foot greater than 30 _____ *Geomys bursarius*

Genus *Thomomys*

The name *Thomomys* is from the Greek words *thomos*, meaning "heap" (referring to the mounds of soil the animal pushes up in the course of digging), and *mys*, meaning "mouse." The dental formula is 1/1, 0/0, 1/1, 3/3, total 20. The genus is found over almost the entire western half of the United States, in a large area of southwestern Canada, and over much of Mexico. Although at least nine species are currently recognized in the genus, only one is found on the Northern Plains.

Thomomys talpoides:
Northern Pocket Gopher

Name. The specific name *talpoides* is from the Latin *talpinus* and means "molelike." In the vernacular name, "pocket" refers to the paired fur-lined external cheek pouches, and "gopher" is from the French *gaufre* meaning "honeycomb," a reference to the series of subterranean tunnels dug by the animal.

Distribution. The northern pocket gopher is the most widespread and geographically

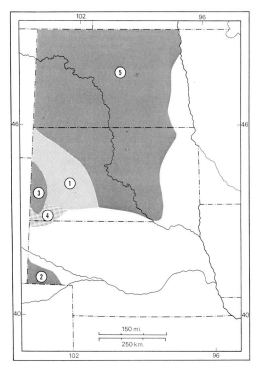

89. Distribution of Thomomys talpoides:
(1) T. t. bullatus; (2) T. t. cheyennensis;
(3) T. t. nebulosus; (4) T. t. pierreicolus;
(5) T. t. rufescens.

variable species in the family. It occurs from Alberta and Saskatchewan to northern Arizona and New Mexico, and from Washington to western Minnesota. On the Northern Plains, the species is present in nearly all of North Dakota, in much of South Dakota, and in the Panhandle of Nebraska. With few exceptions, the general ranges of the genera *Geomys* and *Thomomys* are mutually exclusive.

Five subspecies are found in the tristate region. *Thomomys talpoides rufescens*, one of the larger races, has a wide distribution, covering more than three-quarters of North Dakota and the central half of South Dakota. *Thomomys talpoides bullatus*, another large subspecies, is found along the western edge of North Dakota and in northwestern South Dakota, south at least to Badlands National Park. To the south of *bullatus*, in the Pierre soils of southwestern South Dakota and the northwestern tip of Nebraska, is the medium-sized *Thomomys talpoides pierreicolus.* The remaining two subspecies have small geographic ranges. *Thomomys talpoides nebulosus* is endemic to the Black Hills of South Dakota and the adjacent Bear Lodge Mountains of Wyoming; it is a small to medium-sized, dark-colored gopher. *Thomomys talpoides cheyennensis,* a small, pale race, is found in the southwestern corner of the Nebraska Panhandle, south of the North Platte River.

Description. Thomomys talpoides is the smaller of the two pocket gophers found on the Northern Plains and is a slightly less efficient burrower than the more powerful *Geomys.* Both are, however, well adapted for fossorial life: the body is fusiform, the eyes and ears are greatly reduced, and the fur is medium to short and rather smooth. The upper incisors and the long, sharp claws on the forefeet are used for digging, aided by powerful neck and shoulder muscles. The tail is sparsely haired and has a sensitive tip, which is used when the animal travels rapidly backward in the narrow burrow system. Exterior fur-lined cheek pouches open adjacent to the mouth on each side and extend to the shoulders. The pouches are used for transporting food to storage areas but not for carrying soil, as was formerly supposed. As in other pocket gophers, the upper incisors pierce the lips, allowing them to be closed behind the teeth. This modification enables the gopher to use the incisors for digging while excluding soil from the mouth.

The five subspecies of *T. talpoides* on the Northern Plains exhibit gradients of size and color. Dorsal pelage is some shade of dark brown to pale buffy gray, and the venter ranges from dark to pale gray. Of the five, *T. t. nebulosus* of the Black Hills is by far the darkest. The adjacent *pierreicolus* is markedly paler, and *bullatus* and *rufescens* are of intermediate color. *T. t. cheyennensis* is the palest. The color is somewhat variable in each subspecies because of seasonal changes in pelage.

Molts in this species are complex. Summer pelage first appears in early spring and proceeds from nose to rump in a series of

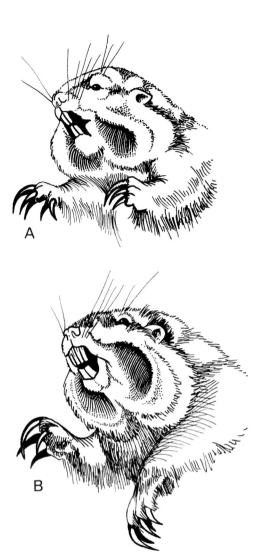

90. *Head and front feet of pocket gophers found on the Northern Great Plains:* (A) Thomomys talpoides; (B) Geomys bursarius *(after Wrigley 1971).*

"waves," several of which may be seen as distinct lines in the fur at any one time. Molt to winter pelage occurs in September or later and usually is less conspicuous. Sometimes, especially in breeding females, the two are telescoped, and autumn molt begins before spring molt is completed.

Externally, *bullatus* is the largest subspecies, *rufescens* and *nebulosus* are slightly smaller, *cheyennensis* is considerably smaller, and *pierreicolus* is the smallest. Cranially, differences are found in size of the bullae, length and width of the rostrum, and proportions and angularity of the skull as well as in overall cranial size.

Male pocket gophers usually are significantly larger than females, though this is much more pronounced in *Geomys bursarius* than in *Thomomys talpoides*. Average and extreme external measurements of three male and 10 female *T. t. bullatus* from Billings County, North Dakota, are, respectively: total length, 235.7 (232–240), 226.3 (210–242); length of tail, 70.0 (63–75), 64.1 (42–76); length of hind foot, 29.3 (27–31), 29.4 (27–32); length of ear, 9.0 (8–11), 7.7 (7–9). Weights of the above specimens averaged 157.0 (143–172) grams for males and 143.5 (117.8–174.5) for females. Comparable measurements for six male and 10 female *T. t. pierreicolus* from Dawes County, Nebraska, are: 190.8 (175–205), 189.3 (170–202); 57.7 (50–65), 55.6 (48–63); 27.0 (26–28), 26.3 (25–28); 5.3 (5–6), 5.8 (5–6). Weights of these specimens average 74.4 (65.8–100.5) grams for males and 76.2 (61.0–87.8) for females.

Average and extreme cranial measurements for the above specimens of *T. t. bullatus* males, followed by those of females, are: greatest length of skull, 41.0 (39.3–42.6), 40.5 (32.7–43.5); zygomatic breadth, 27.2 (one individual only), 25.0 (22.8–26.3). Comparable cranial measurements of six male and nine female *T. t. pierreicolus* are, respectively: 33.6 (32.0–37.3), 33.8 (32.5–35.4); 20.4 (19.4–22.8), 20.3 (18.6–22.3).

Thomomys talpoides differs from *Geomys bursarius* in having smaller and lighter body proportions, no grooves on the upper incisors, six rather than three pair of mammae, lighter and more slender claws on the front feet, and longer, less glossy fur.

Natural History. In addition to being highly variable morphologically, the northern pocket gopher is, throughout its range, found in the greatest variety of habitats of any gopher, ranging from cultivated fields to alpine meadows. In the tristate region it is restricted to relatively thin and rocky upland soils or to heavily compacted clay

91. Northern pocket gopher, Thomomys talpoides *(courtesy U.S. Fish and Wildlife Service).*

soils when it occurs in the same general area with the larger *Geomys*, as is true in southwestern Nebraska, south-central South Dakota, and perhaps North Dakota. However, in extreme northwestern Nebraska and throughout much of the Dakotas, *T. talpoides* is the only pocket gopher present, and there it exhibits a preference for deep, soft, easily worked soils, such as moist sandy loams, in which to construct its extensive burrow systems.

The burrow system often reaches lengths of 400 to 500 feet and is constructed on two levels. The lower level, sometimes reaching a depth of six feet or more, contains the nest, food storage areas in short lateral pockets, and blind tunnels for deposit of feces. The upper level is shallower (six to eight inches below the surface) and is used when foraging. Short lateral tunnels are constructed off the main tunnel, primarily for the disposal of soil from digging nearby. Other tunnels at this level also are used for deposit of feces; when these are filled, they are sealed off. All openings to the surface are kept tightly plugged except when the burrow system is "aired" on a sunny day. Often the mounds of pocket gophers are confused with those constructed by moles, but they differ in that a mole approaches the surface perpendicularly and deposits the ejected soil in concentric rings (a sort of miniature "volcano cone"), whereas a gopher approaches obliquely and deposits the soil in a fan-shaped mound.

Pocket gophers are decidedly solitary, and

each burrow system is usually occupied by only one individual, except during the breeding season and for a short period while the dependent young are still in their mother's burrow. Although pocket gophers do not hibernate, they do undergo brief periods of inactivity in winter and midsummer. They are most active in burrowing in spring and autumn when the soil is neither too dry nor frozen. Digging may take place at any time, but peaks of activity are at dawn and dusk. *T. talpoides* often makes burrows in snow in the late autumn and early winter before the ground freezes. These snow burrows may be used for travel and foraging. They also are used for deposit of soil from underground burrowing; thus, when the snow melts, networks of cylinders of frozen soil are exposed on the surface.

Food consists entirely of vegetation—succulent roots of broad-leafed plants (forbs), cacti, and grasses, underground stems, bulbs, tubers, and occasional leaves of broad-leafed plants. The latter are obtained only if they are within close range of the opening of a burrow. Forbs and pricklypear cactus are the staples of the diet. Food is placed in the cheek pouches for transport to storage chambers. Occasionally this stored material spoils and is sealed off by a soil plug.

Mating in the northern pocket gopher takes place from March to mid-June, depending on weather and latitude. After a gestation period of about 19 or 20 days, four to seven young are born in a grass- or leaf-lined nest within the burrow system. The young remain with their mother until able to forage for themselves. Weaning may occur at about 40 days of age, and at two months or so the young are forced to disperse from the natal burrow to begin a solitary life of their own. Dispersal may be either above or below ground. Juvenile pocket gophers do not approach adult size until they are about six months old. There is never more than one litter per year, and both sexes are reproductively mature at one year of age. Adult longevity is about two years, but maximum life span may exceed five years.

Badgers and great horned owls probably are the chief predators of northern pocket gophers, but coyotes and weasels also are important in this region. Other species known to prey on *Thomomys* include red foxes, bobcats, skunks, other owls, and rarely hawks and snakes. However, weather and nutrition are also of major importance in determining mortality rates in northern pocket gopher populations. Spring flooding, in particular, may cause heavy losses locally. Lice, fleas, mites, and ticks have been recorded as infesting the northern pocket gopher; internal parasites include trematodes, cestodes, nematodes, and at least one protozoan.

Home range in pocket gophers is basically the burrow system. While it may be long, it is often tortuous, and the average home range may be 100 to 200 feet long and 30 to 40 feet wide. Population densities vary with habitat quality—on poor rangeland there may be less than one individual per acre, but numbers may exceed 50 per acre in the most favorable habitat.

Selected References. General (Turner 1974); natural history (Andersen 1978; Gettinger 1975; Reid 1973; Tryon 1947; Vaughan 1967); systematics (Swenk 1941; Thaeler 1980).

Genus *Geomys*

The name *Geomys* is derived from the Greek *geo*, meaning "earth," and *mys*, meaning "mouse." The name "gopher" is frequently misapplied to ground squirrels of the genus *Spermophilus*, and the southeastern pocket gopher (*Geomys pinetus*) often is referred to as a "salamander." The dental formula is 1/1, 0/0, 1/1, 3/3, total 20. Of about six nominal species in the genus, only one is found on the Northern Plains.

Geomys bursarius: Plains Pocket Gopher

Name. The name *bursarius* is derived from the Latin word *bursa*, or "pouch."

Distribution. The range of the plains pocket gopher includes much of central North America, from Texas north to Manitoba, and from Indiana west onto the High Plains. Three subspecies are recognized in the tristate region: *Geomys bursarius bursarius,* a northeastern subspecies extending as far west as eastern North and South Dakota; *Geomys bursarius majusculus,* ranging over the eastern half of Nebraska and into the southeastern corner of South Dakota; and *Geomys bursarius lutescens,* which occurs in the western half of Nebraska and in the southern part of western South Dakota. The ranges of *majusculus* and *lutescens* are known to come into contact in only one place, in Antelope County, Nebraska, but other points of contact may exist. At this place hybrids between the two races are known but rare, and gene exchange between them may be extremely limited.

Locally, the distribution of plains pocket gophers is determined primarily by the composition and condition of soil. The species prefers moist, deep, sandy loam and avoids continuously cultivated fields.

Description. The plains pocket gopher has a broad, flat head, compact body, short snout, sparsely haired tail that is used as a sensory organ, and large front legs and feet bearing large claws on all toes. A fringe of hair along the edge of each toe assists in digging and handling dirt. The skin is rather loose, and the hairs are arranged so as to permit flexibility when the animal moves forward or backward in its subterranean tunnels. Large external cheek pouches (used for carrying food) lined with fur extend back to the shoulders. The incisor teeth are large, yellow, and conspicuously grooved on the front surface; they protrude from the mouth and are enclosed at their base by the lips. Females have only three pair of mammae, one pectoral, one abdominal, and one inguinal.

The plains pocket gopher is the larger of the two species of gophers found on the Northern Plains. *Geomys bursarius bursarius* and *Geomys bursarius majusculus* are darker than *Geomys bursarius lutescens,* usually liver brown to chestnut brown above with somewhat paler underparts. The dorsal fur often has a distinct sheen. Seasonal differences in pelage are slight. As in *Thomomys,* there are two adult molts per year, the spring molt in May being more conspicuous than the molt to winter pelage in September. Hair replacement occurs in a series of waves that move from nose to rump, several of which may be visible at one time.

G. b. lutescens has pale yellowish brown dorsal fur and is considerably smaller than the two eastern races. Average and extreme external measurements of four males and 10 females, respectively, from extreme western Nebraska are: total length, 281.0 (269–300), 248.6 (235–263); length of tail, 82.8 (79–86), 72.4 (68–82); length of hind foot, 34.3 (33–35), 31.7 (30–33); length of ear, 6.0 (5–7), 5.2 (4–7). Two males from the same locality weighed 285 and 305 grams; the 10 females averaged 197.9

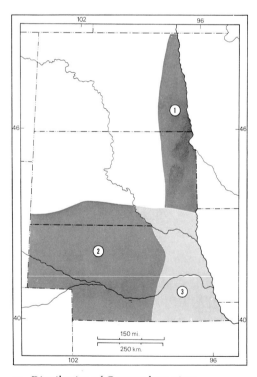

92. *Distribution of* Geomys bursarius: (1) G. b. bursarius; (2) G. b. lutescens; (3) G. b. majusculus.

(170.8–225.0). *G. b. bursarius* and *G. b. majusculus* are similar in size, although *majusculus* is slightly the larger. Females are consistently smaller than males, as in *G. b. lutescens.* Average and extreme external measurements of five adult males followed by those of three adult females of *G. b. bursarius* from eastern South Dakota are: total length, 287.0 (274–295), 270.0 (268–272); length of tail, 77.4 (76–81), 68.0 (62–72); length of hind foot, 37.4 (36–39), 36.3 (35–37); length of ear, 8.4 (8–9.5), 8.3 (7.8–8.5). Corresponding mean and extreme external measurements of 48 males and 50 females of *G. b. majusculus* from eastern Nebraska are: 319.6 (294–352), 290.2 (278–316); 95.2 (75–107), 88.6 (77–102); 38.7 (36–43), 35.4 (33–39); 6.0 (5–7), 5.6 (4–7). Weights varied from about 200 grams for the smallest female of *G. b. bursarius* to nearly 420 grams for the largest male of *G. b. majusculus.*

As in *Thomomys*, the heavily built cranium is flat and broad, with widely spreading zygomatic arches, and the mandible is powerful and heavy. The auditory bullae are relatively small. Both cheekteeth and incisors are unrooted and grow continuously. Average and extreme cranial measurements for five adult males and three adult females, respectively, of *G. b. bursarius* from eastern South Dakota are: greatest length of skull, 53.2 (51.2–54.6), 49.7 (48.0–51.2); zygomatic breadth, 32.6 (31.1–36.3); 29.5 (28.5–30.1). For seven male and three female *G. b. majusculus* from eastern Nebraska these measurements are: greatest length of skull, 58.7 (56.9–61.5), 50.5 (49.4–51.8); zygomatic breadth, 36.4 (35.1–38.9), 30.4 (29.6–31.5). Four male and 10 female *G. b. lutescens* from extreme western Nebraska measured: 48.9 (46.9–51.2), 42.6 (41.6–43.8); 30.9 (29.6–32.8), 26.3 (25.2–27.0).

Natural History. Soils suitable for the plains pocket gopher occur mainly in meadows, at forest edges, along rivers and streams, and on the higher terraces of floodplains. The species also occurs in sandy soils in some localities. This gopher is adapted to a narrower range of soil condi-

93. *Plains pocket gopher*, Geomys bursarius (courtesy D. C. Lovell and R. S. Mellott).

tions than the northern pocket gopher, which tends to occupy drier upland soils; *Thomomys* also has greater tolerance for colder, wetter soils and wooded areas than does *Geomys*. Intensive cultivation restricts plains pocket gophers to field margins, but they favor alfalfa or clover plantings.

This gopher is the most highly specialized for digging of any North American rodent, and it lives underground throughout most of the year. Although rarely seen, it makes it presence known through mounds of dirt pushed to the surface, often in a somewhat linear fashion. These fans of soil indicate a burrow system like that described for *Thomomys* except that it is more extensive, with a slightly larger burrow diameter and perhaps a slightly shallower nest cavity (two to four feet deep). The lateral tunnels are generally only six to 10 inches below the surface, but the main tunnel, which connects all the mounds, is generally deeper, and in the area of the nest chamber it usually is below the frost line. Besides the nesting chamber there are special tunnels for storage of food and deposit of feces. Dirt is excavated with the teeth and strong claws of the front feet; after sufficient dirt has accumulated the gopher turns around in the burrow and pushes the dirt with its front feet, head, and chest to a surface opening where it fans out over the mound. These animals may employ their teeth in burrowing to a greater extent than *Thomomys*. This species does not produce the network of earth cores beneath the snow that are so characteristic a sign of winter activity in *Thomomys*. Gophers mix and deepen soils, and this benefits their plant and animal associates. Their mounds cover surface vegetation, thus incorporating surface plant material into the soil, and their feces and carcasses, as well as uneaten plant material, also enhance soil fertility. The tunnels let water penetrate the soil and are especially valuable in absorbing rapid runoffs of early melting snows or rainstorms. Abandoned tunnels provide ready-made burrows for many other small mammals.

The food of *G. bursarius* usually consists of roots and underground stems of forbs. It is essentially like that of the northern pocket gopher, except that different plants are available, but the species eats a greater proportion of grass than does *T. talpoides*. Green leaves of forbs and grasses are gathered from around tunnel entrances, and at night these animals sometimes forage aboveground short distances away from their tunnels, thus exposing themselves to predators. Large amounts of food are stored in the burrow system and used during dry summer months. In winter *Geomys* apparently depends on food caches to a greater extent than *Thomomys*, because it forages on the ground surface beneath the snow to a lesser extent.

Breeding occurs from February to April or even later in the plains pocket gopher, somewhat earlier than in *Thomomys talpoides*. After a gestation of unknown length, one to six (average three to four) young are born, hairless, pink, wrinkled, and with eyes and ears closed. Only one litter per year is produced in our region. By five weeks of age the cheek pouches and eyes are open, and a few weeks later the young are weaned, when they are approximately half-grown. Dispersal patterns are poorly known, but young may remain with the mother until almost fully grown at a year old. Males and females reach sexual maturity at one year. Adult longevity is generally two to three years, but maximum longevity may exceed five or six years.

Natural enemies of pocket gophers are relatively few; the principal predators are badgers, weasels, and great horned owls. Bull snakes are able to dig through the dirt plug and enter the burrow, and owls capture gophers at tunnel entrances or when they forage aboveground. Other predators include large carnivores such as coyotes and foxes. Parasites include fleas, ticks, mites, lice, trematodes, cestodes, nematodes, and a protozoan.

Home ranges of plains pocket gophers, like those of other species of gophers, are essentially linear. If one attempts to calculate the area traversed by the burrow system, estimates of about 1,500 to 5,000 square feet are obtained, similar to those for *Thomomys*. Males tend to range more widely than females. Population densities appear to be lower on the average than in the smaller *Thomomys*, ranging up to about 15 per acre in preferred habitat.

Pocket gophers are of great benefit to the soil. Soil turnover of an individual *G. bursarius* has been calculated to be two and a half tons or 52 cubic feet per year. In agricultural areas, however, these animals are generally regarded as pests, especially in crops such as alfalfa and clover. Not only do they feed on the crops, but their mounds may damage mowing machinery. Trapping with commercial spring-steel gopher traps is the most effective method of control.

Selected References. General (Kennerly 1959; Vaughan 1961); natural history (Downhower and Hall 1966; Miller 1964; Vaughan 1962); systematics (Hart 1978).

Family Heteromyidae: Heteromyids

The heteromyid rodents constitute a distinctive group of North American origin, although one subfamily now ranges from Texas southward to northern South America. The family includes five living genera, of which the most numerous and widespread—*Perognathus* and *Dipodomys*—have representatives on the Northern Great Plains. Despite diversity in size and general body plan, all members of the family have external fur-lined cheek pouches (as do the closely related pocket gophers) and a light, almost papery skull. The dental formula is 1/1, 0/0, 1/1, 3/3, total 20. There is an evolutionary trend within the family toward bipedal locomotion, best developed in the genus *Dipodomys*. The geological record of the family extends back to the Oligocene, some 20 to 25 million years ago.

Key to Heteromyids

1. Tail bushy at tip; total length more than 240; greatest length of skull more than 35; cheekteeth high-crowned, with single lake of dentine surrounded by enamel; hind foot more than 35
   _____*Dipodomys ordii*
1.' Tail not bushy at tip; total length less than 240; greatest length of skull less than 35; cheekteeth low-crowned, with two lateral rows of cusps; hind foot less than 30 _____2
2. Total length more than 180; greatest length of skull more than 30; pelage somewhat harsh to the touch
   _____*Perognathus hispidus*
2.' Total length less than 150; greatest

length of skull less than 25; pelage not harsh to the touch _____3
3. Buffy patch behind ear distinct; interparietal bone conspicuously narrower than interorbital breadth (fig. 94)
   _____*Perognathus flavus*
3.' Buffy patch behind ear indistinct or lacking; width of interparietal bone about same as interorbital breadth (fig. 94) ___4
4. Auditory bullae generally not meeting anteriorly at midline; dorsal color olivaceous _____*Perognathus fasciatus*
4.' Auditory bullae generally meeting anteriorly at midline; dorsal color buffy to brownish _____*Perognathus flavescens*

Genus *Perognathus*

The genus *Perognathus* includes about 25 nominal species, arrayed in two subgenera. The subgenus *Perognathus* has three representatives on the Northern Great Plains; the subgenus *Chaetodipus* has one (*Perognathus hispidus*). Species of *Perognathus* range across western North America, from the Mississippi River to the Pacific Coast and from British Columbia and the Prairie Provinces of Canada to central Mexico. The name *Perognathus* is compounded from the Greek *pera*, meaning "pouch," and *gnathos*, meaning "jaw," a reference to the fur-lined cheek pouches.

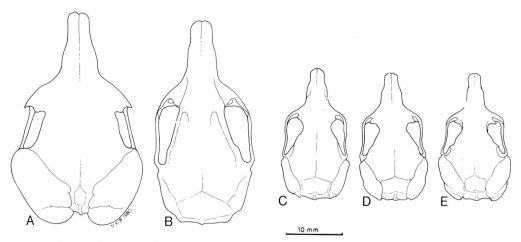

94. *Dorsal view of crania of the five species of heteromyid rodents found on the Northern Great Plains:* (A) Dipodomys ordii; (B) Perognathus hispidus; (C) P. fasciatus; (D) P. flavescens; (E) P. flavus.

Perognathus fasciatus:
Olive-backed Pocket Mouse

Name. The specific name *fasciatus* is a
Latin adjective meaning "banded," a reference to the distinctive yellowish buff lateral line. Other vernacular names applied to
this species include banded pocket mouse
and Wyoming pocket mouse.

Distribution. This is the only species of
pocket mouse for which the Northern
Great Plains and adjacent open country are
the principal range. These mice are found
from Alberta, Saskatchewan, and Manitoba
southward into the Wyoming Basin in
northeastern Utah and northwestern Colorado, and along the Colorado Piedmont east
of the Rocky Mountains. They occur in
northwestern Nebraska and across most of
North and South Dakota, except for the
tall-grass prairie region in the east. At present, a single subspecies is recognized on
the Northern Great Plains, *Perognathus
fasciatus fasciatus.*

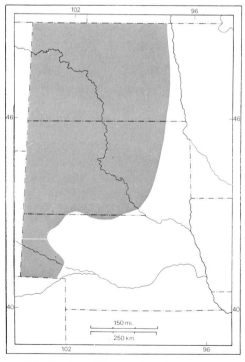

95. *Distribution of* Perognathus fasciatus.

Description. This is one of three small,
silky-haired species of *Perognathus* on the
Northern Plains. It can be distinguished
from the plains pocket mouse (*Perognathus
flavescens*) by the presence of a middorsal
band of black and olivaceous hairs, and in
that the lower premolar is nearly or quite
as large as the first molar and the bullae do
not meet anteriorly. *P. flavus,* the silky
pocket mouse, is smaller and has longer,
more silky pelage, prominent buffy spots
behind the ears, and a much narrower interparietal bone.

This is a beautiful little mouse by any
standard. The basic color is buffy, but the
dorsum is marked by a prominent band of
black and olive hairs. There is a pronounced yellowish buff lateral stripe, and
the belly is pure white. Variation (apparently clinal) in darkness of the dorsal patch
is evident across the Northern Plains. The
animals are darkest (and least buffy) in the
northeast and palest in the southwest.

The sexes are not significantly different
in size. Mean and extreme external and
cranial measurements of 12 individuals
from Medora, North Dakota, are: total
length, 134.7 (125–142); length of tail, 61.7
(57–68); length of hind foot, 17.2 (16–18);
length of ear, 7.5 (7–8); weight, 12.2
(10.8–13.1) grams; greatest length of skull,
22.7 (22.3–23.6); zygomatic breadth, 11.9
(11.3–12.3).

Natural History. The olive-backed pocket
mouse inhabits open country, often sandy
soils with good plant cover. Specimens
have been taken in *Bouteloua-Stipa* grassland, thickets of wild rose and sagebrush,
in alfalfa fields and grainfields, on grazing
lands, and in patches of cactus and yucca
among ponderosa pine.

These animals are active burrowers. Burrows are found in groups of two or three,
usually on slightly elevated ground, the
entrances usually not marked by thrown-
out soil. A tunnel system may cover an
area 20 feet across. Entrances usually are
plugged during the daytime, maintaining
high humidity and thus helping the mice to
conserve water. Tunnels may extend six
feet below ground. Seeds are stored in spe-

96. *Olive-backed pocket mouse,* Perognathus fasciatus *(courtesy R. E. Wrigley).*

cial chambers off the main tunnel, and there may be separate summer and winter quarters. The animals are not known to hibernate, but they may become torpid during cold weather.

Home range and movements of this species are little studied. Movements of 26.5 and 65.7 meters (87 and 217 feet) between live traps on a grid have been reported. Densities have been calculated by various means at about 0.5 to two per acre. As in other species of *Perognathus*, individuals are solitary.

Clues to food habits have been gained by inspecting the contents of cheek pouches, which in animals taken in southwestern North Dakota included seeds of needlegrass, croton, butterfly weed, and knotweed. Seeds in the cheek pouches represented only 14 percent of the plant species in the study area, suggesting some selectivity of diet. Seeds reported from elsewhere on the Northern Plains include lamb's-quarters, pigweed, Russian thistle, and June grass. Most of these plants are considered weeds, suggesting that these mice are of economic benefit. Wheat seed has been recovered from winter stores, but *P. fasciatus* is better adapted as a gleaner than as a harvester of cultivated grains. Probably it also eats some insects.

No intensive study of reproduction in this species has been made. Gestation takes about four weeks. Records of pregnancy extend from mid-May to July, but there seems to be a lull in reproductive activity in June. Counts of fetuses range from two to nine, with a mean of about five. The long and apparently bimodal period of reproductive activity and the presence in a female of 12 placental scars suggest that olive-backed pocket mice may have two litters of young in a season. The young are altricial. Nothing is known of their development. There are three pair of mammae, two inguinal and one pectoral.

As in other species of *Perognathus*, active locomotion in *P. fasciatus* is by quadrupedal hopping, although the mice sometimes walk when moving slowly. When gathering food, the animals rest on hind feet and tail and harvest seeds with their forepaws, stuffing them into the cheek pouches. Pouches are emptied by pushing against the outside of the pouch with the shoulder or paw, not by reaching inside.

Probably these pocket mice are preyed upon by any carnivorous animal that happens upon them, but they are too small and too widely dispersed across the habitat to serve any predator as a staple. Perhaps owls are the most important enemy. Ectoparasites include mites, ticks, and fleas.

Molt patterns have not been described in detail. An individual captured in mid-June showed new pelage beginning behind the ears, progressing to the middorsum and laterally onto the sides and cheeks.

Selected References. Natural history (Lampe et al. 1974; Pefaur and Hoffmann 1974, 1975; Turner and Bowles 1967); systematics (Williams and Genoways 1979).

Perognathus flavescens: Plains Pocket Mouse

Name. The Latin *flavus,* "yellow," and *escens,* "becoming," are combined to form the descriptive specific name.

Distribution. The plains pocket mouse is mostly a mammal of the Central Plains, extending from northern Mexico to east-central North Dakota and adjacent Minnesota, and from the Mississippi River to Colorado and Utah. Two subspecies are recognized on the Northern Great Plains, *Perognathus flavescens flavescens* in the south and west, and *Perognathus flavescens perniger* in the north and east.

Description. For comparisons with sympatric, small pocket mice, see the account of *P. fasciatus.* This is a rather nondescript species, with a grayish buff dorsum variously washed with blackish hairs. There often is an indistinct buffy spot behind the ear (less pronounced than in *P. flavus*) and a buffy lateral line (less pronounced than in *P. fasciatus*), and the venter is white. The blackish dorsal wash varies geographically, being heaviest in the north and east and least prominent in southwestern parts of the Northern Plains.

Average and extreme external measurements of 10 specimens from the Sand Hills of Nebraska are: total length, 121.5 (114–128); length of tail, 59.9 (52–65); length of hind foot, 16.8 (16–17); length of ear, 6.5 (6–7); weights vary from 7 to 12 grams. Cranial measurements of six males from the vicinity of Kearney, Nebraska, are: greatest length of skull, 21.1 (20.2–21.7); zygomatic breadth, 10.8 (10.3–11.3). *P. f. flavescens,* which these measurements describe, is slightly smaller than *P. f. perniger.*

Natural History. Plains pocket mice occur in areas of sandy soils, with cover of grasses or grasses mixed with sagebrush or

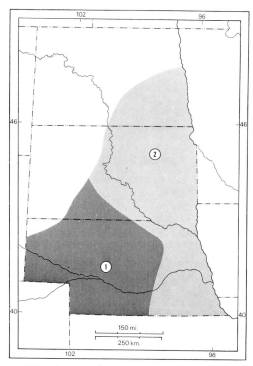

97. *Distribution of* Perognathus flavescens: (1) P. f. flavescens; (2) P. f. perniger.

yucca. Often they are found in grainfields. The grazing lands of the Nebraska Sand Hills seem to be a center of abundance. Little is known of the natural history of this species. Their burrows are similar to those of other small pocket mice, with several entrances (which are plugged during the day) and with separate chambers for nesting and food storage. Food habits have

98. *Plains pocket mouse,* Perognathus flavescens *(courtesy B. L. Clauson).*

not been studied in detail, but the animals are mostly granivorous, supplementing the diet with occasional insects in season. There is no evidence of hibernation. Seeds are stored for winter. Animals maintained in captivity at 5 to 15° C became lethargic, suggesting a capacity for winter torpor.

Owls probably are a principal predator of pocket mice, which are too small to provide a useful resource base for less skillful carnivores. A study of vertebrate remains in castings of great horned owls from central Nebraska found that plains pocket mice made up 2.7 percent of their diet.

Reproduction is known to occur from May to August; counts of two to five fetuses have been reported.

Selected References. Natural history (Beer 1961; Czaplewski 1976); systematics (Williams 1978).

Perognathus flavus: Silky Pocket Mouse

Name. The specific epithet *flavus* is a Latin adjective meaning "yellow."

Distribution. The silky pocket mouse is a species of the semidesert grasslands of the Central and Southern Great Plains and the Mexican Plateau. It reaches its northern limits in western Nebraska and southeastern Wyoming and ranges southward to central Mexico.

A single subspecies, *Perognathus flavus piperi*, has been documented in our area, although should specimens become available from southwestern Nebraska they may be found to represent the paler southern race, *Perognathus flavus bunkeri*.

Description. This is the smallest of pocket mice in our region; indeed, it is, on average, the smallest member of the genus. Mean and extreme external measurements of four individuals from Hooker County, Nebraska, are: total length, 110.0 (107–117); length of tail, 52.0 (48–57); length of hind foot, 15.7 (15–16); length of ear, 6.5 (6.4–6.8). Cranial measurements of two of these individuals are: greatest length of skull, 20.2, 19.6, and zygomatic breadth,

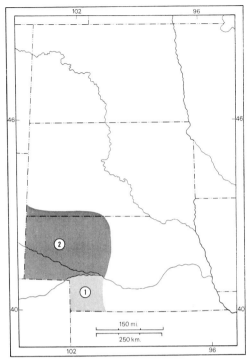

99. *Distribution of* Perognathus flavus: (1) P. f. bunkeri; (2) P. f. piperi.

10.9, —. Weights of silky pocket mice range from about 6 to 10 grams.

This is a handsome little mouse, with remarkably soft, silky fur. It is yellowish to reddish buff, with a pronounced wash of blackish hairs among northern populations. A patch of buff-colored hairs behind the ear is distinctive. The lateral stripe is less well defined than in other local species of *Perognathus.*

Natural History. On the Northern Great Plains, this is the least widespread of the heteromyids, and within this range it seems to be the least abundant. Hence little is known of its natural history there. In eastern Wyoming silky pocket mice were found to be most abundant on loamy soils with a cover of grasses (grama, needlegrass, three-awn) and minimal bare soil. They exhibit much more habitat specificity than does *P. flavescens.*

Species of *Perognathus* generally do not breed in captivity; hence details of re-

100. Silky pocket mouse, Perognathus flavus *(courtesy T. H. Kunz).*

production are scanty, based on incidental field reports. Records of two to six young per litter are available. There seems to be one litter per year, with most breeding in spring. There are three pair of mammae, two inguinal and one pectoral. Gestation probably takes about four weeks.

The food of silky pocket mice is composed almost entirely of seeds, which they shell carefully and eat only the rich endosperm. In periods when native seeds are abundant, this and other pocket mice are hard to attract to artificial baits and thus are not often trapped.

Juvenile pelage is a dull gray, soft and remarkably thin. Mice wear this coat when they first begin to forage aboveground. They retain it until about the time the permanent lower premolar begins to erupt. The postjuvenile molt begins just behind the forelimbs and spreads posteriorly over the back as a wave of bright yellowish buff hairs with black tips. This molt often is not completed until after the mouse attains sexual maturity and morphological adulthood. After this there is a single annual molt, in mid- to late summer, which begins dorsomedially and progresses fore, aft, and laterally. This pattern may break down in older mice, molt becoming erratic.

Selected References. Natural history (Forbes 1964).

Perognathus hispidus: Hispid Pocket Mouse

Name. The specific name *hispidus* is a Latin adjective meaning "bristly," a reference to the distinctive pelage of this mouse, the only member of the subgenus of spiny pocket mice on the Northern Plains.

Distribution. The hispid pocket mouse ranges across the Central and Southern Great Plains, from southernmost North Dakota to central Mexico, and from the Missouri River to the foot of the Rocky Mountains. Two subspecies are recognized in our region, *Perognathus hispidus spilotus* in southeastern Nebraska and *Perognathus hispidus paradoxus* in the drier country to the west.

Description. This is much the largest of local pocket mice. Rubbing the harsh fur from back to front allows unequivocal identification. The animals are ochraceous buff above, mixed with blackish hairs. There is a distinct lateral line of buffy hairs, and the venter is mostly white. The tail has a dark middorsal stripe and buffy sides and is white below. Geographic variation on the Northern Plains follows a pattern seen in other pocket mice. There is a darker sub-

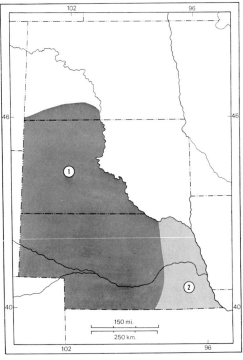

101. Distribution of Perognathus hispidus: (1) P. h. paradoxus; (2) P. h. spilotus.

species (*P. h. spilotus*) in the east and a paler race (*P. h. paradoxus*) in the west. Size does not distinguish the two populations, nor do sexes differ significantly in external or cranial dimensions.

Mean and extreme external measurements of 10 individuals from the Nebraska Panhandle are: total length, 220.0 (203–237); length of tail, 106.6 (93–114); length of hind foot, 26.4 (23.5–29.5); length of ear, 12.9 (12–14); weights of seven specimens averaged 47.8 (39.9–59.6) grams. Cranial measurements include: greatest length of skull, 32.0 (30.7–33.9); zygomatic breadth, 16.5 (15.6–17.2).

Natural History. Hispid pocket mice occupy a variety of upland habitats, usually on loamy soils with considerable bare ground. They are not common on dune sands but do occur in rocky prairie areas. Vegetation of mid- and short-grasses, shrubs, forbs, cacti, and yucca characterizes their habitat.

The diet is mostly seeds, which seem to be selected rather than simply gathered at random. A thorough study of food habits in west-central Texas showed seeds of the following plants to supply the bulk of winter foods: mesquite, sunflower, sagebrush, cacti (both *Mammalaria* and *Opuntia*), sorghum,

and millet. A few leaves and insects also were eaten. In spring the diet shifted to include seeds of blanketflower and bluestem. Insects (mostly carabid beetles) constituted almost 14 percent of the spring diet. Other foods reported include seeds of evening primrose, poppy mallow, buckwheat, noseburn, pecans, persimmons, legumes, and plums.

Hispid pocket mice are solitary. Their burrows have two or three entrances, which usually are plugged. Unlike our other pocket mice, *P. hispidus* often leaves a conspicuous mound of earth about the burrow entrance, reminiscent of that of a pocket gopher but much smaller. Sometimes individuals burrow into banks or beneath shrubs, and such work looks like that of kangaroo rats. Older mice have more elaborate burrow systems than do younger animals. Often there is a short tunnel just inside the main entrance. When disturbed, the mouse enters this antechamber, plugging it behind itself. Food is stored in a separate chamber. No fat is stored against winter, and there is no indication of hibernation, the animals subsisting on their ample stores of seeds during that time.

Enemies include most nocturnal carnivorous mammals and birds. A study of food habits of great horned owls in central

102. Hispid pocket mouse, Perognathus hispidus *(courtesy R. J. Baker).*

Nebraska revealed that *P. hispidus* is included in the diet. Ectoparasites include ticks and other arthropods.

Populations seem never to be especially dense. Thus estimates of population density, home range, and movements are not available. Reports on reproduction are incidental and rare. The number of young per litter has been reported to vary from two to nine. Females taken in August in western South Dakota and Nebraska each had six fetuses, and one captured in Nebraska in early September had five. Adult females probably bear two or more litters annually, because the reproductive season extends from spring through late summer in the tristate region. As for other pocket mice occurring on the Northern Great Plains, few details are available on some aspects of natural history, such as molt patterns, growth and development of young, and longevity.

Selected References. Natural history (Alcoze and Zimmerman 1973; Blair 1937); systematics (Glass 1947).

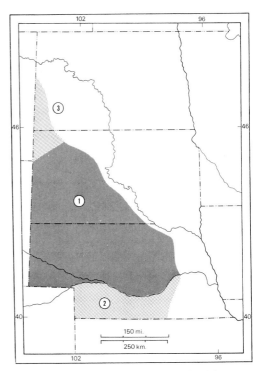

103. Distribution of Dipodomys ordii: (1) D. o. luteolus; (2) D. o. richardsonii; (3) D. o. terrosus.

Genus *Dipodomys*

The genus *Dipodomys*, with a single representative on the Northern Great Plains, includes more than 20 species, some of which are island endemics. Unlike members of the subfamily Heteromyinae, the Dipodomyinae have rather high-crowned cheekteeth. The genus shows the culmination of the trend through the family Heteromyidae toward bipedal locomotion. The name is a compound of the Greek *di*, "two," *podo*, "foot," and *mys*, "mouse," an apt allusion to the peculiar stance and gait.

Dipodomys ordii: Ord's Kangaroo Rat

Name. Pioneer American zoologist George Ord (1781–1866) is commemorated in the specific epithet *ordii*.

Distribution. This kangaroo rat has the widest range of any member of the genus, from southern Saskatchewan and Alberta to the valley of Mexico, and from eastern Nebraska to the eastern flanks of the Sierra Nevada. On the Northern Great Plains the animals occur in extreme southwestern North Dakota, the western half of South Dakota, and the western two-thirds of Nebraska. Three subspecies are recognized in the tristate region, *Dipodomys ordii richardsonii* south of the Platte River in Nebraska, *Dipodomys ordii luteolus* northward in Nebraska and South Dakota, and *Dipodomys ordii terrosus* in northwestern South Dakota and adjacent North Dakota.

Description. Ord's kangaroo rat is unmistakable among mammals of the Northern Plains, with its long, bushy-tipped tail and huge hind feet. The upper parts are a rich yellowish buff with a variable blackish wash, more pronounced in *D. o. richardsonii* than in the other subspecies. The tail has a dark stripe above and below (the ventral stripe is more or less continuous to the tip of the tail in *D. o. richardsonii* but is restricted to the proximal two-thirds in *D. o. luteolus* and *D. o. terrosus*). The sides of the tail, venter, and flank stripes are white.

Average and extreme external and cranial measurements of 10 individuals (seven males and three females) of *D. o. luteolus* from near Valentine, Nebraska, are: total length, 266.0 (245–276); length of tail, 147.6 (139–156); length of hind foot, 41.8 (40–44); length of ear, 13.5 (12–15); greatest length of skull, 39.2 (38.0–40.6); mastoidal breadth, 23.8 (23.3–24.9). Weights of eight individuals from the same series averaged 65.7 (55.9–78.1) grams. Of the three subspecies in the Dakotas and Nebraska, *terrosus* is the largest, *luteolus* is intermediate in size, and *richardsonii* is the smallest.

Natural History. Ord's kangaroo rats occupy areas of sandy soils. They are remarkably abundant on the Sand Hills of Nebraska and increase in response to overgrazing and consequent formation of "blowouts" on the plains. Sandy soils not only allow them to burrow readily but provide a substrate for their frequent and characteristic dust baths. Vegetation of the habitat varies from grasslands to shrublands, usually with considerable bare soil. Burrows often open at the base of shrubs or bunchgrasses, or in banks or cuts. Burrows are dug actively in the sandy soil and may become quite complex. The forelimbs are used to pile sand beneath the body, and the pile then is propelled backward with the hind feet. The rats sometimes gnaw at hard materials in the course of burrowing. Nest chamber and food storage are separate. Defecation is indiscriminate except that the nest is not fouled. Urination takes place in one corner of the burrow. The viscous urine crystallizes quickly in the air.

The nest is built of plant fibers. The rats cut pieces of stem into lengths convenient for transport in the cheek pouches. In the burrow, they shred them by grasping an end in the incisors and then quickly jerking the head back.

Kangaroo rats are mostly granivorous. A study of food habits of the species in west-central Texas showed that in winter there was little food preference, in that seeds were gathered roughly in proportion to their abundance in the habitat. Among important seeds were paspalum, mesquite, croton, sorghum, ragweed, and sunflower. Insects, including caterpillars and ants, made up about eight percent of the diet, and four percent consisted of the fungus *Endogone*. The spring diet, by contrast, showed dietary selection. Seeds of mesquite and blanketflower were selected strongly, and those of bluestem and ragweed also were important. Insects—especially carabid beetles and ant lions—and other arthropods were important in the spring diet, accounting for 18.3 percent of the food. A high degree of dietary overlap with *Perognathus hispidus* was observed.

In Idaho, Ord's kangaroo rat was found to eat seeds of halogeton, shadscale, peppergrass, pricklypear, grasses, and tansy mustard, and leaves of sweet clover, sagebrush, grasses, shadscale, Russian thistle, tansy mustard, summer cypress, and saltsage, as well as arthropods. Studies of food habits on the Northern Plains mostly are anecdotal but concur that seeds of "weedy" plants are dietary staples, although cultivated small grains also are eaten.

A volumetric study of stomach contents of Ord's kangaroo rats from northeastern Colorado showed that seeds made up 85 to 93 percent of the diet. Foliage was most important (to 30 percent) in late spring and early summer. Animal matter—chiefly adult beetles and larval beetles, moths, and butterflies—was most important in winter (10 percent), spring, and early summer. The rats gather seeds with their forepaws, often by sifting the sand. They then stuff them into the capacious cheek pouches and carry them to the burrow, where they are removed by the pressure of the shoulder against the pouch. Seeds are husked before they are eaten. The rat grasps the seed coat in its incisors and strips it from the seed by a jerk of its head.

Kangaroo rats can live indefinitely without free water, supplying their water needs from "metabolic water" produced in the oxidation of food, especially fats, and conserving it by means of an extremely effi-

104. Ord's kangaroo rat, Dipodomys ordii *(courtesy T. H. Kunz).*

cient kidney and by avoiding the heat of the day (and resultant demand for evaporative cooling) by remaining in the humid confines of the burrow. When water is available they take it readily, scooping it up with the forepaws and lapping it with the tongue. Dew also is utilized as available.

Reproductive patterns in Ord's kangaroo rats seem to be highly variable geographically. It is impossible to know how applicable are data from elsewhere over the broad range of the species to understanding the situation on the Northern Great Plains. In Nebraska, 13 females taken over the period May–September averaged 3.1 (two to five) fetuses, whereas 10 taken in southern South Dakota in June and July carried an average of 2.3 (one to three). In northeastern Colorado, studies of pregnancy and placental scars suggested a reproductive season extending from February to August. Testicular regression began in July. Recently weaned young (with body weights less than 40 grams) were found from March to August. Fetus counts from 31 females ranged from two to four (mean 2.87). Farther south, in the Oklahoma and Texas panhandles, reproduction occurred mostly from August to March and only rarely from

April to July. A peak in reproductive activity was seen in September and October. Mean embryo count was about 3.2. In Utah the species was found to have two distinct breeding periods, one from January to March and the other from late August into October. Environmental or physiological correlates of geographic variability in reproductive patterns have not been elucidated.

Kangaroo rats are solitary animals. As in most such species, copulation is preceded by considerable "courtship" activity. Studies of precopulatory behavior in captives showed that there is considerable reciprocal chasing, the male nudging the genitalia and thorax of the female. During copulation the male grasps the skin on the neck of the female with his incisors. Multiple copulations occur, increasing in duration from one-half minute to two minutes. Gestation takes about 29 to 30 days. Just before parturition, nest-building is initiated. The female helps deliver the altricial young with incisors and forepaws. After birth she grooms them with paws, teeth, and tongue, and she eats the placenta, thereby recovering some nutrients.

Young kangaroo rats are highly altricial;

indeed, with their short tails, relatively short hind feet, and lack of cheek pouches, the hairless, pinkish neonates are hard to identify out of context. The body is covered sparsely with hair in about one week, ears open at about 10 days, and at two weeks the eyes open. The cheek pouch is obvious as a fold of skin at 10 days and first is used at about three weeks of age. Adult size is attained at five to six weeks of age.

The locomotion of kangaroo rats is distinctive among North American rodents and involves several modes of progression. Occasionally they move in a quadrupedal gait, although the locomotive power comes from the hind feet and the small forelimbs lend little more than support. The basic unhurried gait is bipedal, with leaps little longer than the rat's body. Between hops the animals may lean forward to gather seeds with the forelimbs, with the weight over the feet, counterbalanced by the tail. Sometimes they engage in a bipedal walk, moving the two hind feet alternately. Fast locomotion is of two sorts. The forefeet are not involved at all, but are carried on the chest. In one kind of fast progression, when not pursued, the animals move straight ahead with one- to two-foot leaps, about eight inches above the ground. It is when pursued that kangaroo rats use their remarkable erratic bipedal locomotion, making leaps six feet or more long and two feet high, with frequent midair changes in direction. They have been known to careen off rocks and ridges of earth. The tail is essential to these locomotor patterns. In slow movement it is used as a prop, a sort of third leg. In fast, straightaway locomotion it serves to counterbalance the body upon landing. In erratic flight the tail is essential for balance through the tortuous changes in direction. In a fall the tail flails the air to right the animal. A kangaroo rat that has lost its tail will somersault as it attempts to land from a long jump.

This evasive action, coupled with the flash of the white-striped flanks and brush-tipped tail, enables the rats to avoid many pursuing predators. But, still, owls take a toll, badgers and coyotes dig the animals from their burrows, and larger snakes follow them quietly into their dens and probably also capture some as the rats forage aboveground at night.

Kangaroo rats show considerable active intraspecific agonism. Fighting bouts are strikingly reminiscent of those of their marsupial namesakes. There may be pushing, tussling, and clawing with the forelimbs. Butting with the shoulders is common. The most characteristic offense is a bipedal kick with the hind feet, the animal partly standing on its tail. Some vocal signals occur with fighting, including growls and squeals. The most characteristic sound of arousal, however, is a drumming of the hind feet and a chattering of the teeth.

Grooming behaviors include wash-licks, scratching and combing with the forepaws, and violent scratching with the hind feet. The cheek pouch linings are stretched out for brushing. In sand bathing the animals dive into the sand and wriggle along with weasel-like motion, the forefeet on the chest, pushing the body along with the hind feet.

Surprisingly little is known about molt in kangaroo rats. There is a distinct juvenile pelage in *D. ordii*, but whether there is a single postjuvenile molt or several is in question. Adults probably molt only once a year, in late spring or summer depending on climate, reproductive condition, and other variables, but this too is uncertain. Similarly, there are few data on longevity. Adults have been known to live in captivity for five years or more, but the average life span in the wild among animals that survive into a second year of life probably does not exceed two years.

Selected References. Natural history (Alcoze and Zimmermann 1973; Eisenberg 1963; Flake 1973, 1974; Schmidt-Nielsen and Schmidt-Nielsen 1952); systematics (Kennedy and Schnell 1978; Setzer 1949).

Family Castoridae: Beavers

The family Castoridae includes only the single genus *Castor*, with but two living species, one in Eurasia and the other in North America. The fossil record, which extends from early Oligocene to Recent in North America and from late Oligocene to Recent in Eurasia, includes 14 extinct genera, among them a giant form, *Castoroides*, during the Pleistocene. Beavers have no close relatives among modern rodents, but the lineage has been traced to a common ancestry with the squirrels, which have a comparable arrangement of musculature in the jaw.

Genus *Castor*

The name *Castor* is based directly on the Greek word for beaver. The dental formula is 1/1, 0/0, 1/1, 3/3, total 20. The origin of the vernacular name is the Old Anglo-Saxon *befer*.

Castor canadensis: Beaver

Name. The specific name *canadensis* was applied to the North American beaver because the specimen on which the original description was based came from the vicinity of Hudson Bay, Canada.

Distribution. Castor canadensis ranges in suitable habitat throughout North America, southward to northern Mexico. Because its existence depends on permanent water and a supply of woody vegetation, the beaver is absent from the Arctic Slope of northern Alaska, parts of northern Canada, and a sizable area in the arid regions of the southwestern United States; it also is absent from peninsular Florida. Two subspecies occur on the Northern Plains, *Castor canadensis canadensis* in northeastern North Dakota and *Castor canadensis missouriensis* in the Missouri River drainage to the west and south.

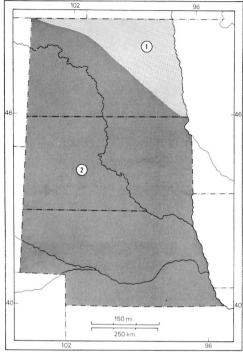

105. Distribution of Castor canadensis: (1) C. c. canadensis; (2) C. c. missouriensis.

Description. The beaver is the largest rodent in North America north of Panama. Large adults may weigh 60 pounds or more. The animals are highly modified externally for an aquatic existence—the feet are webbed, the eyes are small and protected by a nictitating membrane, and the nostrils and the pinnae of the ears are valvular, capable of being closed underwater. Anal and urogenital openings share a cloaca, hence there is no obvious external sexual dimorphism. The tail is scaly, flattened, and nearly naked. Paired anal scent glands called "castors" are present in both sexes but are better developed in males.

Size and color distinguish the two subspecies of *Castor canadensis* of the Northern Plains. *C. c. canadensis* is darker brown and slightly larger than *C. c. missouriensis*.

Average and extreme external measurements of five females (*missouriensis*) from Nebraska are: total length, 1,098.2 (940–1,212); length of tail, 328.0 (290–376); length of hind foot, 182.4 (170–192); the ear usually is about 35 mm long. Adult weight averages about 40 pounds. There are two annual molts in adults, one in spring and a second in autumn. Albinos have been reported.

The skull is robust and is characterized by a heavy dentition, a broad, deep rostrum, and a relatively narrow braincase. The large incisors are ever-growing, as in other rodents. The zygomatic arch is deep and heavy. Representative cranial measurements of six animals (subspecies *missouriensis*) are: greatest length of skull, 130.4 (125.2–140.2); zygomatic breadth, 94.5 (90.4–104.1).

Natural History. The curious and specialized habits of the beaver make it one of the most interesting of our native mammals. Its engineering feats have been the subject of a vast amount of popular literature, some of it exaggerated, for beavers probably have a more profound effect on their immediate environment than any other mammal except man. The greatest environmental change results from their construction of dams, barriers of sticks and logs covered with mud built across shallow streams. These are carefully maintained, and a dam in disrepair probably has been abandoned. The sound of running water stimulates dam-building behavior.

Ponds formed behind dams protect lodges or bank dens from predators, provide a trafficway for the movement of materials and food, and allow a swimming area beneath the ice in winter. In many areas such dams cause local flooding. Many species of trees cannot survive in the resulting waterlogged soil, and their death allows the spread into the area of species that are adapted to permanently saturated ground. The bark of such trees, including willows, cottonwoods, and alders, forms the major source of food for beavers. Thus damming of streams not only protects the animals and allows easy

movement throughout the year, but creates a water table favoring a plant community dominated by preferred food species. Beaver ponds also provide habitat for other kinds of wildlife that favor still water, especially muskrats and shorebirds.

The well-known beaver lodge is built in a natural pond or in a pond formed behind a beaver dam. The lodge, erected on a foundation such as a natural elevation on the bottom of a pond, in water three to six feet deep, is begun with a ring of sticks that forms a door. A hollow is maintained inside throughout the construction of the lodge. Limbs and logs are added to the walls, and smaller material is used to narrow the walls and close the top. The entire house is then plastered with mud and sod. The floor inside is built up a few centimeters above the water. This nest chamber is lined with grasses or shredded bark. There are usually two or more underwater entrances.

In some situations, bank dens are built with access only from the water. Where the bank is high the entrance is unmarked by building material, but on a low bank a house of logs and sticks plastered with mud may be present; these, however, usually are smaller than typical lodges. In some instances beavers also build canals through

106. Beaver, Castor canadensis *(courtesy G. Latham).*

which logs are floated to the pond. Water levels in such canals may be maintained with dams.

The industry of beavers and their intricate construction projects have made the animals the subject of much folklore, but care must be taken not to ascribe to superior intelligence activities that are based on instinct or chance. It is frequently claimed, for example, that a beaver can fell a tree in any desired direction; but the fact that trees often fall toward the water is actually related to their tendency to grow toward the lighter open area above a stream and hence to have a natural "lean" in the proper direction.

The social structure of a beaver colony is based on a nuclear family consisting of a pair of adults, yearlings, and kits. A colony consists of one or more such families. Beavers disperse (usually as two-year-olds) from crowded areas to form new colonies, which may begin with a single animal. The average number of individuals occupying a lodge or bank den is about five.

The breeding season begins in January, and, after a gestation period of approximately 120 days, one to six (usually three to four) young are born. A single litter is produced annually. Larger litter sizes correlate positively with good nutrition in the female and with mild winter weather. The young are born at a fairly advanced stage, fully furred, eyes partially open, and incisors erupted, at weights of 350 to 650 grams. Lactation lasts about three months, although weaning begins at about two months, and the young eat vegetation even earlier. There are four pectoral mammae. Females do not breed until the second season after their birth, or even later in large colonies that contain several older females. Full growth is not attained for several years. Longevity frequently is 10 years or more, the record in the wild being 21.

Beavers are vegetarians, specific foods varying somewhat with the season. Grass and other greens are important during the warmer months, and at this time such bark as is eaten is incidental to preparing construction material; in autumn, branches are stored for winter food. In most areas

willows form the bulk of the woody diet and provide preferred building material, but cottonwoods and aspens are favored when available. Hardwoods and some conifers occasionally are cut for building material but rarely are eaten.

Communication is visual, auditory, and olfactory. Grunts, hisses, and tooth-chatters are aggressive sounds. The well-known tail slap warns the colony of danger but also may frighten would-be predators. Castoreum, the secretion of the anal glands, is deposited on mounds of mud, presumably to mark colonial territory.

The beaver is protected from predators by its aquatic habits. On land these animals are vulnerable to predation by coyotes, domestic dogs, bobcats, lynx, and mountain lions. Mink may enter the lodge or den and prey on the young. Beavers also are susceptible to various diseases, including rabies and tularemia, and to infestation by parasites. The most important controls on the beaver population, however, are the direct or indirect effects of man and the effects of extremes in weather. Trapping by man kept populations low at one time, but today destruction of suitable habitat is probably a more important control. Both drought and floods have serious consequences for established beaver colonies, and they may have been the most important natural control of numbers before the disturbance of wide areas by man.

The former abundance of the beaver over much of North America and the commercial value of fur were responsible for much of the early exploration of large sections of the continent. Where modern resource management is practiced, the animals still are of commercial value (averaging about twenty-four dollars per pelt in Nebraska in the 1979–80 season). The remarkable behavior of these animals makes them a unique and invaluable aesthetic resource. Beaver ponds form a desirable habitat for other kinds of wildlife including muskrats, mink, water- and shorebirds, and fish.

Beavers may do considerable damage to man's interests in certain circumstances. Their dams may impound irrigation water, alter watercourses, or raise water levels to the detriment of low-lying woodlands or croplands. Occasionally they cut valuable hardwoods or fruit trees in the course of gathering construction materials. Where the interests of a landowner are jeopardized by beavers, they may be harvested or trans-planted to more favorable sites by conservation personnel.

Selected References. General (Jenkins and Busher 1979 and included citations).

Family Cricetidae: Cricetids

The cricetids are the most diverse family of rodents and the most widespread, both on a global scale and on the Northern Great Plains. There are more than 650 species in the family, arranged in nearly 130 genera. These animals range in size from harvest mice, with a weight of about 10 grams, to the muskrat, 150 times as heavy. Cricetids occupy a spectrum of habitats from salt marsh to desert to alpine and arctic tundra, and every conceivable habitat in between. They occur nearly worldwide, except for Australia, Malaysia, Antarctica, and some islands.

There are three principal subfamilies of the Cricetidae. Two of them, the Cricetinae and Microtinae, are important on the Northern Great Plains. Each is represented there by eight species in five genera. A third subfamily (Gerbillinae—an Old World group) is familiar to many, for it includes the Mongolian "gerbil," an increasingly common household pet and laboratory animal.

The fossil record of cricetids is traceable back to Oligocene times in Eurasia and North America. It is increasingly common for specialists to merge the cricetids with the closely related Old World rats and mice into an inclusive family Muridae, but this arrangement is not yet universally accepted and is not followed here.

Key to Cricetids

1. Cheekteeth with two longitudinal rows of cusps (except *Neotoma*) _____2
1.′ Cheekteeth with distinctive prismatic occlusal pattern _____9

2. Total length more than 200; greatest length of skull more than 28 _____3
2.′ Total length less than 200; greatest length of skull less than 28 _____5
3. Upper parts harsh to the touch, dark brown; tail scaly; cheekteeth nearly equal in size, square and low-crowned; claws brown _____*Sigmodon hispidus*
3.′ Upper parts soft to the touch, grayish to yellowish brown; tail not noticeably scaly; first cheektooth nearly twice as long as last, cheekteeth high-crowned; claws pale, translucent _____4
4. Tail bushy, dark gray; diastema one and a half times length of cheektooth row _____ *Neotoma cinerea*
4.′ Tail not bushy, brownish above and whitish below; diastema approximately same length as cheektooth row _____ *Neotoma floridana*
5. Tail less than half length of body; coronoid process of mandible prominent _____*Onychomys leucogaster*
5.′ Tail nearly as long as body or longer; coronoid process of mandible not especially well developed _____6
6. Greatest length of skull more than 19; length of hind foot more than 17; upper incisors not grooved (fig. 111) _____7
6.′ Greatest length of skull less than 19; length of hind foot less than 17; upper incisors grooved (fig. 111) _____8
7. Tail distinctly bicolored; length of hind foot usually 21 or less; first upper cheektooth trapezoidal, with anterior lobe _____*Peromyscus maniculatus*
7.′ Tail usually not distinctly bicolored; length of hind foot usually 21 or more; first upper cheektooth rectangular,

lacking anterior lobe _____*Peromyscus leucopus*

8. Dorsal stripe on tail relatively narrow, about one-fourth diameter of tail; mid-dorsal stripe well defined; rostrum relatively short; greatest length of skull 20 or less _____*Reithrodontomys montanus*

8! Dorsal stripe on tail relatively broad, about one-half diameter of tail; middorsal stripe poorly defined; rostrum relatively long; greatest length of skull more than 20 _____*Reithrodontomys megalotis*

9. Tail naked, laterally flattened, length more than 100; greatest length of skull more than 50 _____*Ondatra zibethicus*

9! Tail well furred, not flattened, length less than 100; greatest length of skull less than 50 _____10

10. Tail approximately same length as hind foot; upper incisors with distinct groove on anterior surface (fig. 126) _____*Synaptomys cooperi*

10! Tail at least slightly longer than hind foot; anterior surface of upper incisors smooth (at most, a shallow, faint groove present) _____11

11. Tail short, less than 10 longer than hind foot; dorsal pelage pale gray; sole of hind foot densely haired; auditory bullae large, extending backward beyond occipital condyles and thus visible in dorsal view _____*Lagurus curtatus*

11! Tail usually more than 10 longer than hind foot (except *Microtus pinetorum*); dorsal pelage reddish, brownish, or grayish; sole of hind foot not densely haired; auditory bullae not extending backward beyond occipital condyles and not visible in dorsal view _____12

12. Dorsal stripe reddish to reddish brown, contrasting with gray sides (except in some juveniles); rear of palate shelflike (fig. 126); cheekteeth rooted in adults _____*Clethrionomys gapperi*

12! Dorsal pelage lacking distinct reddish stripe contrasting with gray sides; rear of palate a sloping median ridge bordered laterally by pits (fig. 126); cheekteeth rootless, ever-growing _____13

13. Tail less than 29 (usually less than 25); pelage soft and smooth, reddish brown

dorsally; skull wide and flat; third upper molar with two closed triangles (fig. 126) _____*Microtus pinetorum*

13! Tail more than 26 (usually more than 29); pelage relatively coarse, some shade of grizzled brownish, grayish, or blackish; third upper molar with three closed triangles or, if only two, skull high and narrow _____14

14. Tail 25 to 35 percent of body length; underparts usually buffy to ochraceous; third upper molar with two closed triangles (fig. 126) _____*Microtus ochrogaster*

14! Tail more than 35 percent of body length; underparts grayish to whitish; third upper molar with three closed triangles (fig. 126) _____15

15. Tail usually more than 50 percent of body length; second upper molar with four closed triangles and no posterior loop (fig. 126) _____*Microtus longicaudus*

15! Tail less than 50 percent of body length; second upper molar with distinct posterior loop that often appears as a fifth triangle (fig. 126) _____*Microtus pennsylvanicus*

Genus *Reithrodontomys*

The genus *Reithrodontomys* includes 19 species, which range from southern Canada to Panama. The dental formula is 1/1, 0/0, 0/0, 3/3, total, 16. The incisors are distinctive among local cricetines in that they are grooved on the anterior face. The name *Reithrodontomys* alludes to this character; it is a compound of Greek roots meaning "channel-toothed mouse."

Reithrodontomys megalotis: Western Harvest Mouse

Name. The specific name is a compound of two Greek roots, *megal,* "large," and *otis,* "ear."

Distribution. The western harvest mouse is the most widespread member of its genus, ranging from southern British Columbia, Alberta, and Saskatchewan to the Isthmus of Tehuantepec. The species extends east-ward to western Indiana and westward to the Pacific Coast. A single subspecies, *Reithrodontomys megalotis dychei*, occurs on the Northern Great Plains.

Description. For this larger of our two har-vest mice, mean and extreme external mea-surements of 27 individuals (19 males and eight females) from Cherry County, Nebraska, are: total length, 134.7 (122–155); length of tail, 62.8 (56–73); length of hind foot, 17.4 (17–18); length of ear, 12.8 (11–14); 10 males weighed 13.3 (11.7–14.7) grams. Average and extreme cranial measurements of 20 adults are: greatest length of skull, 21.0 (20.4–22.1); zygomatic breadth, 10.9 (10.0–11.3). The dorsal color is grayish brown, somewhat darker on the midline. The tail is distinctly bicolored, and the venter is white. For com-parison with the plains harvest mouse, see the account of that species.

Natural History. The western harvest mouse ranges widely and occupies a fairly

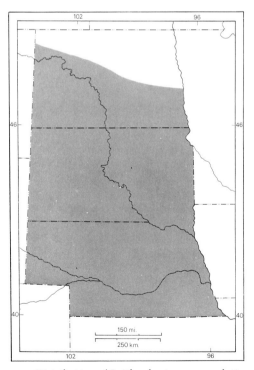

107. Distribution of Reithrodontomys megalotis.

wide variety of habitats, seemingly avoid-ing only dense forest and xeric uplands. Typical habitats include dense patches of tall grass, shrublands (including sagebrush), yucca-grass associations, brushy riparian habitats, cattails, alfalfa fields, and borrow pits with weedy vegetation. Often these mice are found in association with *Micro-tus ochrogaster* or *M. pennsylvanicus*, uti-lizing the runways of the voles. Being strictly nocturnal, harvest mice overlap minimally with the microtines, which are mostly crepuscular but may be active at any time of day. The habitats of *R. mega-lotis* and *R. montanus* usually are mutually exclusive (see account of the latter species for comparison).

These harvest mice build globular nests like those of birds, at or near ground level in heavy vegetation. The outer nest is loosely woven of fibrous plant materials and is about three inches across. A half-inch entrance hole toward the bottom of the nest leads to an inner chamber about an inch and a half in diameter, lined with "down" from thistle, milkweed, or cottonwood.

The diet consists mostly of seeds, al-though some insects and leaves are taken. Fibrous and woody plant materials are not eaten. Seeds seem to be taken as available; those of grasses, mustards, and legumes often are reported. The animals also eat small grains such as oats, barley, and wheat, but probably populations never are high enough to take much of a toll on row crops. Over half of stomachs examined in a study in Idaho had remains of insects, bee-tles, weevils, and larval moths and spiders. Food storage has not been reported.

Western harvest mice are active the year around. Fat content is about 20 percent of dry body weight, depleted in summer and accumulated during autumn to a maximum in winter. These tiny animals have un-favorable ratios of radiative surface to vol-ume, and one mechanism to conserve heat is communal huddling. They are as tolerant of each other as any mammals known. Not only can individuals be housed together, but females with newborn young tolerate males in the nest, and strangers introduced

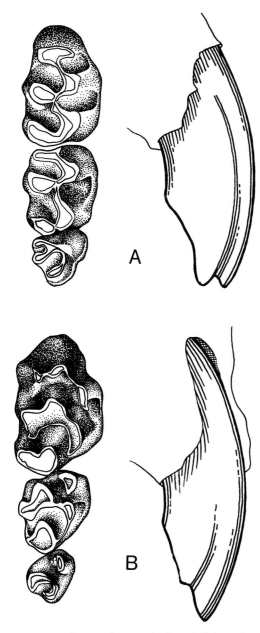

108. *Ventral view of upper cheekteeth* (left) *and lateral view of upper incisors* (right) *of* (A) Reithrodontomys megalotis *and* (B) Mus musculus.

huddling further lowered the rate by 28 percent. (Mice huddling in a ball-shaped bundle have a more favorable surface-to-volume ratio.) The habits of these animals thus can be seen as part and parcel of their small size. They are nocturnal to avoid overheating (and predation, perhaps), but activity at cooler nighttime temperatures also expends extra energy. In harvest mice in California, the cost of nocturnality was estimated to be about three and a half grains of wheat daily.

Harvest mice are more leisurely in their activities than most mammals of their size, spending much of their time in the nest. Activity is confined to the period from just after sundown to just before sunrise. There is some evidence that this pattern is innate, because it persists in well-lighted laboratories. Homing has been recorded from distances up to 1,000 feet, suggesting a maximum area with which an individual is familiar. Movements of 200 to 300 feet have been recorded across a live-trapping grid.

Reproduction takes place throughout the warmer months. Litter size usually is about three to seven (mean about five), though fetus counts from one to nine have been recorded. Laboratory-reared females are remarkably fecund. One gave birth to 14 litters, producing 57 young (mean litter size 4.1, range two to five) over an 11-month period; another produced 58 young in 14 litters (mean 4.1, range two to six) in 12 months. Such fecundity would not be expected in the wild, but obviously the combination of a long breeding season, a three-

109. *Western harvest mouse,* Reithrodontomys megalotis *(courtesy T. H. Kunz).*

to this blissful commune are readily accepted. Huddling with house mice and deer mice has been reported in laboratory situations.

In trials at 1° C, the nest contributed to lowering metabolic rate by 24 percent, and week gestation period, postpartum estrus, and moderately large litter size contributes to a high biotic potential. In addition, harvest mice have a high reproductive efficiency, with the weight of litters averaging more than 50 percent of the weight of the female.

Birth weight of the altricial young is about 1.5 grams, total length about 33. During days one to three postpartum, there is weight gain with little outward morphological change, except that the dorsum darkens as the pelage begins to develop. During the period of days four to 12, weight gain slows, but the body and extremities elongate. The ears unfold and the pinnae develop, the toes separate and become functional, the grayish pelage emerges, the incisors erupt, and the eyes and ears open. Solid food is taken at about two weeks of age, and the young are weaned when three weeks old. The sex ratio at birth is about 53 to 47, biased in favor of males. Young females bred in the laboratory as early as four weeks of age, males at about five and a half weeks.

As in cricetines in general, there are two maturational molts. The juvenile pelage is grayish brown and rather sparse. At about the time the young begin to leave the nest, there is a postjuvenile molt, and a subadult pelage appears. Although more brownish than the juvenile pelage, it is less bright than the adult pelage, and the dark hairs of the dorsum are less pronounced. Seasonal molt in adults occurs in late spring and early summer (later for reproductive females than for males) to summer pelage, and in autumn to the denser, longer, somewhat paler winter pelage.

Predation by various avian and mammalian carnivores probably is the principal check on populations. Owls, hawks, jays, weasels, skunks, badgers, coyotes, and foxes prey on *R. megalotis.* Communal nesting leads to a rather high ectoparasite load, although less heavy than in the deer mouse. Ectoparasites recorded on specimens from the Black Hills are ticks, mites, chiggers, fleas, and a louse.

Selected References. Natural history (Fleharty, Krause, and Stinnett 1973; John-

son 1961; Pearson 1960); systematics (Jones and Mursaloğlu 1961).

Reithrodontomys montanus: Plains Harvest Mouse

Name. The specific name *montanus* is a Latin word referring to "mountain," but it is a misnomer. The animals occur near the Southern Rockies but never in mountainous situations.

Distribution. The plains harvest mouse is mostly an animal of the Central and Southern Great Plains, occurring from the Missouri River westward to the Rockies and from western South Dakota south to northern Mexico. Two subspecies are recognized on the Northern Great Plains, *Reithrodontomys montanus griseus* in southeastern Nebraska and *Reithrodontomys montanus albescens* elsewhere.

Description. This is the smallest of the cricetine rodents on the Northern Plains. Mean and extreme external measurements

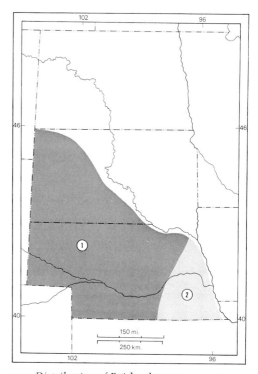

110. *Distribution of* Reithrodontomys montanus: (1) R. m. albescens; (2) R. m. griseus.

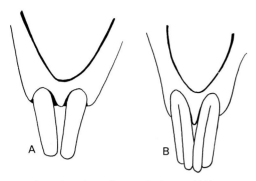

111. Anterior view of upper incisors of (A) Peromyscus *and* (B) Reithrodontomys.

of eight specimens from Cherry County, Nebraska, are: total length, 124.5 (113–134); length of tail, 54.7 (48–60); length of hind foot, 16.5 (16–17); length of ear, 12.5 (12–13). Weights range from about 10 to 12 grams. Mean and extreme cranial measurements of six animals from central Nebraska include: greatest length of skull, 19.4 (19.0–19.8); zygomatic breadth, 10.6 (10.1–10.8).

The dorsum is grayish brown, with a well-defined dark medial stripe. The dorsal stripe of the bicolored tail is narrow, the width being less than one-fourth the diameter of the tail. The underparts are white. The eastern subspecies, *R. m. griseus*, is darker (more brownish and less grayish) than *R. m. albescens* to the west.

In addition to the characters mentioned in the key, this small mouse differs from the sympatric western harvest mouse as follows: smaller external and cranial size; paler dorsal and lateral color overall, but with a distinct blackish middorsal patch; narrower dorsal stripe on tail; buffy patch behind ear; relatively shorter rostrum; margins of braincase more nearly vertical.

Natural History. The plains harvest mouse is a species of well-developed grasslands, usually on uplands. Typically there is less than 50 percent bare soil in the habitat. Weedy situations, including patches of pricklypear, are occupied, and the species occasionally is found in tall-grass prairie in the eastern parts of its range. Often it is broadly sympatric with *R. megalotis*, the

western harvest mouse, but the two species exhibit different habitat preferences, the western harvest mouse usually occurring in areas of somewhat greater cover, including shrubs. Habitat segregation is similar to that seen in the two species of *Peromyscus* in the tristate region, with *R. montanus* being the analogue of *P. maniculatus* and *R. megalotis* the analogue of *P. leucopus*. Typical mammalian associates are *Perognathus hispidus, Spermophilus tridecemlineatus, Cynomys ludovicianus, Onychomys leucogaster,* and *P. maniculatus*. Often plains harvest mice have been reported from stony pasturelands, where they nest beneath rocks. The food probably consists mostly of seeds.

These mice seldom seem to be abundant. Home ranges of about one-half acre have been reported, with individual movements of more than 500 feet across a live-trapping grid.

Litter sizes of two to five young (mean about three) have been reported. In Oklahoma the breeding season is from February to November. Gestation takes 21 or 22 days. The young are highly altricial, weighing 1.0 to 1.3 grams at birth. They are pink, hairless, and helpless, and the ears, eyes, and anus all are closed. They kick feebly within hours of birth; in a day or so they become more active, and the dorsum begins to darken. The mother carries the young about in her mouth, grasping them by the nape, back, or belly. Birth weight has doubled by day six, tripled by day 10, and quadrupled by the time of weaning,

112. Plains harvest mouse, Reithrodontomys montanus *(courtesy J. L. Tveten).*

about day 14. The lower incisors erupt at day six, and at about day eight the eyes and ears open. A captive female gave birth to her first litter at about 12 weeks of age.

Harvest mice are remarkably tolerant of each other. Males and females caged together show no obvious agonistic behavior, despite the presence of young.

Selected References. Natural history (Goertz 1963; Hibbard 1938; Hill and Hibbard 1943; Leraas 1938).

Genus *Peromyscus*

By almost any criterion, the genus *Peromyscus* is one of the most successful groups of New World mammals. These mice are known as fossils as early as the late Pliocene, and since then they have diversified into nearly every terrestrial habitat, from tropical forests to alpine tundra. Whatever area one studies in temperate and subtropical North America, some species of *Peromyscus* is likely to be the most abundant mammal. Nearly 50 species are recognized today (several of which are insular taxa), and they range from Panama to tree line in the Canadian arctic. Mexico is the center of diversity in the genus.

The name *Peromyscus* is derived from two Greek roots, *pero,* "defective," and *myskos,* "mouse," certainly a misnomer, because these animals are the quintessence of "mouseness." The dental formula is 1/1, 0/0, 0/0, 3/3, total 16.

Peromyscus leucopus: White-footed Mouse

Name. The specific name *leucopus* is Greek for "white-footed," a name that is descriptive but hardly diagnostic, because nearly all members of the genus are white-footed. Wood mouse is another vernacular name sometimes used in reference to this species.

Distribution. The white-footed mouse occurs from Nova Scotia and Alberta southward to the Yucatan Peninsula. Toward the western limits of its range, the species has

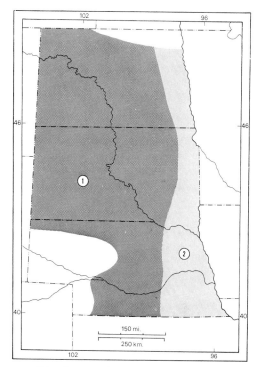

113. Distribution of Peromyscus leucopus: (1) P. l. aridulus; (2) P. l. noveboracensis.

an increasingly dendritic distribution, following deciduous woodlands along watercourses. Isolated populations occurring on wooded uplands—the Killdeer Mountains in North Dakota, the Slim Buttes and Black Hills in South Dakota, and the Pine Ridge of northwestern Nebraska—probably are relicts of the warmer and wetter postglacial Climatic Optimum. Two subspecies are recognized on the Northern Great Plains, *Peromyscus leucopus noveboracensis* in the east and *Peromyscus leucopus aridulus* in the central and western parts of the region.

Description. White-footed mice sometimes are difficult to distinguish from deer mice, unless one has at hand material that allows comparison of animals of similar age. Such comparisons show *P. leucopus* in the tristate region to be larger overall, with an absolutely and relatively longer tail; generally duller, more grayish dorsally; and with a tail less sharply bicolored (except occasionally in winter-taken specimens).

Average and extreme external measurements of 25 specimens (16 males, nine females) from the vicinity of Valentine, Nebraska, are: total length, 173.2 (161–186); length of tail, 74.8 (68–81); length of hind foot, 22.2 (21–23); length of ear, 15.8 (15–17); weights of eight animals averaged 30.5 (27.5–35.6) grams. Cranial measurements of 23 animals from the same area include: greatest length of skull, 27.3 (26.2–28.3); zygomatic breadth, 14.5 (13.7–15.2).

The animals are buffy to dark grayish brown above, paling along the sides, and white on the venter and feet. The tail is indistinctly bicolored, being brownish above and white below. Specimens from the western part of our region (P. l. aridulus) are distinctly paler than those from the east (P. l. noveboracensis).

Natural History. On the Northern Great Plains, white-footed mice almost invariably are found in woodlands or in brushy areas such as shelterbelts and fencerows (whereas the closely related deer mouse inhabits more open country). The only usual exception to this rule is dispersing young, which are rather commonly taken in open country between patches of woody vegetation. In the western part of its range, the white-footed mouse occurs mostly in riparian woodland, and its distribution thus is dendritic, following major watercourses. That the species does not follow rivers such as the Platte farther west suggests that riparian woodland was not so nearly continuous in the recent past as it is at present. Studies of microhabitat indicate that patches of habitat within a tract of woodland are used in proportion to their shrub cover. The animals function well in the three-dimensional environment of thickets, climbing readily both to forage and to escape, and they occasionally climb high into trees. Home ranges vary considerably, in part related to habitat and population density, but rarely exceed an acre. Ranges of males usually are larger than those of females. There evidently is a well-developed sense of "homing," because individuals have returned to their original home ranges after having been displaced up to a mile away.

A nest seven inches or so in diameter, made of plant fibers, the "cotton" of tree and weed seeds, leaves, and shredded grasses, is built beneath a stump, fallen log, or rock, or sometimes in the hollow of a log or tree or even in the abandoned nest of

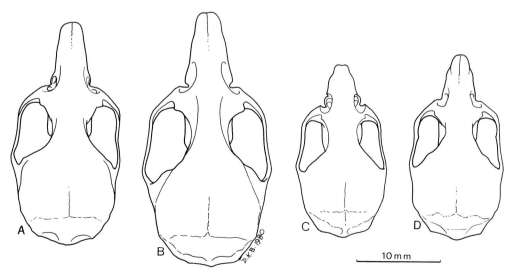

114. Dorsal view of crania of four species of cricetine rodents found on the Northern Great Plains: (A) Peromyscus maniculatus; (B) P. leucopus; (C) Reithrodontomys montanus; (D) R. megalotis.

115. White-footed mouse, Peromyscus leucopus *(courtesy T. H. Kunz).*

a squirrel or bird. About three-quarters of the time in winter is spent in the nest, and about one-fourth of the time in summer is spent there.

The daily energy budget for nonlactating animals is about 0.602 kilocalories per gram per day in winter, and roughly half that amount in summer; lactation increases this budget by about 25 percent. Foods consist mainly of berries and seeds but include a wide variety of other items such as fruits, nuts, insects and other invertebrates, and even occasionally small vertebrates.

On the Northern Great Plains, reproduction is restricted mostly to the warmer months, with peaks in March–April and September–October. In the heat of midsummer reproduction slows down, and testes of many males regress; in winter reproduction ceases. Females are polyestrous, a single cycle being five days long. Postpartum estrus occurs. Gestation in nonlactating females takes 22 to 23 days, but in lactating females it may take as long as 28 days, the lengthening being due, perhaps, to delayed implantation. The highly altricial, pinkish young are delivered while the female is in a sitting position; she assists the birth with her forepaws and bites through the umbilical cord. The young arrive at intervals of about 10 minutes. Mean litter size is about 4.3, with a known range of one to seven.

Neonates have been recorded as weighing between 1.4 and 2.4 grams; the total length of one measured 47, the tail 12, and the hind foot 7. Sparsely distributed hairs are visible on the venter by the second day of life, and dorsal hairs are evident by day five or six. The juvenile pelage, which is grayish, is complete by the third week. The ears open between days nine and 13, and the eyes, on the average, on day 13 or 14.

As in *P. maniculatus,* weaning usually takes place between three and four weeks of age. Females may reach sexual maturity as early as 28 days, but normally it is not attained until they are about seven weeks old; males mature somewhat later.

Virtually all carnivorous animals take a toll of white-footed mice. They are a staple for woodland snakes and owls and an important resource for weasels, skunks, bobcats, and foxes. *P. leucopus* harbors fleas, mites, ticks, botfly larvae, or warbles

(*Cuterebra*), and a variety of endoparasites. The botfly larvae are large enough to interfere with locomotion and may lower the ability of the rodents to escape predation. These warbles usually are found in the abdominal wall, and usually a mouse hosts a single larva (although infestations with two or three larvae have been observed). Warbles may infect a third or so of the individuals in a given population.

As in other small rodents, mortality among juveniles is high, and of animals that reach adulthood only a small percentage survive into a second year of life. One individual is known to have lived 22 months in the wild. Males have outnumbered females in most studies thus far conducted of free-living populations, averaging about 55 percent of the mice captured.

As in the deer mouse, there are two maturational molts, one from juvenile pelage to subadult pelage and another from subadult to adult pelage. Adults probably molt twice annually (although this is not certainly known), in spring or early summer and again in autumn, the winter coat being somewhat paler and noticeably longer and more luxuriant than the summer coat.

Selected References. Natural history (Baar and Fleharty 1976; Miller and Getz 1969; Svendsen 1964); also see references in account of *Peromyscus maniculatus*.

Peromyscus maniculatus: Deer Mouse

Name. The specific epithet *maniculatus* is a diminutive adjective derived from the Latin word for "hand."

Distribution. The deer mouse is the most widespread of our cricetine rodents, and in many places over its broad range it is the most abundant species as well. From the tree line in northern Canada southward to the Isthmus of Tehuantepec and from the Pacific Coast to New England, the animals occupy all but the wettest habitats. In the United States, only the Southeast has no deer mice; a closely related species, the oldfield mouse, replaces it there.

Three subspecies of *P. maniculatus* are recognized on the Northern Great Plains:

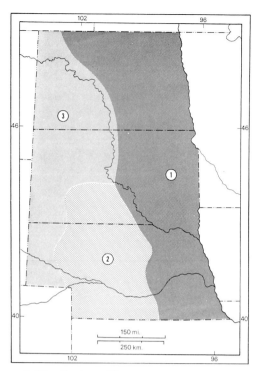

116. Distribution of Peromyscus maniculatus: (1) P. m. bairdii; (2) P. m. luteus; (3) P. m. nebrascensis.

Peromyscus maniculatus bairdii in the eastern part of the region, *Peromyscus maniculatus luteus* on the Nebraska Sand Hills and in adjacent areas, and *Peromyscus maniculatus nebrascensis* to the west.

Description. This is a beautiful little mouse, varying in dorsal color geographically from nearly black in the eastern part of the tristate region, through yellowish buff, to grayish brown in the west. The venter and limbs are white, and the tail usually is sharply bicolored, white below and laterally and dark brown to black above.

Geographic variation in *P. maniculatus* is pronounced. *P. m. nebrascensis* is larger than either *P. m. bairdii* or *P. m. luteus. P. m. bairdii* is decidedly darker than either *P. m. nebrascensis* or *P. m. luteus.* For comparison with *P. leucopus,* see the account of that species.

Average and extreme external measurements of 30 specimens of *P. m. nebrascen-*

sis from Sioux County, Nebraska, 30 specimens of *P. m. luteus* from Cherry County, Nebraska, and 17 specimens of *P. m. bairdii* from Richardson County, Nebraska, are, respectively: total length, 160.7 (151–177), 147.9 (130–164), 147.5 (125–164); length of tail, 63.6 (52–77), 59.6 (50–72), 58.5 (43–66); length of hind foot, 19.9 (18–21), 19.3 (18–20), 18.7 (16–20); length of ear, 16.5 (15–19), 14.5 (13–15), 14.0 (13–15). Weights of mice from those samples were: 25.0 (19.8–27.9), 21.1 (18.7–24.9), and 21.4 (20.0–24.0) grams. Mean and extreme cranial measurements of those series include: greatest length of skull, 25.6 (24.3–26.5), 24.5 (23.3–25.5), 24.4 (23.1–24.8); zygomatic breadth, 13.6 (13.0–14.3), 13.3 (12.6–13.7), 13.1 (12.5–13.7).

Natural History. P. maniculatus occurs in a greater variety of habitats on the Northern Great Plains than any other species of mammal. It is unlikely that there is a square mile of rural upland in the three-state area that does not support at least a few deer mice. Open grasslands, brushy country, badlands, cliffs, coniferous woodlands, hedgerows, shelterbelts, pasturelands, and croplands all provide suitable habitat. Deer mice usually are not seen in wetlands, and they are rare or absent from woodlands and riparian situations within the range of *Peromyscus leucopus*. However, when *P. leucopus* is absent, *P. maniculatus* readily inhabits wooded areas. In isolated areas where *Mus musculus* is absent, deer mice will set up as commensals of man, raiding stored provisions and nesting in furniture, clothing, or haystacks.

117. *Deer mouse,* Peromyscus maniculatus *(courtesy D. C. Lovell and R. S. Mellott).*

Deer mice live in burrows, sometimes of their own making, perhaps more often borrowed from other animals. They also den in eroded root channels, soil cracks, or beneath rocks, stumps, fallen logs, or debris from human settlements. A cup- or ball-shaped nest is built of finely shredded plant materials, including dry grass and shredded bark, and fur—their own or that shed by other animals. The nest is used for sleeping during the daylight hours, protection against the winter cold, and rearing the young. Some food is stored in the burrow, especially in winter.

Deer mice are polyestrous, and in the laboratory they breed throughout the year. In the wild, breeding is concentrated in spring and early autumn, with a drop in activity in mid-summer (although pregnant or lactating females have been taken in every month of the year). The gestation period is about 23 days for nonlactating females, slightly longer for nursing mothers. Litter size averages about four, with a range of one to nine. There seems to be seasonal variation in litter size at a given locality, although no widespread pattern is discernible. On the Black Hills, largest litters (mean 5.2, range four to seven) were in late spring, smallest (4.3, range three to six) in early autumn. In southwestern North Dakota, greatest numbers of placental scars were counted in mid-June (6.8) and early August (6.7), with a range of four to eight in each period; scars were fewer in late June and July; embryo counts there were highest in August (mean 5.9, range three to seven). There is a postpartum estrus.

The female assumes a bipedal sitting position while giving birth, assisting the process with paws and snout. She snips the umbilical cord with her incisors and eats the placenta. She grooms the young with her teeth and tongue during and immediately after birth. They arrive at intervals of several minutes, and between deliveries there is much stretching of the mother's body. Parturition usually takes place during daylight hours. When disturbed, the female may kill and eat the young, but she may tolerate the male in the nest during parturition.

Birth weights average 1 to 2 grams, about

half the weight of a penny. The neonates are highly altricial, being pinkish, naked, and blind (the eyelids not even thickened); the pinnae are folded and fused to the head, and the digits, although visibly clawed, are fused. Suckling is almost continuous. Pigmentation begins to develop within 24 hours. The pelage first arises dorsoanteriorly. The gray juvenile pelage at first seems fluffy and luxuriant, but as the animals grow it becomes sparser. The pinnae elevate on about the third day, and the eyes open at about two weeks. The incisors appear (lowers first) at about day five, and the yellowish color of the faces of the incisors develops at the time of weaning, at about four weeks of age. The cheekteeth also emerge at that time. The young may stay with the female after weaning, especially in autumn, when the mother ceases to reproduce.

The average age at first estrus is about seven weeks, although females as young as five weeks have been reported to conceive young. Males become sexually mature at about eight weeks of age. With this early maturation, young of spring litters enter the reproductive pool by autumn and hence contribute their own young to the overwintering population.

Deer mice undergo two developmental molts, postjuvenile and postsubadult. The juvenile pelage is gray. At 30 to 45 days of age, the animals commence a postjuvenile molt to a brownish subadult pelage, which is completed by about day 85. The molt to rich, brownish adult pelage takes place at about 15 weeks of age. Thereafter there is a seasonal molt, in spring or early summer, that begins on the head and moves tailward, and apparently another in the autumn to the longer, paler winter pelage.

Population density varies widely from place to place, from year to year, and from season to season within a given year. Highest densities usually are seen in autumn, when the season's recruits are in the trappable population and females of spring litters have produced their own young. Estimates of 6.8 mice per hectare (about three per acre) were derived from live trapping on moderately grazed pasture in south-western North Dakota. This is a rather average population estimate, although local variation on either side of the mean by a factor of nearly 10 may be expected. Home ranges vary with habitat, density, sex, and age, from 0.1 acre to 10 acres. Distance between points of capture on live-trapping grids ranged from 15 to about 190 meters (50 to 625 feet). Shortest movements are by adult females, longest by dispersing young.

Deer mice are omnivorous. Food varies with seasonal and local availability. The dental adaptation is to a diet of seeds, but 20 to 60 percent of the summer diet may be insects: adult and larval beetles, caterpillars, grasshoppers, and leafhoppers. Seeds, especially those of weedy species, provide the bulk of the winter diet and are important year round. Green plant material, such as leaves and stems, furnishes a small proportion of the diet (usually less than 10 percent), but this may be an important source of moisture. Deer mice drink readily when water is available, and they lap dew from plants, but they can survive quite well on the moisture present in their varied diet.

Deer mice are active throughout the year. Seeds are stored against lean times, as is body fat. The fat content of wintering animals ranges to about 20 percent of dry weight. Sometimes tracks of deer mice are seen on the snow, but wherever possible winter activity is confined beneath the insulating blanket of the snowpack. There is some suggestion that these mice become torpid in the coldest weather, for there are times that they seem not to enter traps, even those set beneath the snow. Torpor can be induced in the laboratory under conditions of starvation.

Being abundant and nearly ubiquitous, deer mice are a rather predictable prey item for a variety of predators. Foxes, weasels, skunks, badgers, coyotes, bobcats, and grasshopper mice all prey on them, as primary or secondary sources of food depending on the size of the predator, the number of predators, and the condition of other prey populations. The two species of *Peromyscus* accounted for 44.8 percent of the contents of owl pellets in a study in

Nebraska; locally this toll could be even higher, especially for small owls foraging over uplands. Nocturnal habits protect deer mice to some extent from diurnal avian predators, but hawks and shrikes still exact a toll. Snakes can follow deer mice into their nests, and even reptiles as small as garter snakes prey on these rodents, especially nestlings. Weather is another check on populations. Breeding begins before the spring thaw and the spring rains; thus, local flooding often kills early litters in their nests.

Numerous parasites afflict deer mice, including ticks, mites, lice, fleas, and botflies. Scabies (a scab-producing mange mite) affects many individuals but seems seldom to be fatal.

Selected References. Natural history (King 1968 and included citations); systematics (Jones 1964).

Genus *Onychomys*

The genus *Onychomys* includes three species, restricted to western North America. The fossil record extends back only to the late Pliocene. The name *Onychomys* is a compound of the Greek *onycho*, "clawed," and *mys*, "mouse."

Onychomys leucogaster: Northern Grasshopper Mouse

Name. The specific name *leucogaster* is derived from the Greek *leuco*, "white," and *gaster*, "belly."

Distribution. The northern grasshopper mouse ranges across the Great Plains and the Great Basin, from Saskatchewan to northeastern Mexico and from northwestern Iowa to northeastern California. Four nominal subspecies have been recognized on the Northern Plains: *Onychomys leucogaster arcticeps* in the southwest, *Onychomys leucogaster breviauritus* in the southeast, *Onychomys leucogaster leucogaster* in the northeast, and *Onychomys leucogaster missouriensis* in the northwest.

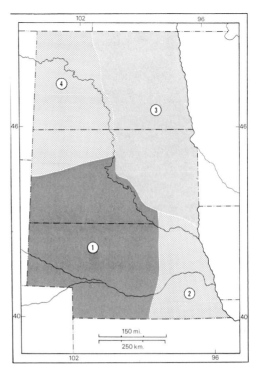

118. Distribution of Onychomys leucogaster: (1) O. l. arcticeps; (2) O. l. breviauritus; (3) O. l. leucogaster; (4) O. l. missouriensis.

Description. This heavy-bodied, short-tailed mouse is of distinctive appearance. The dorsal pelage is buffy to reddish brown in adults, graying in old age to become again the color of subadults. The venter is pure white. The short tail is grayish to buffy above basally, white below and distally. The claws are notably long.

Mean and extreme external measurements of 13 individuals from western Nebraska are: total length, 140.5 (135–146); length of tail, 39.8 (37–44); length of hind foot, 20.8 (20–21); length of ear, 15.3 (14–17); weights of adults average about 40 grams. Measurements of greatest length of skull and zygomatic breadth of adults from southwestern Nebraska averaged 28.0 (27.5–28.9) and 14.8 (14.4–15.6), respectively. Mice of more northerly populations are somewhat larger. External measurements of seven individuals from North Dakota are: total length, 157.0 (141–168); length of tail, 43.1 (38–47); length of hind foot, 21.7 (21–22); length of ear, 14.1 (13–15).

As noted above, four named subspecies of *O. leucogaster* have been recognized on the Northern Great Plains. The names *O. l. leucogaster* and *O. l. breviauritus* were applied to dark-colored mice of the tall-grass prairie of the eastern part of the region, whereas *O. l. missouriensis* and *O. l. arcticeps* were applied to paler races of the more arid country to the west. The northern populations (*O. l. missouriensis* and *O. l. leucogaster*) average slightly larger in size than do grasshopper mice from the area to the south. A recent study of geographic variation in this mouse from the Central Plains showed that variation in color is broadly clinal; as a result, *O. l. breviauritus* was regarded as a synonym of *O. l. arcticeps*. A thorough modern study is needed of geographic variation in this species throughout its broad range.

Natural History. The northern grasshopper mouse is something of a curiosity among local rodents, because it is adapted in anatomy and behavior to a carnivorous existence. It is hard to know whether a more apt analogy is with the shrews or the carnivores; it has convergent adaptations with both, as study of its natural history suggests.

These are animals of semiarid grasslands and shrublands. A prime characteristic of habitat is sandy or silty soils for these mice seem to be obligate dust-bathers. They may burrow actively, or they may usurp the burrows of other small rodents. A common habitat in modern times is fencerows, which tend to support high numbers of their preferred insect prey and also provide the cover of eroded areas around fenceposts.

Grasshopper mice are not strictly carnivorous, but a greater portion of the diet is animal matter than in other rodents. A thorough study in northeastern Colorado found that animal matter constituted over three-quarters of the diet (by volume) as a year-round average, ranging from above 80 percent in spring and early summer to a low of about 60 percent in midwinter. The most common food items were adult and larval beetles, grasshoppers, caterpillars, and spiders. Elsewhere, crickets, scorpions, flies, and true bugs are reported in the diet. Ants seem not to be taken in any great numbers. If given a number of grasshoppers, a captive mouse will decapitate them all and then eat them at leisure, leaving behind a neat pile of heads and legs. In addition to arthropods, vertebrates are eaten. They sometimes take flesh as carrion, but grasshopper mice are active predators, quite capable of killing rodents the size of cotton rats (which may be three times their size). Cannibalism is not rare; males kill and eat intruders in their territories. Rodents are almost invariably killed by a bite at the base of the skull. The incisors are well suited to this task, being long and sharp. The claws are long and curved and serve to grasp the prey firmly. The mandible has a notably high coronoid process, which serves for attachment of the powerful jaw musculature. Internally, the large intestine is shorter than is typical of rodents, again an adaptation to a diet of meat, for proteins are digested mostly in the small intestine.

Burrows are of three types and usually are excavated in open areas, well away from cover. The U-shaped nest burrow is about one to two inches in diameter and descends some six inches below ground to an enlarged, cylindrical nest chamber. The tunnel then rises to an exit, which usually is plugged. The male excavates the nest burrow (the female assisting by moving soil to the surface), using his forefeet to accu-

119. Northern grasshopper mouse, Onychomys leucogaster *(courtesy T. H. Kunz).*

mulate a pile of soil, which he then kicks back with his hind feet. The nest burrow is the center of activity for a mated pair and is used for resting, retreat, some feeding, and rearing the young. Solitary mice live more simply, beneath rocks or clumps of vegetation.

Retreat burrows are used for cover when foraging away from the nest burrow. Short tunnels also are dug as food caches. These are plugged. Short (two-inch-long) latrine tunnels are used for defecation. About the perimeter of the territory are sand-bathing points, which seem to retain scent from anal glands, thus posting the territory against intruders.

Sand bathing is a characteristic behavior. The animal places the side of its head against the soil, pushing with its hind feet while rolling over on its back and neck. The pattern is established in the young by the age of 14 days. Other forms of grooming help to keep the animals impeccably clean. Self-grooming involves scrubbing the face and head with the forepaws and nibbling at the fur of the belly and limbs. The face is cleaned frequently while feeding. The mother grooms the young about the neck and back, and mated pairs engage in mutual grooming, which seems to reinforce the pair bond and has a courtship role.

These are rather noisy mice, with vocalizations of several kinds. A "squeak" is given by neonates and sometimes by adults when fighting. A "chirp" serves as an alarm note. High, piercing notes (which may be double or single) about a second in duration seem to aid in intraspecific localization. Also, they reveal the presence of these mice to the nocturnal naturalist. Some have likened this call to that of the coyote or wolf.

As is true of many carnivorous mammals, courtship is rather complex, involving mutual chasing, nasonasal and nasoanal contact, grooming, and somersaults. Odor seems to be a cue to sex recognition. Nesting material used by an estrous female stimulates the male to initiate courtship.

In the laboratory, grasshopper mice will breed throughout the year, with a peak from late spring to early autumn. In the wild, reproduction is limited to the warmer months. Both females and males are capable of reproduction at about four months of age, but in the wild they seem not to breed during the summer of their birth. Gestation is about four weeks for nonlactating females and about five weeks for those that are nursing. There is a postpartum estrus, and three to six (maximum 10) litters per year have been reported in the laboratory, where females have a two- or three-year reproductive lifetime. Number of young per litter ranges from one to six, with a mode of four and a mean of about 3.5. There are six mammae.

The young are born blind, pink, and hairless, but during the first postnatal day the dorsum begins to darken. The mystacial vibrissae are white and silky at birth, and the digits are fused and clawless. Mean birth weight is about 2.8 grams. By day four the pinnae are free of the head, and by day six the fur becomes apparent on the back. Claws form by day seven, and the velvety juvenile pelage completely covers the young on day eight. Lower and upper incisors erupt on days nine and 11, respectively. The eyes open about day 18, and the young are weaned in about three weeks. The male helps his mate rear the young, an unusual behavioral pattern in rodents. Both parents bring parts of insects to the nestlings, but the mother does most of the grooming. The young play like young puppies, fighting, digging, and mounting each other.

Juvenile grasshopper mice have the pale gray, sparse pelage typical of young cricetines. The venter is white. At about nine weeks of age they molt to a dense, grayish brown subadult pelage. This pelage is replaced, usually in autumn or winter, by a more richly colored buffy to cinnamon pelage. Old animals—those that have survived to a second autumn—assume a grayish brown pelage similar in color to that of subadults.

Grasshopper mice seem to be highly territorial. Active agonism establishes a dominant-subordinant relationship among individuals, after which nonviolent encounters predominate. During active agonism (and in hunting), the aggressor pursues its quarry, pounces on it, and holds on with the forefeet. In the defensive posture (which denotes submission), the mouse stands on its hind feet with its back straight and its forefeet raised against the chest. The mouse that fails to assume this posture will be killed by a bite to the base of the skull. Strange individuals that wander onto the territory of a pair of grasshopper mice may be summarily dispatched by the male. The boundaries of the territory are marked clearly with scent posts, and violators are not tolerated. Size of territories in nature is unknown, but home ranges are large for rodents of this size, about six acres. What controls populations of grasshopper mice is not known in detail, although surely their active and sometimes deadly territorial defense is a factor. Populations seem never to be large; when several grasshopper mice are taken in an area, juveniles invariably predominate in the sample. These mice fall prey to owls and other nocturnal carnivores, but they seldom are common in scats or castings. No predator can afford to make a staple of another predator, because the latter are too far up the food chain. Efficiencies along the food chain average about 10 percent. Hence it takes 10 pounds of grass to make one pound of grasshoppers to make a tenth of a pound of grasshopper mouse (to make a hundredth of a pound of owl). Obviously, the owl makes an easier living by eating the herbivores directly.

Among ectoparasites reported from *O. leucogaster* are mites, lice, fleas, and ticks. Ectoparasite loads generally are high, as is typical of meat-eating mammals with permanent nest burrows. Endoparasites include protozoans, spiny-headed worms, tapeworms, flukes, and roundworms.

Selected References. General (McCarty 1978 and included citations); systematics (Engstrom and Choate 1979).

Genus *Sigmodon*

The genus *Sigmodon* is known in the fossil record as early as the late Pliocene and includes about eight modern species, most of which occur in tropical America. The name *Sigmodon* is a compound of two Greek words, *sigma*, "S-shaped," and *odon*, "tooth," a reference to the pattern of the occlusal surface of the cheekteeth. The dental formula, as in other cricetines, is 1/1, 0/0, 0/0, 3/3, total 16.

Sigmodon hispidus: Hispid Cotton Rat

Name. The specific name *hispidus* is of Latin origin, meaning "bristly," and like the vernacular name it refers to the harsh texture of the dorsal pelage.

Distribution. This cotton rat is a tropical and subtropical animal that has only recently inhabited the Northern Great Plains. It occurs across the southern United States, from the Colorado River Valley in California eastward to Florida. Northernmost records are in Nebraska, Iowa, and Virginia. A single subspecies, *Sigmodon hispidus texianus*, occurs in our region.

The northward expansion of the range of the hispid cotton rat in recent decades has

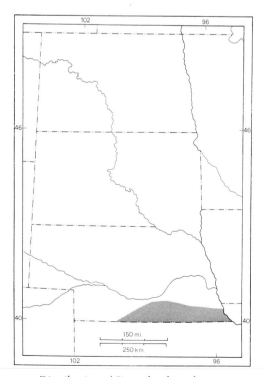

120. Distribution of Sigmodon hispidus.

been monitored rather carefully. Between 1933 and 1947 the animals moved northward through Kansas at about seven miles per year. They first were captured in Nebraska in 1958, in the extreme southeastern corner of that state. By 1965 they were known from as far north as Adams County, and they first were taken in the vicinity of Kearney in 1966. This species occurs north of the Missouri River in northwestern Missouri and adjacent southwestern Iowa, so there is no reason to suspect that the rivers of the Northern Plains will limit distribution. Rather, the northern limits of the range seem to vary with the severity of winter, especially the prevalence of freezing rain and extended periods of subfreezing temperatures. There is some evidence that northern "pioneers" are altering their habits to occupy burrows well below the frost line.

Description. Cotton rats superficially resemble large, long-tailed voles. They are smaller than Norway rats and woodrats and have a shorter tail. The long, coarse, almost spiny, dorsal pelage is distinctive. The tail is sparsely haired, and the ears are nearly concealed by the long hair of the head. The dorsal pelage is grizzled yellowish brown, frequently with a "salt and pepper" appearance. The ventral pelage is grayish, sometimes washed with buff. Albino animals have been reported.

External and cranial measurements (extremes in parentheses) of three males and 12 females, respectively, from central Kansas are: total length, 239.6 (225–265), 251.5 (202–277); length of tail, 102.3 (95–111), 102.5 (87–110); length of hind foot, 30.5 (29.5–32.0), 31.6 (29–35); length of ear, 17.0 (16–18), 17.5 (16–20); greatest length of skull, 34.6 (31.0–38.6), 31.7 (29.1–33.2); zygomatic breadth, 19.3 (17.5–21.0), 18.3 (16.5–19.8). Adults generally weigh between 50 and 150 grams—an extreme that probably includes pregnant females.

Natural History. Hispid cotton rats occupy a wide variety of relatively mesic habitats: thickets, woodland borders, riparian ecosystems such as sloughgrass and cattails, as well as weedy margins of fields and moist

pastures—especially stands of sunflowers and summer cypress—and tall-grass stands in roadside ditches. A common denominator of suitable habitat is nearly complete ground cover, although marginal areas are occupied when populations are high.

The animals clip sizable quantities of vegetation, especially in autumn and winter, which they pile at intervals along a maze of runways (broader than those of *Microtus*, and often through nongrassy vegetation). Piled clippings amounted to about a half of one percent of net primary production on a study area in western Kansas. Most of this material never is eaten. Leaves and stems form the bulk of the diet, which is similar to that of the sympatric prairie vole (*Microtus ochrogaster*).

The diet is somewhat selective. A thorough study in western Kansas showed that less than half of the available plant species actually was eaten. Species, stage of growth, palatability, and time of year all seem to influence choice of food. Summer cypress and wheat proved to be staples in Kansas, but dropseed, sweet clover, switchgrass, brome, goldenrod, wild rose, shepherd's purse, dock, grama, and dandelions also were eaten. Seeds, especially of wheat, were common in the diet, and invertebrates were eaten in small quantities. Some vertebrate flesh is eaten, and cotton rats may prey actively on young voles. (Indeed, such predation, rather than competition, may be the basis for suppression of prairie vole populations in areas where numbers of cotton rats are high.) Cotton rats also eat eggs of ground-nesting birds, and they have been implicated as an important influence on the abundance of quail in the Southeast. The nutritional efficiency is relatively high in that about two-thirds of calories ingested are used.

Cotton rats are mostly crepuscular and nocturnal, although the pattern of activity is highly variable. Light and high temperatures suppress activity in the laboratory, and food shortages and sexual behavior shift activity to daylight hours. Young animals are essentially arhythmic.

Generally cotton rats nest in cup-shaped hollows beneath logs, rocks, or debris. Sometimes they den in shallow burrows of

121. Cotton rat, Sigmodon hispidus *(courtesy J. K. Jones, Jr.).*

ground squirrels or other rodents or small carnivores or excavate their own burrows. Most burrows are less than 10 inches deep, but recent reports of denning in deep burrows on the High Plains of western Kansas, well below the frost line, suggest a change in behavior that possibly has allowed for successful invasion of the Northern Great Plains, where abandoned tunnels of active burrowers are available. The nest itself is a sphere of shredded plant fibers.

Cotton rats seem not to be territorial, although there may be some mutual avoidance. Huddling may take place in the cold of winter, though adult males and females are solitary during the breeding season. Home ranges of 0.97 acres (0.14–4.88) for males, and 0.54 acres (0.12–1.69) for females were calculated in western Kansas, but home ranges about twice as large were calculated during population "lows" in eastern Oklahoma. Home ranges of males increase during the breeding season; those of females decrease. Movements as great as 900 feet in 24 hours have been recorded. Densities of four to nine individuals per acre were reported in western Kansas, but populations as dense as 500 per acre have been noted. Some animals displaced experimentally showed an ability to "home" from as much as 1,500 meters from the point of original capture. "Homing" generally was successful with displacements of up to 300 meters (when rats seemed to be moving in familiar territory), and then suc-

cess decreased with distance. Animals released in habitat poor for cotton rats generally "homed" better than those released in good habitat.

Populations vary from year to year and season to season, being low in spring and high in autumn. Where abundant, this species may cause serious losses to crops of grain, vegetables, or cotton.

In the southern United States hispid cotton rats breed throughout the year, but on the Northern Plains breeding occurs only in the warmer months. No pregnant females were found from November through March in a recent study of this rat in northeastern Kansas, for example. Litter size varies clinally in that litters of northern populations average larger than those from southern latitudes. Geographically, it varies also with size of the female and time of the year. It has been reported that the number of fetuses per multiparous female averaged 9.9 in northeastern Kansas, whereas primiparous females (first pregnancy) carried an average of 7.7 fetuses. There are usually eight mammae, but there may be as many as 12.

The gestation period is 27 days, and there is a postpartum estrus. There is little variation in gestation period for lactating females. Cotton rats are rather excitable animals; females are good mothers, however, and seem not to cannibalize young even when disturbed. In the laboratory, females tolerate males in the nest with neonates.

Young are rather precocial for cricetines, being large (birth weight ranging from about 6.5 to 8 grams), well haired, and active. Their eyes open within 24 hours of birth. Gaining weight at the rate of a gram per day, the animals are weaned at 10 days of age. Young are first active outside the nest as small as 10 to 12 grams, which would make them only about four days old. At about 20 grams the young begin to show hispid guard hairs, and by the time they weigh 40 grams they have lost their juvenile appearance. Females may breed as young as about six weeks of age, at only about half of adult weight. Longevity in the wild is not known, but most animals live less than a year. Less than 20 percent of a population is likely to survive the winter. Surprisingly, apparently little is known of the molt pattern in this species.

Hawks, owls, and a wide variety of mammalian carnivores (including foxes, skunks, weasels, coyotes, and domestic cats and dogs) help keep populations in check. Remains of cotton rats have been found in pellets of burrowing owls in Nebraska. A variety of ectoparasites as well as cestodes and nematodes infest *S. hispidus.*

Selected References. Distribution (Farney 1975); natural history (Cleveland 1979; Fleharty and Choate 1973; Fleharty and Mares 1973; Kilgore 1970; McClenaghan and Gaines 1978).

Genus *Neotoma*

The genus *Neotoma,* entirely North American in distribution, has a geologic record extending back to late Pliocene times. There are some 19 living species, of which all but two are restricted to the western United States and Mexico. The name *Neotoma* is a neologism compounded of two Greek roots, *neo,* "new," and *toma,* "cut." The dental formula is 1/1, 0/0, 0/0, 3/3, total 16.

Neotoma cinerea: Bushy-tailed Woodrat

Name. The specific name *cinerea* is from a Latin adjective meaning grayish or ashen. Other vernacular names that have been applied to this species are trade rat and pack rat.

Distribution. The bushy-tailed woodrat is mostly a species of the western mountains, occurring from the Yukon to Arizona and from the western parts of the Dakotas and Nebraska to the Pacific Coast. Two subspecies are recognized on the Northern Great Plains, *Neotoma cinerea rupicola* in the badlands of the western Dakotas and Nebraska, and *Neotoma cinerea orolestes* on the Black Hills.

Description. Bushy-tailed woodrats are large, handsome rodents. Mean external

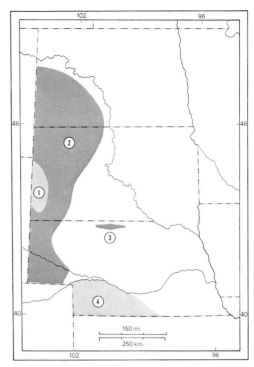

122. Distribution of (1–2) Neotoma cinerea *and* (3–4) Neotoma floridana. *Subspecies of* N. cinerea: (1) N. c. orolestes; (2) N. c. rupicola. *Subspecies of* N. floridana: (3) N. f. baileyi; (4) N. f. campestris.

and cranial measurements of a large series from the Black Hills of South Dakota are: total length, 359.1; length of tail, 148.0; length of hind foot, 40.6; length of ear, 33.4; greatest length of skull, 48.3; zygomatic breadth, 25.8. Weights of individuals in the same series averaged 272.5 grams.

The fur of the dorsum is pale grayish to buffy, with a pronounced blackish wash in *N. c. orolestes.* The sides are pale buff. The hairs of the venter and feet are white to their bases. The pronounced bushiness of the tail, the basis for the vernacular name, is best developed in adult males. Females and subadult males superficially resemble *N. floridana,* but the eastern woodrat is smaller, and the white hairs of the belly are gray at the base. The tail of *N. cinerea* is dark gray to blackish above and white below.

From *N. c. orolestes,* the subspecies *N. c. rupicola* differs in slightly smaller size and paler (more grayish, less reddish) dorsal pelage, which also is less heavily suffused with black. The several local populations of *N. c. rupicola* on the Northern Plains are isolated from one another in patches of discontinuous badlands habitat. Whether the morphological similarity among the populations is due to genetic similarity or to common physiologic responses to similar habitats is not known.

Natural History. Bushy-tailed woodrats are found almost exclusively in rocky areas (or in buildings nearby). They build large stick houses in crevices, cracks, and caves, among boulders in talus slides, or in abandoned man-made structures. In mines and buildings, they commonly take up residence on shelves or atop rafters or timbers, man-made analogues of their natural den sites. The houses incorporate just about any object in the vicinity that the rats can carry—such things as sticks, bones, conifer cones, bits of rope and leather, bottles, cans, hardware, pieces of dung, shotgun shells, owl pellets, feathers, bits of hide, paper, shingles, and wire. Their highly developed collecting instinct has earned them the vernacular names "pack rat" and "trade rat," the latter name referring to the habit of leaving one object in exchange for another. This is due not to any particular sense of justice, but to the fact that a rat can haul only so much debris at once. If it comes upon a new treasure while carrying another, it may put down the first to pick up the second.

The "house" that results from all this dealing in refuse provides protection from weather and enemies and is the center of activity, a territory defended strenuously against other woodrats. Bushy-tailed woodrats are solitary; the dispersing young build their own houses or remodel those abandoned by rats of an earlier generation. The nest itself is a globular or cup-shaped affair within the house, made of dry grass, bark fibers, moss, and leaves.

Bushy-tailed woodrats are active the year around. In winter they feed on vegetation stored during the previous autumn. The diet is varied, emphasizing foliage of shrubs

123. Bushy-tailed woodrat, Neotoma cinerea *(courtesy R. W. Barbour).*

and forbs, but no particular species. Needles of conifers, mushrooms, fruits, and seed heads are gathered for storage or immediate consumption. A study in Colorado suggested that fully 75 percent of available plant species were utilized to some extent. Big sagebrush (*Artemisia tridentata*), however, seems to be avoided.

The breeding season spans the warmer months. Some females have two litters per year. The gestation period is about five weeks. Three or four altricial young make up a typical litter. Juveniles commonly are trapped from mid-June to August.

Maturational molt, as in other species of woodrats studied, begins as a midventral band and spreads laterally, then fore and aft toward the limbs. Molt lines from opposite sides first meet dorsally on the rump and lower back. A wave of new hairs moves forward to the shoulders, neck, and head. There are postjuvenile and postsubadult molts as well as an annual molt (in spring or early summer) in adults. Annual molts are less regular in pattern than maturational molts.

Important predators include coyotes, long-tailed weasels, bobcats, and owls, although their "houses" (and their rugged and inaccessible location) protect the rats somewhat from predator pressure. Ectoparasites include ticks, fleas, and botflies. Among internal parasites are protozoans, flukes, tapeworms, and nematodes. For animals reaching adulthood, longevity in the wild frequently is three or four years.

Selected References. Natural history (Finley 1958; Turner 1974).

Neotoma floridana: Eastern Woodrat

Name. The name *floridana* is derived from Florida, whence the species—the only woodrat in eastern North America—first was described early in the nineteenth century.

Distribution. This woodrat occurs in the eastern and central United States, from Connecticut south to Florida, and westward to Nebraska and eastern Colorado. On the Northern Great Plains, two subspecies occur in two distinct areas—along the Niobrara River and its immediate tributaries in north-central Nebraska (*Neotoma floridana baileyi*), and along the Platte and Republican rivers and their tributaries in southwestern Nebraska (*Neotoma floridana campestris*). The northern population may be a relict of a warmer, moister Climatic Optimum period some 4,000 to 9,000 years ago (see fig. 122).

Description. These woodrats are handsome rodents, hardly deserving the emotion-laden name "rat." The upper parts are brownish, paling on the sides. The belly has pale grayish to white hairs, dark gray at their bases. The tail is bicolored, but not sharply so, being brownish above and white below; although sparsely haired, it is not scaly like that of the introduced Norway rat. Eastern woodrats from northern Nebraska are darker than those from farther south on the Northern Plains. Average and extreme external measurements of 13 adults from Cherry County, Nebraska, are: total length, 376.8 (350–398); length of tail, 166.1 (136–180); length of hind foot, 39.4 (38–41); length of ear, 25.5 (25–26). Weights range from 300 to 400 grams. Cranial measurements of greatest length of skull and zygomatic breadth for 10 individuals from the same area are 49.2 (48.2–50.6) and 26.2 (25.6–26.9), respec-

tively. Cranially, *N. f. campestris* differs from *N. f. baileyi* in having distinctly longer incisive foramina and generally greater size.

Natural History. Eastern woodrats are restricted on the Northern Great Plains primarily to riparian woodlands, frequently in areas with rocky cliffs. Such situations provide shelter and materials for the "houses" these animals typically construct, the architecture of which is varied enough to defy description. A house may be situated beneath a rock shelf, among downed timber, on a shelf or rafter in an abandoned building, or occasionally at the base of a shrub or tree or even in the low crotch of a tree. There may be an excavation beneath houses built on the ground. The house itself is constructed mostly of sticks, but a remarkable variety of materials—virtually anything the rats can carry—may be found in the walls or gracing the roof: bones, pieces of stone, dung of livestock or native mammals, owl pellets, shed snakeskins, feathers, pieces of glass, paper, metal, and the like. These homes usually are dome-shaped, 18 to 24 inches across (and slightly shorter than they are wide, but they may be three times this size). There are two or more entrances. Where eastern woodrats occur, one often finds houses that are temporarily vacant. Spiderwebs across entrances and a lack of fresh-cut vegetation

124. Eastern woodrat, Neotoma floridana *(courtesy R. W. Wiley).*

nearby indicate an unoccupied den. Such a vacancy soon will be rectified when the season's young disperse. Serial occupancy of dens is the rule. The vast accumulations of cemented feces and urine outside some houses suggest habitation by many generations of woodrats spanning several decades.

The house is the center of activity. It affords protection and a pantry for food storage. The construction of a good, old home is tight enough that the interior remains dry in all but prolonged rainy spells, and the cactus joints, rose clippings, hackberry branches, and other thorny materials incorporated in the walls are enough to withstand the efforts of any but the most determined predator.

These rats store food in the upper part of the den, above the nest chamber. Leaves of shrubs, trees, and forbs are staples. They seldom eat grasses, although they use them to line the nest chamber. Roots, fruits (especially legumes), and nuts (particularly acorns) are eaten, but dry seeds seem not to be taken to any great extent. Although they have rather catholic diets, woodrats are picky eaters, sorting through large quantities of forage before choosing just the right piece (which, to the human observer, looks like all the rest). When foraging outside the den, they take plants in rough proportion to their occurrence in the environment. Palatability may have something to do with moisture content. In dry seasons, leaves are a source of water; hence cacti are a common food where available. While feeding, the rats rest on the hind feet and tail, manipulating the food with the forelimbs.

Within the house, somewhere near the bottom (or even below ground level), is a cup-shaped nest chamber, about six inches in diameter, with thick walls of shredded bark and other plant fibers. Woodrats are mutually intolerant, and the nest is occupied by a single individual or by a female and her young.

The breeding season extends from February to about August in eastern Kansas, with a lull in July, when the testes of males regress. The estrous cycle is about a week long. Gestation takes about five weeks.

Females have been known to breed while lactating. Males and most females breed first as yearlings, but some females of early litters do reproduce in late summer of their first season. Litters of one to five young have been reported, with three or four being the mode. During parturition, the female sits on her hind feet and tail with her back arched. She assists the birth with her forepaws.

The young are born blind and nearly helpless and weigh about 10 grams. They are slate gray above and pale below, with pinkish muzzles and feet. White hairs are apparent on the chest and venter, and with the aid of a hand lens minute grayish hairs can be seen on the dorsum. The pinnae are folded over the auditory canal. Claws are visible though small, and the incisors have erupted. Locomotion is by uncoordinated wriggling, but this is enough to allow the young to reach the nipples, to which they attach firmly, grasping them, as in other rodents, with widely spaced deciduous incisors. The grip is sufficiently tenacious that a mother with young in tow will readily climb shrubbery or trees to escape disturbance.

Weight gain is rapid to about 90 days of age—about 1.5 grams per day. The pinnae are erect by the second day after birth. The eyes open at about two weeks, and the first solid food is taken at about the sixteenth day. Postjuvenile molt begins at about five weeks of age, starting on the venter and progressing to the dorsal midline, whence it moves anteriorly and posteriorly.

These woodrats are territorial, defending the homesite against other individuals. The home range around the den seems not to be marked, and woodrats are separated while foraging mostly by mutual avoidance. Obvious feeding trails usually are within 30 feet of the house, and all foraging is within about 75 feet. In a study in eastern Kansas, the mean distance between captures of males was 345 feet (range 0–1,080); that for females was 143 (0–650). The longer movements may have been to establish new dens. Homing has been reported across distances of about a fourth of a mile in suitable habitat. To speak of densities of woodrats is somewhat misleading because their habitat often is linear and frequently is decidedly patchy. Under excellent conditions one may locate active dens at intervals of 30 to 60 feet along a gully or hedgerow.

Woodrats run with the head and chest close to the ground. They climb well but seldom leap from branch to branch. The relaxed posture is bipedal, with some of the weight on the tail. When alert the animals crouch. Fighting is preceded by tail rattling, grinding of the teeth, and a characteristic thumping of feet. Agonistic encounters involve sparring with the forefeet and bites to the neck, chest, and muzzle, accompanied by much squealing.

Woodrats are meticulous about the condition of their pelage. When grooming, they assume a bipedal stance and lick the fur. The front feet are used to pull folds of skin within reach. The face and vibrissae are "scrubbed" with the paws, a behavior that begins before weaning, at about 15 days of age. The tail is pulled through the paws for cleaning. Dust bathing is common, and there often is a patch of bare soil in front of the house to accommodate this behavior.

Predators of eastern woodrats include coyotes, weasels, skunks, bobcats, owls, occasional hawks, and large snakes. These animals frequently are observed with pieces of tail missing. The skin of the tail is loose and easily pulled off. Once it is removed, the bare bones wither and fall away. This may be an important defense against predators.

Numerous ectoparasites live on *N. floridana*—fleas, mites, ticks, and lice. A common and conspicuous parasite is the warble, or botfly larva. The adult fly lays eggs on the entrance to the house, which then stick to the hair of a passing woodrat. The larva passes through the skin, often on the throat, neck, chest, or groin, and grows there at the expense of the host. These are large insects, with pupae 20 mm or more long, and one would imagine that their presence would incapacitate the host, but they seldom seem to be lethal. Once the adult fly has emerged, the wound heals rapidly, and within two weeks it is no longer visible externally. Among endoparasites are protozoans, roundworms, and flatworms.

In addition to their parasites, woodrats host a variety of commensals in their dens, which provide a favorable microclimate and some physical protection. Among these commensals are numerous kinds of insects and spiders, salamanders, toads, turtles, snakes, lizards, opossums, small rabbits, and white-footed mice. The mice sometimes help themselves from the host's pantry.

Selected References. General (Wiley 1980 and included citations).

Genus *Clethrionomys*

The generic name is derived from two Greek words, *kleithrios*, meaning "alder wood," in reference to the boreal forest and woodland habitat of some species, and *mys*, meaning "mouse." The preferred vernacular name for the group is red-backed vole (voles are short-tailed, short-eared mice), and the members of this genus usually exhibit a reddish dorsal coloration, at least as adults. The dental formula is 1/1, 0/0, 0/0, 3/3, total 16.

Clethrionomys is found in the northern parts of both North America and Eurasia, in tundra, boreal coniferous forest, northern deciduous forest, and montane habitats. Some seven species are recognized in the genus, but only three are found in North America, one of which occurs in the tri-state region.

Clethrionomys gapperi: Southern Red-backed Vole

Name. The specific name *gapperi* was bestowed in honor of a nineteenth-century naturalist named Gapper who, in 1830, wrote a short essay on Canadian mammals to which was appended the description by Vigors of *C. gapperi*. The full name and background of Gapper are unknown to us. The vernacular name distinguishes the general distribution of this vole from that of its North American relatives, the northern and western red-backed voles.

Distribution. Clethrionomys gapperi is widely distributed in the transcontinental coniferous forest, south of the tree line from northern British Columbia to Labrador. The species ranges southward through the Atlantic seaboard states to Chesapeake Bay and in the southern Appalachians to North Carolina, through the upper Great Lakes states in the Midwest, and throughout much of the Rocky Mountains and Pacific Northwest. Two of the approximately 29 recognized subspecies are found on the Northern Plains. *Clethrionomys gapperi loringi* is known from the extreme northeastern part of South Dakota, and through North Dakota generally to the north and east of the Missouri River; an apparently isolated population is known south of the Missouri on the Killdeer Mountains. *Clethrionomys gapperi brevicaudus* is an isolated race found only in

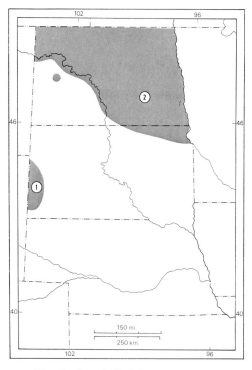

125. *Distribution of* Clethrionomys gapperi: (1) C. g. brevicaudus; (2) C. g. loringi.

the Black Hills region. Locally, distribution is determined by the presence of suitable densities of trees and shrubs and probably by the occurrence of free water as well.

Description. This species is volelike in general appearance, but the head and ears are larger, the ears are more conspicuous, and the tail is relatively longer than in many other small voles of the Northern Plains. The tail is bicolored and about one-third the body length. The most conspicuous external feature is the broad band of bright reddish chestnut hairs extending dorsally from the forehead to the base of the tail. The sides of the body are grayish, intermixed with buffy or ochraceous, and the underparts are whitish to buffy. Summer pelage is darker than that of winter, but the fur in all seasons is quite variable in color. There are two seasonal molts in adults each year; summer pelage is attained by mid-May, and winter pelage is complete by November. Molt in breeding females may be postponed. Juveniles are grayer than adults and undergo two molts between days 33 and 90 to 120.

Average and extreme external measurements for a sample of 10 females and six males of *C. g. loringi* from Marshall County, South Dakota, are: total length, 133.7 (125–144); length of tail, 35.3 (32–40); length of hind foot, 18.7 (18–19); length of ear, 14.3 (12.5–15.5). Weights of these individuals averaged 24.3 (19.2–30.5) grams. Comparable measurements of nine females and eight males of *C. g. brevicaudus* from Lawrence County, South Dakota, are: 133.4 (119–143); 34.3 (27–39); 18.2 (17–19); 14.1 (12–16). The average weight was 28.0 (20.6–35.2) grams. The two subspecies thus are about the same size externally.

The skull generally has a "weak" appearance in comparison with that of other voles, being smooth and relatively free of ridges and angularity. This and the absence of an arched rostrum often are presumed to be evolutionarily primitive features. The skull is relatively narrower than those of other small voles, and the straight posterior border of the palate distinguishes *C. gap-*

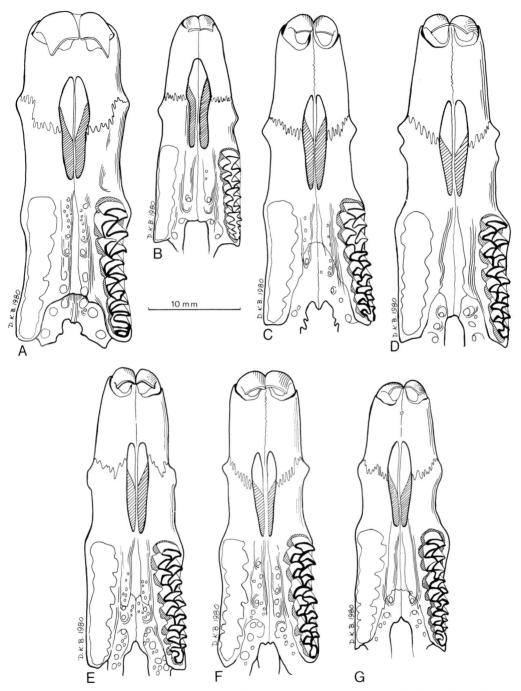

126. *Ventral view of palates and upper teeth of some microtine rodents of the Northern Great Plains:*
(A) Synaptomys borealis; (B) Clethrionomys gapperi; (C) Microtus pinetorum; (D) M. ochrogaster;
(E) M. longicaudus; (F) M. pennsylvanicus; (G) Lagurus curtatus.

peri from other species in the region (fig. 126). The incisors are ever-growing, as is true of all rodents, but the molar teeth of *Clethrionomys* become rooted as the individual matures. Possession of rooted molars, in which tooth growth ceases and wear patterns become evident on the grinding surfaces, is one of the principal distinctions between this and other voles found on the Northern Plains. The lower molars also are diagnostic in that the inner and outer reentrant angles are of about equal size.

Average and extreme cranial measurements of the six female and four male *C. g. loringi* from Marshall County, South Dakota, are: greatest length of skull, 23.5 (22.7–24.2); zygomatic breadth, 12.5 (11.8–13.5). Comparable measurements of eight male and seven female *C. g. brevicaudus* from Lawrence County, South Dakota, are: 24.0 (22.4–25.9); 13.0 (12.2–13.8).

Natural History. Red-backed voles are basically forest-dwelling mammals, although they sometimes are found in relatively dry, open woodlands, and they also live in brushy riparian habitats. *Clethrionomys* seems to require moisture and therefore selects a microhabitat that meets this need. In dry, upland woods in Connecticut, red-backed voles were found only in stone fences, which had cooler temperatures during summer days than the surrounding woodlands. More typical habitat for this species is moist lowland forest with uprooted trees and moss-covered fallen logs. Occasional flooding is tolerated, but desiccation is not; laboratory studies have shown that this vole has a high daily water requirement—about 0.5–0.6 gram of water per gram of body weight per day. The usual restriction of *C. gapperi* to moist areas thus is related to the availability of free water.

In relatively dry ground, *C. gapperi* does not build runways or tunnels, though individuals often use those of other small mammals. However, in damper habitats these voles build shallow tunnels in sphagnum moss, needle litter, or in peat-loam, especially around rotting stumps and logs.

Wide places in the tunnels are used for nests or for storing small food items. The voles often establish such tunnel systems in the "middens" of red squirrels, and they may rob the squirrel of its hoard of conifer seeds.

Staple food items include an array of fruits, nuts, and seeds, supplemented with the bark of small trees and shrubs when necessary. Stomach contents have revealed that the food of some individuals consists mostly of the mycelia of the fungus *Endogone* and other subterranean fungi. Principal summer and autumn foods are fungi, fruits, and some seeds, whereas winter staples are conifer seeds, roots, and the inner bark of deciduous trees. Caching of seeds may occur in autumn. Though basically an herbivore, the red-backed vole does eat some insects and other invertebrates when they are seasonally abundant. These voles have been trapped in trees and shrubs, so they may at times be sufficiently arboreal to exploit plant parts above ground level. Because of their varied diet and habitat preferences, these attractive small mammals cannot be considered a major threat to man's interests.

Red-backed voles are active at all hours of the day and night, but daytime foraging probably constitutes only 20 to 40 percent of their overall activity. They tend to be diurnal in winter but nocturnal in warmer seasons. These animals are not colonial or gregarious; groups consist only of families of a female and her young. The male is tolerant of his offspring, but he does not assist in their care. When adult red-backed voles of the same sex encounter one another, they usually react with aggressive, defensive, or avoidance behavior. They also interact competitively with voles and mice of other species during the breeding season, and red-backed voles often displace both deer mice and meadow voles from forest habitats. Conversely, meadow voles may interfere with red-backed voles in grassland habitats. Outside the breeding season, interspecific aggression is reduced, and coexistence in the same habitats may occur.

One study reported that seven times as many red-backed voles were caught in live

127. Southern red-backed vole, Clethrionomys gapperi *(courtesy E. C. Birney).*

traps as in snap traps. *C. gapperi* has been described as being of a nervous temperament and is known to be likely to die when caught in live traps. Individuals sometimes die when being handled by an investigator for routine marking and weighing in the field. These characteristics mark this species as considerably different from the other voles in the Northern Plains region.

Litters of red-backed voles are born from late winter through late autumn after a gestation period of 17 to 19 days. In a five-year study in Minnesota, the number of fetuses varied from three to 10, with 6.07 the average of 107 pregnant females (females have eight mammae). Only 1.24 percent of embryos were calculated to have died after implanting in the wall of the uterus, an exceptionally low proportion for a small mammal. In another study in Michigan, litter size ranged from two to five, with an average of 4.16. As in many other voles, litter size probably varies with the length of the breeding season, being larger in the north and at higher elevations where that season is shorter. Postpartum mating occurs, usually within 12 hours, and a second gestation period follows. Newborn mice are helpless, weighing about 2 (1.7–2.3) grams. In a related subarctic species, *C. rutilus*, growth is linear for 20 days and weight increases an average of about half a gram per day; a similar growth rate probably occurs in the southern red-backed vole.

At one week of age the incisors are erupting, and the nestlings, though usually firmly attached to the nipples of the female, can crawl about. If disturbed while nursing such young offspring, the mother may flee with the young firmly attached to her mammae, thus removing them from danger. Eyes open about day 12 or 13, after which the young voles run about freely and begin to eat solid food. Weaning may not occur until three weeks in some instances, but if another litter is born sooner the older young are weaned abruptly. Females reach sexual maturity at about two months of age. The juvenile pelage begins to be replaced by a distinct subadult pelage at about 33 days of age. Molt to the adult pelage is completed between 90 and 120 days of age.

In common with other higher vertebrates, the young of *Clethrionomys* have lower body temperatures than do adults, average 35.4° C for 10-day-old animals as opposed to 39.3° for adults. Temperature variance is much greater in young animals and does not stabilize at the adult level until they reach 23 or 24 days of age. Experimental exposure of young to freezing environmental temperatures for 60 to 90 minutes showed that 11- to 12-day-old voles responded by reducing body temperatures to 25° C or less. These low temperatures caused no visible ill effects. By 14 days of age, young voles showed great improvement in thermoregulation, and by the eighteenth day all experimental animals did so successfully. However, because young become free-living at about 18 days of age (weight approximately 11 grams), they do so before attaining adult body temperature.

Maximum longevity in the wild is probably less than three years, and most individuals live only a year or less. As in other small voles living in the Northern Plains region, a significant proportion of the population fails to survive a second winter, so the probable life expectancy of a newly weaned *Clethrionomys* is considerably less than a year.

Population densities are fairly low in the few studies that have been reported, on the order of four or five per acre. However, one study estimated eight individuals per acre in spruce–white birch forest and 14 per acre in hemlock–yellow birch forest. These relatively high densities suggest the possibility of increased intraspecific competition, resulting in a higher rate of dispersal of young animals from the home territory. Using a forested dune (50 to 80 yards wide) on the south shore of Lake Manitoba, a Canadian mammalogist studied the possibility that adult males drive young from the home territory. Data from the three-year study showed that the presence of adult males seems to affect the age at which young animals enter traps, but no clear evidence for intraspecific strife was found. Home ranges in *Clethrionomys* usually are about 0.25 to 0.70 acre, but sometimes they are much larger. Winter home ranges beneath the snow tend to be larger than those in warmer seasons. The extent to which this species undergoes the great fluctuations in density seen in some other microtines is uncertain, and fluctuations seem to vary from place to place. However, high densities are attained in at least some populations, followed by sudden "crashes."

As might be expected of mammals with a great reproductive capacity and relatively low densities, many individuals are lost to predation. Small carnivores, such as foxes and weasels, raptors, and snakes take a portion each season. In one study in Wisconsin, remains of red-backed voles were found in a small percentage of scats from timber wolves.

This species harbors a variety of parasites, both internal and external. Tapeworms, as well as lungworms and intestinal roundworms, have been reported, as have various protozoan parasites. These voles also are hosts to many species of fleas, mites, ticks, chiggers, and lice.

Selected References. General (Merritt 1981 and included citations).

Genus *Microtus*

Microtus is derived from two Greek words, *mikros*, "small," and *otos*, "ear." The general vernacular name for *Microtus* and its relatives is "vole," an Old English word originally derived from a Scandinavian dialect. Voles are small, short-eared, usually short-tailed mice that, together with their relatives the lemmings, are widely distributed in the Northern Hemisphere, from the northern arctic tundra zone south to southern China and Central America. Most species inhabit open landscapes, such as grasslands, but a few are found in forests or in forest edges and openings. There are about 45 species in the genus *Microtus* as strictly defined, of which about 19 occur in North America. However, there is widespread disagreement over the limits and content of this genus. The Northern Plains harbor four species, two of which are peripheral and two of which are widespread throughout the region. The dental formula is 1/1, 0/0, 0/0, 3/3, total 16.

Microtus longicaudus: Long-tailed Vole

Name. The specific name *longicaudus* is a combination of two Latin words meaning "long" and "tail" and refers to the fact that this species possesses one of the longest tails among voles.

Distribution. The long-tailed vole is a mammal of the western mountains—in the Rocky Mountains from east-central Alaska south to southern New Mexico and Arizona, and in the Coast and the Cascade–Sierra Nevada ranges as far south as northern and southern California, respectively. The species also is widely distributed in the various interior ranges and the eastern outliers of the Rockies, including the Black Hills, where it just penetrates the Northern Great Plains from the west. This population represents the widespread subspecies *Microtus longicaudus longicaudus* but appears to be isolated from other populations farther to the west.

Description. The most distinctive feature of this vole is its long tail, which is often about half as long as the head and body. Otherwise its small, rounded ears, brownish gray dorsal coloration, and grayish white belly do not distinguish it from other voles. It does, however, have a diffuse reddish tinge to the dorsal fur that

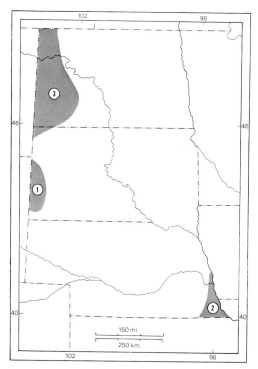

128. *Distribution of* (1) Microtus longicaudus,
(2) Microtus pinetorum, *and* (3) Lagurus curtatus.

Natural History. This is one of the least
known voles in North America, even
though it is widely distributed and occasionally not uncommon. The species is
found in an unusually wide variety of habitats, from sea level to at least 12,000 feet,
and from dense coniferous forest to rocky
alpine tundra and sagebrush semidesert.
The particular habitat in a given area seems
to be strongly influenced by whether other
species of *Microtus* occur in the same
place. Where it is the only species in an
area, it often occurs in moist meadows and
may attain fairly high densities. Where it is
found with another species, *M. longicaudus*
frequently appears to be "displaced" to
shrubby or forest-edge habitats. However, a
number of exceptions have been described,
and the habitat relationships of coexisting
voles require more study.

On the Black Hills, this species ranges
from 4,200 feet, in the semiarid foothills,
up to boreal forest at the crest. At lower
elevations it is restricted to moist riparian
environments, where it occurs with the
meadow vole, but at higher elevations it
ranges into aspen and spruce forest in company with red-backed voles. This species
appears to be subordinate to other con-
generic voles when they occur together,
and vigorous chasing and fighting have
been observed, with the other species of
Microtus the aggressor.

Except where they occupy grassy habitats, long-tailed voles usually do not make
well-defined "runway" systems through
vegetation, as do most other *Microtus,* and
they appear to be solitary in their habits.
Burrow systems and nests have not been
described in detail, but the nest, of grass
and other plant fibers, often is under a
fallen log or in an underground chamber.

Like voles generally, *M. longicaudus* is
herbivorous, principal dietary choices being
green plant parts as long as these are available. Seeds, berries, and underground fungi
also are eaten when abundant, especially in
autumn. In winter, when other foods are
scarce, these voles may feed heavily on the
inner bark of shrubs and small trees.

Breeding in long-tailed voles is poorly
known; the reproductive season presum-

may help set it apart from other *Microtus*
on the Northern Plains (not to be confused
with the reddish dorsal stripe of the red-
backed vole). Molt probably is similar to
that described for *Microtus pennsylvanicus.*
On the Black Hills, adults molt to summer
pelage in late June and early July. The size
of long-tailed voles varies greatly from
place to place; large individuals are known
from the northwest coastal region and central Montana, whereas small animals are
found along the Oregon and California
coast and in the Southern Rocky
Mountains.

A series of six males and five females
from Pennington County, South Dakota,
had the following average and extreme
measurements: total length, 172.0
(148–191); length of tail, 56.3 (48–66);
length of hind foot, 21.0 (20–22); length of
ear, 15.5 (13–17); greatest length of skull,
26.4 (23.8–27.4); zygomatic breadth, 14.7
(13.4–15.2). Weights of these specimens
averaged 38.2 (29–43) grams.

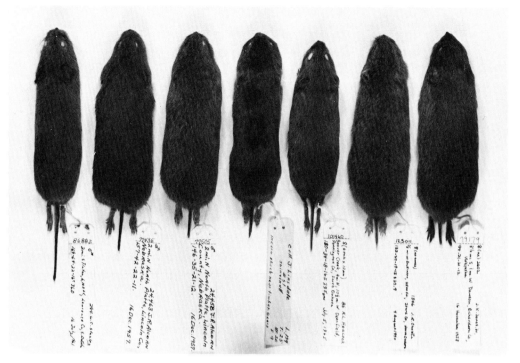

129. *Dorsal view (left to right) of museum skins of microtine rodents (except muskrat) of the Northern Great Plains:* Microtus longicaudus; M. pennsylvanicus; M. ochrogaster; M. pinetorum; Clethrionomys gapperi; Lagurus curtatus; Synaptomys cooperi. *Note general size and texture of pelage as well as length of tail.*

ably varies with altitude and latitude, as it does in other voles. On the Northern Plains, breeding probably begins in May and continues through August. The average litter size was 4.7 (range four to six) in 15 females captured in July on the Black Hills of South Dakota, but litter size probably varies seasonally. In adjacent Wyoming, two to seven (average 4.8) young have been reported. The female has four pair of mammae. The small, naked, helpless young are born after a gestation period of about three weeks. The young grow rapidly, developmental events probably following the same general sequence as in the meadow vole (see below); the young begin to be weaned, and become active outside the nest, at about two weeks of age.

Reproductive maturity may be reached at three weeks of age, although usually a bit later, but this too varies seasonally. Thus a female born early in the breeding season may herself produce several litters that

same summer. An adult female that has safely overwintered might produce four consecutive litters in the course of a single breeding season, if she lives long enough. Mortality rates in voles are high, however, and longevity seldom exceeds one year, usually being considerably less. This is partly a matter of the toll exacted by predators, because many prey heavily on voles, including coyotes, foxes, bobcats, weasels, hawks, and owls. The long-tailed vole may, in brushy and forested habitats, be somewhat less suceptible to predation than grassland species, but data are scanty. *M. longicaudus* also may be relatively more sensitive to water deprivation than many other species, and drought conditions may have a particularly detrimental effect on survival. Ticks, lice, and trematodes have been found infesting this species.

Population densities of long-tailed voles usually are relatively low, but may occasionally build up to 100 or more per acre.

Regular fluctuations in numbers occur in some places, but other populations seem to exhibit stable low densities; again, the pattern found at a particular place may depend in part on the other "vole neighbors" of this species. Where densities become high, local damage may result to crops, orchards, and forest plantations. However, most long-tailed vole populations exist outside agricultural areas, and they cannot be considered serious agricultural pests.

Selected References. General (Turner 1974); natural history (Conley 1976; Randall 1978).

Microtus ochrogaster: Prairie Vole

Name. The specific epithet of this vole was coined by combining two Greek words, *ochra*, referring to "ochre yellow," and *gaster*, meaning "belly." The yellowish buff color of the belly sets it apart from other voles of the Northern Plains.

Distribution. The prairie vole is mostly endemic to the great central grasslands of North America, from the central prairie provinces of Canada south to Oklahoma, and from the Rocky Mountains east to the edge of the eastern deciduous forest and beyond. The species originally followed the Prairie Peninsula eastward into Indiana and Ohio, and clearing of primeval forest in the past 200 years has allowed the species to increase in distribution and abundance along the eastern margin of its range. Prairie voles also occurred on the Southern Great Plains in late Pleistocene times, and a relict population survived at several localities in eastern Texas and western Louisiana until at least 1905. This population now may be extinct, as efforts (since 1934) to find this vole have been to no avail.

Three subspecies are found on the Northern Plains: *Microtus ochrogaster minor* occurs north and east of the Missouri River in North Dakota and in the northeastern quarter of South Dakota; *Microtus ochrogaster similis* is restricted in the region to the vicinity of the Black Hills; and the remainder of the area is occupied by *Microtus ochrogaster haydenii.*

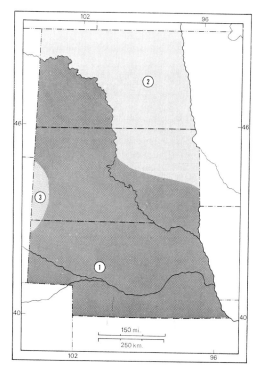

130. Distribution of Microtus ochrogaster: (1) M. o. haydenii; (2) M. o. minor; (3) M. o. similis.

Description. This small vole is stocky and compact-bodied, with short legs and tail. The small eyes and ears are inconspicuous on the relatively large head. The dorsal fur is grayish brown with a pronounced grizzled appearance imparted by the mixture of blackish to ochraceous buff tips on the guard hairs. The sides are paler, and the venter is ochraceous to grayish white. The feet are grayish tan, the tail dark above and paler below. Pelage color is similar in winter and summer, but fresh winter fur is slightly darker and the belly is usually more buffy. The intensity of color fades with time, and by spring the pelage is pale. The sexes do not differ in color.

This species differs from the meadow vole (*Microtus pennsylvanicus*) in that the dorsal fur is more grizzled and paler, and the underparts are usually ochraceous rather than grayish. The prairie vole also has a shorter tail. Characters that distinguish it from the woodland vole (*Micro-*

tus pinetorum) may be found under the account of that species. There appears to be a single, prolonged annual molt in adults during summer. Juveniles undergo two molts, to subadult and then to adult pelage. There are usually only five plantar tubercules on the sole of the hind foot, and females possess three pair of mammae (one pectoral, two inguinal). Flank glands are poorly developed. These as well as cranial and dental differences have led some systematists to place *M. ochrogaster* in a separate genus, *Pitymys*, along with *M. pinetorum*, or even to place it in its own monotypic genus, *Pedomys*.

Average and extreme external measurements of 26 adult *M. o. haydenii* from north-central Nebraska are: total length, 165.6 (153–186); length of tail, 37.0 (33–45); length of hind foot, 21.1 (20–23); length of ear, 12.3 (11–15). Average weight of 11 specimens from this locality was 60.1 (49.6–70.5) grams. *M. o. similis* from the Black Hills is somewhat smaller and darker than *haydenii*. The subspecies north and east of the Missouri River, *M. o. minor*, is much smaller and darker than the other two. Average and extreme external measurements of nine adult *M. o. minor* from Sherburne County, Minnesota, are: total length, 128.4 (120–135); length of tail, 30.0 (27–33); length of hind foot, 16.2 (16–17). Weight averaged only 23.0 (20.0–25.5) grams.

The skull of the prairie vole is high and narrow. Average and extreme cranial measurements of 15 to 16 adult *M. o. haydenii* from north-central Nebraska are: greatest length of skull, 28.9 (27.6–30.8); zygomatic breadth, 16.4 (15.5–17.5). Comparable measurements of eight adult *M. o. minor* from Sherburne County, Minnesota, are 23.3 (22.3–24.1) and 12.3 (11.9–12.6).

Natural History. The prairie vole typically inhabits upland prairies, although the species also may occur under certain conditions in swales and riparian grasslands—habitat that usually harbors the meadow vole, *M. pennsylvanicus*. This is especially true in western and southern Nebraska in those places where the meadow vole does not occur. Elsewhere, when the two species are found in the same areas, the prairie vole usually is restricted to drier habitats unless the population density of *M. pennsylvanicus* is low, as happens periodically. Prairie voles construct and maintain extensive runway systems in their grassland habitat by clipping grass stems and other vegetation along routes of travel. With continued use, runways become quite distinct, and all vegetation may be removed in places, leaving the bare soil exposed. These voles tend to be more fossorial than meadow voles, particularly in short-grass habitats, living in subterranean burrows, with runways between burrow entrances (and with foraging runways extending beyond the burrow system). In the eastern part of the tristate region, vegetative cover is taller, and prairie voles are less fossorial.

On the High Plains, these voles often are "patchy" in distribution and form small colonies where they occur. Unlike the meadow vole, prairie voles are highly social and tolerant of conspecifics. They also may form monogamous pairs, and both parents cooperate in rearing the young and maintaining the burrow and runway system. It also has been discovered that, unlike the meadow vole, growth and reproductive maturation of young is suppressed so long as they remain with their siblings and parents. Dispersal from the natal area is therefore essential if subadult prairie voles are to become reproductively active adults.

The reproductive cycle is complex, variable, and influenced by many factors. Some reproduction may occur in any season in different times and places, but the usual pattern in the southern part of the tristate region is for breeding activity to peak in March and April and again in September and October, with less, or no, activity in midsummer and midwinter. As one moves progressively northward on the Northern Plains, the initiation of spring breeding becomes later and the cessation of fall breeding earlier, and the magnitude of lull diminishes until a unimodal breeding season is established, at least in years when the summer is not too hot and dry.

Sexual maturity in *M. ochrogaster* is at-

131. Prairie vole, Microtus ochrogaster *(courtesy D. C. Lovell and R. S. Mellott).*

tained when both males and females are still in subadult pelage and weigh about 20 to 30 grams. Females are about a month old or less when they first become sexually receptive (given normal dispersal—see above). Ovulation is induced in this species by the stimulus of breeding and occurs nine or 10 hours after copulation. Following a gestation period of about three weeks, the young are born in a globular nest of dry grass, usually underground but sometimes in a depression on the surface. The litter size varies from year to year, seasonally, and sometimes with the age of the female; the range is one to seven, and the mean is between three and four. Studies in Kansas and Indiana showed that largest litters were associated with peaks of breeding activity in spring and autumn, and also with peak population densities that occur in some years. Farther north, litter size might be expected to be greatest during midsummer breeding peaks when and where they occur.

The naked young weigh about 2.8 grams and are born with eyes and ears closed, but both open in about eight days. Growth is rapid for the first three weeks of life, during which the juvenile pelage develops, and weaning begins during the third week. After three weeks of age, growth slows down somewhat, and postjuvenile molt begins. Young females and males both can breed by day 30, when they are approx-

imately two-thirds the size of adults. In young voles born late in the breeding season, growth usually is slowed during winter, and they remain at subadult size (about 30 to 35 grams). Growth resumes in March, however, and adult size is quickly attained.

Female prairie voles often exhibit postpartum estrus and thus may produce litters at three-week intervals during the breeding season. However, mortality rates in voles are high, and most females do not survive long enough to produce more than a few litters in their lifetime. In fact, mortality rates are so high that they must be expressed in time shorter than a year, for few prairie voles live as long as one year. The situation is further compounded by the great variation in mortality rates from year to year, seasonally, and among different sex and age classes. During the breeding season, adult survival rates usually range between 30 and 80 percent per month; among adults, males tend to have slightly higher mortality rates than females at this season. Thus the typical adult female may live little more than a month, long enough to produce one, or at most two, litters if she first became pregnant as a subadult. Subadults often exhibit somewhat higher mortality rates than adults in the same population, and juveniles have the lowest survival rate of any age class, often less than 20 percent per month. In subadults and juveniles, as in adults, males tend to have higher mortality rates than females. The situation in the winter nonbreeding season differs; young voles born in late summer and autumn may have higher survival rates, equal to or better than those of the adult component of the population, and overall survival rate in the population often is better than 50 percent (and may be as high as 90 percent) per month.

Generalizations on mortality rates in prairie voles are further confused by large changes between years, the causes of which are still obscure. Periodically in this and other vole populations, mortality in juveniles and subadults becomes exceptionally high, and the survival rate may fall to about five to 10 percent per month, leading to an abrupt decline, or "crash," in population density. These crashes are followed by an interval during which survival improves, and population density increases, only to terminate in another crash decline. In some circumstances these oscillations in population density are regular enough to have been termed the "microtine cycle." Cycles in the prairie vole are not so pronounced as in the meadow vole and seem not to occur in many places. The amplitude of fluctuation in populations appears to be strongly influenced by the amount of vegetative cover in the habitat, and only if cover is adequate are peak densities of about 30 to 50 per acre attained periodically (the cycle in *M. ochrogaster* lasts two years according to most studies). If cover is sparse, as it is over much of the Northern Plains, vole densities may remain below two to four per acre, and density fluctuations are not detectable. Even though dense cover may be necessary for high-density peaks to occur, crash declines nevertheless occur even though habitat quality as measured by vegetative cover remains good.

No single factor seems identifiable as the cause of cyclic declines. Quantity of food appears unchanged, although recent studies suggest that food quality may be important. Predators take many voles, but it has been difficult to prove that they are the cause of the surge in vole mortality that leads to a crash. More subtle, intrinsic mechanisms such as physiological or genetic changes within the vole populations themselves also have been investigated; while many interesting correlations between population density and these intrinsic changes have been found, the ultimate causal mechanism of vole cycles is still not understood.

In spring and summer prairie voles feed almost exclusively on green vegetation, principally grasses and sedges, forbs such as legumes (clover and alfalfa), and composites. The younger growing leaves are sought, and to obtain them the vole cuts through the base of the plant, then cuts off successive sections 40 to 50 mm long until the top of the plant is reduced to ground level. Some of the sections, or "cuttings," subsequently may be eaten, but often they are left in a pile in the runway—testimony to the feeding activity of the vole. Food plant selection is partly influenced by availability, but palatability and nutrient value (determined in part by growth stage) appear more important; relatively few of the species of plants in the habitat are eaten, and, of those food species, the voles consume only a fraction of the available plants. From late summer onward, green vegetation progressively disappears, and these microtines turn to alternative foods such as roots, tubers, fruits, seeds, and bark. They cache quantities of such food, as well as grass, in underground chambers for winter use; one such chamber was 250 mm wide, 400 long, and 200 high, and it contained almost 3,000 seeds of the Kentucky coffee tree, a legume. There is some evidence that prairie voles also may eat a significant amount of insect food, consuming the soft body parts and discarding the hard exoskeleton. More study of this is needed, however.

Foraging activity probably is restricted to the immediate vicinity of runways in summer, but this is not certainly known. Runways are 40 to 50 mm wide, as are the connecting burrows. The system of a family group may cover from less than 50 to more than 150 square yards, and may contain several burrows from two to four inches deep, and nest chambers eight to 10 inches underground.

Prairie voles may be active in their runway and burrow systems both day and night. Patterns of activity are affected by weather and food availability, and also by the presence of cotton rats (*Sigmodon hispidus*) in the habitat. The voles shift their daily activity patterns, and also their use of space within the habitat, to avoid contact with the much larger cotton rat. Members of the genus *Sigmodon*, particularly breeding adults, act aggressively toward prairie voles and may be a significant source of vole mortality in southern Nebraska where the two species occur together.

Other predators include coyotes, foxes, weasels, skunks, bobcats, and raccoons. The opossum is reported to eat prairie voles on occasion, although an active adult vole probably can elude one. Short-tailed shrews also are potential predators, especially of nestlings and young weanlings. A number of raptors, both hawks and owls, also have been observed to prey on prairie voles, particularly when vole densities are high. The marsh hawk, red-tailed hawk, rough-legged hawk, long-eared owl, and short-eared owl are the most frequent vole predators. Finally, a number of species of snakes may dine on this vole, and *M. ochrogaster* clearly is a most important prey species for many predators on the Northern Plains.

As if this were not enough, many parasites afflict the prairie vole. Common external parasites are fleas, lice, chiggers, mites, and ticks; internal parasites include several species of trematodes, cestodes, and nematodes, as well as a coccidian protozoan.

Home-range size in prairie voles varies with habitat, being larger where the vegetative cover is sparse, as in the drier western half of our region. In tall-grass prairie, home ranges average about a tenth of an acre, whereas in short-grass prairie the mean may be three-tenths of an acre. Home ranges of adjacent families often overlap, but the vicinity of the nest probably is defended. Intraspecific aggression is highest outside breeding periods, especially in winter, and may serve to disperse individuals; whether there is differential tolerance related to sex or kinship as in colonial ground squirrels and marmots is not known. More studies are needed before social structure and use of space in this species are well understood.

Selected References. General (Jameson 1947); natural history (Batzli, Getz, and Hurley 1977; Birney, Grant, and Baird 1976; Gaines and Rose 1976; Meserve 1971); systematics (Choate and Williams 1978; Severinghaus 1977).

Microtus pennsylvanicus: Meadow Vole

Name. Both the vernacular and the scientific names of this widespread vole refer to the fact that the species was described by George Ord from a specimen taken from a meadow near Philadelphia, Pennsylvania.

Distribution. It is difficult to generalize about the geographic range of this rodent, which has the widest distribution in North America of any species of *Microtus.* It occurs from tree line or slightly beyond in the north down through the transcontinental boreal forest region, into the northern half of the eastern deciduous forest, across the Northern Great Plains, and in the Rocky Mountains; the southernmost population is in the Mexican state of Chihuahua. However, it is associated neither with forest nor with plains habitats, but rather with the moist to wet meadows that are scattered in patchy fashion throughout this vast region. On the Northern Plains, meadow voles occur throughout the region except for most of southwestern Nebraska, where suitable habitat is apparently too infrequent to support the species.

Four subspecies occur in the Dakotas and Nebraska. *Microtus pennsylvanicus finitus* is found only in the southwestern corner of

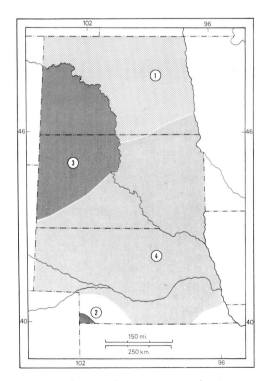

132. Distribution of Microtus pennsylvanicus: (1) M. p. drummondii; (2) M. p. finitus; (3) M. p. insperatus; (4) M. p. pennsylvanicus.

Nebraska. Its range is disjunct from that of other meadow voles farther north in the state, which belong to the nominate subspecies *Microtus pennsylvanicus pennsylvanicus.* This race occurs throughout most of Nebraska and the southern and eastern two-thirds of South Dakota, and it intergrades east of the Missouri River with *Microtus pennsylvanicus drummondii* and west of the river with *Microtus pennsylvanicus insperatus.* Like the prairie vole, this species occurred farther south on the Great Plains (as well as in other parts of North America) during the most recent glacial period.

Description. This is a rather large vole that, although its color varies geographically and to some extent seasonally, usually is fairly dark in general appearance. The adult summer pelage is, in *M. p. pennsylvanicus,* dark chestnut brown dorsally, with guard hairs that possess pale subterminal bands and black tips of variable length, imparting a grizzled appearance. The sides are paler brown than the middorsal region, and the venter is dusky gray. The tail is moderately long for a vole (longer than in *ochrogaster* and *pinetorum,* shorter than in *longicaudus*), dark brown above and gray below. The feet usually have six plantar tubercles (rather than five as in *ochrogaster* and *pinetorum*) and are dark gray; the ears are short and rounded, barely projecting above the fur. Females possess four pair of mammae, two pectoral and two inguinal, one pair more than in *ochrogaster* and two more than in *pinetorum.*

Descriptions of the annual molt of this species are unsatisfactory; it is apparently irregular and prolonged, occurring between May and September. Winter pelage thus is complete in the autumn; when fresh it is longer and fuller than the summer fur, as well as duller and slightly darker in appearance; the tail is more distinctly bicolored. By the end of winter, however, the fur is quite faded. Juvenile pelage is "woolly," dark brownish black above and grayish black below; the feet and tail are blackish. Postjuvenile molt leads to a subadult pelage that is paler than juvenile pelage, inter-

mediate in color between the latter and the pelage of adults; texture approaches the adult condition. A second molt then results in adult pelage, which is complete as the animal approaches adult body size. At least one author has suggested that there may be two seasonal molts annually in adults.

M. p. finitus is larger and quite a bit darker than the other subspecies in the tristate region. The nominate subspecies also is large and dark, whereas *M. p. drummondii* is slightly paler and more buffy and is much smaller. *M. p. insperatus* is only slightly smaller than *M. p. pennsylvanicus,* but it is the palest of the races on the Northern Plains.

Average and extreme external measurements of nine males and 11 females of *M. p. finitus* from Dundy County, Nebraska, are: total length, 167.4 (155–187); length of tail, 42.1 (35–50); length of hind foot, 22.0 (21–24); length of ear, 12.5 (11–16). Nine adults—four males and five females—from this locality weighed an average of 57.2 (42.9–75.0) grams. Average and extreme external measurements of 11 males and six females of *M. p. drummondii* from eastern North Dakota are: total length, 157.3 (150–170); length of tail, 45.7 (40–52); length of hind foot, 19.2 (18–20); length of ear, 12.4 (11–14). Average weight of these specimens was 39.3 (30–47) grams. Average and extreme cranial measurements of nine male and nine female *M. p. finitus* from the above locality are: greatest length of skull, 28.8 (27.2–31.6); zygomatic breadth, 16.0 (15.1–17.5). Comparable cranial measurements of *M. p. drummondii* are 25.5 (24.4–27.1) and 14.2 (13.2–15.0).

Natural History. The meadow vole typically inhabits moist to wet meadows, usually living in lush, dense vegetation of grasses, sedges, and rushes. Such meadows and marshes are frequently found only in riparian situations on the Northern Plains, particularly to the west, and *M. pennsylvanicus* often is closely restricted to stream banks and lakeshores. However, there are local exceptions, and meadow voles are sometimes found in relatively dry upland habitats so long as cover is sufficient. Habitat segregation between *M. ochrogaster* and *M. pennsylvanicus* typically restricts the former to drier, less densely vegetated habitats and the latter to mesic areas of better cover where the two species are sympatric. On occasion, though, both may be found in the same habitat.

Habitat segregation reduces the intensity of potential competition for space and food between these two species. Both feed primarily on green vegetation, but, because they tend to occupy different habitats, the plant species they consume usually are different. Moreover, neither species is periodically so abundant that food availability is grossly affected, although more subtle quantitative effects may occur. There is, however, dietary overlap—both species fed heavily on bluegrass (*Poa*) in an Indiana study (15 percent of the diet of *ochrogaster* in the course of a year and 32 percent of that of *pennsylvanicus*). Meadow voles eat more leaf and stem material of graminoids, whereas prairie voles eat more roots and broad-leafed plants, especially legumes. This difference reflects not only the differential abundance of plant species in their respective habitats, but also the less fossorial habits of meadow voles.

M. pennsylvanicus lives mainly in surface runways in dense grass or sedge cover. Meadow voles are excellent swimmers, and their runways may lead in and out of the water. Although burrows and underground nest chambers are sometimes constructed, the soil in typical habitat often is too waterlogged to permit burrowing. Conse-

133. Meadow vole, Microtus pennsylvanicus *(courtesy E. C. Birney).*

quently, nests are usually on the ground surface or in a tussock of grass. The nest is spherical, five to six inches in diameter and three or four inches deep, constructed of dry stems and leaves with a lining of finely shredded plant material. Home ranges of meadow voles vary in size—from less than a twentieth of an acre up to about two-tenths of an acre in one Michigan study. Hence home ranges are, on the average, smaller than in the prairie vole. Males usually occupy larger home ranges than do females, especially in the summer breeding season. Home ranges of both sexes overlap to some degree, those of females more so than those of males. However, intraspecific aggression seems to keep the animals apart most of the time, and territorial defense probably occurs in the vicinity of the nest. Unlike the prairie vole, this species has not been reported to form monogamous pairs, and each adult in a population occupies a separate nest. The meadow vole is considerably more aggressive intraspecifically than is the prairie vole and also may be behaviorally dominant to the latter species in most cases, although the evidence for this is equivocal.

Experiments have demonstrated that displaced meadow voles usually can return home successfully from up to 200 yards away, although this ability declines when the displacement distance is farther. "Homing" ability probably is based on a combination of things, including familiarity with features of local terrain and use of sun position to orient in the correct direction. Although both sexes can use sun-compass orientation, only females consistently employed it in one study, perhaps because the female has more incentive to return to a nest containing her young.

Unlike the prairie vole, this species shows no evidence of social organization, and the breeding system may be termed promiscuous. When a female vole enters estrus and becomes receptive, male voles in the area probably become aware of this through smell and are attracted to her. Ovulation is induced by the stimulus of copulation, occurring 12 to 18 hours later, and pregnancy lasts 21 days. Litter size varies depending on a number of factors, as is generally true of *Microtus*. Latitude is one important component in that litters tend to increase in size with higher latitude, a situation perhaps inversely correlated with average length of breeding season. For example, mean litter size estimated from counts of fetuses in southern Indiana ranged from 4.1 to 4.9, whereas in southern Minnesota the average was 5.7, and in North Dakota, 6.4; the known range in litter size also is large, from one to 11. As might be predicted from the differences in the number of mammae (eight and six, respectively), *M. pennsylvanicus* has larger litters than does *M. ochrogaster* in places where both species occur.

The altricial young usually weigh between 2 and 3 grams at birth, and they grow rapidly. Body hair begins to appear about day five, and eyes and ears open on day eight. Weaning begins on the tenth day, and the young may achieve independence at between two and three weeks of age. The postjuvenile molt usually begins between days 20 and 25; young females may achieve sexual maturity about this time, although this depends upon the season of the year (males become sexually mature about a week after females). Young meadow voles are not inhibited in their growth and development by the presence of siblings, as is the case in the prairie vole. In any event, they disperse from their natal nest in the third or fourth week of life. This dispersal may be triggered by maternal abandonment, inasmuch as females usually undergo postpartum estrus and the next litter is due to be born about three weeks later. At this time the female may leave the weaned young of the earlier litter in the nest and seek another place to give birth to the next, although probably still close to, or within, her original home range. Such early dispersal of young and the shifting of adults in the population probably is of adaptive advantage to a species with a patchy optimal habitat, one in which long-distance dispersal thus may be necessary to colonize isolated meadows where chances of survival may be better for a young vole.

Such dispersal is, of course, hazardous,

and the disappearance rate of juveniles is sometimes as high as 80 to 90 percent per month. After subadulthood is achieved in the second month of life, survival improves and remains relatively constant as the vole becomes an adult in its third month. Mortality rates fluctuate greatly, both seasonally and from year to year, but there is no apparent pattern related to age after the first month. Survival rates ranged from about 30 to 90 percent per month in one study, and from 30 to 60 percent in another; as in prairie voles, males suffer higher mortality than females. Longevity is consequently low; average life expectancy at birth is less than a month, and mean survival time varies from about six to 10 weeks once a vole becomes sexually mature.

Thus most females do not live long enough to produce more than one or two litters, even though the breeding season may be long. In a Minnesota study, breeding extended from late March to mid-November in most years. However, during a winter with heavy snow cover, breeding continued through the winter months as well, although only a few females participated. On the other hand, winter breeding is more common during an "increase" phase of the population cycle.

In most circumstances, *M. pennsylvanicus* exhibits much more pronounced cycles of population density than does the prairie vole. Peak densities range from about 60 to a reported 250 voles per acre. The pattern of fluctuation varies, with the "increase" phase sometimes occurring in a single breeding season or sometimes over several. Similarly, the peak density period may be brief or prolonged over about 18 months. The decline is usually rapid, however, and may occur at any season. During the decline, mortality rates increase and recruitment into the population is reduced. After a "crash," the population density may be only about one vole per acre, and survivors are found in small pockets in optimal marshy habitat.

Although many predators may take advantage of the peak in a meadow vole population to feed on them, predation and disease seem not to be causes of the abrupt decline, nor does inclement weather per se, although a hard winter (cold, little snow cover) or a series of early spring storms may cause much mortality. The ultimate cause(s) of microtine cycles are still unexplained.

Probably more species of predators have been enumerated for this vole than for any other North American mammal. One work, published a half century ago, listed 12 snakes, 27 birds, more than 20 species of mammals, the bullfrog, and the snapping turtle as predators of *M. pennsylvanicus*, as well as bass and several species of pike. External parasites include many species of fleas, lice, chiggers, and mites; internally are found trematodes, nematodes, cestodes, and a number of protozoans. Botflies may be more important parasites on the Northern Plains than in many parts of the range of this vole, infestation rates of 15 to 25 percent having been reported.

Selected References. Natural history (Ambrose 1973; Getz 1960, 1972 and included citations; Keller and Krebs 1970; Krebs, Keller, and Tamarin 1969; Madison 1978; Zimmerman 1965); systematics (Anderson 1956).

Microtus pinetorum: Woodland Vole

Name. The name *pinetorum* means "of the pine grove," being derived from the Latin word *pinetum.* This is actually a misnomer, because the preferred habitat of this species is deciduous forest and woodland. Other vernacular names that may be encountered are pine vole and "mole" mouse, the latter a reference to the fossorial habits. This species formerly was placed in a different genus, *Pitymys,* which includes a number of species in Eurasia. It seems to be closely related to the prairie vole, however, and for this reason is provisionally retained in *Microtus.*

Distribution. The woodland vole is found throughout most of the eastern deciduous forest, from southern New England and the Great Lakes states south to northern Flor-

ida and eastern Texas, and there is an isolated population in south-central Texas and a Pleistocene record from the Edwards Plateau in the central part of that state. The species barely reaches the southeastern corner of the tristate region in extreme southeastern Nebraska (fig. 128). The subspecies found on the Northern Plains is *Microtus pinetorum nemoralis*, which has been regarded by some specialists as a full species, distinct from typical *M. pinetorum* farther to the east.

Description. Like all voles, the woodland vole is a small rodent with a cylindrical body and relatively short legs and tail. Its eyes are small, and the small, rounded ears are mostly hidden beneath the fur. Both the upper incisors and the brownish gray forefeet are used for excavating soil. The upper parts are dull reddish brown to chestnut brown with a wash of dark-tipped guard hairs; the sides are paler, and the underparts are grayish buff. The tail is faintly bicolored, brownish above and dark gray below. The pelage is dense and velvety to silky in texture. The hair structure is adapted to both forward and backward movement in a tunnel, as in pocket gophers and moles. Sexes are alike in size and color, but juveniles are darker and more grayish than adults. The woodland vole can be differentiated from the other species of *Microtus* on the Northern Plains by the distinctly reddish color of its dorsal hair.

Two annual molts are reported for this species. The darker winter pelage is replaced in April or May by the bright reddish summer pelage, and molt to fresh winter pelage takes place in September and October. Young woodland voles undergo postjuvenile and postsubadult molts before attaining adult pelage, as is typical of microtines.

The degree of relationship of this species to the prairie vole is unclear. Characters shared by both include: five (rather than six) plantar tubercles; pattern of occlusal surfaces of molar teeth; structure of the penis; and fewer than four pair of mammae. *M. pinetorum* has only two pair of mammae (situated inguinally), in contrast to

134. Woodland vole, Microtus pinetorum (courtesy R. R. Patterson).

three in *ochrogaster* and four in *longicaudus* and *pennsylvanicus*.

Average and extreme external measurements of six males and three females from southeastern Nebraska are: total length, 134.0 (127–140); length of tail, 25.1 (20–28); length of hind foot, 19.2 (18–20.5); length of ear, 11.9 (11–12.5). Weights of three males and two females from northeastern Kansas averaged 35.0 (27.3–44.3). Average and extreme cranial measurements of three males and two females from southeastern Nebraska are: greatest length of skull, 25.9 (25.3–26.5); zygomatic breadth, 15.9 (15.4–16.2).

Natural History. Unlike most North American voles, this species is most at home in forest and woodland habitats, where it burrows beneath leaf litter and into the duff and soil of the forest floor. It does not make runways as other voles do, and its burrows are inconspicuous because they usually are concealed by leaf litter. The species inhabits a variety of soil types and forest associations in various parts of its range, but in southeastern Nebraska it is found in oak-hickory forest along the Missouri River and its tributary streams. Well-drained slopes with dense ground cover appear to be preferred. These voles also have been trapped occasionally in brushy pastures some distance from the edge of the

nearest woodland; locally they may occur with the prairie vole and southern bog lemming, and they probably use the runways of these species. In eastern populations, at least, openings leading to the ground surface from the burrow system are frequent, and conical piles of fresh dirt are associated with recently excavated burrows, reminiscent of molehills.

Few studies have been made of woodland voles, especially *M. p. nemoralis*. The animals appear to be tolerant of conspecifics and may live gregariously in burrow systems or may form monogamous pairs that share a burrow system with their offspring. The second speculation is based on analogy with the prairie vole but lacks supporting documentation. In any case, several adult voles usually are associated with a system of burrows; small colonies are discontinuously distributed in suitable habitat, and the general density of the population frequently is low. High densities build up locally from time to time, but the cause(s) of these outbreaks are not known, and woodland voles do not exhibit the marked cyclic fluctuations seen in many other species of *Microtus*. Although in most cases population density is 10 per acre or fewer, it may reach 50 per acre during periods of local increase, and there is one unconfirmed report of 500 woodland voles taken from two acres of orchard in New York.

Because these voles exhibit some degree of social organization, it is not suprising that individual home ranges show extensive overlap. A study in Oklahoma revealed the average size of home ranges of males and females to be 0.58 and 0.42 of an acre, respectively. These values are somewhat larger than those reported for the species in New York and Michigan. As in other voles, home-range size varies with habitat; in an area of favored woodlands, as judged by number of voles present, home-range size was smaller than in a marginal habitat of grass and brush—about four-tenths versus seven-tenths of an acre—in the Oklahoma study. Females seem usually to have home ranges 15 to 30 percent smaller than those of males.

Little is known about breeding behavior

in woodland voles, but what may be courtship, in which the female takes the initia-
tive, has been described. Copulation
induces ovulation; the gestation period has
been determined as 24 days, slightly longer
than in other species of the genus. The
young at birth weigh less than 2.5 grams,
are naked, and have eyes and ears closed.
By day seven the head and dorsum are well
covered with glossy brown fur; the incisors
begin to erupt on day five. The ears open
on day eight, and the eyes open between
nine and 12 days of age. Weaning commences early in the third week, and the
young become active outside the nest at
that time. Postjuvenile molt begins in the
fourth week; the subadult pelage subsequently is molted beginning in about week
seven so that adult pelage is acquired at
two months of age at an average weight of
20 to 25 grams. Juvenile woodland voles
may not disperse from their nests as quickly as other young *Microtus*. Sexual maturity is reported at day 40, but data are
scanty and variation undoubtedly exists.

The breeding season in the woodland
vole is about nine months long, from January or February through September or October. Winter breeding also has been reported
to occur in some years, and in Oklahoma a
summer lull was found, but what factors
may influence these occurrences are not
known. Litter size is smaller than in most
other species of the genus. The average
number of fetuses carried by 11 females of
M. p. nemoralis from Nebraska, Kansas,
and Oklahoma was 2.4, the range being
only two to three per female; in a study in
Oklahoma, 26 females yielded an average of
2.6 (two to five) fetuses. These samples
came from throughout the year, and the
low range of variation suggests that there is
little seasonal or geographic change in litter
size, in contrast to most other voles. Eastern subspecies may have larger litters, up
to eight young having been reported.

Postpartum estrus occurs regularly, and
the average interval between litters is about
25 days. In one study, woodland voles captured as adults subsequently lived an average of 2.3 months more, so a breeding
female may, on the average, have two or

three litters in her lifetime, somewhat
more than in most *Microtus*. Also in contrast to *M. ochrogaster* and *M. pennsylvanicus*, adult females had a higher
mortality rate than adult males, and juveniles did not seem to suffer the high mortality characteristic of other voles, perhaps
because of the protection offered by their
semifossorial habits.

The spherical nest in which the young
are raised is about 100 mm in diameter and
often is at the end of one of the subterranean tunnels. It consists of dry grasses and
leaves and is lined with finer plant fibers.
Surface nests are rarely used, but nests
frequently are found under pieces of flat
bark that lie on the forest floor or under
logs or stumps. The social nature of this
vole is highlighted by the finding of three
separate litters of young of different ages in
one nest, together with a single pregnant
and lactating female. The published interpretation of this was that both the oldest
litter (estimated age 18 to 20 days) and one
of the two younger litters (estimated ages
10 to 12 and eight to nine days) had been
produced by the adult female that was
found in the nest, but given the known
gestation period of the species this is not
possible, and the three litters must have
been the offspring of three females sharing
a communal nest.

This species is active both by day and by
night, and, although its surface activity
may be reduced in daytime, particularly in
summer, underground activity probably is
maintained during that period. Subnival
foraging in winter is frequent. Foraging behavior has not been described in detail, and
whether it is primarily on the surface or, as
seems more likely, mostly from the shallow burrows is not known. The bulk of
food items of this species consists of underground plant parts—roots, tubers, corms,
bulblets, and the like. According to one
study, woodland voles subsist mostly on
roots and stems of grasses in summer, on
fruits and seeds in autumn, and on roots,
inner bark of woody stems, and food caches
in winter. Large caches are sometimes
made, up to a gallon in volume, containing
seeds, sections of stems of grasses and

forbs, tubers, roots, hickory nuts, hazelnuts, and acorns. Caches are stored in underground chambers as much as 18 inches below the surface. In its habit of caching, this species resembles *M. ochrogaster.*

Because they eat bark in winter, woodland voles may become a serious pest in orchards, where they girdle fruit trees at their bases, usually just below the soil surface, sometimes killing a significant number.

Most authors agree that *M. pinetorum* is somewhat protected from predation by its semifossorial habits. However, at least six species of owls and four species of hawks are known to feed on it, and mammalian predators include the opossum, raccoon, mink, gray fox, and red fox. Short-tailed shrews (*Blarina*) are common in the habitat of woodland voles throughout their range and may kill them. Cotton rats (*Sigmodon*) also may suppress woodland vole populations where the two are sympatric.

Selected References. General (Benton 1955; Hamilton 1938); natural history (Gentry 1968; Goertz 1971; Kirkpatrick and Valentine 1970); systematics (van der Meulen 1978).

Genus *Lagurus*

These small voles are found only in western North America and central Eurasia. The name *Lagurus* is derived from the Greek words *lagos*, meaning "hare," and *ura*, meaning "tail." Members of this genus have short tails, that of *L. lagurus* of the Eurasian grassland being, relative to its body size, as short as that of a hare or rabbit. The vernacular name applied to the group is steppe vole or sagebrush vole. These animals are primitive in many respects, and their relation to other voles is enigmatic. One species occurs in the Old World, and another occurs in North America. The dental formula is 1/1, 0/0, 0/0, 3/3, total 16.

Lagurus curtatus: Sagebrush Vole

Name. The specific name *curtatus* refers to the short tail and possibly to the limbs of this vole. It is derived from the Latin adjective *curtus*, meaning "short," which in turn gave rise to an adjective, *curtal*, that specifically refers to short-tailed or cut-tailed; hence the English word "curtail."

Distribution. This vole has two principal centers of distribution in North America. The largest, and the one where the species is most common, includes the arid Great Basin and the adjacent Snake River Plains and Columbia Plateau of southern Idaho, and eastern Oregon and Washington. The second center includes the Northern Plains, from southern Alberta and Saskatchewan south to northern Colorado, then west into northeastern Utah. Whether these two centers are connected by populations in the intervening areas across mountain ranges, or whether they are effectively separated, is not certain. On the Northern Plains *Lagurus* is known only from western North Dakota and the extreme northwest corner of South Dakota (fig. 128). However, it is a rare and elusive little vole and may eventually prove to occur elsewhere in South Dakota or western Nebraska. The subspecies in the region is *Lagurus curtatus pallidus.*

Description. The extremely short tail of this vole does not extend much beyond the hind feet when they are pulled back. This and the long, rather fluffy, grayish pelage (pale gray to brownish gray dorsally and whitish to pale brown below) serve to distinguish it from other voles on the Northern Plains. There are two molts each year in adults, resulting in distinctive summer and winter pelages, the former being shorter and slightly darker. The soles of the feet are relatively hairy, in contrast to those of other voles, and the size is small. Females have two pair of pectoral and two pair of inguinal mammae. An adult male from southeastern Montana measured: total length, 139; length of tail, 19; length of hind foot, 16; length of ear, 12; greatest length of skull, 25.9; zygomatic breadth, 15.5. Two adult females from western North Dakota measured: total length, 137, 133; length of tail, 24, 27; length of hind foot, 18, 19; length of ear, 9, 12; greatest length of skull, 24.7, 25.1; zygomatic breadth, 14.6, 15.3. Weight of the male was 38.2 grams, and the two females weighed 27.9 and 32.3.

Natural History. Sagebrush voles usually are found in temperate, dry, shrub-steppe habitat, where grasses are interspersed with clumps of sagebrush (*Artemesia*) or rabbit brush (*Chrysothamnus*). Brush cover may be tall (three to four feet) to short (less than a foot), and grass cover may be sparse or dense; however, *Lagurus* usually is most abundant where vegetative cover is fairly dense. Poorly defined runways are formed in grass adjacent to or beneath rocks and at the bases of shrubs. An unusual habit of sagebrush voles is to hollow out dry droppings ("chips") of cattle (and previously probably those of bison). These chambers are used for temporary shelter from both inclement weather and predators and as feeding stations, but not as nest sites.

Sagebrush voles are more social than most of their relatives and live in small colonies. The burrow and runway system of a colony forms an interconnected cluster. Nests often are situated beneath rocks or

135. Sagebrush vole, Lagurus curtatus *(courtesy M. L. Johnson).*

the roots of shrubs at a depth of three to 10 inches. They are built of finely shredded grass and sometimes of bark, paper, or feathers, and they are from four by four to eight by 10 inches in width and length. Burrows are shallow, not penetrating more than a foot below the surface.

Groups of sagebrush voles of differing age and sex may share nests. Females in captivity even have been known to nurse young from another litter. Aggression between males has been observed, however, and at high densities serious fighting may occur.

Another unusual feature of the behavior of this vole is the mobility of colonies. In winter these are concentrated in areas where snow cover is deepest and most continuous, and the animals live underneath the shelter it provides. Such areas often are in valleys and on the lee slopes of hills. With the onset of spring and snowmelt, colonies shift to more elevated south slopes, where new plant growth is most advanced. As the summer progresses they move to lower elevations, as the more elevated sites become hot and dry. Colony movements also appear to be triggered by local depletion of food supplies. No reports of home-range size or population density appear to have been published for this species.

Sagebrush voles seem to be strict herbivores, but they feed on a wide variety of plant species, principally the leaves, stems, and flowers or immature fruits of grasses and forbs. Leaves of sagebrush and rabbit brush also are eaten, but it appears that *Lagurus* depends on deer mice to climb up into those shrubs and clip off the terminal twigs and leaves, whereupon the sagebrush voles steal the accumulated stores. This species also is reported to eat the inner bark of shrubs in winter and may resort to the undigested plant residues in cow "chips" during seasons when other foods are scarce. These voles appear to derive all their water from the foods consumed.

Foraging goes on at all hours, but there are peaks of activity in the hours around sunrise and sunset. Neither temperature nor cloud cover appears to have much affect upon activity, but strong wind greatly reduces surface movements. It has been suggested that this is because sagebrush voles are more dependent on hearing to detect predators than are other voles, and that wind reduces the sharpness of this sense. Certainly in their relatively barren shrub-steppe environment they often are exposed to avian predators. The most important of these is the burrowing owl, which hunts *Lagurus* by swooping down from a perch on the top of a shrub, or by waiting on the ground next to a burrow entrance. Like some other rodents of arid lands, sagebrush voles have enlarged auditory regions, which apparently are associated with their ability to hear the wingbeats of an approaching raptor. Mammals such as coyotes, badgers, bobcats, and long-tailed weasels, and probably foxes and grasshopper mice, also prey on *L. curtatus*, as do several species of snakes.

Sagebrush voles have a high reproductive rate. Unlike most other small mammals, they do not have a restricted breeding season, at least in Oregon and Washington, where they have been studied most intensively. Pregnant females may be found at any time of year, as may juveniles. There is some evidence, however, that reproductive effort may be reduced during periods of severe summer drought or winter cold. Females mature at about one to two months of age, as do males. The gestation period averages 25 days, and a wide range of embryos has been reported—from one to 13. Average litter sizes for different populations have varied from 4.4 to 6.1 per female, perhaps depending on the season when the sample was taken or on the amount of rain and resulting green vegetation in that year.

Newborn sagebrush voles are naked, with eyes and ears closed, and weigh about 2 grams. Between days 11 and 14, the eyes of most young open, and they move about readily. By about 18 days of age, they begin to feed on solid food and are weaned shortly thereafter. In contrast to other vole species, young *Lagurus* scatter from the nest if they are disturbed, and they are later re-trieved by their mother when the source of the disturbance is gone. This behavior may be a means of avoiding nest predation by mammals.

Like other small mammals, sagebrush voles are exploited by many parasites. Those described include 10 species of fleas (some accidental), as well as lice, mites, and ticks. Tapeworms have also been reported. *Lagurus* is an important reservoir for sylvatic plague in Washington.

Selected References. General (Carroll and Genoways 1980 and included citations).

Genus *Ondatra*

Whereas most scientific names are derived from classical roots, *Ondatra* is based on a native American name for the animal, from the Huron language, which passed into use in French Canada. Although the word ends in *a*, which would be feminine in Latin or Greek, it is masculine, and therefore the specific name must carry the appropriate masculine ending (-*us*). The genus is monotypic and found only in North America; the vernacular name alludes to the "musk" secreted by the large perineal glands. The dental formula is 1/1, 0/0, 0/0, 3/3, total 16.

Ondatra zibethicus: Muskrat

Name. Musky-smelling is the meaning of *zibethicus*, which refers to the product of the scent glands, as does the vernacular name. A similar-sounding name, "musquash," is rarely used; it is said to be derived from the Cree name for the animal.

Distribution. The muskrat is widely distributed in North America, from the tree line in Alaska and Canada south to the coast of the Gulf of Mexico (excluding Florida), the Rio Grande Valley in western Texas, and the lower Colorado River Valley. It was originally absent from most of California, but it was introduced into the Pitt River Valley and is now widespread throughout the Central Valley and else-

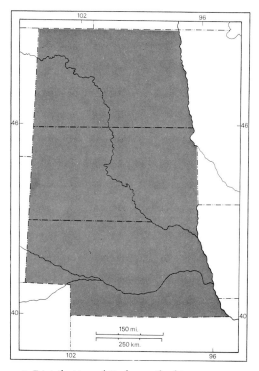

136. Distribution of Ondatra zibethicus.

a yellowish tinge and fewer guard hairs. The ventral fur is silvery gray, shading into white on the throat. The feet are dark brown, and the toes have a dense fringe of stiff hairs. Young are more grayish and duller than are adults. Juveniles undergo a molt to subadult pelage, followed by a molt to adult pelage. Adults seem to have a single annual molt. Musk glands are present in both sexes, although larger in males, and are situated at the base of the tail. Females have four pair of mammae, two pectoral and two inguinal, and the presence or absence of these nipples is the best way to differentiate males from females, because the sexes otherwise are quite similar externally.

Average and extreme measurements of eight adults (four females and four males) from southeastern Nebraska are: total length, 544.0 (515–578); length of tail, 242.9 (225–252); length of hind foot, 76.6 (71–81). The ear usually measures 20 to 25. Average weights of a large series (198 males, 215 females) from north-central Nebraska were reported to be 2.6 (1.6–3.4) pounds for males and 2.4 (1.7–3.2) pounds for females. Average and extreme cranial measurements of seven males and six females from western Nebraska are: greatest length of skull, 62.7 (60.5–64.8); zygomatic breadth, 38.9 (36.3–41.1).

Natural History. The muskrat is the most aquatic of microtine rodents. Whereas many voles may enter water and some spend a good deal of time there, the muskrat requires that a large part of its habitat be submerged in at least a few inches of water. This species is most common in marshland, such as in the Sand Hills of Nebraska or the lake region of North Dakota, where the combination of large expanses of shallow water and abundant emergent vegetation, such as cattails and bulrushes, provides both optimal security and food. *O. zibethicus* also inhabits lakes, rivers, and streams, although population densities are lower in these more open, deeper-water habitats.

The muskrat is an excellent swimmer and often is noticed by the V-shaped ripples

where. The species is found throughout the Northern Plains wherever suitable streams, lakes, or marshes occur. Humans have favored muskrats by creating impoundments ranging from small farm ponds to large reservoirs and by planting row crops near these newly created aquatic habitats. Only one subspecies, *Ondatra zibethicus cinnamominus*, is found in the region.

Description. The muskrat is the largest of the voles. It has several morphological adaptations, including dense, waterproof underfur, a laterally compressed, long, nearly naked tail, and large, partly webbed hind feet that adapt it to a mostly aquatic mode of life. Its heavyset body is supported by short legs, and its relatively massive head has small ears and eyes. The lips close behind the front teeth, which permits the muskrat to secure food underwater. The dorsal fur is blackish brown, darker on the head, with long black guard hairs that give it a glossy sheen. The sides are paler, with

137. Muskrat, Ondatra zibethicus *(courtesy V. B. Scheffer).*

it makes as it swims with its head partly above water and body submerged. Sometimes it may be seen feeding on a beach, riverbank, or platform in the water, in its typical "humpbacked" posture. Tracks in mud or sand along the edge of water, with the drag mark of the long tail, are another frequent sign of this rodent.

lodge

Other positive evidence of muskrats is the conspicuous conical piles of damp vegetation that are heaped in shallow areas of ponds or marshes. These "houses," which are from one to five feet high with a cavity in the center, are of two kinds—nest areas and feeding areas. Nest houses usually are the larger, as much as 20 feet in circumference, and contain a dry nest chamber well above the water level. Several entrances, termed "plunge holes," lead in and out of the chamber. Feeding houses or platforms are smaller and do not always contain an enclosed chamber; there usually are several within 100 feet of each occupied house. In summer one family occupies a single nest house and defends a limited territory around it.

As winter approaches, more vegetation and more mud are added to the house for insulation, and a larger group of muskrats may occupy the structure. The body heat of the group, held inside the house by the heavily insulated walls, raises the temperature within by an average of 20° C above that of the cold outside air. When warm weather returns, the house is "opened" to regulate the temperature of the nest chamber. In the hottest part of the summer, muskrats may move to a summer home with a completely open nest.

In streams and lakes with abrupt banks, muskrats may not find suitable habitat for houses. Here they construct burrows and nests. Bank burrows are excavated underwater and curve upward until they terminate in a dry nest chamber above the water level. Access is directly from deeper water to avoid ice blockage of the underwater entrance in winter. Sometimes the burrow entrance is inland, connected by a shallow canal to an adjacent beach or mud flat. Receding water necessitates elaborating the canal system, and shallow water between two bodies of deeper water frequently is deepened by a regularly used trail or canal.

Although they may be somewhat social in winter, muskrats disperse as the breeding season approaches. Temporary bonds are formed between pairs, and they either reoccupy old nest houses or build new homes. The area around the nest, to a distance of 100 to 200 feet, is the exclusive home range of the family. Inasmuch as the home ranges of adjacent families ordinarily do not overlap, it is assumed that the boundaries are defended or marked, and that the home ranges thus are territories.

territory

Within the territory, shallow platforms of vegetation may be used for resting or feeding, as well as for depositing scent marks and feces. Logs, rocks, and elevated sites along the edge of the water also are used as scent posts.

In winter, when ponds and marshes are frozen, the home range of muskrats occupying a nest house or bank burrow is marked by a series of "push-ups" radiating out from the nest in more-or-less straight lines. Push-up is the name given to a hole or crack in the ice into which a muskrat has "pushed up" a mass of vegetation. Within the mass a cavity sometimes is formed, with a shelf just above the water level. An average of 100 to 150 feet separates the winter lodge from the nearest "push-up," and these feeding houses are themselves spaced that far apart. Muskrats can swim at least 180 feet underwater without surfacing to breathe, so that spacing between "push-ups" and feeding houses probably represents the longest distance they can swim and forage beneath the ice and provides safe access to all parts of the winter home range while minimizing exposure to cold water.

Foods include a wide variety of aquatic plants, including roots, bulbs, stems, and leaves. Particularly common in the diet are cattails, bulrushes, sedges, lilies, and pondweeds. Leaves and stems are summer staples, but these become unavailable in winter, and roots are more heavily utilized. Muskrats use some animal food as well, particularly in winter, when they eat clams, snails, fish, frogs, and crayfish. There is no storage of food, but the inner walls of the house may be consumed in winter emergencies.

The muskrat is a prolific breeder and produces two or, more often, three litters per year (two to nine young per litter) after a gestation period of 29 to 31 days. Both sexes attain breeding condition in late March, although mating behavior appears to be controlled to some extent by the spring breakup of ice. On the Northern Plains this results in the first litters being born beginning in late April. Litter size varies geographically and from year to year, being higher in the north, where the breeding season is shorter, and in years when habitat conditions are good. The average litter size for our region usually is between six and seven. Females may undergo postpartum estrus, and second litters are produced in early June. Many adult females, and occasional juvenile females born in early spring, give birth to litters later in the summer as well.

The eyes of the newborn are closed, and the body is hairless; the young weigh about 20 to 25 grams at birth. In one week hair appears, and in 14 to 16 days the eyes are open and the 90-gram young begin to swim and feed. In three or four weeks they are weaned, disperse from the house, and build new summer nest houses within the family home range. In six months they attain adult size, and in a year or less they reach sexual maturity. The life span may be three or four years, but mean life expectancy is much less, owing to the high mortality rate of young.

As the breeding season draws to a close in late August and September, there is a general shifting and redistribution of individuals, especially young born that summer. This has been termed the "fall shuffle." Young muskrats moving out of their natal home ranges and attempting to establish home ranges elsewhere engage in intraspecific strife, the severity of which is directly related to density. Moreover, late summer drought, resulting in falling water levels, may force adult muskrats to seek new homes at this time as well. In such circumstances, many young animals may be killed in direct combat with adults or else fall victim to predators or exposure as they wander. Another period of movement occurs in spring when the ice breaks up and breeding begins. Males are principally involved, because most females stay within the home ranges they established during the previous winter. Males simply may disperse within the confines of a marsh or lake, or they may undertake cross-country migrations to new habitats. This latter course especially exposes wanderers to predation and intraspecific strife. Pair bonds are probably restructured for the coming season during this period.

The mink generally is regarded as the most important predator on the muskrat. Adult *Ondatra* within their home ranges are large enough to defend themselves successfully from this mustelid, but in unfamiliar territory, especially on land, they are vulnerable. Young muskrats are taken under many conditions, but especially when they are dispersing. Other mammalian predators, such as the coyote, red fox, and otter, are significant causes of mortality in certain times and places, as are eagles and the larger owls and hawks. Occasionally young muskrats are eaten by snapping turtles or large predatory fish such as northern pike.

Muskrats harbor a large number of parasites. Internally they host as many as 24 kinds of trematodes, as well as cestodes, nematodes, and several protozoans. External parasites include fleas and mites. This species also is susceptible to tularemia, a bacterial disease transmissible to humans, which is known to cause sizable muskrat mortality at times. Other diseases specific to muskrat populations include a fungal skin infection, to which juveniles are particularly subject, and a poorly understood hemorrhagic disease that occasionally may cause severe losses.

Muskrat mortality also is induced by fluctuating water levels. Flooding may drown nestlings and damage food plants, and drought exposes muskrats to predators and, by concentrating animals, may result in local exhaustion of food resources and ultimate starvation.

Population densities fluctuate in response to changing water levels and other climatic conditions as well as other factors. Muskrats sometimes become extremely abundant in a marsh and consume most of the emergent aquatic vegetation. Such "eatouts" are followed by a sharp decline in numbers, as might be expected, and the marsh environment may take several years to recover. However, the causes of most muskrat population fluctuations are obscure.

Densities of 25 *O. zibethicus* per acre may be maintained in optimal habitats, but in most marshes there are about 15 animals per acre. Populations along streams and rivers seldom exceed 15 or 16 per mile of bank, and in many places average densities are about one muskrat per acre. Better habitats support an important fur industry, and many farmers and other landowners control water levels in wetlands to encourage muskrats.

Trappers in Nebraska in 1949–52 averaged about $264,000 per year for the value of the muskrat harvest. During the winter of 1977–78, more than 46,000 muskrats were harvested in Nebraska and South Dakota (nearly 41,000 in Nebraska alone), with a value of about $168,000. The trapping season is restricted to the winter period (usually December through March) when the fur is "prime" (that is, when molt to winter pelage is complete). Muskrat meat is palatable but not much used. On the negative side, muskrats may damage crops such as corn when fields are near water, and their burrowing activities may weaken dikes, dams, and drainage ditches and cause them to leak or collapse. Local control, combined with regular harvesting of populations, usually results in muskrats being one of the most economically valuable of all Northern Plains mammals.

Selected References. General (Willner et al. 1980 and included citations).

Genus *Synaptomys*

The generic name for these primitive lemmings is derived from two Greek words, *synaptein*, meaning "to unite," and *mys*, meaning "mouse." At the time the genus was named, this rodent was regarded as intermediate between the middle-latitude voles and the high-latitude true lemmings. The use of "bog" in the vernacular name is a partial misnomer in that these animals often are found in habitats other than bogs. The two species in the genus are, however, lemmings rather than voles, and they are primitive in a number of traits. Although fossil *Synaptomys* are recorded from Eurasia, the genus now is restricted to North

America, one species, the northern bog lemming (*S. borealis*), being found in the transcontinental northern coniferous forest, or taiga, and a second species farther to the south, including a part of the Northern Great Plains. The geographic ranges of the two species overlap in northwestern Minnesota and southeastern Manitoba as well as in upper New England and southern Quebec. The dental formula is 1/1, 0/0, 0/0, 3/3, total 16.

Synaptomys cooperi:
Southern Bog Lemming

Name. William Cooper (1797–1864), the American zoologist who collected the first specimen, lent his surname to this species.

Distribution. Although occasionally local populations become abundant, the southern bog lemming is not often encountered over much of its range. In Nebraska the species was unrecorded until 1947, but subsequent collecting has yielded a number of records in the eastern part of that state, and the species probably occurs also in extreme southeastern South Dakota. The subspecies there is *Synaptomys cooperi gossii.* Much of the natural habitat in this part of the tristate region has been converted to farmland, thereby restricting these lemmings to fencerows, damp corners of cultivated fields, swales, grassy riparian communities, and bogs. The distribution of the species coincides in a general way with that of the eastern deciduous forest and the mixed deciduous-coniferous forest that occurs from the Great Lakes region east through New England. Within this large area, bog lemmings apparently occur sporadically, probably in response to the distribution of suitable habitat. An isolated population, *Synaptomys cooperi relictus,* is known only from the vicinity of cold-water springs draining into Rock Creek, Dundy County, in southwestern Nebraska.

Description. The bog lemming resembles the woodland vole in having a short, thick body and short tail, but it differs externally in having rougher, coarse-textured, longer,

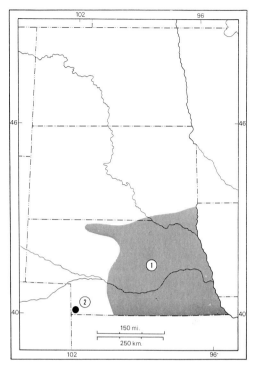

138. *Distribution of* Synaptomys cooperi: (1) S. c. gossii; (2) S. c. relictus.

darker-colored hair on the back and long grayish and silvery hairs on the venter. The dorsal pelage is dull brown to golden brown, mixed with black- and gray-tipped hairs. Juveniles have darker, slate gray fur. At about 34 to 37 days of age, a post-juvenile molt replaces the gray juvenile coat with a subadult pelage similar in color to that of adults. In Ohio the molt of adults to summer pelage is not completed before August. Winter pelage, attained by November, is longer, softer, and more grayish than that of summer. Animals captured in midsummer generally have a dark brown summer coat, but as hairs wear the dark appearance is replaced by a bright ochraceous tawny cast. Molt patterns in *Synaptomys* are similar to those in other microtines, beginning in the middorsal region and radiating in all directions. Other differences from woodland voles include six (rather than five) plantar pads, three (rather than two) pair of mammae in females, and whitish hairs in the region of the hip glands, especially in breeding males. The tail is short, nearly the same length as the hind foot. Prairie and meadow voles have longer tails and also differ in dorsal color (see accounts of those species). The upper incisor teeth of *Synaptomys* possess grooves along the outer edge that are unique among the several microtines of the Northern Plains.

Average and extreme external measurements of five males and three females of *S. c. gossii* from eastern Nebraska are: total length, 142.0 (136–152); length of tail, 22.2 (18–25); length of hind foot, 21.1 (20–26); length of ear, 11.7 (10–13). One male weighed 41.3 grams. Specimens of *S. c. relictus* do not differ in external size from *S. c. gossii* but are slightly paler in comparable pelages and have smaller teeth and shorter molariform toothrows. The sexes are colored alike. Average and extreme cranial measurements of four males and one female of *S. c. gossii* from eastern Nebraska are: greatest length of skull, 30.0 (29.6–30.6); zygomatic breadth, 18.2 (17.7–18.5).

Natural History. The habitat of *Synaptomys cooperi* appears to be quite variable. In the northern part of their range in the Great Lakes region and to the northeast, southern bog lemmings are found in spruce bogs in summer. During other seasons they move into adjacent upland areas. In the southern parts of the range, populations often are restricted to sphagnum bogs where daytime temperatures are several degrees lower than in nearby areas. Over much of the range of the species, however, bog lemmings are found in grasslands, often in bluegrass associations, and in damp areas where sedges (genus *Carex*) predominate. They appear to favor stabilized communities with a thick mat of dead grass covering the ground and with stems of grasses or shrubs forming a dense, high canopy overhead. The margins of permanent springs, damp to wet grasslands, and marshes are favorite refuges. From these stable areas the mammals move out onto upland grasslands and other adjacent communities. They do not avoid grassy areas containing shrubs and trees as meadow voles may do. Bog lemmings also may inhabit forest and woodlands that have a dense ground cover of herbs and forbs, and they may share this habitat with woodland voles. Finally, while they often occur with prairie voles, they are not found in the dry short-grass habitats that the latter species can tolerate.

Bog lemmings use runway systems built by other microtines and also build and maintain systems of their own, although these tend to be less "tidy" and well-defined than those of *Microtus*. Nests three and a half to eight inches in diameter are constructed along principal trails in the system, usually just below ground level or under logs. The inner cavity of the nest is lined with shredded grasses or sedges. In spruce and sphagnum bogs, nests are built within grass or sedge clumps or hummocks, above the water level. In adjacent forested situations, runways are built under leaf mold and litter and may be mistaken for those of the red-backed vole or woodland vole.

S. cooperi is active in both summer and winter and, because of dense overhead cover, individuals frequently are active during the day. In fact, they probably are equally active during daytime and dark, although daytime activity may be reduced during summer months. In winter these microtines have been trapped in temperatures well below freezing. Lemmings probably

139. Southern bog lemming, Synaptomys cooperi *(courtesy R. R. Patterson).*

leave their runways and tunnels only to cut grasses and sedges, their principal foods. Leaf and stem cuttings accumulate in runways, and often entire leaves are pulled into holes. The bulk of the diet consists of fleshy leaves and stems of sedges, which are finely ground before being swallowed. Rootstocks of sedges and grasses are also dug out and consumed, and fruits (such as raspberry, huckleberry, and blueberry) are an important part of the summer diet. Occasionally fungi, mosses, and bark are eaten. The presence of *S. cooperi* often can be detected in runways by the bright green feces typical of this species, combined with piles of uneaten sections of leaves and stems of about equal length (one to three inches long). Other voles tend to void darker feces, and their piles of cuttings usually are smaller and less regular.

Except for a study in southern New Jersey, little is known about reproduction and development in bog lemmings. Breeding may occur throughout the year, but greatest activity is from early spring to late autumn in both New Jersey and Kansas. Gestation lasts 21 to 23 days, postpartum matings occur, and litter sizes range from one to seven, averaging about three young per litter.

The litter size is lower than that of many voles of similar size living in similar habitats, and the breeding season of voles usually is not so extended. It is likely that a female bog lemming produces three or four litters in the course of a year, if she survives. Birth weights ranged from 3.1 to 4.3 grams (mean 3.9) in one study, and development was found to be slightly faster in smaller litters. The pinnae unfold by the third day, and hair begins to appear by the fifth day. Incisors erupt between the sixth and eighth days, the eyes open on day 11 or 12, and the juvenile pelage is completed by the time young bog lemmings are two weeks old. Weaning begins about that time and continues for a week or so, depending on how soon the mother gives birth to another litter.

Young weigh about 12 grams when weaned, but their rapid growth continues until adult pelage is acquired, when weights approach 25 grams in laboratory-reared animals. Sexual maturity probably is attained at about this time. Growth then slows abruptly; adults generally weigh between 30 and 40 grams. Animals captured in juvenile pelage are probably two to five weeks old, weighing from 9 to 22 grams. It thus seems that young bog lemmings begin to forage at an early age under natural conditions. Young born early in the spring have a low probability of living through the following winter, whereas those born in autumn have a more favorable chance of surviving through the winter and into the next year. Adults generally have a lower mortality rate than young, and survival over winter is usually higher for older ages. One year is generally considered average maximum longevity. Known predators of bog lemmings include coyotes, red foxes, raccoons, long-tailed weasels, and several species of owls. They harbor tapeworms, ticks, mites, lice, and fleas.

Estimates of the home range of the southern bog lemming vary greatly, probably owing to small sample sizes and the different trapping regimes employed in different studies, but such ranges probably are less than one acre, and more often a tenth to a half an acre. Males have, on the average, larger home ranges than do females, and they overlap those of females as well as those of adjacent males. There is some evidence, however, that home ranges of females tend not to overlap, suggesting that they may defend territories. Estimates of populations are few, but densities probably are much lower than are usually reported for the genus *Microtus*, perhaps on the order of five to 15 per acre. One estimate of 35 per acre probably represents an uncommonly high density. Populations fluctuate from year to year, but whether they exhibit a typical "microtine cycle" is not agreed upon. Bog lemmings also show long-term changes in population density that are not yet understood. For example, *S. cooperi* was considered quite rare in northeastern Kansas through most of this century, with the single exception of the period 1924 through

1928. However, it began to increase in numbers again in the late 1960s and remains fairly common at present. Perhaps such changes result from the response of a primarily northern species to subtle climatic changes at the southern edge of its range. Southern bog lemmings have little or no adverse effect on humans; there are no references to economic loss attributable to them.

Selected References. General (Connor 1959); ecology (Gaines, Baker, and Vivas 1979); systematics (Wetzel 1955).

Family Zapodidae: Jumping Mice

North American jumping mice are near relatives of the jerboas (family Dipodidae) of the Old World deserts and steppes and are even more closely related to the birch mice (genus *Sicista*) of the Palaearctic, which also are included in the family Zapodidae. The dental formula is 1/1, 0/0, 1/0, 3/3, total 18, in *Zapus*, although the small upper premolar occasionally is missing, as it regularly is in the woodland jumping mouse (*Napaeozapus insignis*) of the eastern United States and Canada. A third kind of jumping mouse, *Eozapus setchuanus*, is confined to the remote and inaccessible mountains of southwestern China. Two of the three recognized species of the genus *Zapus* occur on the Northern Great Plains.

No key is provided to distinguish between the two jumping mice found in the tristate region, because they so closely resemble each other where they occur together on the Northern Plains that no truly "key" characteristics can be used in identification. Rather, a combination of traits must be employed, as explained in the account of *Z. hudsonius*.

Genus *Zapus*

Jumping mice typically are found in deciduous and coniferous forests and in meadows and riparian vegetation along streams. In the mountains they extend up into the alpine zone. On the Northern Plains, however, they usually occupy marshy areas and moist riparian habitats in the more open landscape of the prairie, even venturing into relatively dry grasslands in some places. The name of the genus is derived from two Greek elements, *za*, which is a prefix added for emphasis that may be translated as "exceedingly," and *pus*, meaning "foot," in reference to the large hind feet. The vernacular name alludes, of course, to the leaping ability conferred on these mice by their long hind legs and feet.

Zapus hudsonius: Meadow Jumping Mouse

Name. The specific name *hudsonius* refers to Hudson Bay; the specimen from which the species was first described was reported to have been captured along the southwest shore of the bay. The vernacular name refers to preference for meadows.

Distribution. The meadow jumping mouse occupies a broad transcontinental range, from Labrador to Alaska south of the tree line in the northern coniferous forest, or taiga. The species also occupies the eastern deciduous forest region as far south as Georgia and Alabama and occurs westward across much of the Northern Great Plains.

The subspecies *Zapus hudsonius pallidus* occurs in the southern part of the Northern Plains, throughout much of northern and central Nebraska, as well as in the southeastern part of that state, and in southcentral and eastern South Dakota. It is the smallest, palest, and most brightly colored of the races in the tristate region. Larger, duller-colored jumping mice occur to the north and west. *Zapus hudsonius intermedius* inhabits eastern and northern South Dakota and virtually all of North

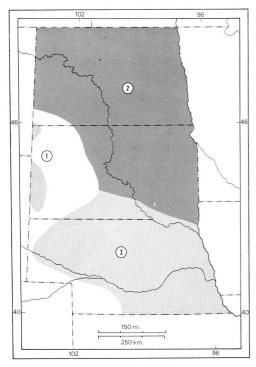

140. Distribution of Zapus hudsonius:
(1) Z. h. campestris; (2) Z. h. intermedius;
(3) Z. h. pallidus.

dorsal stripe. The paler sides are yellowish
orange, with fewer guard hairs, and the
underparts are white. The bicolored tail is
dark brown above and yellowish white below, and its slender tip is dark. The ears
also are dark, edged with buff or white, and
extend beyond the fur of the head. The feet
are pale yellow gray to white. Spring hair
loss in May and June is diffuse and incomplete, but the molt to winter pelage in
late August and early September is complete and well defined, and jumping mice
often are regarded as having only one molt
annually. Young animals are duller than
adults, and their fur is softer. They undergo
a single postjuvenile molt in the fourth
week of life. Jumping mice can be differentiated from kangaroo rats and pocket
mice by their lack of external cheek pouches;
neither do they possess internal cheek
pouches as has sometimes been reported.
Average and extreme measurements of nine
Z. h. pallidus from Cherry County,
Nebraska (five males, four females), are:
total length, 206.6 (194–220); length of tail,
119.7 (112–131); length of hind foot, 29.1
(28–30); length of ear, 12.6 (12–13.5).
Weight of five males and a female averaged
18.8 (16.5–22.1) grams.

In eastern North Dakota and extreme
northeastern South Dakota, the meadow
jumping mouse is found in the same places
as a similar species, the western jumping
mouse, *Zapus princeps* (see below). The
two taxa are difficult to identify, but comparison of individuals from continuous habitat in North Dakota indicated that the
following features are useful in distinguishing between them: *princeps* is paler (having
a somewhat drab appearance caused by
many white or grayish hairs in the dorsal
stripe), the edge of the ear is bordered by
whitish hairs, and the tail is not sharply
bicolored; *hudsonius* is darker, with more
orange on the sides and fewer white or
grayish hairs in the dorsal stripe, usually
has no conspicuous border on the ear (if a
border is present it is composed of yellowish hairs), and the tail is sharply bicolored.
Additionally, *princeps* tends to be the larger
of the two. Average and extreme cranial
measurements of six males and four

Dakota and intergrades broadly with the
smaller southern populations. *Zapus hudsonius campestris* is found on the Black
Hills and in adjacent isolated mountain
ranges of Wyoming and Montana. It is the
largest of the races of meadow jumping
mice and is darker in general coloration.
Integradation with *intermedius* and *pallidus* probably is restricted by the scarcity
of suitable habitat over much of the western part of the region. Even in the more
mesic parts of the Northern Plains the distribution of the species is local and erratic.

Description. Zapus hudsonius is one of the
most graceful and colorful of small mammals, characterized by a tapering, sparsely-
haired tail that is longer than the head and
body. It also possesses large hind legs and
feet adapted for jumping, which contrast
with its small front legs and feet. Upper
parts are clothed in long, coarse yellowish
brown hair. Black-tipped hairs form a dark

females of *Z. h. pallidus* from Nebraska are: greatest length of skull, 22.9 (22.4–23.2); zygomatic breadth, 11.4 (10.8–11.8).

Natural History. The meadow jumping mouse usually is found in meadows, in moist, brushy abandoned fields, in dense vegetation along the edges of marshes, ponds, and streams, and even in woodland habitats and on the forest floor where the ground cover is dense. Adequate herbaceous or grassy ground cover is essential for the species. On the Northern Plains this usually results in its restriction primarily to riparian habitats.

These mice are essentially solitary in their habits at all seasons. Although home ranges overlap, there is little evidence for intraspecific antagonism, at least in captive individuals. Home ranges vary greatly in size, from 0.14 acre to more than four acres depending on habitat quality. In better habitats, population densities are higher (11.9 per acre in one Minnesota study), and home ranges are smaller, with those of both sexes being about the same size. In poorer habitats, such as those usually found on the Northern Plains, densities are lower (1.8 to 3.6 individuals per acre), home ranges are larger, and there is some indication that males occupy larger areas on the average (2.7 acres), than do females in the population (1.57 acres). There also is evidence that home ranges of jumping mice are less stable than those of other small mammals; for example, marked individuals suddenly may abandon an area they have been occupying and reappear elsewhere. Equally unstable are population densities, which fluctuate

considerably from year to year. The causes of these shifts and changes are not yet understood.

Territorial behavior has not yet been described for *Z. hudsonius*, but if it occurs it probably is restricted to the area around the nest. Summer nests are "neat little balls of fine grass with a tiny opening at one side and a soft lining in the central chamber," according to one observer. They may be situated in a clump of grass on the ground, beneath a rock or log, or at the end of a short, shallow burrow dug by the mouse, usually no more than about six inches below the surface.

Jumping mice, unlike many other small rodents, are true hibernators. To survive a long period of winter torpor, they must avoid prolonged exposure to freezing temperatures. Winter nests therefore are deeper underground, below the frost line in a well-drained site, usually at least a foot (and as much as 27 inches) below the surface. The nest itself is a ball of dead grass or leaves about four inches in diameter, within which the torpid jumping mouse is rolled up in a ball. Its head is tucked between its hind feet, its nose rests against the lower abdomen, the forefeet are pressed to the chest, and the long tail is curled completely around the head and body.

Most adults enter hibernation in late September or early October, probably depending on the availability of autumn foods necessary for fat accumulation. The latter is a rapid process. It begins about the end of August or first of September and probably results from a reduction in the basal metabolic rate rather than an increase in food consumption. The accumulation of fat

141. *Meadow jumping mouse,* **Zapus hudsonius** *(courtesy W. J. Hamilton, Jr.).*

probably is controlled by day length and mediated via hormone production, but much work remains to be done on this aspect of the physiology. Adults begin accumulating fat before juveniles, increasing the average weight from 16 to 19 up to 26 or more grams before becoming torpid. Jumping mice born in the same summer at first lag behind adults in fat deposition by at least two weeks, but ultimately they attain nearly the same prehibernation weight.

Once hibernation has begun, these jumping mice are not active aboveground until late April or early May; the total period of torpor thus usually exceeds seven months. While in the winter nest, the animals may periodically "break" torpor, but for most of this time the body temperature is between 35° and 40° F, and both heart and breathing rates are greatly reduced, as is true of other torpid mammals. Periodic arousal, when the body temperature increases to near normal, may be related to eliminating accumulated metabolic wastes by urination. All jumping mice appear to have sensitive control over metabolic rate, so that emergence and reentry into torpor during the hibernation period may be thought of as "voluntary." This control also is reflected in the fact that an individual experimentally exposed to air temperature below freezing increases its metabolic rate to prevent its body temperature from falling to the freezing point, which would, of course, result in death.

Males tend to emerge from hibernation in spring before females. Mating at the time of emergence (which may vary geographically and from year to year), followed by a gestation period of 18 to 19 days, results in the first litters of young being born as early as late May, but most are born in the first two or three weeks of June. Some females do not breed upon emergence but delay mating, and their litters do not appear until July. Postpartum estrus has not yet been demonstrated in this species, and it is possible that females do not breed immediately after the birth of their first litters but wait until they have nursed their young for a week or two. The gestation period of lactating females may be extended to 20 to 21 days. Most adult females probably have two, or rarely three, litters in the course of a summer breeding season; birth peaks occur in early June, mid- to late July, and mid-August, but a few have been recorded as late as mid-September. Juvenile females in early litters (May and June) may attain sexual maturity at six to eight weeks of age, breed, and produce August litters, but it is not certain how frequently this occurs in the wild.

Litter size averaged 4.5 (range two to eight) to 5.7 (range four to seven) in different studies but probably varies depending on age of the female and timing of the litter; however, no data are published on this point. Adult females have two pectoral, four abdominal, and two inguinal mammae.

Newborn *Zapus hudsonius* weigh about 0.8 gram (0.7–1.0); their eyes and ears are closed, and their bodies are hairless. Dorsal hair is present by the ninth day, and the incisors erupt on day 13. The ears begin to open on day 19, and the eyes open on day 22. By the end of the third week the juvenile pelage is complete, but almost immediately there is a molt into an "adult" pelage, which is attained by the end of the fourth week. Weaning is completed by then and the young are independent, with a body weight of 8 to 11 grams. Growth continues more slowly after weaning; at day 60 body weight is 14 to 15 grams, and at day 90, 18 to 20. This relatively slow growth rate has important implications for attainment of fat stores adequate for successful hibernation. A young meadow jumping mouse born early in summer (by mid-June) has by mid-September reached adult weight and can accumulate fat rapidly in the next two weeks and enter hibernation. Young born later in the summer must delay correspondingly their entry into hibernation, accounting for late October and November records of active jumping mice. A juvenile born as late as mid-August probably cannot attain suitable prehibernation condition before late November; the later the birth date of juveniles, the less likely they are to survive to enter hibernation, or, having entered, to survive the winter.

Winter mortality is probably the most important cause of death in jumping mice less than a year old and may be important in adults as well (see account of *Z. princeps*). Predators also take a toll during the five months or so these mice are active, owls, hawks, weasels, snakes, foxes, coyotes, and skunks being the principal predators. Jumping mice seem to try to avoid predators by making a few jumps of about a yard in length, varying the direction at each jump, then suddenly "freezing" motionless to avoid detection. They also may take to the water if it is near, being excellent swimmers and divers. It is not known whether mortality due to predation is less than encountered by other small mammals, but the long period of relative safety during hibernation makes this likely.

Perhaps because they are solitary, meadow jumping mice also seem to support fewer parasites than other small mammals. Nevertheless, the list is impressive—flukes, flatworms, and roundworms internally, and ticks, chiggers, mites, and fleas externally.

Zapus hudsonius is omnivorous, its diet varying seasonally in response to availability of preferred foods. In spring, animal matter (mostly caterpillars, ground beetles, and weevils) makes up about half the food. As the warm season continues and seeds ripen, these become more important, and insects diminish in the diet. Grass seeds are particularly favored; these are used as they ripen over the summer. Seeds of other herbs also are eaten. Fruit is extensively consumed in late summer and autumn, as is the soil fungus *Endogone*. Early reports suggested that jumping mice might store food for winter use, but no recent observations support this.

Selected References. General (Whitaker 1972 and included citations).

Zapus princeps: Western Jumping Mouse

Name. The vernacular name of this species alludes to its occurrence in western North America, contrasting it with the meadow jumping mouse of the north and east (see above). The scientific name *princeps* is a Latin word meaning "the first" or "chief," bestowed on this large and handsome mouse as a laudatory appellation.

Distribution. This mouse occurs in western North America. The Rocky Mountains, from the extreme southern Yukon south to central New Mexico and Arizona, are the main range of this species, but it also occupies suitable habitat westward across the Snake River Plains, Columbia Plateau, and northern part of the Great Basin to the Sierra Nevada and eastward across the northern Great Plains. It is found in the tristate region in northern and eastern North Dakota southward to extreme northeastern South Dakota.

Description. Z. princeps is difficult to distinguish from the meadow jumping mouse. Characters useful in telling the two species apart are mentioned in the account of the latter. Timing of molt is probably the same in both species, although the patterns of

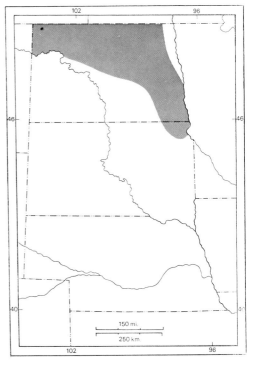

142. Distribution of Zapus princeps.

hair replacement are slightly different. *Zapus princeps minor*, the subspecies occurring on the Northern Plains, is small for the species, thus approaching *Z. hudsonius* in size where the two occur together and rendering immediate identification difficult.

Average and extreme external measurements of 11 males and 14 females from eastern North Dakota are: total length, 215.2 (204–230); length of tail, 126.4 (118–138); length of hind foot, 28.2 (27–30); length of ear, 12.5 (11–14). Weights of 22 of these specimens averaged 20.3 (18–24) grams. Average and extreme cranial measurements of 10 males and 12 females are: greatest length of skull, 23.6 (23.0–24.4); zygomatic breadth, 11.9 (11.3–12.4).

Natural History. Like *Z. hudsonius*, this species is found most often where grass and herbs grow tall and lush, providing a dense ground cover. On the Northern Plains this habitat usually is found along banks of streams and ponds and in marshes. No obvious differences exist between the habitat preferences of the two species where they occur together, but this has not yet been studied carefully, and subtle differences may occur.

Like the meadow jumping mouse, the western species seems to be solitary, but individuals are tolerant of overlap of home range with their neighbors. The few summer nests and burrows that have been described are like those of *Z. hudsonius*, as are the general patterns of behavior. Home-range size—0.76 acre for males and 0.58 acre for females in one study—is within the range of that determined for the meadow jumping mouse. The shape of the home range often is elongated, stretching along the bank of a stream or lake in suitable habitat.

Zapus princeps eats a mixed diet, with insects and other invertebrates an important component in spring, followed by a major shift to seeds in midsummer; there is some evidence that seeds are of even greater importance in the diet of *Z. princeps* than in that of *Z. hudsonius*.

The breeding season in the western jumping mouse is perhaps shorter than in *Z. hudsonius*, beginning later and terminating earlier. A study in eastern Wyoming revealed that males first emerged from hibernation between mid-May and mid-June, depending on elevation and spring thaw, and that females emerged nine to 12 days later. Breeding occurs within the first week after the females emerge, and the young are born following a gestation period of 18 days. Most young thus are born in late June or early July, and one litter per year may be the rule rather than two or three as in *Z. hudsonius*. The average litter size was 5.4 in the Wyoming study, with a range of four to eight, and in another study it was five (range two to seven). Development of the young has not been described, but it probably resembles that found in *Z. hudsonius*. Lactation extends for at least a month, and the young continue to nurse after they have begun to eat solid food. Young of both species of jumping mice, then, have a rather prolonged period of dependence on the mother compared with that found in most other small rodents.

The pattern of fat accumulation in preparation for hibernation differs in some respects between the two species. Rapid prehibernation fat deposition begins seven

143. Western jumping mouse, Zapus princeps, *in hibernation (courtesy V. B. Scheffer).*

to seven and a half weeks after emergence and occupies four weeks in *Z. princeps*, about twice as long as in *Z. hudsonius*. Its initiation is coincident with seed-set in many grasses and herbs, at which time the mice switch over almost exclusively to this new food source. The young, born in early summer, are completely weaned at that time, and they too fatten rapidly on a seed diet. The more prolonged period of fat deposition results in accumulation of reserves up to two-thirds of body weight, compared with up to half the body weight in *Z. hudsonius*. Body weight just before entry into hibernation in early or mid-September averages above 35 grams. Both species of jumping mice, when fully "prepared" for hibernation, begin to lose weight; this may represent a last "fine tuning" and alteration of metabolism accompanied by emptying of the gastrointestinal tract so that torpor can be achieved.

Once hibernation has begun, the first phase, about three months long, is characterized by arousal from torpor an average of once every nine days, and weight loss is fairly rapid. Frequency of arousal is higher at the beginning of this phase and gradually declines; eventually the mouse enters a second phase of torpor in which arousals are much less frequent, averaging once in 21.3 and 38.0 days in two separate studies. Weight loss during this longer second phase is much reduced and may be due to more efficient heat conservation associated with improvement in hibernating posture (see account of meadow jumping mouse above); at the start of hibernation, western jumping mice simply are too fat to curl up in a tight ball. The greater fat reserves of this species permit it to hibernate somewhat longer than is typical for *Z. hudsonius*; the active period of *Z. princeps* averaged only about 87 days in a study in Utah. Thus this species has less than a quarter of the year to breed, rear young, and accumulate fat in preparation for the coming winter.

Little margin of safety exists in such a tight schedule, and some mice fail to meet it. A jumping mouse, either adult or juvenile, that fails to put on enough fat in the autumn will either be caught by the onset of bad weather and perish or else be forced to enter hibernation with perilously low fat reserves. If the latter occurs, and body weight falls to the critical level of 18 or 19 grams (corresponding to about 15 percent body fat) before it is "time" to wake up, the mouse either will fail to arouse, will continue to lose weight and die, or else will arouse and leave the winter nest to face a hostile late winter or early spring environment in which its chances of survival are low.

How does a jumping mouse know when it is "time" to arouse? The cue apparently is the gradual increase in the temperature of the soil surrounding the winter nest. As the spring sun melts late winter snow and warms the surface, this heat penetrates to the lower levels, and when it reaches a temperature of 8.5° to 9° C there is a virtually synchronous arousal of all hibernating *Z. princeps* that survived the winter. Elevation, slope exposure, and year-to-year changes in snowfall and in spring weather all may affect the rise of soil temperature, so that the time of arousal varies from year to year, and thus the hibernation period may be longer or shorter. This is critical to the survival of young jumping mice that are hibernating for the first time, because they are smaller and can carry less body fat than adults. During a long hibernation period imposed by a snowy winter and a cold spring, as many as 70 percent of the hibernating juveniles may not survive. During a period shortened by little snow accumulation or a warm spring, mortality is lower (40 percent in one study), though still higher than that in adult mice. Being larger, adults usually can accumulate sufficient fat and consequently are less sensitive to the length of the hibernation period. Even so, about 25 percent of all adults that enter hibernation fail to reappear the following spring. Western jumping mice that survive the hazards of their first winter are likely to live longer than many other species of small mammals. Unfortunately, few data are available on this point.

Inasmuch as the gradual warming of the soil around the winter nest is the signal for arousal to hibernating jumping mice, the nest must not be so deep that the effect of spring sun and warmer air temperature is unduly delayed. The average depth of 26 hibernacula studied in Utah was about two feet and the range of variation fairly small (plus or minus six inches), although two found in Wyoming were less than a foot underground. All nests were in mounds or banks elevated above the surrounding ground level and thus were well drained. During the cold season (December to May) the soil temperature at the average depth of winter nests remained near 4.6° C, well above the freezing point. The nest itself, like that of the winter nest of *Z. hudsonius*, is composed of vegetation gathered in the vicinity and shredded to form a soft, highly insulative core. A single burrow three to four feet long leads to the nest chamber. When the jumping mouse enters hibernation, it plugs the first six to 10 inches solidly with dirt, and the next 20 inches may contain loose soil. The final portion of the burrow leading into the nest chamber is open, and it is here that the jumping mouse retreats to urinate when it periodically arouses from torpor. Like the meadow jumping mouse, the western species does not cache food in or near its winter nest.

Because they spend so much time in a secure underground chamber, western jumping mice are less vulnerable to predators than are some other rodents, although hawks, owls, weasels, and garter snakes are known to prey on them. Relatively few parasites have been recorded, either, but records include two species of tapeworms and several roundworms.

Selected References. General (Krutzsch 1954); ecology (Brown 1967a; Clark 1971; Cranford 1978).

Family Erethizontidae: New World Porcupines

The Erethizontidae are characterized externally by quills, which are hairs modified into sharp spines with imbricate barbs. The feet are broad and plantigrade, modifications for arboreal life. The skull has prominent auditory bullae, and the infraorbital canal is larger than the foramen magnum. Fossil erethizontids are known from the Oligocene to the Recent in South America, and from the late Pliocene to the Recent in North America.

The geographic range of the family extends over much of North America, except for the southeastern and south-central United States, southward to northern Mexico, and from southern Mexico to northern Argentina east of the Andes. The family consists of 11 living species in five genera.

Genus *Erethizon*

The genus *Erethizon* includes a single species, *E. dorsatum*, the second largest native rodent of the Northern Plains. The dental formula, as in other erethizontids, is 1/1, 0/0, 1/1, 3/3, total 20; the incisors grow continuously, but the cheekteeth are rooted. There are two pair of mammae, and a baculum is present in males.

The name *Erethizon* is derived from a Greek word meaning "stir" or "arouse," in reference to the supposed irritability of the porcupine.

Erethizon dorsatum: Porcupine

Name. The name *dorsatum* is derived from the Latin word for "back." Porcupines occasionally are referred to as "hedgehogs" (although true hedgehogs are Old World insectivores), "quill pigs," or "porkies." Porcupine is derived from the French *porc espin* ("thorny hog").

Distribution. Porcupines are most frequently animals of coniferous forests, but they may range well out into grasslands

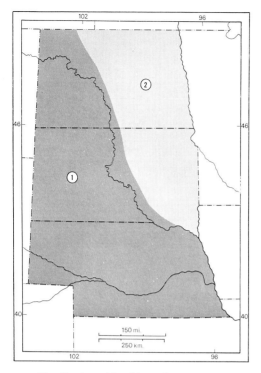

144. Distribution of Erethizon dorsatum:
(1) E. d. bruneri; (2) E. d. dorsatum.

along wooded riparian communities. The presence of such woodlands determines the local distribution of the animals over much of the Northern Plains, where two subspecies have been recognized. *Erethizon dorsatum bruneri* occupies the area south and west of the Missouri River, whereas *Erethizon dorsatum dorsatum,* typically an animal of the northeastern boreal forest, occurs north and east of the Missouri River. Additionally, *Erethizon dorsatum epixanthum* may enter the extreme northwestern corner of North Dakota; the range of this subspecies centers on the Rocky Mountains.

Description. Porcupines are large, thickset rodents; large adults may weigh more than 30 pounds. Their size is accentuated by their heavy, specialized pelage. Porcupines have three kinds of hairs. The underfur is thick, fine, and dark-colored distally. Long (to four inches) guard hairs occur over the underfur, being most abundant on the back and tail. Hollow quills occur on the dorsum, sides, limbs, and tail. These quills, up to three inches or more in length, are black at the tips and white to yellowish white proximally. Quills are longest on the rump and shortest on the cheeks, and a single individual may bear 30,000 or more. Albinos and other color variants are known.

The three nominal subspecies on the Northern Great Plains are separable by color and size. *Erethizon dorsatum dorsatum* has white-based quills and white-tipped guard hairs, whereas *E. d. bruneri* and *E. d. epixanthum* have yellowish quills and yellow-tipped guard hairs. Of the latter two subspecies, *bruneri* generally is paler.

E. d. bruneri generally is larger than *E. d. epixanthum,* and both are larger than *E. d. dorsatum.* Average and extreme external measurements of adult females (three *bruneri,* four *dorsatum,* and four *epixanthum,* respectively) are: total length, 851 (820–872), 732.0 (731–735), 765.5 (727–790); length of tail, 233 (175–285), 186.7 (175–195), 239.0 (205–285); length of hind foot, 113 (110–116), 101.2 (95–115), 103.0 (95–120); length of ear, 34 (30–37), 28.3 (20–35), —.

The skull of the porcupine is heavy and tends to become rugose with age. Average and extreme cranial measurements of seven female *bruneri,* three female *dorsatum,* and four female *epixanthum* are, respectively: condylobasal length, 104.1 (91.6–115.5), 99.4 (98.2–100.1), 98.8 (96.4–105.1); zygomatic breadth, 72.7 (63.1–80.6), 71.6 (70.6–73.7), 70.2 (68.7–71.5).

Natural History. Despite their large size and apparent clumsiness on the ground, porcupines are able climbers, well adapted to their arboreal life. The digits bear strong curved claws that allow the animals to climb with remarkable speed. Once in a tree they are difficult to dislodge. The powerful tail is used as a "fifth foot," and porcupines may be observed on seemingly precarious perches, balanced on the tail and hind limbs, using the forefeet to handle food.

On the ground porcupines move with an awkward gait, accentuated by the movement of the long pelage. Their eyesight is poor, but their sense of smell is keen, and they continually sniff the air, the nostrils moving constantly. Hearing also is well developed.

Porcupines are not territorial and generally are solitary, although pairs sometimes are seen during the breeding season, and two or more may den together, particularly in extremely cold weather. Movements center on the den, for which any available cavity will suffice. Shelters beneath overhanging rocks or in hollow logs are common, but dugouts, abandoned outbuildings, and similar structures also are used. The dens are generally unimproved but are often recognizable by the prevalence of numerous characteristic, oval- to crescent-shaped feces. Experienced observers can locate porcupine dens by their odor.

The animals generally move from the den at dusk to feed, but they may be seen during the day, particularly when disturbed. Frequently there is a well-worn path from the den to the feeding site. During extensive movements from the home den, sec-ondary dens may be established. Porcupines do not become dormant in cold weather, being active throughout the year, but they may retire to dens in severe weather. The home range seems to be about 25 to 35 acres. There is a single annual molt in adults, in late spring or summer.

Food varies with the season and with local vegetation. During the growing season, herbs may form the bulk of the diet, but woody plants are eaten in the colder months. Preferred food is the cambium of coniferous trees, but where conifers are absent porcupines thrive on broad-leafed trees. Leaves, buds, young twigs, roots, berries, seeds, and flowers are eaten in addition to bark. Local preferences for food species may develop, so that porcupine damage to forest trees becomes selective. A pound or so of food may be ingested per day. The cecum has a symbiotic microflora that digests cellulose. The ever-growing incisors (the annual growth may be as much as 300 mm) are kept sharp by continuous wear against each other.

The maternity den is slightly more elabo-

145. *Porcupine,* Erethizon dorsatum *(courtesy J. L. Tveten).*

rate than the typical resting den and may be lined with boughs and roots. Breeding occurs in November or December. Females may come into estrus repeatedly if fertilization does not take place at the first ovulation. A single young (rarely twins) is born in spring after a gestation period of about seven months. At birth porcupines are well developed; the eyes are open, and the thick pelage includes quills, which harden quickly on exposure to the air. Females have four mammae. Vegetation is added to the diet of the young animal within two weeks, and the young grow rapidly, at a rate of about 450 grams per month.

When molested, a porcupine retreats to the safety of a tree or resists the attacker by erecting the quills and flailing the air with the powerful tail. It is probable that this species is not subject to systematic predation by any of the carnivores of the Northern Plains, although the fisher, *Martes pennanti*, is an able climber and reputedly is a habitual porcupine-killer where the two occur together. Perhaps any large carnivore will kill an occasional porcupine, particularly when other prey is scarce. Records are available of porcupines having been killed by coyotes, red foxes, wolves, wolverines, martens, fishers, mink, mountain lions, lynx, and bobcats. Individual carnivores may develop the ability to kill porcupines by exposing the unarmed belly, but mostly they are attacked in the head region. Weasels reportedly feed on por-

cupine carcasses. Parasites include roundworms, flatworms, fleas, lice, ticks, and mites. The life span may exceed 10 years.

Porcupines affect man's interests primarily by damaging forests and crops and injuring unwary domestic animals. Damage to crops is usually negligible, although occasional individuals may develop a taste for alfalfa or grain. Fruit trees and ornamental trees may be damaged or destroyed. Injury to forest trees may be spectacular in some places but often is overemphasized. The loss of cambium to porcupines breaks the flow of nutrients in the tree, and affected parts die; if it is completely girdled the entire tree will die. Such severe damage is not common, however, and trees are more often disfigured than killed outright. Porcupines often return to the same tree to feed on several successive nights. Timber loss attributable to porcupines is small indeed compared with losses to fire, insects, fungus, and wasteful utilization practices.

Where porcupines are abundant, domestic dogs and livestock may be injured. The barbed quills lodge tightly in the tender skin of the mouth and nose, and if not removed quickly they work deep into the flesh and may cause permanent damage or death. Quills are best removed with pliers. Twisting the quill slightly as it is drawn out makes removal easier.

Selected References. General (Costello 1966; Woods 1973 and included citations).

Order Carnivora

Carnivores are distributed nearly worldwide, occurring on all major continents, although the wild dog of Australia, the dingo, was introduced by early man. They are absent, however, from most oceanic islands except where introduced (principally domesticated dogs and cats and the mongoose). The order is well represented by a long fossil history. Carnivores encompass 12 modern families, among which more than 100 genera and about 270 species generally are recognized. There is current controversy, however, concerning carnivore taxonomy, particularly at the generic level. Five families occur on the Northern Great Plains.

As the ordinal name implies, these mammals typically are flesh-eaters. Among North American families, this tendency is highly developed in the cats and some mustelids, but members of all other groups regularly include vegetable matter (such as fruit) and a variety of invertebrates in their diets, and the bears and the raccoon are truly omnivores. All terrestrial carnivores have a well-developed pair of generally conical canine teeth in both the upper and the lower jaw. In North America the number of premolars varies from 4/4 to 2/2 in various combinations, and the molars vary from 2/3 to 1/1. A special combination of large shearing teeth, the last premolar in the upper jaw and the first molar below—the so-called carnassial pair—is typical of carnivores; among Northern Plains species these specialized teeth are best developed in cats and least developed (but still evident) in bears and procyonids. The feet are five-toed or four-toed, the toes bearing sharp, curved claws. The stance is plantigrade to digitigrade.

Key to Families of Carnivores

1. Claws retractile, completely concealable in fur; molars 1/1, premolars 2/2 or 3/2 _____Felidae
1.' Claws not retractile, not concealed in fur; molars 1/2 or more; premolars never less than 3/3 _____2
2. Tail conspicuously ringed; molars 2/2 _____Procyonidae
2.' Tail not ringed; molars 1/2 or 2/3 _____3
3. Size small to medium (*Gulo* largest); molars 1/2; premolars 4/4 (*Gulo, Mar-*

tes), 4/3 (*Lutra*), or 3/3 _____ Mustelidae
3.' Size medium to large (*Vulpes* smallest); molars 2/3; premolars always 4/4 _____4
4. Size large; feet plantigrade; large cheekteeth flattened, only slightly cuspidate, and lacking conspicuous cutting edges _____Ursidae
4.' Size medium; feet digitigrade; large cheekteeth not flattened, cuspidate, and with conspicuous cutting edges _____Canidae

Family Canidae: Dogs and Allies

The family Canidae is a close-knit assemblage of some 35 living species, arranged in 11 genera (although some authorities would refer all Northern Plains species to the genus *Canis*). An ancient group, canids first are known from the fossil record of Eocene times. They range on all continents except Antarctica (although the wild dogs of New Guinea and Australia arrived there as commensals of humans). The native fauna of the Northern Great Plains includes five canids, of which one (the gray wolf) was extirpated within the past half century. The following key includes the domestic dog (*Canis familiaris*) as well as native canids.

Key to Canids

1. Total length more than 1,050; condylobasal length more than 150; postorbital process thickened and convex dorsally —————————————2
1.′ Total length less than 1,050; condylobasal length less than 150; postorbital process thin, concave dorsally ——————4
2. Total length usually more than 1,400; length of lower first molar more than 25; condylobasal length more than 200 —————————————*Canis lupus*
2.′ Total length usually less than 1,400; length of lower first molar less than 25; condylobasal length less than 200 ——————3
3. Upper parts brownish gray; orbits not rising abruptly above rostrum; dewclaw absent —————————*Canis latrans*
3.′ Upper parts variable in color; orbits usually rising abruptly above rostrum; dewclaw frequently present ——————*Canis familiaris**
4. Dorsum of skull flat from nasals to orbits; temporal ridges prominent ——————5
4.′ Dorsum of skull rising abruptly at orbits; temporal ridges generally not prominent ——————————*Canis familiaris**
5. Upper parts reddish; ears with black tips; condylobasal length more than 125; length of maxillary toothrow more than 55 —————————————*Vulpes vulpes*
5.′ Upper parts not reddish; ears not black-tipped; condylobasal length less than 125; length of maxillary toothrow less than 55 —————————————6
6. Upper parts grizzled grayish; fur of cheek reddish; temporal ridges (fig. 148) lyre-shaped — *Urocyon cinereoargenteus*
6.′ Upper parts yellowish brown; fur of cheek white; temporal ridges (fig. 148) not lyre-shaped ——————*Vulpes velox*

Genus *Canis*

The genus *Canis* includes about seven species of medium-sized to large carnivores that are adapted by their long, lean limbs and digitigrade feet for cursorial locomotion. The fossil record extends back to the early Pliocene. Members of the genus occur on all continents, although the Australian dingo probably arrived there as a commensal of man. Two species of *Canis*, the coyote and the gray wolf, once occurred throughout the Northern Great Plains, but only the coyote remains. The dental formula is 3/3, 1/1, 4/4, 2/3, total 42. *Canis* is the Latin word for "dog."

Canis latrans: Coyote

Name. The specific name *latrans* is a Latin word meaning "barking." In early literature the coyote often was referred to as the prairie wolf or brush wolf, in contradistinction to the timber wolf, *C. lupus,* for which the vernacular name gray wolf is preferred. Coyote is a Spanish word derived originally from the Aztec *coyotl.*

Distribution. Coyotes are distributed from Costa Rica to the North Slope of Alaska and from coast to coast in the United States. Their range seems to be increasing, perhaps in part a response to the demise of the gray wolf (*C. lupus*) and the red wolf

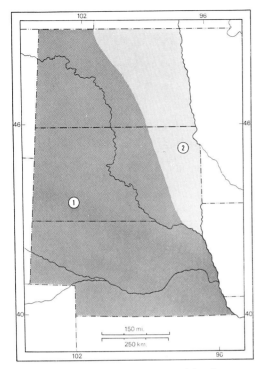

146. Distribution of Canis latrans: (1) C. l.
latrans; (2) C. l. thamnos.

(*C. niger*). Coyotes of the Northern Great
Plains are separable into two subspecies,
Canis latrans latrans throughout much of
the region and *Canis latrans thamnos* in
the northeast.

Description. Coyotes are about the size of a
small shepherd dog. External measurements
of three males and two females from
Nebraska are: total length, 1,302, 1,250,
1,234, 1,190, 1,171; length of tail, 368, 337,
346, 340, 362; length of hind foot, 219, 212,
208, 190, 200. The ear measures about 100
from notch to tip. Weights of six males
averaged 33.7 (30.0–42.8) pounds; two
females weighed 27.5 and 25.1 pounds.
Average and extreme cranial measurements
of five males and 10 females from Nebraska
are, respectively: condylobasal length, 192.8
(191.0–196.0) and 177.3 (168.0–181.5), and
zygomatic breadth, 102.2 (99.2–105.3) and
94.3 (87.7–99.2).

The upper parts are grayish buff to
brownish gray, paler in the long, soft winter pelage. The back of the ears is buffy to
reddish. The tail is bushy and tipped with
black. The underparts are whitish. There is
a distinct mane of rather stiff hairs over the
shoulders. *C. l. thamnos*, the subspecies of
the northeastern part of our region and
adjacent states, is darker than *C. l. latrans*.
Some crossbreed domestic dogs (especially
German shepherd–husky) look much like
coyotes. Living coyotes are most readily
distinguished from such dogs by their habit
of holding the tail low—almost between
the legs—when running.

Coyotes also differ from dogs in having a
more elongate track. Furthermore, the upper canine teeth are longer; they extend
below a line drawn between the anterior
mental foramina. In addition, the skull of
most coyotes is relatively longer than that
of dogs. The ratio of length of upper molariform toothrow to palatal width in
coyotes is 3.1 or more, whereas in domestic
dogs it is 2.7 or less.

From gray wolves, coyotes differ in
smaller size. The skull of a coyote is 220
mm or less in greatest length, whereas that
of the wolf is 230 or more. The coyote's
track is shorter, and the footpad is 32 mm
or less (compared with 38 or more). The
coyote, however, has longer ears.

Natural History. If coyotes had no other
virtues, their adaptability would be enough
to command respect. These canids occur
nearly everywhere on the Northern Great
Plains, sometimes in rather thickly settled
areas, but mostly in more remote country.
They succeed in open grassland, brushy
country, badlands, and woodlands. Elsewhere over their increasingly broad range,
they prosper in such diverse habitats as
tundra and tropical deciduous forest.

The daily peak of activity is in early
evening, but coyotes may be seen foraging
at any time of the day or night. Young
animals are especially likely to be active
diurnally. These are wide-ranging animals,
with females moving over an area of about
four square miles and males using areas
three or four times larger. Home ranges of
females tend not to overlap, suggestive of
territoriality, but those of males overlap

broadly. Daily movements average three or four miles, with long movements—well beyond 100 miles—being reported as animals migrate to new home ranges or when young disperse from the parental den. As coyotes move through the home range, they urinate and defecate frequently. The odor of urine and feces probably serves to mark the boundaries of the home range. Densities of coyotes vary widely with habitat quality, degree of persecution, and other factors, but one per square mile or two is perhaps a reasonable guess at an average.

Coyotes observed usually are solitary, but they may hunt as packs in autumn and winter. Such packs frequently are found to be family groups before the dispersal of young; in any event, packs are not as typical as they are of wolves or feral dogs.

The diet varies widely in that coyotes are opportunistic and tend toward omnivory, although mammalian flesh usually is a staple. Rabbits, hares, mice, rats, pocket gophers, and ground squirrels are important dietary items. Birds and eggs are eaten, as are insects and a wide variety of fruits.

147. Coyote, Canis latrans *(courtesy U.S. National Park Service).*

Frogs, lizards, snakes, turtles, fish, and crustaceans also are taken. In winter carrion is important over much of the range and may include such big-game mammals and domestic livestock as deer, pronghorns, sheep, cattle, and poultry. Large animals in good health can evade or chase off coyotes, but young or sick deer, for example, or those hampered by heavy snows or caught on ice, become rather easy prey, as do lambs born on open range. Just how great are the losses of livestock to coyote predation is almost impossible to determine. Starvation, infection after docking, and abandonment all result in lamb mortality. Coyotes may eat the carcasses, but what killed the animals is frequently impossible to determine. Basic research on the biology of predation eventually may prove more valuable in the solution to (and definition of) "the coyote problem" than applied research on methods of extermination. Today the most effective method of preventing coyote losses around a farmstead seems to be careful disposal of dead poultry and livestock so that it will not attract scavengers.

Coyote-dog hybrids are found occasionally on the Northern Plains. These "coydogs" are best documented in the tristate region in southeastern Nebraska. There is some evidence that coydogs are more likely to attack livestock in that area than are coyotes.

Coyotes are of substantial monetary value in some years when fashion dictates a taste for long-haired furs. In the winter of 1979–80, for example, some 12,000 coyotes were taken in Nebraska and sold for an average price of about thirty-nine dollars.

Female coyotes—unlike domestic dogs—are monestrous, the period of heat lasting two to five days. Ovulation occurs near the midpoint of estrus, which takes place in winter or early spring (January to March). Courtship may begin two to three months before estrus. The gestation period is about nine weeks, and average litter size is six, increasing with age of the mother and ample food supplies. The young are born blind and nearly helpless. They are nursed in a den excavated by the female, often beneath a downed tree, rock, or cutbank or in a thicket. The den may be newly constructed or may be remodeled from an existing excavation. Dens generally are 15 inches or so in diameter and from about five to 20 feet or more in length. Two or more entrances are common. If the den is disturbed, the young may be moved to another. Occasionally dens are shared by two females, suggesting that some males are polygynous; such combined litters led some early workers to report exceptionally large numbers of young per female.

The young grow rapidly; they weigh about 250 grams at birth and gain about 300 grams per week before weaning at week seven. The male provisions the nursing bitch. Both parents, beginning in about the third week, regurgitate semisolid food for the pups. The eyes open at about two weeks, and the deciduous dentition erupts at about the same time. Soon thereafter the pups first emerge from the den to play. Severe fighting among pups three to four weeks old establishes a dominance hierarchy. Position in the hierarchy seems to have something to do with the probability of dispersal. Coyotes usually live less than 10 years in the wild, but captives have lived as long as 19 years.

The vocal repertoire is less complex than that of the wolf, but more complex than those of less social canids. There are doglike barks and squeals, and also a characteristic howl, higher in pitch than that of the wolf, but still an awe-inspiring serenade on a crisp, moonlit autumn evening on the open prairie. Visual signals are much like those of domestic dogs. Dominant animals exhibit a stiff-legged walk, stiffened tail, and erect hairs on the mane and tail. Submissive animals "grovel," urinate, whine, and roll over to expose the belly.

Endoparasites of coyotes include tapeworms, roundworms, lungworms, heartworms, hookworms, pinworms, and spiny-headed worms, and ectoparasites include fleas, ticks, mange mites, and lice. Microbial diseases include rabies, distemper, tularemia, and plague. Wounds inflicted by prey are a factor in mortality, as are hunting, trapping, and poisoning by man.

Selected References. General (Bekoff 1977 and included citations); natural history (Gier, Mahan, and Andelt 1977; Mahan 1977; Mahan, Gipson, and Case 1978).

Canis lupus: **Gray Wolf**

Name. The specific name *lupus* is the Latin word for "wolf." Gray wolves also are commonly referred to as timber wolves.

Distribution. The gray wolf once ranged over virtually all of North America, from Greenland, the Canadian Arctic, and Alaska southward to the Valley of Mexico. A Holarctic species, it also occurred across Eurasia, southward to India and the Arabian Peninsula. In the United States, the animals were absent only from parts of California and Arizona and an area in the Southeast occupied by the related red wolf (*C. niger*).

Today, gray wolves have been extirpated from much of their native range. In the United States they are mostly restricted to the northern Great Lakes region, parts of Montana and Idaho, and perhaps western Texas and other areas of the Southwest, and they also occur in adjacent Mexico. Single wolves were killed in Jerauld and Hanson counties, South Dakota, in 1970, and another was taken in Sargent County, North Dakota, in 1973. There is some evidence that each of these animals may have been reared in captivity, however. In 1981 a wolf—apparently from wild stock—was taken in extreme northeastern South Dakota. This animal probably entered the state from Minnesota. Occasional dispersing wolves also may enter North Dakota from Manitoba or Montana.

The subspecies *Canis lupus nubilus* once occurred throughout the tristate region except for the Black Hills; this race now is extinct. Wolves of the Black Hills apparently belonged to the subspecies *Canis lupus irremotus*, which once ranged through the Northern and Central Rocky Mountains. *Canis lupus lycaon* is the subspecies that now may be moving onto the Northern Great Plains from adjacent areas. Because of the uncertain status of the gray wolf in the Dakotas, its distribution is not mapped.

Description. Wolves are pale gray in general color, the underfur mixed with buff, and the dorsum, tail, upper legs, and backs of ears variously overlain with black. Although northern populations average paler than those in the south, local variability tends to obscure geographic variation. Published external measurements of the extinct *C. l. nubilus* from the Northern Great Plains are rare. An individual of unknown sex from eastern Nebraska was about 1,645 mm in total length, of which the tail accounted for 330. An old female taken in 1833 along the Missouri River in North Dakota was 1,435 mm long, tail, 368. Adult females weigh from 45 to 120 pounds, adult males, 45 to 175. Mean and extreme cranial measurements of seven males from Nebraska, followed by measurements of three individual females, are: greatest length of skull, 251.0 (243.7–258.4), 228.7, 231.0, 233.5; zygomatic breadth, 136.2 (133.7–140.9), 119.4, 130.5, 126.2. From *C.*

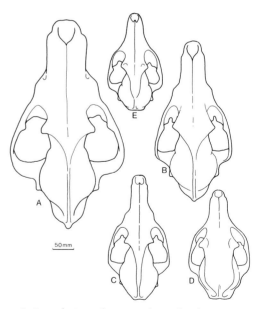

148. Dorsal view of crania of canids of the Northern Great Plains: (A) Canis lupus; (B) C. latrans; (C) Vulpes vulpes; (D) Urocyon cinereoargenteus; (E) Vulpes velox.

l. irremotus of the Northern and Central Rockies, *C. l. nubilus* differed in having grayer pelage, with less tendency to white coloration.

Natural History. By nearly all reports, wolves once were abundant throughout the Northern Great Plains, both on the prairie (where they followed herds of bison) and in forested areas (where they preyed mostly on deer). Apparently it was only in badlands and brushlands that coyotes outnumbered wolves. But the Great Plains have changed. The great migratory herds of bison were broken by the railroad-building, the United States army, and market hunting. With them went the wolves, most of which were gone by 1890. Deprived of their native prey, wolves turned to domestic livestock. Their carrion-feeding habits made them susceptible to extermination. A carcass of a bison or cow laced with strychnine was highly effective. Other wolves were trapped or shot for the fur trade or for "sport." In Nebraska they were rare by the turn of the century and extirpated by 1920. Vernon Bailey, writing of North Dakota in 1926, noted: "At the present time there are probably a few wolves left in the least-settled parts of the rough Badlands region west of the Missouri River, but it is to be hoped that a very few years will see the last of these destructive animals in this state." With that as the opinion of a federal wildlife official, the hope was soon realized. Today, occasional individuals wander into the Pembina Hills, along the Manitoba border. The last wolf was taken on the Black Hills of South Dakota in 1934. The subspecies of the Northern Plains is extinct, one of very few taxa of animals deliberately driven to extinction by humans.

Given present-day land use practices, there is no possibility of reestablishing populations of wolves widely on the Northern Great Plains. There are, however, localities on the plains that have been set aside for preservation of native ecosystems and their fauna. Reintroduction of wolves in some of these areas could prove of inestimable scientific, educational, and aesthetic value.

In general habits, wolves are much like

149. *Gray wolf,* Canis lupus, *male (left) and female (courtesy R. Peterson).*

coyotes, their smaller, more resilient relatives. Dens are established in earth banks, gullies, and hillsides; there a female brings forth a litter of six to 10 fluffy black pups, usually in March, after a gestation period of 60 to 63 days. The male (or other pack members) provisions the nursery den. During the breeding season the den is the center of activity, and it may be used by a pair year after year. Month-old pups, now gray, play at the mouth of the den. When about three months old and half-grown, young wolves begin to take part in hunting, and such packs show remarkable cooperation in taking prey—as large as bison in the old days, and moose, elk, deer, and caribou over their northern and western range today. The diet also includes rabbits and hares, beaver and smaller rodents, and some berries and other fruits. Carrion is important in the diet, especially in winter.

Packs range widely through a home range, often an elongate area bounded by natural landmarks such as ridgetops. Hunting territories are known to range from about 10 to 95 square miles or more, and neighboring territories are mostly mutually exclusive. Ranges of "lone wolves," dispersing individuals or remnants of depleted packs, may be interspersed between hunting ranges of existing packs. Wolves are "top" carnivores, and adults are subject to attack by no other animal (although occasionally a grizzly bear must stumble on a den of young). Injuries incurred during hunting larger prey or in fights with other wolves may cripple or kill individuals, and an occasional wolf starves to death when incapacitated by a mouth full of porcupine quills. Rabies and distemper are known. Ectoparasites include fleas, ticks, and mange mites. Helminth parasites are common, as in other carnivores that feed on carrion. In lean years some pups starve to death, as do dispersing young. The greatest threat to wolves today over much of their remaining distribution is man, armed with rifle, trap, or poison and carried into the vestiges of wilderness by airplane or snowmobile.

In behavior, wolves are highly reminiscent of domestic dogs (to which wolves gave rise in Paleolithic times). Adults are solicitous of young. Pups play together violently. Juvenile behavior—rolling on the back or crouching to urinate, for example—shows submission and halts the attack of a dominant individual. Males urinate frequently on prominent objects in the environment. Probably this indicates to other wolves a territorial boundary. Like other social canids, wolves are highly vocal. Barks and yips maintain cohesion among groups. Deep growls express threat. Whines bring solicitous behavior. The howl may express restlessness or it may be a call to assemble. Whatever its function among wolves, it will raise the hair of the most hardened wilderness traveler. It is a "call of the wild," stilled now on the Northern Plains.

Selected References. General (Mech 1974 and included citations); natural history (Young and Goldman 1944; Van Ballenberghe, Erickson, and Byman 1975).

Genus *Vulpes*

Some 11 species of the genus *Vulpes* range across North America, Eurasia, and most of Africa. The geologic record of the genus extends back to the late Miocene. Two species occur on the Northern Great Plains. The dental formula is 3/3, 1/1, 4/4, 2/3, total 42. The generic name *Vulpes* is the classical Latin name for this animal; as an adjective, *vulpes* meant "cunning."

Vulpes velox: Swift Fox

Name. The name *velox* is the Latin word for "quick," an allusion to the supposed speed of this animal. Actually, the swift fox is probably no more swift than any other fox, but the long fur does give the animal a fluid, graceful appearance while moving and creates an illusion of remarkable quickness. Swift foxes sometimes are mistakenly referred to as kit foxes. The kit fox (*Vulpes macrotis*) is a closely related species that occurs to the south and west of the Rocky Mountains.

Distribution. The swift fox is a mammal of the Great Plains, ranging from Alberta and Saskatchewan southward to northern Texas. The species once occurred nearly throughout the Dakotas and Nebraska, and seemingly it was fairly common until the early 1900s. Then it was reduced to the verge of extinction, probably through the

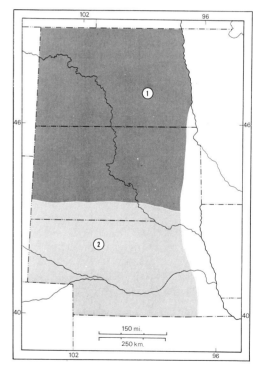

150. Distribution of Vulpes velox: (1) V. v. hebes; (2) V. v. velox.

indiscriminate use of traps and poison baits to control coyotes. The swift fox is not at all trap shy and also is easily captured by domestic dogs.

Since about 1950 there has been increasing evidence that swift foxes are making a comeback on the Northern Plains, especially in the western part of the tristate region. It is not known whether the present population represents expansion from some local refugium (such as the Nebraska Sand Hills), where swift foxes persisted unnoticed for more than half a century, or whether it represents invasion from elsewhere on the Great Plains.

Two nominal subspecies have been ascribed ranges on the Northern Great Plains, a southern race (*Vulpes velox velox*) and a northern race (*Vulpes velox hebes*). Available specimens are too scant to critically evaluate the status of these subspecies or to accurately delineate their distribution.

151. Swift fox, Vulpes velox *(courtesy K. Steibben, Kansas Fish and Game Commission).*

Description. V. velox is a beautiful little fox, much the smallest native canid of the Northern Plains. The pelage is long and lax, especially in winter, and the tail is notably bushy. The general color is pale, from buffy to reddish gray above, slightly paler on the sides. The legs, neck, and back of the ears are clear orangish buff. The tail has a prominent black tip, and there are black patches on each side of the muzzle. The venter is pale buff.

Mean and extreme external measurements of 19 males and 17 females, respectively, from the Central and Southern Plains are: total length, 800 (735–880), 788 (702–850); length of tail, 284 (254–350), 280 (253–350); length of hind foot, 126 (114–132), 120 (113–128); length of ear, 65 (56–75), 62 (58–66). Weights of adults generally range from 4 to 6.5 pounds. Cranial measurements of three males and one female from Colorado are: condylobasal length, 114.5, 108.7, 109.7, 113.3; zygomatic breadth, 63.9, 63.6, 64.9, 63.0.

Natural History. Swift foxes are animals of the open prairie and, perhaps as a consequence, are strongly subterranean in their

habits. Unlike many canids, which occupy dens mostly in the breeding season, swift foxes use dens throughout the year. These may be remodeled burrows of other animals, but more often they seem to be excavated by the foxes themselves, because the entrances are too small for coyotes or even badgers. Shallow temporary dens with a single entrance are common, but more complex systems of tunnels, with four to nine entrances, seem to be used over extended periods, particularly by vixens with young. Some tunnels are three feet or more deep, but most are about twenty inches in depth. The longest tunnel reported in an unpublished study in northwestern Nebraska was about 15 and a half feet long, although many burrows, particularly those not containing young, are much shorter. Entrances are about eight inches across. Sometimes a nest chamber with a floor of dried grasses is observed, although other tunnels are free of litter. Dens are marked on the surface by low ridges of pale subsoil. The dens may be on hillsides or ridgetops, in flat rangeland, or even in cultivated fields. Of forty located in Sioux County, Nebraska, all were in short-grass prairie, in sandy loam to loam soils. Abandoned dens form a sheltered habitat on the open prairie favorable for other animals, including striped skunks, burrowing owls, deer mice, rattlesnakes, toads, and numerous invertebrates.

Swift foxes usually mate for life, but if one of a pair dies the partner will take a new mate. Some observations indicate that occasionally males are polygynous. Females are monestrous, breeding once each year, in late winter. The gestation period is about seven weeks long. The young are born blind and helpless in late April or early May. Litter size seems to vary from about three to six. The male has been reported to provision mother and young, but this recently has been questioned. Eyes and ears open at about two weeks of age, and the young are weaned at about six weeks. The juvenile pelage is wavy and fine; guard hairs are most noticeable on the rump. By August the young, now four to five months old, are of adult size and have adult pelage. In early autumn the young disperse, to mate and breed in their first winter.

The diet of swift foxes consists mostly of animal matter. Mammals, especially cottontails, jackrabbits, and mice, constitute the bulk of the diet. Birds, including both songbirds and quail, are taken, as are lizards and a wide variety of insects and other arthropods, as well as some plant material. Some of the mammalian flesh is taken as carrion. Swift foxes are mostly nocturnal, although in winter they may become at least partially diurnal, sunning themselves at the burrow entrance. In northwestern Nebraska, home ranges of males fitted with radio collars averaged about seven square miles, and those of females averaged about five. The nightly hunting circuit may be 16 miles long.

Like other mammals that use established dens as shelter, the swift fox harbors a heavy parasite load, especially fleas. Intolerable ectoparasite loads may force abandonment of dens. Parasitic ticks, flatworms, roundworms, and protozoans also have been reported.

The principal enemy of the swift fox is man. The animals seem to have little fear of humans, sometimes denning within a few hundred yards of human habitation or busily used roadways. They are not wary of traps and apparently have succumbed in large numbers to indiscriminate campaigns to trap and poison coyotes and (especially in the late nineteenth century) wolves. Vehicular casualties are frequent, both on highways and in cultivated fields. There has been no regular market for the fur, but the dietary habits of the swift fox make it a firm ally of the farmer. We hope that swift foxes will recover from near extirpation to resume their historical role in the short- and mid-grass ecosystems of the Northern Plains.

Selected References. General (Egoscue 1979 and included citations); natural history (Hines 1980; Hines, Case, and Lock 1981).

Vulpes vulpes: Red Fox

Name. Vulpes is the classical Latin name for this animal. Red foxes of atypical color phase are known as cross foxes and silver foxes. Typical red foxes frequently are referred to simply as "fox" throughout much of the range of the species.

Distribution. The red fox is a Holarctic species. In North America, it occurs across Canada and the United States, except in the desert Southwest and the driest parts of the Great Plains. A single subspecies, *Vulpes vulpes regalis,* is recognized on the Northern Plains. The red fox seems to be extending its range westward at present, perhaps as a response to the increased "patchiness" of plains environments caused by permanent settlement and agricultural practices, especially planting of shelterbelts, stabilization of riparian woodlands, and impoundment of irrigation water. In the eastern United States, its spread was helped in colonial days by the clearing of forest for agriculture.

Description. The red fox is a beautiful little canid, with a full reddish yellow dorsal pelage, great bushy tail tipped with white, and black feet, nose, and backs of ears. The

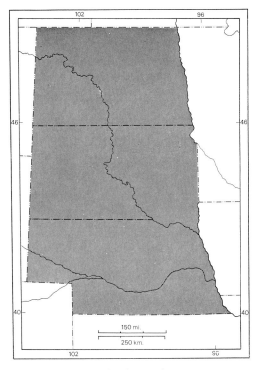

152. Distribution of Vulpes vulpes.

underparts are pale, ranging from grayish to white. Over its range, the species occurs in several color phases. Melanistic individuals are known, and some otherwise black individuals have white-tipped guard hairs that lend the dorsal pelage a silvery appearance; hence the name "silver fox." The "cross fox" is another color phase, marked by a cross of brown hairs over the shoulders and down the dorsal midline. The various color morphs all have white-tipped tails. All seem to be more common in the northern parts of the range and in the mountains of the west, and all are highly prized in the fur trade. Reliable observations of any of these alternative color phases on the Northern Great Plains would be of interest.

Red foxes are considerably smaller than their ample pelage makes them appear. External measurements of three males and two females from southeastern Nebraska are, respectively: total length, 1,024, 965, 946 (males only); length of tail, 381, 298, 370, 349, 368; length of hind foot, —, 152, 160, 149, 156; length of ear, 76, —, 85, 73,

82. Weights of 16 males and five females from northeastern Kansas averaged, respectively: 10.1 (9.0–12.5) and 8.6 (8.2–9.1) pounds. Condylobasal length of eight males and 10 females from eastern Nebraska averaged 139.3 (133.8–146.0) and 132.9 (128.2–137.1), respectively, and zygomatic breadth averaged 75.6 (73.7–77.5) and 71.2 (67.8–74.3).

Natural History. The red fox inhabits a wide variety of habitats, from "edge" communities adjacent to deciduous or coniferous forest to riparian communities in semidesert country. These foxes prosper in areas altered by man, although they seldom den close to settlements where domestic dogs run loose. It is in bushy successional areas—where the small mammalian prey that are their dietary staple are most abundant—that red foxes reach their greatest numbers. The animals seldom are found far from permanent water, either streams or ponds.

The diet consists mostly of animal matter, with the proportion of prey species depending on availability. Over most of the range, rabbits and hares make up a third to half of the diet. Another third consists of mice—*Microtus* or *Peromyscus,* depending upon the character of the habitat available for foraging—or other rodents, such as woodrats, cotton rats, woodchucks, pocket gophers, harvest mice, jumping mice, or muskrats. Opossums, moles, shrews, and young raccoons also are taken. The remainder of the diet is composed of birds—ducks, songbirds, quail, pheasants, and grouse—as well as insects, fruit, fish, and crayfish. Red foxes eat carrion, and, where man's carelessness makes it profitable, they also take poultry and young pigs. The diet may vary with seasonal abundance of prey—with increasing dependence on insects in summer, fruit and mice in autumn, and nestling birds in spring. Some food is cached—buried in a depression under a cover of soil or leaves.

A hunting fox is a study in silent grace. The animals move quickly over their home ranges, nose to the ground, slowing frequently to investigate clumps of grass, patches of brush, or fallen logs. Foxes seldom pursue prey; their strategy is one of surprise. When a fox catches the scent of a prey species, it stops, takes a single bound, and either dispatches the animal with a bite to the back or pins it under a foot. Except when a vixen has young in the den, food is eaten on the trail. Foxes are solitary hunters except during the breeding season, when they follow a pair of parallel trails through the hunting range. Tales of cooperative hunting are legion in the older literature. Unfortunately, some of this lore is probably unreliable, but foxes are "clever" hunters, and cooperative hunting deserves study. A common observation is that hunting pairs almost always encircle brush piles along their foraging route. Also, one of a pair of foxes will station itself at the end of a culvert while the other crawls through it from the other end, driving any resident rodents to their doom. A hunting fox frequently will pause motionless atop a prominence such as a log, stump, or grassy hummock to survey the surroundings for sight, sound, or smell of a meal. It has been calculated that an average fox eats about a pound of food a day. Foxes usually are crepuscular or nocturnal hunters, but they sometimes are active by day as well.

The home range of the red fox varies from one to three square miles, depending on the productivity of the habitat and the density of the fox population. Home ranges of adjacent families tend to be contiguous and nonoverlapping, and they often have natural boundaries such as a stream, a ridgetop, or a roadway. Individuals cover much of the home range in the course of a round of foraging. Probably the boundaries of the home range are marked by scent. Active territorial defense is observed mostly in the immediate vicinity of nursery dens. This is ritualized and seldom involves actual combat. Adults return to rest in the same vicinity day after day but seem not to reuse the same bedding place. Even in winter, adult foxes rest mostly in the open, beneath brush or downed timber rather than in a den, curling the bushy tail over the muzzle and feet.

A den is the center of activity during the

153. *Red fox,* Vulpes vulpes *(courtesy K. Steibben, Kansas Fish and Game Commission).*

breeding season, which begins in December or January. At this time a male and female hunt as a pair. Occasionally a third adult, male or female, is involved in the breeding group. The den usually is on a sunny, well-drained slope, often on a bank or other prominence, where the adults can set up a "look-out" post at the entrance. Sometimes a burrow is actively excavated, but more often it is remodeled from one formerly occupied by a woodchuck, skunk, or badger. Hollow logs, culverts, and small rocky caverns also are used. The tunnels are six to 12 inches in diameter and usually are three or four feet long. A den frequently has two or more entrances.

Mating takes place in January or February. Females are monestrous, the period of receptivity lasting one to six days. Gestation takes 53 days, and one to 12 (mode four or five) young are born in March or April. The neonates are blind but completely covered with grayish black hair, except for the white-tipped tail, and weigh about 100 grams. At one week of age the eyes open. The young first emerge from the den at about five weeks of age, when they are about 500 mm long and weigh about 650 grams. Older pups play in front of the den with bones and bits of feathers. The young may be moved from one den to another several times during their first two months of life. There are reports of partly grown litters being split between two dens. The male provisions the female and brings the pups their first solid food. At eight weeks of age the young accompany their parents on foraging trips. The permanent dentition erupts at about four months.

In early autumn the family unit breaks up. Young males disperse first, then females, and then the parents separate (often only until the onset of the next breeding season). By this time the young are nearly fully grown. Subadult males disperse farther (12 miles on the average in a study in Iowa) than females (five miles). Dispersal distances of more than 50 miles have been recorded. Dispersal seems not to result from agonism by the parents, although in autumn all individuals exhibit mutual avoidance, and some adults relocate their home ranges at that time. Individuals of both sexes may breed during their first winter.

Red foxes molt annually in spring and early summer. The male attains fresh pelage before the lactating vixen. Worn, bleached pelage falls out in unkempt tufts as the new pelage emerges beneath. In autumn the pelage is augmented by additional underfur. The pelt is prime from December to February, all hairs being full grown and tightly rooted.

The luxuriant pelage of red foxes provides habitat for a variety of ectoparasites, including ticks, mites, lice, and fleas, some of which are common to domestic dogs and cats. The varied diet, including carrion, assures a heavy load of endoparasites as well, among them roundworms, stomach worms, heartworms, and tapeworms. Young foxes sometimes fall prey to hawks or owls and to coyotes. By far the greatest threat to the red fox, however, is man. Adults and young are hunted for "sport" or because of their real or imagined depredation on domestic and game mammals and birds. When fashion dictates, foxes are in demand for fur. In the trapping season of 1979–80, about 2,000 red foxes were taken in Nebraska, pelts bringing an average price of nearly fifty dollars. Also, traffic takes a toll. In a thorough study in Iowa, humans accounted for more than 90 percent of mortality in red foxes.

Selected References. Natural history (Korschgen 1959; Pils and Martin 1978; Rue 1969; Sargeant 1972; Stanley 1963; Storm et al. 1976).

Genus *Urocyon*

Urocyon seems to be a genus of Neotropical origin. The known fossil record extends back only to the Pleistocene. There are two Recent species, one of which occurs from northern South America northward onto the Northern Great Plains. The name *Urocyon* is a compound of the Greek *uro*, "tail," and *cyon*, "dog," calling attention to the curiously maned tail of this fox. The dental formula is 3/3, 1/1, 4/4, 2/3, total 42.

Urocyon cinereoargenteus: Gray Fox

Name. The specific name *cinereoargenteus* is a compound of two Latin adjectives, *cinereus*, "grayish" or "ashen," and *argenteus*, "silver," and is an apt description of the color of this beautiful canid.

Distribution. The gray fox has a curious distributional pattern. From northern South America, it extends northward through the eastern United States to extreme southern Ontario and Quebec. To the west it occurs northward through the desert Southwest, along the Sierras to the Columbia River, and along the Rockies to northern Colorado. Thus these animals avoid the drier, more open parts of the Great Plains. In recent years they seem to have been extending their range westward, along the Platte River in Nebraska and the Smoky Hill and Saline rivers in Kansas; and in 1961 a gray fox was taken in Butte County, South Dakota, 130 miles west of the Missouri River. The Missouri Valley formerly had been taken to be the western limit of the distribution of this species. A single subspecies, *Urocyon cinereoargenteus ocythous*, occurs in the tristate region.

Description. The gray fox is somewhat less striking in appearance than its red cousin, but it is a handsome animal. The general color of the back is grizzled "salt and pep-

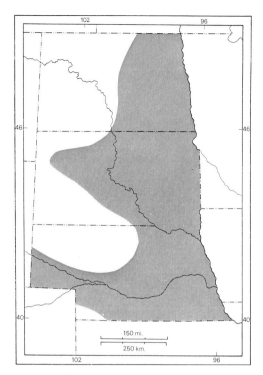

154. Distribution of Urocyon cinereoargenteus.

per," silvery gray to buffy, with an over-layer of blackish hairs, especially pronounced on the dorsum. The throat is white, as are the ears inside; the muzzle is gray, and the legs, neck, and ears, and underside of tail all are reddish buff. The tail has a dorsal mane of stiff black hairs and a black tip.

Gray foxes are somewhat smaller than red foxes. A male from southeastern South Dakota had the following external measure-ments: total length, 964; length of tail, 343; length of hind foot, 140; length of ear, 75. Cranial measurements of two males from northeastern Nebraska are: condylobasal length, 120.5, 123.3; zygomatic breadth, 69.3, 70.5.

Skulls of *Urocyon* are readily distinguish-able from those of *Vulpes* because the tem-poral ridges are lyre-shaped and the lower border of the jawbone has a distinct step below the angular process. Adults average about 8 pounds in weight.

Natural History. Gray foxes are animals of forest, woodland, or rocky and brush-cov-

ered country, unlike the red fox, which prefers meadowland and forest edge. In con-trast to other canids, gray foxes are some-what arboreal, taking refuge in trees, where they hunt squirrels, birds and eggs, and fruit. The animals simply run up sloping trunks to perch daintily on a limb. They also "shinny up" trees in a distinctly un-doglike motion. A home range up to two and one-half miles in diameter is occupied through most of the year. Densities of three or four gray foxes per square mile have been reported in the eastern United States.

Dens seldom are in burrows, but rather are under rock piles, in brushy thickets, in hollow logs, beneath wind-thrown timber, or even in abandoned beaver lodges. Dens are used as daytime retreats as well as nurseries. These foxes seldom are abroad during daylight, preferring to hunt at dawn and dusk and throughout the night. Like other canids, they are active throughout the year.

Food habits differ in proportion from those of red foxes and partly reflect dif-ferences in habitat and the arboreal pre-dilections of gray foxes. A study in Iowa showed that rabbits made up about a fifth of the diet and rodents a fifth (but no microtines—animals of meadowlands— were taken). Birds accounted for about a fourth of the menu, with three times as many passerines eaten by gray foxes as by red foxes. The real difference in diets, how-ever, is in the amount of vegetable matter eaten. Plant material—including fruits, ber-ries, and corn—accounted for nearly 30 per-cent of the diet of *U. cinereoargenteus*. In winter, of course, the foxes rely less on plant products and eat proportionately more rabbits.

The reproductive pattern is similar to that in the red fox. Females are mon-estrous, and most reproduce in their first winter, from January to March (or even later). After a gestation period of 50 to 60 days, a litter of one to seven young is born, well-furred but blind. The neonates weigh only 75 grams but grow quickly, the eyes opening at the end of the first week. At weaning, eight to 12 weeks after birth, the young weigh 1,000 to 1,500 grams. Perma-nent dentition erupts when the young are

155. Gray fox, Urocyon cinereoargenteus *(courtesy K. Steibben, Kansas Fish and Game Commission).*

three-fourths grown (about 16 weeks old). Adult weight is attained at the time of dispersal in late autumn, or early winter, at about 25 weeks of age.

Enemies of this fox include ectoparasites (mites, ticks, fleas, lice), and endoparasites (roundworms, tapeworms, and microbes) as well as some other carnivores (such as coyotes and bobcats, which prey on the young). Humans are the greatest threat to the gray fox, which is hunted for "sport" and (to a much lesser extent than the red fox) for fur or is pursued simply because, as a predator, it must be bad. Because the fur is only about half as valuable as that of the red fox, only a few hundred gray fox pelts are sold from the Northern Plains annually. The probability of an individual surviving the first year is only about 30 percent; thereafter, the expectancy of reaching each successive year-class is about 50 percent. A six-year-old fox is an old animal.

Selected References. Natural history (Hatfield 1939; Lord 1961; Scott 1955; Sheldon 1949; Wood 1958).

Family Ursidae: Bears

The family Ursidae is closely related to the Canidae and seems to have arisen from a canid stock in Miocene times. Bears occur extensively in North America and Eurasia (including the Malay Peninsula), the Atlas Mountains of North Africa (formerly), and the South American Andes. The taxonomy of bears is in flux. Currently most authorities recognize five Recent genera, including about eight species. The ursids include the largest of extant terrestrial carnivores.

Key to Ursids

1. Claws of front and hind feet of approximately equal length; length of maxillary toothrow less than 110
   _____ *Ursus americanus*
1.' Claws of front feet obviously longer than those of hind feet; length of maxillary toothrow greater than 110
   _____*Ursus arctos*

Genus *Ursus*

The genus *Ursus* appears in the fossil record in the early Pliocene and is Holarctic in modern distribution. The dental formula usually is 3/3, 1/1, 4/4, 2/3, total 42, although variation in the number of premolars has been reported in the black bear. The animals are omnivorous and plantigrade. *Ursus* is the classical Latin word for "bear."

Ursus americanus: **Black Bear**

Name. The specific name *americanus* is a Latinized form of America. Black bears of colors other than black are known; in the western part of the range, brown "cinnamon bears" are the most common phase.

Distribution. Black bears once occurred from coast to coast in North America and from the Arctic Slope southward to central Mexico. Extant populations still occur roughly within these limits, but in general they are limited to places remote from civilization. A single subspecies, *Ursus americanus americanus,* occurs in our area of interest. The animals never were as generally distributed on the Northern Plains as was the grizzly bear, being mostly confined to wooded areas. In Nebraska they seem to have been most abundant along the Missouri River but occurred westward along the Loup and the Niobrara, and perhaps other major rivers as well. In North Dakota they also were found along the Missouri but were most abundant along the Red River and in the Turtle Mountains and Pembina Hills. In South Dakota black bears occurred in the eastern part in the early days but now are found only on the Black Hills, where they are rare. Because of its currently restricted range on the Northern Great Plains, the distribution of the black bear is not mapped.

Description. The black bear can be confused with no other native mammal except the grizzly bear, from which it differs in smaller size, usually darker color, markedly shorter front claws, and generally flat to convex, rather than concave, forehead. The color usually is black, but some individuals are brown, ranging from pale cinnamon to chocolate. The muzzle is yellowish brown.

Representative ranges of external measurements of males and females, respectively, are: total length, 1,375–1,780, 1,270–1,475; length of tail, 90–125, 80–115; length of hind foot, 215–280, 190–240; weights, 250–500, 225–450 pounds. The condylobasal length and zygomatic breadth of two males and a female from Colorado were 285, 293, 234, and —, 194.0, 147.8, respectively.

Natural History. Black bears are omnivorous animals of wooded country, occurring mostly in heavy forest. They do occur in riparian woodland but never are found in the open country that the grizzly bear once inhabited. Like the grizzly, black bears are wide-ranging, solitary foragers. When one adult meets another, mutual avoidance is the rule, though direct combat sometimes occurs. Primarily nocturnal and crepuscular, the animals begin a foraging circuit several miles long in late afternoon or early evening and amble along until early morning. The stance is plantigrade, with the entire foot, toes to heel, in contact with the ground. The usual gait is a plodding walk. Bear trails may be obvious where bears are

156. Black bear, Ursus americanus *(courtesy J. L. Tveten).*

common, because each successive user of the trail tends to step in the footprints of the last bear down the trail. The result on soft ground is a decidedly bumpy path, hard for a human hiker to negotiate. Another sure sign of bears is the "bear tree," a tree with deep marks of claws (and sometimes teeth) from five to eight feet above ground level. The function of such "register trees" is not known but may relate to territorial behavior of adult females in some circumstances.

Days are spent in a resting den, a depression lined with leaves and other plant materials and sheltered by a boulder or fallen log. When pressed, black bears can run surprisingly fast, more than 30 miles an hour. They climb readily and well, reaching to their full height, digging in their claws, and pulling themselves up. They also swim well, and they frequent muddy "wallows" in summer, seemingly to escape pesky in-

sects. Although usually quadrupedal, they may stand on their hind legs to observe the surroundings and can even take a few shuffling steps in that position. Generally silent, bears growl or snort when alarmed.

These bears eat virtually anything organic. The folklore surrounding bears has a considerable basis in fact—they do favor both berries and honey. Thus bears can be a genuine economic detriment to apiaries in their range, eating honey, bees, brood, and comb. They eat virtually any fruit in season and can be serious pests in orchards and berry patches. They also eat carrion, fish, rodents, birds and eggs, insects (including ants), and nuts, especially acorns, beechnuts, and hazelnuts. As for meat, probably the mainstay is carrion, but they will kill and eat young pigs and other small livestock when available. Bear scat is readily recognizable because bears tend to feed "wholesale" and frequently make a large meal from a single food item. Thus, the scat, 50 by 150 mm or more in size, is a compressed mass of nutshells or insect parts or whatever. Garbage attracts many mammals—coyotes, skunks, opossums, raccoons—and bears are no exception. An open dump in bear county is an invitation to trouble.

Winter is an unproductive time in the forest. Most plants are dormant and so are many animals, and those animals that are active often are protected by a layer of snow. The evolutionary response to this situation by the black bear is a four- or five-month "nap." These bears do not truly hibernate; their body temperature remains high, although not as high as in summer. They do enter a deep sleep, relying for insulation and energy on a three- or four-inch layer of fat, stored up in the bountiful days of late summer and autumn. Winter dens are not as large as those of grizzly bears and usually are not dug by the bears themselves. A hollow beneath the roots of a wind-thrown tree or beneath a pile of boulders is a typical den site. A bed is constructed of leaves, boughs, and moss. Black bears become dormant in late November or December, or even earlier, and remain asleep until March or April, usually arousing only enough to change position and stretch. In spring, a well-fed bear from the previous autumn will still have sufficient body fat to hold it until new seasonal growth appears, although ants and dead grass are eaten. Other bears emerge from dormancy in an emaciated condition; such animals are dangerously hungry and may turn to domestic livestock if that is the most readily available food supply.

Black bears mate in June or July, but the embryos (blastocysts) do not implant until about November. Thus, although the gestation period is about 225 days, the young—born to a dormant mother in her winter den—are small, only 150 to 200 mm long and weighing only about 250 grams. There are six mammae, two inguinal and four pectoral. The one to five (usually two or three) young are highly altricial, covered sparsely with fine brown hair. At three to four weeks of age the young are well-cov-

ered with a fuzzy brown pelage, and the eyes open a week or so later. By the time the mother leaves the winter den, the cubs are quite active, playing roughly among themselves and with their mother. Black bears are promiscuous breeders, and the male does not participate in rearing the young; indeed, the female tolerates no other bears near her offspring.

When they first emerge from the den, the cubs weigh 6 to 8 pounds. The next spring they will weigh 40 to 75 pounds. Full size and weight are not attained until four or five years of age. The young den with the mother their first winter and disperse the next spring or summer. They first mate no earlier than their fourth summer, at about three and a half years of age, females reproducing every other year thereafter. Maximum longevity in the wild is 20 to 25 years, but the average is much less.

Bears have much in common with the domestic pig. Both are omnivores, seemingly always ravenous. Their cheekteeth are similar, being bunodont grinding tools. They wallow in mud, and mothers in both groups are termed sows. Both have poor eyesight and keen noses, and both are susceptible to the trichina worm, *Trichinella.* Hence bear meat, while edible, must be thoroughly cooked.

Other parasites of black bears include a variety of helminths and ectoparasitic arthropods. Grizzly bears seemingly drive black bears away when the two species come in contact. Porcupines kill bears indirectly, for, with mouth and muzzle swollen with quills, the bear starves to death. Mountain lions are said to prey successfully on cubs, as do wolves. Mostly, however, bears have little to fear in their native haunts except armed humans.

Selected References. Natural history (Burk 1979; Jackson 1961; Seton 1929; Van Wormer 1966).

Ursus arctos: Grizzly Bear

Name. The specific name *arctos* is the classical Greek word for "bear." Grizzly bears may be referred to as brown bears or "silvertips" in the older literature. The grizzly bears of North America are the subject of some nomenclatural confusion. About 90 different specific and subspecific names have been proposed for various populations. The tendency at present is to treat all North American grizzly bears as members of *Ursus arctos.* For the population of the Northern Great Plains, the subspecific name *Ursus arctos horribilis* is available, based on a specimen obtained by members of the Lewis and Clark expedition along the Missouri River in northeastern Montana in 1805.

Distribution. Grizzly bears once occupied virtually all of western North America, from the Red River Valley of Minnesota to the Pacific Coast and from the Arctic Slope to Durango, Mexico. Today the animals are restricted to areas sparsely populated by humans, from the Yellowstone region of northwestern Wyoming northward to Alaska. This is a Holarctic species; the Old World brown bear occurs sparingly today across northern Eurasia, but at one time it ranged southward to the Atlas Mountains of North Africa.

On the Northern Great Plains, grizzly bears once occurred across nearly all of North Dakota and in at least the western parts of South Dakota and Nebraska. The latest report in Nebraska was probably from Keith County, in 1858, and grizzlies apparently never were common in that state. In South Dakota the animals were last seen about 1875 in the vicinity of the Black Hills and about 1890 in the extreme northwest. The last report of the grizzly bear in North Dakota was in 1889, along the Little Missouri. In South Dakota these bears ranged eastward to at least the Missouri Valley, and in North Dakota they were present eastward past Devil's Lake to the Red River Valley. The former distribution of the grizzly bear is not mapped.

Description. Grizzly bears are by far the largest carnivores of the contiguous United States. They have shaggy brown coats, sometimes with a yellowish to silvery or whitish wash on the shoulders and back.

The shoulders are distinctly humped, and the facial profile is concave. The claws of the forefeet are 75 to 150 mm long, obviously longer than those of the hind feet, and clearly visible in tracks. The forelegs are conspicuously longer than the hind legs.

Reliable external measurements of plains grizzly bears are rare in the literature. A three-year-old male from North Dakota was 1,880 mm in total length; a large male taken by Lewis and Clark was reported to be 2,740 mm long. An animal of that size would stand five feet at the shoulder and weigh perhaps 600 pounds. The skull of an adult of unknown sex from extreme western Nebraska had the following measurements: condylobasal length, 330; zygomatic breadth, 195.

Natural History. Grizzly bears have no particular habitat preference. They move through most kinds of ecosystems, following their noses and propelled by their appetite. Once they occurred in and near riparian woodland on the Northern Plains, feeding on the open prairie on the carrion provided by the great herds of bison. Their present occurrence mostly in forested mountains belies the fact that they once occupied more open country, including brushlands, badlands, and semidesert scrub.

Grizzly bears are omnivorous. They are powerful enough to kill any land animal in

157. *Grizzly bear,* Ursus arctos *(courtesy R. L. Palmer).*

North America, but they are too large to be effective in fast pursuit of hoofed game (although they can run 30 miles an hour if pressed). Thus they rely for meat mostly on carrion and fish and such young animals as they stumble across. They may excavate marmots and ground squirrels from their dens, but their great claws are used mostly to dig up roots, to rip apart rotten timber for a feast of grubs, and to dig the winter den. They eat vast quantities of berries and other fruit in season and also eat grass, bark, and mushrooms in quantity. Food may be cached, but winter fuel is stored as body fat in late summer and autumn. Unlike black bears, adult grizzly bears do not climb trees.

The breeding system is promiscuous. Mating takes place in mid-July, and the gestation period, characterized by delayed implantation, is about 184 days. The young—usually twins, but ranging from one to four—are born in the depth of winter. The newborn cubs are highly altricial, blind, and weigh only about a pound and a half. The pinkish skin is covered with fine gray hair. The young first emerge from their den in April or May, weighing about 6 pounds. They may continue to nurse through the summer, but they begin to eat solid food at about four to six months of age, when they weigh 20 to 40 pounds. The young usually den with the mother for their first two winters. Yearling bears weigh 100 to 200 pounds; they attain full growth at eight to 10 years but may reach sexual maturity in three and a half years. Females breed only every two to three years.

Grizzly bears are mostly crepuscular and nocturnal, although young animals have a tendency to forage through the daylight hours except at midday. Most of the time is spent feeding or moving to feeding grounds.

The home range is 10 or 20 to 150 square miles.

In winter, grizzlies enter a den and become dormant. The winter den may be a natural shelter, but usually it is a cavern dug by the bear itself, usually in a shady spot where the snow will lie long and deep. The den is about 50 to 100 inches in diameter and 50 to 70 inches high, at the end of a tunnel some 30 inches in diameter and up to 10 to 12 feet long. Often it is beneath a large tree. A bed of leaves, conifer needles, branches, and grass is constructed. The bears enter their hibernal retreat about the time of the first heavy snow, in October or November. They evacuate the digestive tract before entering the winter den, so that dung is never found inside. They once were thought to hibernate, but this is not the case; the body temperature remains high, and the animals can arouse readily if disturbed. Some bears arouse occasionally during the winter and may leave the den briefly. They emerge in spring in April or May. At this time the range may still be covered with snow, and it is then that the remaining fat reserves laid by in autumn are mobilized.

Mortality in grizzly bears once was due mostly to accidents or to injuries incurred in battle with prey or other bears. Wolves took a few cubs, and many bears probably starved to death in late and lean springs. Now the grizzly has to contend with man. These magnificent animals remain today in only four of the contiguous United States—Montana, Idaho, Wyoming, and Washington. They have been extirpated by man in 13 western states, usually deliberately.

Selected References. Natural history (Schoonmaker 1968); systematics (Merriam 1918).

Family Procyonidae: Raccoons and Allies

Procyonids generally are medium-sized, omnivorous carnivores, related to the bears and dogs. Most species are excellent climbers. All but one are American in distribution, with seven genera and about 19 species—kinkajous, coatis, cacomistles—

mostly in tropical and subtropical areas (only the raccoon and the ringtail occur widely in the temperate zone). Another procyonid is the lesser panda (*Ailurus*) of southern Asia. The giant panda (*Ailuropoda*) has been considered a procyonid by several authorities, but there is a growing consensus that giant pandas are true (if aberrant) bears. The geologic range of the Procyonidae is late Oligocene to Recent.

Genus *Procyon*

The name *Procyon* is derived from the Greek, *pro*, meaning "before," and *cyon*, meaning "dog." The reason for this name is unknown, but it seems likely that it was meant to indicate a primitive, doglike animal. The dental formula is 3/3, 1/1, 4/4, 2/2, total 40. There are seven species (five of them insular) in North and Central America, one of which ranges into South America. Only one species occurs on the Northern Great Plains. The geologic record of the genus dates from the late Pliocene.

Procyon is characterized by a broad head with a robust skull, a pointed muzzle haired across the median line of the upper lip, feet that have smooth naked soles and nonretractile claws and lack well-developed digital pads, and a tail shorter than the body and marked with five to seven rings.

Procyon lotor: Raccoon

Name. The specific name comes from the Latin *lotor*, meaning "washer." The vernacular name is derived from various Indian dialects. This is often shortened to "coon," especially in the southern United States.

Distribution. The raccoon ranges from southern Canada southward throughout most of the United States (except the higher parts of the Rocky Mountains), Mexico, and Central America. There are at present some 25 named subspecies, only one of which (*Procyon lotor hirtus*) is found in the Northern Plains region.

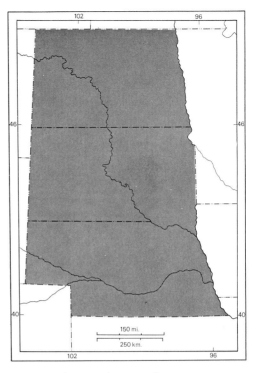

158. Distribution of Procyon lotor.

Raccoons occur primarily in wooded parts of the Plains, especially near water—along streams, lakes, and marshes—and tend to increase in numbers in agricultural areas, where they find a greater food supply (especially corn) and more places to live (such as abandoned buildings). In the more arid, treeless parts of the region, they are generally scarce, and those present use abandoned burrows or other ground dens.

Description. The upper parts are grizzled grayish and blackish, washed with pale yellow, with long, coarse guard hairs and soft, short underfur. The sides are grayer than the back, and the underparts are dull brownish washed with yellowish gray. The face is whitish, with a broad black mask across the eyes and cheeks, and the ears are medium-sized, pointed, and grayish, with black at the base. The skin of the feet is black, the hair whitish or yellowish gray. The front feet are handlike, used in manipulating food, the hind feet plantigrade. The tail is well furred and cylindrical, with five to seven alternating rings of yellowish gray and brownish black. Black specimens are rare, white or albino animals are rarer, and reddish coloration is subject to individual variation. The sexes are similar in color, and there is no seasonal variation beyond the growth of a heavier coat in winter. The annual molt of adults takes place in spring.

Representative ranges of external measurements are: total length, 655–960; length of tail, 200–288; length of hind foot, 105–132; length of ear, 48–58. Weight generally ranges between 13 and 25 pounds, but some individuals are much larger—the record weight is 62 pounds. Cranial measurements of three males and two females, respectively, from Nebraska include: condylobasal length, 119.1, —, 119.8, 110.5, 113.6; zygomatic breadth, 84.0, 83.0, 83.0, 83.2, 75.5, 72.9.

159. Raccoon, Procyon lotor (courtesy K. Steibben, Kansas Fish and Game Commission).

Natural History. Raccoons prefer hardwood-timbered areas, especially the narrow stands along watercourses, lakes, or marshes, but they also can live in treeless areas in ground dens. Oak, elm, and sycamore are preferred where available, but individuals also will use such tree species as locust, osage orange, shagbark hickory, and cherry. Hollows in limbs or trunks are used as dens, but no nest is made, the interior containing only the debris from the tree itself. Old squirrel nests also are used, as well as caves, crevices in rocky outcrops, cavities under tree roots, abandoned mines and buildings, stands of slough grass, cornshocks, haystacks, muskrat houses, stumps, and rotten logs. An individual raccoon has several dens spaced out within its home range and uses them in a random pattern. Raccoons sometimes share larger dens or burrow complexes with opossums and striped skunks, these three species being similar in general habits and ecological distribution.

Raccoons are strongly opportunistic in their food habits, eating a variety of fruits, nuts, grains, insects, crayfish, mollusks, and small vertebrates, including rodents, birds, and carrion. Bottomland inhabitants consume 70 to 80 percent animal food, whereas upland dwellers consume as much plant material. Individuals eat a great deal

in autumn, building up a large fat reserve for winter.

Some young are taken by great horned owls and coyotes, but man is the chief predator of raccoons, hunting them for both fur and sport. Also, they are frequent traffic casualties. Trapping for fur was extensive in the 1920s and 1930s and declined in the 1940s, but since then it has mostly increased. Hunting usually is done on foot, with packs of hounds bred and raised especially for the purpose. Aside from their value as fur-bearers (pelts averaged more than thirty-two dollars each in Nebraska in the 1979–80 season), raccoons have been used for their fat, which is a good lubricant for leather and machinery, and the baculum has been used by tailors as a ripping tool. In addition, this carnivore is also a valuable food source of which more use could be made. A young raccoon is delicious roasted, the meat being rich and dark. Removing the scent glands under the legs and along the spine near the tail before roasting, as well as some of the copious fat, is strongly recommended.

External parasites include a variety of biting and sucking lice, ticks, and fleas. Internal parasites are roundworms, tapeworms, and flukes. Raccoons are subject to tuberculosis, pneumonia, infectious enteritis, canine distemper, rabies, and raccoon encephalitis, of which only the last two are significant in the wild. Like other carnivores, they do not seem to be as subject to bacterial infection as are herbivores and birds.

Individuals may be dormant through excessively cold weather, but, as in bears, there is no true hibernation because they remain responsive to touch and to changes in the weather. Metabolic rate and body temperature remain high, and the heart rate is near the warm-season norm. Stored fat is the principal energy source in cold weather; weight loss over winter may be as high as 50 percent. Raccoons remain dormant for most of the three winter months in southern Canada but are hardly dormant at all in the southern United States. They den up singly or in groups of as many as a dozen or more, waking to forage during warm spells.

P. lotor is primarily nocturnal and rarely is seen during the day except when sunning on a tree limb or squirrel nest in autumn and winter months. Raccoons generally prowl from dusk to dawn, but some may linger in the den until near morning. Females with young tend to forage early at night, whereas single adults and groups of mixed ages tend to forage later. Females with young are given highest social status at the feeding place, with groups of older animals next in dominance. Single adults are shy and move away if a group approaches. The distance covered in a night depends mainly on the amount of food available and on weather. Raccoons, which depend greatly on hearing, do not range widely on windy nights. Adult males have home ranges of 4,800 acres or more, whereas females usually have home ranges of less than 2,400 acres. They rarely travel more than 20 miles, but incidents of individuals traveling more than 175 miles have been recorded.

Raccoons move with a lumbering gait, rarely faster than 15 miles an hour, and could easily be overtaken by a man on foot were it not for their habit of eluding enemies by concealment. They are strong and ferocious fighters when cornered, but they prefer flight. Raccoons are not territorial and do not post sentries. Family groups of five to six animals are found living together outside the nesting season. The young are playful either alone or together, but adults play mostly by themselves, and only rarely in groups. They are good climbers and descend both head- and tail-first, often jumping from a considerable height. They frequent water, both wading and swimming. Sounds uttered vary with age and season and include a chuckling sound when undisturbed; a snarl, growl, or hiss when annoyed or fighting; a soft, low snort in recognition of another individual; a grumbling purr from mother to young; and a sound like a tree frog made by a young animal when it is separated from its mother. In the autumn of the year adults can be heard to utter a long, shrill, tremulous whistle.

The sense of touch is especially strongly developed, with smell and taste less so. This may account for the alleged washing habit so often mentioned in the literature, which may be not for cleanliness, but to satisfy the sense of touch. Eyesight is good, especially at night, and raccoons also hear well. Familiar loud noises are ignored, but an unexpected noise may attract them. They may look out the opening of a den hole, for example, if someone strikes the tree in which it is situated. Curiosity is a strong trait in raccoons, and combined with their intelligence, makes them interesting pets. One taken young is easily tamed, but problems may arise later. These are extremely active animals, and their curiosity and agility lead them into much mischief, so that they must be caged or chained much of the time. In addition, many become truculent and temperamental as they grow older and can be dangerous when thwarted. The rare animal that maintains its good temper as an adult, however, is a captivating and affectionate pet, expressing itself with doglike licking and with catlike purring and rubbing of cheek and chin.

Molt generally occurs from the first of March to the end of May, either as a gradual shedding of hair from all parts of the body or as shedding that begins on the head, proceeding to the sides and underparts and finally to the back. There is a relatively light coat in summer, but in winter, when raccoons have their primary value as fur-bearers, the guard hairs are longer and the underfur thicker. In October and November new guard hairs appear, starting on the venter and proceeding over the back and tail. This process takes about six weeks, and the animals generally have a prime coat by the first of December. Some females that have borne young, however, are not prime until later in the winter, and an occasional animal may not become prime at all.

Male raccoons have a large, recurved baculum, and females possess an os clitoridis. Females have six mammae—two pectoral, two abdominal, and two inguinal. In January and February the males become sexually active, traveling from den to den searching for receptive females, which bar them from the den after mating. Recent evidence suggests that, contrary to older reports, ovulation is spontaneous. Gestation takes 63 to 65 days, and the litter, generally three to four (extremes one to seven), is born in the latter part of April or the first part of May. Males are active sexually until late June or July, breeding with females that did not conceive earlier or that lost a litter. These late matings result in births in August and September, and a few litters have been reported as late as October.

Young raccoons are blind at birth, and their eyes open between 18 and 29 days of age. Neonates weigh from 70 to 85 grams. The face mask is present as a sparsely haired area of pigmented skin, which is fully haired at two weeks. The tail rings are present as pigmented areas in the skin and are fully haired by four weeks. The back and sides are uniformly yellow and gray, sparsely haired at first but well covered by one week. Agouti guard hairs are present by six weeks, and the first molt occurs at seven weeks of age. The kits (also termed pups or cubs) stay in the den from eight to 10 weeks, when they are weaned and begin to forage with their mother. Females occasionally move litters from one den to another, either for safety or for better access to food. Weaning generally takes place in August, and the milk dentition is replaced between then and October. Some young move out of the den in autumn, but most remain with the female until the following spring. About 40 percent are reported to breed in their first year, but young males become sexually active a little later than adults and generally find most females bred by that time. Raccoons in the wild have lived more than 12 years, but the average is far less—two to three years in most studies.

The preservation of large dead or decaying trees is helpful to raccoons, as are restrictions on burning. Hunters should be discouraged from cutting down den trees to get at their prey. In areas where suitable den trees are scarce, the placement of den boxes has been recommended. Prevention of water pollution is essential. Damage done by raccoons is outweighed by their value as insect-eaters, fur-bearers, and game animals; individuals that are offensive can be trapped and moved to distant locations. They may have to be removed several times, inasmuch as some raccoons persist in returning to home territories.

Selected References. General (Lotze and Anderson 1979 and included citations).

Family Mustelidae: Weasels and Allies

This family includes quite different sorts of mammals; to reflect this diversity, four or five subfamilies usually are recognized. The subfamily Mustelinae includes some of the more generalized members of the group as well as specialized taxa. Mustelines encompass the true weasels (*Mustela*), which are found throughout most of the range of the family—in North and South America, Eurasia, Africa, and Southeast Asia—along with about 10 other genera. On the Northern Great Plains, taxa belonging to this subfamily include the several species of *Mustela*, two martens (*Martes*), and the wolverine (*Gulo*). The honey badger (*Mellivora*) of Africa and western Asia is sometimes considered closely related to *Gulo*, but other authorities place it in its own subfamily, Mellivorinae. The badgers, subfamily Melinae, also include both generalized and specialized taxa in four to six genera; badgers occur only in North America, Eurasia, and Southeast Asia, and one species is present on the Northern Plains. The skunks, Mephitinae, restricted in Recent time to the New World, are represented by three genera. The otters, Lutrinae, include four genera, one of which is the aquatic sea otter (*Enhydra lutris*) of the North Pacific Ocean; others are found throughout the range of the family, including one species in the tristate region. The mustelids are known in the fossil record back to the early Oligocene.

Key to Mustelids

1. Feet webbed between toes; tail thick at base, tapering toward tip; hair short, dense; premolars 4/3
   _____ *Lutra canadensis*
1.' Feet not webbed; tail not thickened at base and tapering toward tip; hair not especially short and dense; premolars 4/4 or 3/3 _____2
2. Length of hind foot 96 or more (rarely slightly less in *Taxidea*); greatest length of skull 90 or more _____3
2.' Length of hind foot 95 or less (rarely slightly more in *Martes americana*); greatest length of skull 85 or less _____5
3. Dorsum yellowish gray, with white medial stripe from nose to neck or beyond; premolars 3/3 _____*Taxidea taxus*
3.' Dorsum brownish, lacking white stripe; premolars 4/4 _____4
4. Dorsum dark brown to blackish, with pale brownish band extending from shoulder to rump on each side; tail relatively short, approximately 25 percent of total length; greatest length of skull more than 130 _____*Gulo gulo*

4.' Dorsum grizzled dark brown to blackish; tail relatively long, approximately 40 percent of total length; greatest length of skull less than 130
   _____ *Martes pennanti*
5. Pelage black, with white stripes or spots on dorsum; palate extending posteriorly only to level of last upper molar or slightly beyond _____6
5.' Pelage not black, lacking white stripes or spots dorsally; palate extending posteriorly well beyond last upper molar
   _____7
6. Dorsum black, with four to six white stripes, breaking up into spots posteriorly; skull flattened, deepest over occipital region, greatest length less than 65 _____*Spilogale putorius*
6.' Dorsum black, with two white stripes of variable length that merge with a white spot on the head and neck; skull highly arched, deepest over orbital region, greatest length more than 65
   _____ *Mephitis mephitis*
7. Hind foot more than 80 in males, more than 70 in females; premolars 4/4
   _____ *Martes americana*
7.' Hind foot less than 80 in males, less than 70 in females; premolars 3/3 ____8
8. Total length usually more than 475; greatest length of skull more than 55
   _____9
8.' Total length usually less than 475; greatest length of skull less than 55
   _____10
9. Pelage buffy, with black legs, feet, tip of tail, and mask across eyes; rostrum and mastoid region broad (see fig. 165); distance between canines greater than that between medial margins of auditory bullae _____ *Mustela nigripes*
9.' Pelage uniformly dark brown, except occasional patches of white ventrally; rostrum and mastoid region narrow (see fig. 165); distance between canines about equal to that between medial margins of auditory bullae ____ *Mustela vison*
10. Total length less than 240, usually less than 200; tail less than one-third length of body, lacking distinct black tip (a few black hairs may be present); great-

est length of skull less than 36, usually less than 34 ⎯⎯⎯⎯⎯ *Mustela nivalis*

10.′ Total length more than 190, usually more than 200; tail at least one-third length of body, with distinct black tip; greatest length of skull more than 32, usually more than 34 ⎯⎯⎯⎯⎯11

11. Tail more than 44 percent of length of body, usually more than 110; summer pelage brownish above, yellowish below, with no white line connecting venter and hind toes; greatest length of skull usually more than 44 ⎯⎯ *Mustela frenata*

11.′ Tail less than 44 percent of length of body, usually less than 110; summer pelage brownish above, whitish below, with white line running down inside of hind legs to toes; greatest length of skull usually less than 44 ⎯⎯ *Mustela erminea*

Note: The skulls of small *M. erminea*, especially females, are difficult to distinguish from those of *M. nivalis*. Skulls in the area of overlapping size (basilar length 27.3–30.6) usually may be separated on the basis of relative height of skull. *Mustela nivalis* has a relatively narrower, higher skull (skull height at anterior margin of basioccipital divided by mastoid breadth, 64.6 to 76.3 percent), as compared with the broader, lower braincase of *M. erminea* (height divided by breadth, 59.1 to 68.6 percent). Even in this character there is overlap between relatively higher skulls of young *M. erminea* and relatively flatter skulls of old *M. nivalis*, and in the absence of comparative material such skulls cannot be identified with certainty.

Genus *Martes*

Martes is the Latin name for the Eurasian martens, of which the two species occurring in western Europe were known to the Romans. These large, semiarboreal weasels are Holarctic in distribution; five species occur in Eurasia, one of which extends southward into Indonesia. Two species occur in North America, and both are found on the Northern Great Plains only in

northeastern North Dakota. The dental formula is 3/3, 1/1, 4/4, 1/2, total 38.

Martes americana: Marten

Name. The specific name of the North American marten, *americana,* serves to distinguish it from its Eurasian counterpart, *M. martes.* The North American species is related not only to *M. martes,* but even more closely to the Siberian sable, *M. zibellina,* and the Japanese marten, *M. melampus.* The vernacular name pine marten sometimes is used for *M. americana* as well as for *M. martes.*

Distribution. The several species of martens are associated primarily with the great circumboreal coniferous forest, or taiga, that extends in a nearly unbroken belt around the Northern Hemisphere. *M. americana* occurs from Newfoundland and Labrador west to Alaska, south of the tree line, and in the Appalachian, Rocky, and Cascade mountains, Sierra Nevada, and the Coast ranges as far south as the dense stands of coniferous forest. On the Northern Plains, the marten once barely penetrated the extreme northeastern corner, in the mixed forests of the Red River Valley, Pembina Hills, and Turtle Mountains of North Dakota. Whether any individuals still survive in this area is problematic. The subspecies is *Martes americana americana.* A single individual was captured on the Black Hills (Custer County, South Dakota) in 1930, and it is possible, but not probable, that the species occurred naturally there. The subspecies geographically nearest the Black Hills is *Martes americana origenes.*

Description. The marten is a large, semiarboreal weasel about the size of a domestic cat. It has a long, slender body, as do the weasels generally, and a bushy, cylindrical tail that is about one-third the length of the head and body. The dorsal fur is long, fine, and silky, and the underfur is dense and soft; the color ranges from yellowish brown to tawny brown. The ventral pelage is somewhat shorter and less dense than that of the dorsum and is slightly darker, except

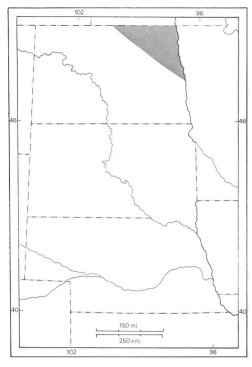

160. Distribution of Martes americana *and* Martes pennanti.

for an irregularly shaped patch of orange to creamy white on the throat, sometimes extending to the upper chest. The legs, although relatively short, are longer in proportion to the body than in other weasels, and this probably represents an adaptation for climbing trees. The legs and tail usually are noticeably darker than the body, and the claws on the five toes of each foot are sharp, recurved, and semiretractile. The head and face usually appear slightly paler in color than the body because of whitish or buffy hairs interspersed with the brown; the ears are prominent, oval in outline, and have pale buffy edges. As in most mustelids, males are larger than females.

One male and two females of *M. a. americana* from northwestern Minnesota had the following measurements, respectively: total length, 660, 559, 508; length of tail, 180, 168, 165. A male from northeastern Minnesota measured: total length, 615; length of tail, 195; length of hind foot, 80; length of ear, 48. Extreme external mea-

surements of adult males from throughout the range of the species are: total length, 500–760; length of tail, 180–250; length of hind foot, 80–98; weight 2.2–3.2 pounds. Comparable measurements for adult females are: 460–580; 170–200; 78–88; weight, 1.6–2.2 pounds. Skulls of a large series of adult male *M. americana* averaged 76.8 (72.1–79.3) in greatest length, and those of a large series of females averaged 69.0 (66.1–73.5); zygomatic breadth averaged 42.3 (37.6–47.6) for males and 37.4 (34.9–41.1) for females.

Natural History. The marten is most common in extensive tracts of dense, boreal coniferous forest of fir, spruce, and hemlock, but it also occurs in mixed coniferous-deciduous forest such as once covered the Red River Valley in northeastern North Dakota and adjacent Minnesota. Unfortunately, the species was sensitive not only to trapping pressure, but also to habitat disturbance and was extirpated in many areas, especially along the southern margins of its range, including the Northern Great Plains and the Great Lakes states. Commercial trapping throughout the nineteenth century and in the early 1900s resulted in near extinction of *M. americana* in this region. With protection and regrowth of cutover forests, martens are increasing in adjacent Minnesota, but they probably have not reappeared in North Dakota.

These animals are solitary inhabitants of the taiga, adults consorting only to mate in midsummer. As in other mustelids, delayed implantation occurs. After the fertilized eggs (zygotes) have divided several times so that each embryo consists of a small number of cells, development ceases, and the blastocysts remain quiescent in the cavity of the uterus and do not implant in the uterine wall as is typical in the development of most mammals. This delay in implantation extends throughout the autumn and winter, until February or March. At that time, in response to increasing day length, hormonal changes in the pregnant female cause the blastocysts to implant in the uterine lining and begin once again to develop. The rate of development is then rapid, and after only 25 to 28 days the two to five (usually three) young are born in March or April. Thus the total gestation period is long, 259 to 276 days, but embryonic growth is suspended during most of this time. The adaptive significance of this arrangement probably is to ensure that both breeding and birth of young occur at the optimal seasons of the year.

At birth the young are naked, blind, and helpless, weighing only 30 grams. Postnatal growth is rapid, however, and by three months of age they are fully furred and active and weigh 600 to 1,000 grams (females and males, respectively), nearly adult size. The young are reared by the female alone, in a nest, usually a cavity in a hollow tree, stump, or log, lined with dry grass, leaves, and moss. When the young are about 35 to 40 days old, their eyes open and juvenile pelage is developing. Weaning begins at about six to seven weeks of age, but the female continues to bring food to the young. When weaned, young leave the natal nest and accompany the mother while she hunts, gradually acquiring the skill.

The onset of estrus in the female in midsummer (July–August) may be the cue for the nearly grown young to become independent. A female in heat may utter a "clucking" sound, and she actively places scent marks in her territory, thus attracting an adult male to mate with her. The pair bond is brief, however.

Home ranges of males are larger than those of females, averaging nearly a square mile as opposed to a quarter of a square mile. Home ranges of adult males tend not to overlap those of adjacent males, and similar lack of overlap is seen between the ranges of adult females. However, home ranges of males often overlap those of several females, and juveniles apparently are tolerated by adults of both sexes. Adult home ranges are stable from year to year; if an adult disappears there are adjustments in home range boundaries by adjacent individuals of the same sex. Scent marking of home range topography may be supplemented by territorial agonistic behavior

161. Marten, Martes americana (courtesy D. Randall).

when two adults encounter each other.

The young do not mature sexually until they are two years old, although they reach adult weight by the end of their first summer. Such immature individuals probably do not establish stable home ranges but wander widely, seeking suitable unoccupied habitat; 65 percent of the martens in one population studied were transients or temporary residents. Females first attain breeding condition as two-year-olds (that is, in their third summer) and usually are bred by a nearby adult male. Having established a stable home range, they then may reproduce annually until they are at least nine years old. Maximum longevity is 15 to 17 years.

Predation on martens is not intense except by humans. Golden eagles prey on them during daylight, but the animals are mostly nocturnal. Fishers, lynx, coyotes, and gray wolves are other potential predators. Lice and fleas have been recorded, and internal parasites include intestinal and nasal roundworms and lungworms.

Martens are active predators on many small mammals of the northern forests. Their climbing ability permits them to pursue red squirrels (*Tamiasciurus*) and flying squirrels (*Glaucomys*) effectively, and they also kill a significant number of snowshoe hares, particularly when the hares reach their cyclic peaks of abundance. However, their staple food is mice and voles, particularly the red-backed vole (*Clethrionomys gapperi*) and meadow vole (*Microtus pennsylvanicus*). Birds are eaten occasionally, and in late summer and autumn fruits, especially huckleberries, blueberries, and strawberries, are an important supplement to the diet.

In the past martens were an important fur resource, and in some places in Canada and Alaska where population densities are relatively high they still are trapped commercially; the value of these northern furs is about double that of pelts originating in the southern part of the range. Densities vary both geographically, according to habitat suitability, and from year to year depending on food availability. High densities may be about three to four individuals per square mile, whereas low densities may be only a tenth as great. Over most of its former range in the contiguous United States, the marten is much reduced in numbers if not extirpated, and at present it is of no commercial significance except in the Northeast, where populations have recovered somewhat. In 1980 a good marten

pelt might have commanded thirty to thirty-five dollars at auction.

Selected References. Natural history (Hawley and Newby 1957; Weckwerth and Hawley 1962); systematics (Anderson 1970).

Martes pennanti: Fisher

Name. This large marten was named in honor of the famous British naturalist Thomas Pennant (1726–98), best known for his work *Arctic Zoology*, which included the mammals of northern North America as then known. It has a number of other vernacular names, including "pekan," an Indian name, and "black cat," but fisher and "fisher marten" are now the only names in general use. The species is not aquatic, and the name appears to be derived from its supposed habit of robbing the caches of frozen fish made by trappers to feed their dogs in winter.

Distribution. The fisher inhabits the northern coniferous forests of North America from Nova Scotia to British Columbia. Unlike the smaller marten, it has no close relatives in the Old World. The fisher's range does not include as much of Alaska, the Northwest Territories, Newfoundland, Labrador, and northern Quebec as does that of the marten, but it has a more extensive distribution, formerly at least, in the Appalachians and the Great Lakes states. Both species occurred over a wide area, including the Pacific coastal forests as far south as San Francisco Bay, the Sierra Nevada, and the Northern and Central Rocky Mountains. Early fur trapping records indicate that the fisher was more abundant than the marten in northeastern North Dakota (see fig. 160) in the early nineteenth century, but both fur-bearers were extirpated there at some unknown time by excessive trapping and clearing of the forest, and the fisher probably was absent from the Northern Great Plains for many years. In December 1976, however, a specimen was trapped in Pembina County, indicating that this mustelid may be reestablishing itself in the

Pembina Hills of North Dakota. The subspecies on the Northern Plains is *Martes pennanti pennanti*. There is no evidence that the fisher ever occurred on the Black Hills.

Description. Like the marten, the fisher is a semiarboreal weasel, but it differs in being larger (about the size of a small fox) and darker. It has a long body, relatively a little less slender than that of the marten, and a long, bushy, cylindrical tail that is more than half the length of the head and body. The dark brown to nearly black dorsal pelage has a sprinkling of buff-tipped guard hairs, heaviest on the head, neck, and shoulders, that give the fisher a grizzled appearance. Males tend to develop this griz-

zling to a greater extent than the more uniformly dark-colored females. The venter and legs are as dark as the dorsum, but there may be a few irregular white spots. Adult males are about 15 percent larger in linear dimensions than adult females. One female of *M. p. pennanti* from northwestern Minnesota had the following measurements: total length, 858; length of tail, 325.

Average and extreme external measurements of nine adult males, followed by the extremes for three adult females, are as follows: total length, 987 (905–1,055), 884–921; length of tail, 372 (341–397), 355–359; length of hind foot, 116 (103–123), 112–115; length of ear, 50 (45–55), 45–49. Weights of these specimens

162. *Fisher,* Martes pennanti *(courtesy San Diego Zoological Society).*

were 3.94 (2.86–5.67) kilograms (about 6.3 to 12.5 pounds) for males, and 2.13–2.15 kilograms (about 4.7 pounds) for females. Sexual dimorphism is also strongly expressed in cranial morphology. Adult males have larger, more massive skulls with strongly developed sagittal crests and widely flaring zygomatic arches. Greatest length of skull for adult males and females from all parts of the range averaged, respectively, 114.7 (100.3–121.0) and 98.6 (92.1–105.9), and the zygomatic breadth averaged 73.1 (61.7–88.3) and 56.1 (51.4–65.1). The female *M. p. pennanti* from northwestern Minnesota measured 103.9 and 64.0.

Natural History. Like the marten, the fisher is mainly a creature of coniferous and mixed deciduous-coniferous forests. Whereas the marten favors dense, mature spruce-fir forest, the fisher seems to be more common in mixed stands of hardwoods and conifers, or even pure hardwood stands, often along streams or lakes. This may explain the early abundance of the animal relative to martens along the rivers of northeastern North Dakota. Such lowland forests are sometimes seasonally wet, in contrast to the favored upland habitat of the marten, and some authors have argued that the vernacular name derives from these animals' feeding on small fish, amphibians, and crustaceans at such times. Certainly the fisher is omnivorous, feeding opportunistically on fruits, nuts, mice and voles, chipmunks, tree squirrels, cottontails, snowshoe hares, woodchucks, and occasionally martens. Carrion also is eaten, and there have been reports of attacks on deer fawns. The aspect of the fisher's diet that has attracted the most attention, however, is its propensity for feeding on porcupines (*Erethizon dorsatum*). The folk explanation for the fisher's ability to deal with this well-protected rodent was that the predator was quick enough to avoid the thrashing of the porcupine's heavily quilled tail and, getting a forepaw beneath its intended prey, would flip it onto its back, exposing the unprotected belly. More recent observations of the fisher's predatory behavior, however, reveal that this low-slung mustelid attacks the porcupine head-on; the porcupine's face is poorly protected, and the fisher inflicts multiple wounds there until the porcupine is overcome. The fisher then eats its prey from the head, throat, and belly inward, avoiding as much as possible contact with the quills of the back and tail. The final result of the encounter is a fairly well-flensed porcupine skin flattened quill side down on the forest floor. Fishers also appear to be relatively immune to the effects of quills that they ingest or that penetrate the skin. Most seem to pass through tissue without causing serious trauma or infection, unlike the case in many other predators.

Relatively little is known of the behavior of the fisher, but the species appears to be solitary except when courting and mating or rearing young. Females usually breed when they are one year old, in their second summer, whereas males attain sexual maturity as two-year-olds. The gestation period of the fisher is longer than that of the marten, being just short of one full year (327 to 358 days). The females become receptive to breeding in March or April. Breeding behavior has not been observed but is likely to resemble that of other *Martes*. Fertilized ova develop to the early blastocyst stage and then become quiescent until February or March, when they implant and develop rapidly for about a month. Females give birth to two to four young (usually three) in March or April, and about a week after parturition a postpartum estrus occurs, and the animals mate again. Postnatal growth of the young is rapid, paralleling that in the marten, and juveniles approach adults in body weight by autumn, when the family unit breaks up and the young begin to forage on their own.

Not enough work has been done on habitat use by the fisher to reveal much about home ranges or territorial behavior. These animals may be more diurnal than the marten, perhaps because their larger size protects them from attacks by diurnal birds of prey. This larger size also dictates that the home range be larger than that of the marten. One estimate of 10 square miles or more may, however, be excessive, and the opinion of trappers that fishers move at random, rather than within home ranges, also is suspect. Dens are presumed to be in holes, stumps, and rock crevices, but documentation is scanty.

Population densities of fishers have been estimated as one animal per 3.6 to 4.5 square miles in good habitat in Maine, and they may have been even higher in mixed or deciduous forests in the southern part of the original range. The decline in distribution and numbers of *M. pennanti* reached its nadir in the 1930s and 1940s, owing both to habitat loss and to overexploitation. Since that time there has been a significant increase in numbers, and part of the original range has been reoccupied. This increase was noticed first in the Northeast, then in the Great Lakes states and Pacific coastal regions. In the Rocky Mountains, the process was aided by restocking. It was observed that increases in fisher density in these areas often were accompanied by a decline in porcupines, and it is possible that predation by fishers results in lower porcupine density. If this postulated cause-and-effect relationship is true, the fisher has an economic worth beyond that of its valuable pelt, for porcupines can be detrimental to timber. At fur auctions, pelts of the darker females currently bring up to one hundred and fifty dollars and those of the more grizzled males up to one hundred dollars.

Fishers harbor fleas and ticks in their pelage, and internal parasites include tapeworms and roundworms. Longevity in the wild may regularly be as much as 10 years and probably is often longer.

Selected References. General (Hamilton and Cook 1955); natural history (Wright and Coulter 1967); systematics (Anderson 1970).

Genus *Mustela*

This genus is most abundant in the Northern Hemisphere, but it contains species that penetrate the tropics to the equator and beyond in South America as well as Southeast Asia. There are about 16 species in the genus, five of which occur on the Northern Great Plains. The latter are distributed among three subgenera. The weasels proper (subgenus *Mustela*) include three species, the ferrets or polecats (subgenus *Putorius*) only one, the nearly extinct black-footed ferret, and the subgenus *Lutreola* contains the semiaquatic mink. The generic name is the Latin word for "weasel;" the vernacular name weasel is derived from the Anglo-Saxon *wesole*. The dental formula is 3/3, 1/1, 3/3, 1/2, total 34.

Mustela erminea: Ermine

Name. The species name is derived from an Old French word for the white winter color phase of this animal, *ermine* (in the brown summer pelage it is referred to as stoat), which is now used as a vernacular name in the British Isles and parts of North America. Another name widely used in the United States is short-tailed weasel.

Distribution. The ermine is a Holarctic species, occurring in a variety of habitats throughout Eurasia and North America, from Greenland and the Canadian and Siberian arctic islands south to about 35° N latitude. On the Northern Plains the species occurs in eastern North Dakota (*Mustela erminea bangsi*) and on the Black Hills (*Mustela erminea muricus*). It is found principally in forest and woodland habitats, whereas the least weasel (*M. nivalis*) tends to occur in open habitats. The two species are relatively easy to distinguish in eastern North Dakota, where they are sympatric, because they are quite different in size and differ in the color pattern of the tail. However, in the Black Hills region a potential problem exists in that the ermine of the Southern Rocky Mountains is nearly as small as the least weasel. Until recently it was thought that the ranges of the two species did not overlap in this area, but specimens of *Mustela nivalis campestris* recently have been taken east of the Bighorn Mountains of Wyoming, and the least weasel also is known from north-

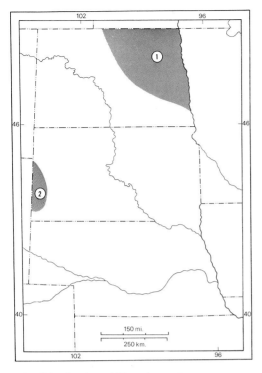

163. *Distribution of* Mustela erminea:
(1) M. e. bangsi; (2) M. e. muricus.

western Nebraska; it is possible that it may be found on the Black Hills, where it would be distinguished from the ermine by its shorter tail, which lacks a black tip, but not by overall size.

Description. Weasels are all short-legged, long-bodied animals with elongated heads and with tails of varying length. Of the three species on the Northern Plains, the ermine is intermediate in both size and relative tail length. All three are seasonally dimorphic in color, being brown dorsally in summer and molting to a white pelage in winter. (For description of molt, see account of *Mustela frenata* below.) The ermine may be told from other species by the following combination of characters: tail about a third the total length or somewhat less, always black-tipped, all white in winter except tip of tail, in summer dark brown above and white below, with white fur extending down inner side of hind leg to ankle. Females have four or five pair of

inguinal mammae. Males are much larger than females, as indicated by the following average and extreme external measurements of 12 males and four females, respectively, of *M. e. bangsi* from Minnesota: total length, 316 (291–340), 249 (240–260); length of tail, 87 (70–101), 61 (55–65); length of hind foot, 43 (40–44), 32 (30–33); recorded ear measurements for the subspecies vary from 15 to 20. Weights of the animals mentioned above ranged from 90 to 170 grams for males and from 43 to 71 grams for females. Average and extreme cranial measurements of 10 males of *bangsi* from central Minnesota are: greatest length of skull, 42.9 (39.7–45.7); zygomatic breadth, 21.8 (19.6–24.0). One female from the same locality measured 36.8 and 17.4.

As pointed out above, *M. e. muricus* is much smaller than *M. e. bangsi*, and the degree of sexual dimorphism is somewhat reduced (see fig. 165 for illustration of skulls). Average and extreme external measurements of seven males and four females, respectively, from Colorado and Wyoming are: total length, 230 (220–243), 206 (200–214); length of tail, 36 (32–39), 35 (28–40); length of hind foot, 29 (28–31), 23 (20.5–26); the ear usually measures from 13 to 16. The females weighed an average of 38.0 (30–44) grams. Average and extreme cranial measurements of six males and four females from the same states are: greatest length of skull, 33.9 (32.7–34.9), 31.9 (31.5–32.2); zygomatic breadth, 17.5 (17.0–18.0); 16.2 (16.0–16.4).

Natural History. A great variety of habitats harbor *M. erminea*, which ranges from arctic tundra to the northern part of the eastern deciduous forest and the Great Plains, including alpine barrens, marsh, and agricultural land. In North America the ermine is most abundant in boreal, montane, and Pacific coastal coniferous forests. Fluctuations in numbers from year to year are common and sometimes extreme, being influenced primarily by prey availability. Ermines feed on a variety of small mammals and birds, ranging in size up to cottontail rabbits. Mice and voles constitute their main food when abundant, but when small

164. Ermine, Mustela erminea *(courtesy L. E. Bingaman).*

mammals are scarce, nesting birds may be prime prey. Large males tend to take larger prey, such as rabbits, than do the smaller females. Both sexes may turn to secondary prey such as shrews, reptiles, amphibians, earthworms, insects, carrion, and berries. Ermines sometimes also kill least weasels, but the significance of this for both species is uncertain. Excess food may be cached, especially in winter.

Ermines live and hunt separately, except when young are being reared. Unlike most mustelids, males assist in raising the young (of the previous mating). Mating takes place in late spring and early summer. Males attain breeding condition in mid-May, shortly after the birth of the litter that was conceived the previous summer. They seek out females in May, June, and July, while they still are nursing their young, and form a temporary pair bond, during which time the pair may hunt together and bring food to the nestlings as they are being weaned. Both adult females and juveniles (about two and a half months old or less) come into breeding condition during this pair-bond period, but whether adults mate before juveniles is unknown. After a brief period of development, the blastocysts become quiescent and do not implant until the following March or April. At that time, in response to hormonal changes triggered by lengthening spring days, implantation occurs. Development is completed in three to four weeks, and four to 12 young are born, the usual number being six to nine.

Newborn ermines weigh an average of 1.7 grams and have a coat of fine, white hair. A prominent brownish mane develops

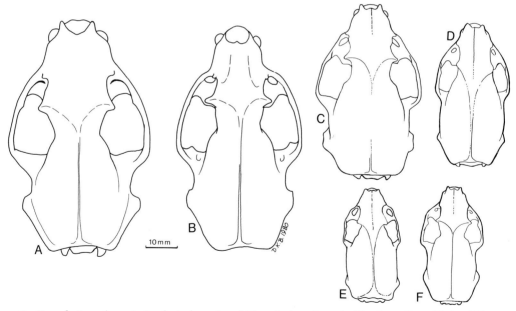

165. Dorsal view of crania (males) of species of Mustela *found on the Northern Great Plains:* (A) M. nigripes; (B) M. vison; (C) M. frenata; (D) M. erminea bangsi; (E) M. erminea muricus; (F) M. nivalis.

on the dorsal surface of the neck soon after birth, which provides the mother with a "handle" to grasp in her teeth when carrying the young. The eyes open in five to six weeks; the brown dorsal pelage has developed by then, and the characteristic black tail tip appears about a week later. Even before their eyes are open, the nestling ermines begin to feed on solid food brought to the nest by the adults, employing deciduous teeth that begin to erupt at about 20 days of age and are in place and functional by day 30. Young may be weaned by the end of the fifth week of life, but usually lactation continues from seven to 12 weeks, with solid food gradually supplementing milk. During the later part of the weaning period, and before they achieve complete independence in the autumn, the nearly full-grown young may hunt with the female, giving rise to reports of hunting packs of ermines. Sometime during the late summer or autumn the adult male leaves the family and resumes his solitary ways.

The nest in which the young are reared may be in the former burrow of a rodent, a rock crevice or pile, or a hollow log or stump. Surface nests of voles also are taken over. The nest typically is thickly lined with rodent fur.

The nest is within a hunting territory that varies in size according to habitat suitability and prey abundance, as well as the presence or absence of neighboring ermines. In good habitat, with abundant vole populations and high ermine density, the territory/home range may be compressed to about eight acres or less (as small as half an acre according to one Soviet study). Usually, however, hunting territories are much larger, up to 500 acres, and may be composed of a series of smaller subterritories (about 20 to 25 acres) that are visited sequentially. However, an ermine may travel a long distance around its territory in the course of hunting in a single day (and night); individuals may be active at all hours. Territorial defense has not been described, but scent marking of boundaries probably is important.

Maximum population densities achieved by ermines are about one per 10 acres but usually are much less. Such high densities are invariably in response to a high vole population; under such conditions ermines feed almost exclusively on voles and may be important in reducing the density of the prey population.

Ermines are in turn preyed on by a variety of raptors, including rough-legged hawks, goshawks, and great horned owls. Mammalian predators of record are coyotes and red and gray foxes. Long-tailed weasels may occasionally kill their smaller cousins. External parasites include lice, ticks, and fleas; helminths include trematodes, cestodes, and several nematodes, the most frequent of which is the nasal roundworm *Skrjabingulus*. Maximum longevity in the wild is about seven years.

Ermines in dense, white winter pelage produce the much sought-after fur of that name. The species is also of great economic benefit by virtue of its rodent-eating proclivities. Careful confinement of domestic rabbits or poultry will prevent ermine depredation on the farm, but the animals are quick to take advantage of a hole in the chicken house wall or floor.

Selected References. General (Corbet and Southern 1977; Hall 1951); natural history (Fitzgerald 1977).

Mustela frenata: Long-tailed Weasel

Name. The long-tailed weasel obtained its vernacular name because among the three species of North American weasels it possesses both absolutely and relatively the longest tail. The specific name is derived from the Latin word *frenum*, "bridle," based on the white facial markings found in some populations (none in the tristate region, however). These white marks, which are variable in extent, generally extend above the eyes and often along the upper lip, giving rise to the fanciful comparison to a bridle.

Distribution. Unlike the other species of weasels found on the Northern Plains, *Mustela frenata* has no close relatives in Eurasia. Its North American distribution

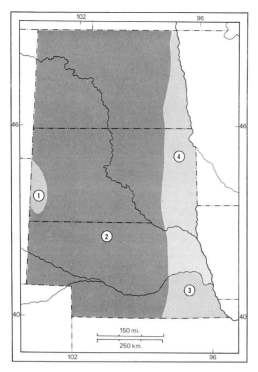

166. Distribution of Mustela frenata: (1) M. f. alleni; (2) M. f. longicauda; (3) M. f. primulina; (4) M. f. spadix.

extends from New England, the Great Lakes region, and the southern provinces of western Canada southward throughout much of the United States, Mexico, and Middle America. The species also is found in northern South America, but the limits of its range in that region are not well known. The long-tailed weasel occurs throughout the Northern Great Plains, where four subspecies have been recorded: *Mustela frenata alleni* is endemic to the Black Hills; *Mustela frenata spadix* is found in the eastern part of the tristate region from the Red River Valley south to northeastern Nebraska; *Mustela frenata primulina* occurs in the extreme south-eastern corner of Nebraska; and the remainder of the three states is inhabited by *Mustela frenata longicauda.*

Description. As in most other mustelids, males in this species are considerably larger than females. Therefore a female *M. frenata*

may approach a male ermine (*M. erminea*) in size, but there is a distinct difference if the same sexes from the same area are compared. The tail of *M. frenata* is relatively longer than that of *M. erminea*, being more than 44 percent of the length of the head and body; moreover, the black tip is shorter in relation to tail length than is usual in *erminea*. Dorsal pelage in summer varies from pale brown to dark brown, being darker along the eastern margin of the Northern Plains than in the more arid western reaches. The chin and upper lips are white, and the rest of the venter ranges from ochraceous to buff. This species molts to a winter pelage that is completely white, except for the black tip on the tail, throughout the tristate region except in southern Nebraska.

Molt begins in October as scattered patches of new white fur on the venter, which gradually coalesce. Once the yellow summer pelage of the belly is completely replaced, molt spreads laterally up the sides, and finally the two molt lines meet middorsally. The entire process of molt takes two months or more. South of the Platte River in Nebraska some individuals have a brown, rather than white, winter coat that is paler and longer than that of summer. Molt back to summer pelage begins in March; new brown hair first appears along the middorsal line and spreads laterally until finally the venter too has molted from white to yellowish fur. Females possess four pair of inguinal mammae.

Average and extreme external measurements of 10 adult male *M. f. longicauda,* followed by those of five adult and subadult females, all from western Nebraska, are: total length, 427.1 (400–459), 362.2 (343–375); length of tail, 155.0 (138–175), 127.2 (115–142); length of hind foot, 47.5 (42–54), 42.6 (39–47); length of ear (five males, two females only), 23.6 (19–27), 17.8 (17.6–18). Two males weighed 275.0 and 315.5 grams; four females averaged 154.7 (130.5–178.6) grams. Average and extreme cranial measurements of seven male *M. f. longicauda* from western Nebraska are: greatest length of skull, 49.7 (48.2–51.4); zygomatic breadth (six individuals only),

28.9 (27.4–30.2). The corresponding measurements for six females of *longicauda* from the Northern Plains (Kansas to North Dakota) are as follows: greatest length of skull, 45.6 (44.5–46.9); zygomatic breadth, 24.8 (24.5–25.3).

The eastern subspecies *primulina* and *spadix* are darker than the western *longicauda* and *alleni*; *primulina* is smaller than the more northerly *spadix* and *longicauda*, and *alleni* also is a smaller race.

Natural History. Like many carnivores, the long-tailed weasel ranges widely and may be encountered in a large number of habitats, both grassland and forest. It has been reported to favor brushy and rocky areas and often is found near watercourses and lakes. Where it is sympatric with the ermine in northeastern North Dakota, there may be some tendency for *M. frenata* to occur in more open habitats, whereas *M. erminea* may be more common in forested areas.

As in other small species of weasels, individuals lead a solitary life except during the breeding season and while young are being reared. Home ranges, which may broadly overlap, probably are larger on the average than those of the smaller ermine, but as in that species they are greatly influenced by the availability of prey. When food is readily available, a home range of 30 to 40 acres is typical for a male long-tailed weasel; females probably have somewhat smaller ranges, except when they are rearing young. At other times the home range may be 200 to 300 acres.

Males also may capture larger prey species than females do, including cottontails, snowshoe hares, tree squirrels, and Richardson's and Wyoming ground squirrels. Young prairie dogs may be subdued, but in most circumstances adults are large enough to defend themselves. Females, being little more than half the weight of males, concentrate on mouse- and chipmunk-sized prey. The several species of voles in the tristate region are staples in the diet of this weasel. Shrews and moles also are consumed, as well as pocket gophers. The elongated body of this weasel is ideally

167. *Long-tailed weasel*, Mustela frenata *(courtesy U.S. Fish and Wildlife Service, photograph by C. N. Hillman).*

adapted to pursuing prey species within their burrow systems, and the animals also tunnel beneath the snow when hunting in winter, something the ermine does not regularly do. Ground-nesting birds, and their eggs and nestlings, also are taken but do not form a major part of the diet. Long-tailed weasels are capable tree climbers and may ravage the nests of birds or pursue tree squirrels or chipmunks. Snakes, frogs, and insects are eaten occasionally.

The technique that weasels, including *M. frenata*, employ in killing smaller prey is to seize the victim in their teeth; then, folding the elongated body around the prey and holding it with all four feet, they bite the dorsal part of the neck at the base of the skull, crushing or severing the spinal cord and brainstem. Large prey such as rabbits also are attacked at the base of the skull, but the weasel cannot kill an adult cottontail quickly in this way, and repeated attacks must be made until the canines, piercing the skin of the neck, sever the muscles that support the head. Once the cottontail is incapacitated, the weasel completes the kill either by strangling the rabbit by gripping its throat or by severing the large blood vessels on the side of the neck.

It is commonly held that weasels "suck the blood" of their prey. The basis for this belief is that, once the prey is dead, a weasel often slits its throat with the canine teeth and laps up the blood that flows from the wound. However, the remainder of the prey also is eaten, either on the spot or, more frequently, after being carried back to the den. Large prey species are cached and may provide food for several days. In consuming their prey, the species of *Mustela* characteristically start with the head, especially the brain. Most or all of the flesh and internal organs are finally eaten, leaving the skin, the tail, and sometimes the rostrum behind.

The den usually is originally excavated by a ground squirrel or pocket gopher and subsequently remodeled by the weasel. Sometimes, however, it is in a rock pile or crevice. The nest, in an enlarged chamber nine to 12 inches in diameter, is constructed of dry grasses and lined with rodent and rabbit fur, as well as the molted hair of the inhabitants. Several burrows radiate from the nest, some used as latrines and others as food caches. However, both feces and animal remains are found in the nest chamber as well.

Whether the sexes form a temporary pair bond is uncertain, but there have been many observations of males and females hunting together and inhabiting the same den, and it is likely that the species is at least seasonally monogamous. Two adult males were seen together on one occasion, but in most cases weasels hunt alone.

The hunting pattern of this species often is described as an elliptical route away from the den and back again. Smell seems to be the sense mainly employed to locate prey. The distance traveled in a single night averaged 346 feet for 10 females and 704 feet for 11 males in Pennsylvania, the ranges being 20 to 1,420 and 60 to 2,535 feet for the respective sexes. A study in Iowa provided similar results, but some individuals may travel two or three miles in a night. In winter some weasels are reported to abandon their dens and embark

upon hunting routes that take from several days to several weeks to traverse, but little is known about such behavior. Males are also known to range widely at the onset of the breeding season in late summer. Because of their low density and long-distance movements, it is difficult to census *M. frenata*; estimates of population densities ranged from a high of 0.15 per acre to a low of 0.03 per acre in Pennsylvania deciduous forest, averaging 0.07 per acre. Other estimates are much lower, down to only one or two weasels per square mile. On much of the Northern Plains, average densities probably are quite low.

In early spring, a female (or a pair) will reestablish a den, if it was abandoned during the winter, in preparation for the birth of the young. There is some evidence that, if a mated pair is involved, the female restricts her movements during the last two weeks of gestation, and it has been speculated that her mate may bring her food during this period. The young are born in April or May after a gestation period of nine months, only the last 23 or 24 days of which are devoted to embryonic development, owing to the long period of delayed implantation. Litter size ranges from four to nine, with modal numbers between six and eight. Newborn young are blind, unfurred, about 65 mm long, and weigh 3 grams. After a day they develop a soft white coat, but the long brownish mane that is characteristic of week-old *M. erminea* is not developed by this species. The juvenile pelage begins to be replaced by adult fur at three weeks of age. At three and a half weeks the deciduous canines and premolars have erupted enough to permit young weasels to feed on small pieces of meat provided by the adults. The eyes open by five weeks of age, at which point weaning may be completed. By the sixth week the difference in size between the sexes is pronounced, males weighing about 100 grams and females about 80. The characteristic musky odor first becomes noticeable in the young at this time. Growth continues until the tenth or eleventh week, by which time the permanent dentition is in place and males weigh about 300 to 350 grams and females about 175 to 225 grams, essentially the same as adults. Probably by the sixth or seventh week the young begin to accompany the female while she hunts, and they apparently begin to disperse when 10 to 12 weeks old.

Young females become sexually mature at three to four months of age, in July or August, and adult females also come into estrus at that time. Males probably locate receptive females by scent. However, vocalization also may be employed; the "trill"—a series of short, rapidly repeated, low-frequency calls—is heard between individuals. After breeding, the fertilized ova develop for approximately eight days, then cease development for about seven and a half months. Following this delayed implantation, embryonic development resumes in the spring. Timing of spring molt and of implantation of embryos both are controlled by hormonal changes in response to increasing day length, implantation occurring about three weeks after the initiation of molt. Testes enlargement in males also is associated with the onset of spring molt, as is development of the baculum from the juvenile to the adult state; juvenile males are sexually mature at 14 to 15 months of age. The enlarged testes descend into the scrotum in May and June, before the breeding season.

Virtually nothing is known of average mortality rates or longevity in the long-tailed weasel, but they probably are similar to those of the ermine. Recorded predators on *M. frenata* include barred owls, great horned owls, rough-legged hawks, goshawks, red and gray foxes, coyotes, and several species of snakes. Other carnivores, such as bobcats, probably kill a few. Fleas are common on *M. frenata*, and other external parasites, such as ticks and lice, also are found. Internal parasites include trematodes, cestodes, and nematodes. A few individuals are taken incidentally by trappers, but *M. frenata* is not now such an important fur species as it once was.

Selected References. General (Hall 1951); natural history (Brown and Lasiewski 1972;

Fitzgerald 1977; Latham 1952; Polderboer, Kuhn, and Hendrickson 1941; Quick 1951).

Mustela nigripes: Black-footed Ferret

Name. The black color of the lower legs and feet in this species contrasts sharply with the pale yellowish buff of the body and provides the basis for both the scientific (from the Latin *nigra*, "black," and *pes*, "foot") and the vernacular names of this species. Ferret comes from the Old French *fuiret*, "thief."

Distribution. The black-footed ferret is restricted to the High Plains and adjacent regions of the Central and Southern Rocky Mountains, including the Colorado Plateau. Its distribution nearly coincides with the range of prairie dogs (*Cynomys*). The similar polecat (*Mustela eversmanni*) of the Eurasian steppes, or grasslands, occurs from eastern Europe to eastern Siberia, and during the last glacial period it also was found in central Alaska; it is possible that the two taxa actually are conspecific. Whereas *M. eversmanni* is still common and widespread in many parts of the Old World, *M. nigripes* has been reduced to a few scattered, relict populations within its former range.

Description. The black-footed ferret is the size of a domestic ferret or a mink. The dorsum is buffy, varying from yellowish to whitish, with the face and venter somewhat paler in color. The face has a black mask across the eyes, and the feet and tip of the tail are black; the body also has a sprinkling of black-tipped hairs. This unique color pattern distinguishes the black-footed ferret from all other mammals in our area. The only species with which it might possibly be confused is the long-tailed weasel, which has a dark dorsum (except in winter) and lacks the contrasting black feet and facial mask. Adult males are about 10 percent larger overall than females; average and extreme external measurements of seven males from throughout the plains (Kansas to South Dakota) are: total length, 519.3 (490–572); length of tail,

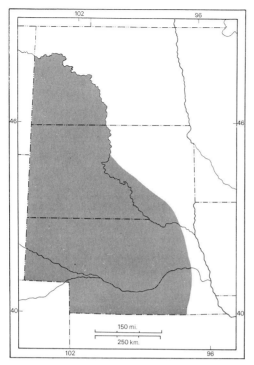

168. Distribution of Mustela nigripes.

126.1 (100–150); length of hind foot, 62.9 (59–68); length of ear, 31.0 (30–32) (three individuals only). Greatest length of skull and zygomatic breadth for eight males from this region are, respectively, 67.3 (65.0–70.5) and 40.6 (38.6–42.1). Average and extreme cranial measurements of six females from Colorado, in contrast, are: greatest length of skull, 64.1 (60.1–67.1); zygomatic breadth, 38.4 (36.3–40.0). No subspecies are recognized.

Natural History. The black-footed ferret has the narrowest range of ecological tolerance of any North American predatory mammal. Its principal prey is the black-tailed prairie dog, and the burrow systems of these colonial rodents also provide shelter. Two other species of prairie dogs, the white-tailed (*Cynomys leucurus*) and Gunnison's (*C. gunnisoni*), also were formerly exploited by ferrets, as are other small rodents to a limited extent. Because prairie dogs are limited to open landscapes—steppe and shrub-steppe—ferrets also are limited

to such grasslands as a consequence of their dietary dependence. Ferret predation seems not to reduce significantly the density of prairie dog populations under present conditions, but ferrets are so rare that this may have been true even before they were nearly exterminated.

Factors contributing to the decline of *M. nigripes* can only be inferred, but it is likely that the destruction of once-extensive prairie-dog colonies through farming and pest control measures were of major importance. Use of persistent biocides to poison rodents, which in turn caused secondary poisoning of ferrets when they ate the rodents, had an effect, the magnitude of which cannot be estimated, although it may have been significant. Direct natural predation is not a major factor, although coyotes, golden eagles, and great horned owls are known to kill ferrets. Viral distemper is also a suspected cause of mortality, because captive individuals are highly susceptible to it.

Females probably first breed at one or two years of age. Unlike most mustelids, they do not exhibit a long period of delayed implantation. Breeding occurs in March or April, and 42 to 45 days later a litter of one to five young (usually three or four) is born. No data on growth and development of the young are available, but probably these do not differ markedly from what is known of Old World ferrets, in which young weigh about 10 grams at birth and are strictly altricial. Eyes open at about 28 to 30 days of age, by which time the juvenile pelage is fully developed. Young black-footed ferrets leave the nest burrow at about two months of age but probably remain with their mothers until autumn. There is some evidence of autumn dispersal of juveniles, and they may be killed while crossing highways at this season.

Because black-footed ferrets are rare, as well as nocturnal and secretive and thus difficult to observe, little is known about their behavior. They are solitary except when mating or with young. Observations of scent marking and agonistic behavior between individuals suggest that they may be territorial. Nothing is known of home-

169. Black-footed ferret, Mustela nigripes *(photographer unknown).*

range size or of frequency of movements within or between prairie dog colonies.

Selected References. General (Hillman and Clark 1980 and included citations).

Mustela nivalis: Least Weasel

Name. The name *nivalis* comes from a Latin adjective meaning "snowy" and reflects the occurrence of this weasel in the boreal parts of the world. Studies of geographic variation in cranial morphology in the North American least weasel and its Eurasian counterpart have led many mammalogists to conclude that the two are conspecific and that the appropriate name is that which was first applied to European populations. However, this view recently has been questioned by workers in Finland, who have evidence that both a larger (*M. nivalis*) and a smaller weasel (*M. rixosa*) occur together there without interbreeding. If additional evidence supports the view that they are different species, the North American least weasel will become *Mustela rixosa,* a name under which it was known for many years. We provisionally apply the specific name *nivalis* to North American populations.

The vernacular name refers to the fact that this is the smallest weasel (in fact, the smallest carnivore) in North America. In Europe *Mustela nivalis* frequently is referred to as the mouse weasel or simply as the weasel.

Distribution. This mustelid is rare to locally common from the Appalachian Mountains and the north-central United States north to Labrador and west to Alaska. Like the ermine, it is thought to be a circumboreal species, reaching the southern limit of its North American distribution in Kansas save for a narrow zone extending southward in the Appalachians to Tennessee and North Carolina. The range on the Northern Great Plains includes the eastern three-fourths of Nebraska and South Dakota (*Mustela nivalis campestris*) and nearly all of North Dakota (*Mustela nivalis rixosa*).

Description. The least weasel is the smallest of the three weasels known to occur on the Northern Plains. Only a small *M. erminea* could be confused with *M. nivalis,*

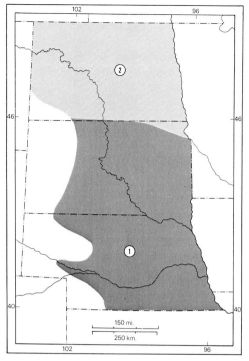

170. Distribution of Mustela nivalis:
(1) M. n. campestris; (2) M. n. rixosa.

but, in the only places in the tristate area where the two species are at present known to occur together (northeastern North Dakota), *erminea* is distinctly larger than *nivalis*. These two species also can be separated by the absence of a black tip on the tail of *nivalis*, which has at most a few black hairs on the end. The tail is also one-fourth or less the length of the head and body. Females are reported to have only three pair of mammae. Distinct white winter and brown summer pelages occur throughout the Northern Plains, but no details are known concerning timing of the molt in this weasel. The summer coat is chocolate brown to umber on the upper parts and sides and white beneath, sometimes with isolated brown patches, whereas winter pelage is completely white; molting individuals taken in spring and autumn are piebald.

Males average larger than females both externally and cranially. Average and extreme external measurements of six males

of the subspecies *campestris* from Kansas, Nebraska, South Dakota, Iowa, and Missouri are as follows: total length, 208.3 (192–216); length of tail, 37.8 (35–44); length of hind foot, 25.0 (23–28); length of ear, 13.0 (10–15). Comparable measurements of eight females from the same region are: 178.5 (166–186); 30.5 (26–37); 21.6 (20–25); 11.1 (9–14). Weights of two males were 54.5 and 62.7 grams, and the average weight of five females was 41.5 (32.2–50.0) grams.

The two subspecies of *Mustela nivalis* on the Northern Plains can be distinguished on the basis of size and color, *M. n. rixosa* being smaller and darker than *M. n. campestris*. Least weasels frequently can be differentiated from *M. erminea*, when the skulls are of equal size, because the mastoid breadth in *nivalis* equals or exceeds the greatest breadth of braincase, whereas the reverse is ordinarily true in subspecies of *erminea* that are as small as *nivalis*, and the latter has a higher skull. Average and extreme cranial measurements for four males of *M. n. campestris* from Kansas, Iowa, and Nebraska are: greatest length of skull, 34.0 (32.6–35.9); zygomatic breadth, 17.6 (16.5–18.7). Comparable measurements for four females from the same region are 29.6 (28.0–31.2) and 15.2 (14.5–15.6).

Natural History. M. nivalis has been recorded from a wide variety of habitats. In the northern part of its range, the species occurs on the tundra and in coniferous forest and woodland. The least weasel is found most commonly on the Northern Plains in meadows and grasslands and reaches its greatest abundance in marshy areas. It is least common in woodlands, where *M. erminea* and *M. frenata* are more likely to occur. Factors affecting the local distribution of the least weasel are not completely understood. Perhaps the presence of a good supply of its principal food, small rodents, is one of the most important parameters.

Voles, deer mice, and harvest mice make up most of the diet. Moles are sometimes eaten, and shrews may be attacked but usually are eaten only when other prey is

171. Least weasel, Mustela nivalis *(courtesy D. L. Pattie). Note paper match for size comparison.*

unavailable. Insects are also eaten when encountered. Small ground-nesting birds are captured, especially when small mammals are scarce, but least weasels are more specialized as predators of small rodents than are ermines, which readily switch to rabbits, birds, berries, or carrion. It is estimated that least weasels require approximately one-third to half their body weight in food each day, so they may have a significant impact on prey populations.

As is usual with carnivores, size of territory varies with prey density. When mice or voles are abundant, hunting territories may be as small as two acres or less, but when prey species are scarcer the size may range from 25 to 65 acres for males. Females have separate, usually smaller, territories. Individual territories do not overlap and appear to be defended against other least weasels, but they may overlap the hunting territories of long-tailed weasels. The hunting technique of the least weasel differs from that of its larger congeners in that this species is small enough to pursue its rodent prey along their own runways and into their burrows and nest chambers. This weasel may hunt by day or night but seems to prefer the latter; prey availability strongly influences activity patterns. Its technique for killing a small mammal involves seizing the victim by the head or neck and wrapping its limbs about the prey. A series of quick bites to the dorsal base of the skull completes the kill. The weasel may feed at once, beginning in the head region, or may cache its prey near the den.

Least weasels follow regular hunting routes, often with great rapidity, and may search their entire territory in a single hunt, covering a mile or more in linear distance. Usually, however, an individual moves only a short distance from its nest in the course of one foray.

The nest of the least weasel is in a shallow burrow made by a mouse, vole, mole, pocket gopher, or other small mammal that the weasel probably has killed as prey, and most often is found underground in a field or meadow. The entrance is about an inch in diameter; the nest is usually five to six or more inches below ground and is four inches in diameter. It is usually constructed of vegetation gathered by the original occupant, then lined with fur gathered from the weasel's prey. In this nest the weasel sleeps, rears its young, and sometimes brings food items to eat.

Unlike many mustelids, least weasels do not exhibit delayed implantation. The breeding season in North America is ill defined, and breeding may occur throughout the year. Females may produce two or three litters per year, although one or two probably is more usual on the Northern Plains. After a gestation period of 34 to 37 days, one to 10 (usually four or five) naked, blind young are born. They average 48 mm in total length and weigh little more than a gram. By the fourth day they are covered with fine, white fur, and the brown summer pelage begins to develop between two and three weeks after birth. The milk teeth erupt between days 11 and 18, and the permanent dentition develops between 30 and 49 days. The young begin to eat solid food at about three weeks of age. Their eyes open between 26 and 30 days, shortly after which weaning is initiated and the young animals begin to leave the den. At this time the female teaches the young to kill prey, and by the time they are six or seven weeks old they are proficient hunters.

The young reach adult size between 12 and 14 weeks of age and probably disperse about that time. The postlactating female may come into estrus again and conceive a second litter. Least weasels become sexually mature at three to four months of age, females earlier than males.

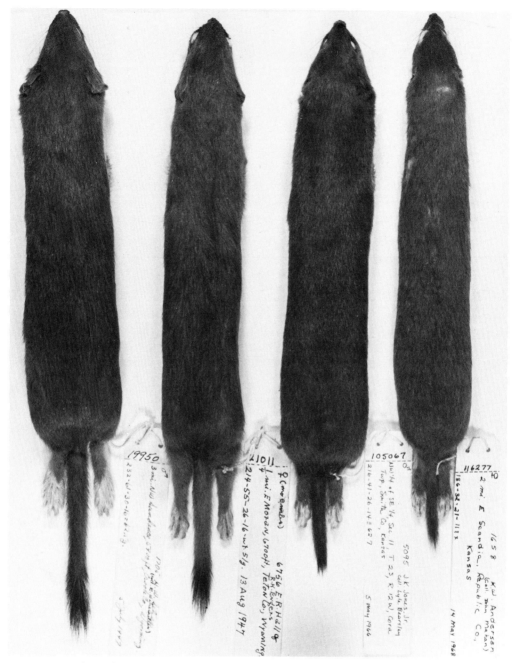

172. Museum skins (dorsal view) of two species of small weasels: left to right, male and female of
Mustela erminea muricus, *and male and female of* Mustela nivalis campestris.

Timing of breeding effort and rate of development of young both appear to be strongly affected by prey availability, season having less influence. It is probable, however, that more litters are produced in summer and autumn, and young of late litters may develop more slowly than those born earlier in the year. Maximum natural longevity is about three years, but captives have lived 10 years.

Many animals are known to prey on *M. nivalis*. Records for the Northern Plains are scarce, but in other states the great horned owl, the barn owl, and broad-winged hawks such as the rough-legged hawk are known predators. Known and inferred mammalian predators include foxes, martens, mink, and house cats; the two larger species of weasels also may prey on the least weasel. Predation thus may be an important control mechanism on populations.

The least weasel is host to a biting louse and to various fleas that usually infest its rodent prey. Several species of roundworms, including the nasal nematode, have been found to parasitize the species. The fur of *M. nivalis* has no commercial value, but the animal is a distinct economic asset because its principal food is small rodents.

Selected References. General (Corbet and Southern 1977; Hall 1951); natural history (King and Moors 1979); systematics (Kurtén and Anderson 1980).

Mustela vison: Mink

Name. The several names associated with this large, semiaquatic weasel are of uncertain origin. The scientific name *vison* is used as a vernacular name in modern French but may be a word originally of Icelandic or Swedish origin. Alternatively, it has been suggested that *vison* is derived from *visor*, an obscure Late Latin word meaning "scout." The English vernacular name possibly is traceable through the medieval English *mynk* to the Swedish *menk*. This species is sometimes referred to as the North American mink to distinguish it from its Eurasian relative, *Mustela lutreola*.

Distribution. The mink has the widest distribution in temperate and boreal North America of any species of *Mustela*, although *M. frenata* is a close runner-up. It has a transcontinental range in the coniferous forest belt and even extends north of the tree line in some places such as Alaska and Ungava. It ranges southward throughout the eastern deciduous forest into subtropical habitats at the southern tip of Florida, and through the Great Plains to

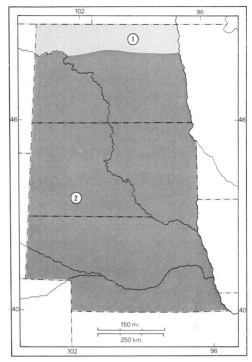

173. Distribution of Mustela vison: (1) M. v. lacustris; (2) M. v. letifera.

the Gulf Coast. It does not occur, however, in the extreme Southern Rocky Mountains, on the Colorado Plateau, in the Great Basin, or in southern California, perhaps because of the aridity and unpredictability of stream flow in those regions.

In Europe and western Siberia there occurs a closely related species of similar habits, the Eurasian mink (*M. lutreola*). In central and eastern Siberia, China, and Japan, a less similar (but probably related) species, *M. sibirica*, occurs. On the Northern Plains, the entire region is occupied by the range of the subspecies *Mustela vison letifera*, except for extreme northern North Dakota, where *Mustela vison lacustris* is found.

Description. The mink is a large, dark brown weasel and shares with its relatives a long, slender body and relatively short legs. Its tail is moderately long and more bushy than those of other North American *Mustela*, resembling that of the marten. The glossy, water-repellent fur is dense and lustrous, the underfur being thick, soft, and gray brown, whereas the long, shiny guard hairs range from pale chocolate to nearly black. The underparts are only slightly if at all paler than the dorsum, but there are usually white blazes on the chin and throat and sometimes streaks or blotches on the chest or belly. The tail is dark brown to nearly black and has an indistinct blackish tip. The ears are short and rounded, and the toes of the feet are partially webbed, both features probably adaptations to semiaquatic life. Spring molt usually occurs in March or April, and the shorter, thinner, and slightly paler summer coat is then worn until September, when molt into the winter pelage occurs. Females possess three pair of mammae and are about 10 percent smaller than males when mature.

Average and extreme external measurements of seven males from Nebraska are: total length, 630.7 (594–686); length of tail, 200.3 (185–217); length of hind foot (five individuals only), 71.0 (68–76); length of ear (three individuals only), 26.3 (25–27). Weights of three of these males averaged 2.6 (2.25–3.0) pounds. Comparable measurements for three females, two from Nebraska and one from South Dakota, are: 536.7 (491–560); 180.7 (154–203); 61.0 (57–64); 22 (one individual only). Average and extreme cranial measurements of two males from Nebraska and one from South Dakota, followed by comparable measurements of three females from the same areas, are: greatest length of skull, 70.7 (70.0–71.7), 63.3 (63.2–63.4); zygomatic breadth, 39.3 (38.2–40.5), 34.8 (33.8–36.0). *M. v. lacustris* from northern North Dakota averages larger and darker than *M. v. letifera*.

Natural History. The mink, although widely distributed geographically, is narrowly restricted ecologically. The species, as noted above, is semiaquatic and normally does much of its foraging in water. It usually is found, therefore, along the banks of lakes and rivers or in marshes. It is particularly partial to watercourses or ponds where stumps, logs, or muskrat houses break the surface, or where the banks have driftwood or brushy cover. Occasionally, however, mink may travel overland for considerable distances or forage far from water.

Mink hunt along the water's edge for frogs, snakes, and crayfish and other invertebrates; such small prey species are easily handled with the teeth and forefeet. Mink also forage in and under the water and swim skillfully enough to capture fish. They also may prey heavily on muskrats and occasionally take aquatic birds, including ducks. When attacking large prey, their killing technique probably is much like that of the long-tailed weasel (see account of that species). Summer foods are derived primarily from the water and include large numbers of crayfish and muskrats; about half the diet may be composed of one or the other of these favorites in that season. In some circumstances, however, waterfowl, including adult and young grebes, coots, and ducks, may constitute the main summer diet, flightless young and molting adults being especially vulnerable. At still other times meadow voles may be a summer staple.

Winter foods reflect less dependence upon crayfish and perhaps also more time spent foraging away from water. Muskrats often are still important, and fish may increase in relative frequency, but birds decline. Some cottontails are taken and, on occasion, squirrels. In a Missouri study based on stream rather than marsh populations, frogs were the principal winter food, and muskrats formed only a small fraction of the diet. This variability in food habits points to the opportunistic nature of the mink's predation. Studies in Iowa have indicated that healthy adult muskrats are able to escape, or defend themselves, from mink. It is mostly when muskrat populations are overcrowded, when disease epizootics break out, or when drought reduces habitat quality that muskrats become extremely vulnerable to mink predation.

The mink, like other species of *Mustela*, may kill more prey than it can consume at one time. Extra food then may be cached in or near the den, although caching may not be so regular as in the smaller terrestrial weasels. One record cache, found in a den

174. Mink, Mustela vison *(courtesy W. D. Zehr).*

in Illinois, consisted of 13 freshly killed muskrats (age and sex were not recorded), two mallards, and a coot. It was made by an adult male in January, and caching may be more common in winter.

The den of this male was in a hollow ash log on the shore of a marsh. Dens usually are close to water and most frequently are in old muskrat burrows or houses, although one was found in an old Franklin's ground squirrel burrow 200 yards from the edge of a marsh. Mink dens typically have one and sometimes more entrances close to the shoreline, with several other entrances on higher ground. In one Minnesota study, the average number of entrances was 2.9 (two to five) in eight excavated dens. Males and females maintain separate dens. Adult females restrict their activities in winter to a relatively few den sites, whereas males appear to wander more widely and use more dens, although few with regularity.

In late winter (mid-January to early March), a male and female share a den briefly during the breeding season. The female, after she has given birth to her litter, further restricts her activities to the vicinity of the natal den. The nest itself is 10 to 12 inches in diameter and lined with dry grasses, feathers, and fur. After the young are weaned, the family makes use of a series of dens within the home range, and after dispersal in mid- to late summer juveniles continue to wander from den to den.

Home ranges of mink tend to be linear because they follow shorelines. Those of adult females vary from about 20 to 50 acres, whereas those of adult males are much larger—in fact, they are too large to estimate accurately but may be six or seven times as large as those of females. Home ranges of males are reported not to overlap, and this may be true also for ranges of females; at least one fight between two mink of the same size has been observed that may have been territorial defense. Adult males appear tolerant of juvenile males in their territories, however, and their ranges may overlap those of several adjacent females.

Late summer and autumn dispersal of juveniles may take them far from the area in which they were born. In a study done on the Madison River in Montana, six young males moved an average of 11.2 miles, ranging from two to 28. Two young females dispersed 14.5 and 18.5 miles. Occasionally an adult also will disperse; one adult female from the same study area was found 24 miles away, but no adult males left the area. Most dispersal does not involve such long movements, however; it is usually up to three miles for juvenile males and less than two miles for juvenile females.

Mink become sexually mature at about 10 months of age, so all individuals in the population may take part in the annual breeding season, although yearling males may not be successful in competing with older males for access to females and thus may not breed until they are 22 months old. Breeding begins as early as mid-January in the southern part of the tristate region and as late as early March in the north. Males are attracted to estrous females, and more than one male may mate with a single female. The length of gestation is variable, because the period of delayed implantation is more variable in this species than in other mustelids that exhibit the phenomenon. The delay is also shorter, so that the total gestation period ranges from 40 to 75 days (average 51), of which about a month is devoted to growth of the implanted embryos. This developmental period is about a week longer than in the long-tailed weasel. The annual litter is born in April or May; the number of young varies from three to six, but most often there are four.

The kits are naked, with eyes closed. Their deciduous teeth erupt between 16 and 49 days of age, and their eyes open at 25 days, at which time they are covered with short reddish gray pelage. At this time the female begins to bring them solid food. They are weaned between the ages of five and six weeks, and the permanent dentition is probably in place toward the beginning of the third month. By eight weeks of age the young accompany the female while she hunts. There is some speculation that the male may help the female raise the off-spring, but this has not been verified.

Population densities ranging from two to 22 mink per square mile have been reported from various parts of the range of the species. These vary, naturally, with habitat quality; on the Northern Plains mink are more common in the better-watered eastern half, especially in areas such as the Nebraska Sand Hills or the glacial lake region in the Dakotas, where marshes and muskrats are common. Populations exhibit fairly high mortality rates, dispersing juveniles in the autumn and first winter of life being particularly susceptible. Most adults do not live much more than three years, but potential life span in captivity is up to 16 years.

Coyotes, red foxes, and bobcats may prey on mink, and great horned owls capture a few. Probably as many mink are killed, however, by other mink as by all other natural predators combined, and intra-specific strife is thought to be one of the main controls on mink populations. Others are starvation and disease—tularemia and rabies may be serious at times, and many species of trematodes, cestodes, nematodes, and protozoans have been described, the effect of which undoubtedly is debilitating. External parasites include fleas, ticks, lice, and botfly larvae.

A great deal is known about the parasites and diseases, as well as the nutrition, physiology, and genetics of captive mink, for this species has been bred for its fur since the middle of the previous century, and mink ranching is an important industry in the tristate region. Many color mutants have been developed from the original wild-type pelage, some of which are in great favor in the world of fashion. Pelts of wild-trapped individuals also remain important in the fur trade. The estimated harvest in Nebraska in the 1979–80 trapping season, for example, exceeded 7,000 mink at an average price of about twenty-two dollars per pelt.

Selected References. General (Jackson 1961); natural history (Enders 1952; Errington 1943; Mitchell 1961; Sargeant, Swanson, and Doty 1973).

Genus *Gulo*

Gulo is a Latin word meaning "glutton" or "gluttonous," in reference to the legendary food habits of the wolverine. The genus is boreal in distribution and now is considered monotypic, although the American population formerly was treated as specifically distinct from the Eurasian population. The dental formula is 3/3, 1/1, 4/4, 1/2, total 38.

Gulo gulo: Wolverine

Name. The derivation and meaning of the vernacular name wolverine is unknown.

Distribution. The wolverine at one time inhabited most of Canada and ranged southward into the western mountain states and portions of the Great Plains and Central Lowlands. Assorted records from the early and mid-1800s indicate it was present in central and northern Minnesota and northern North Dakota, southward to western Nebraska. With the advent of civilization, the wolverine retreated northward, leaving only scattered populations in mountainous regions of the West. With the decline of trapping, northern Canadian populations have stabilized. Reports from western Montana show favorable increases. Recent records from South Dakota and Iowa, however, well may represent imported animals. Thus, although the wolverine was at one time an occasional member of the fauna, it evidently no longer occurs regularly on the Northern Great Plains. The subspecies formerly represented in the tristate region is *Gulo gulo luscus*. The former distribution is not mapped.

Description. The heaviest of the North American Mustelidae, the wolverine somewhat resembles an elongate bear cub. The body is heavily built, and the animal has a short, bushy tail, small and broadly rounded ears, and short, muscular forelimbs. The back is arched high, and the head and tail are carried low. The thick, kinky underfur is topped by stiff, straight guard hairs that reach a length of four inches on sides and hips and six inches on the tail, giving a shaggy appearance. Color varies from black to dusky brown with characteristic chestnut or brownish white bands on either side, extending from shoulders to rump. Pale gray patches are present on the top and sides of the head, and irregular patches of pale brown are found on the throat and chest. Coloration pales with age as gray hairs become present.

A male and three females from California had the following external measurements: total length, 1,002, 932, 914, 895; length of tail, 260, 250, 210, 210; length of hind foot, 180, 161, 155, 165; length of ear, 55, 50, 57,

The feet are heavily furred above and below except for naked pads on the ventral side of the five digits. Nonretractile claws, present on all digits, are one and a half to two inches long. Anal scent glands produce a yellowish brown, rather thick fluid with a potent musky odor. Abdominal scent glands also are present.

The skull is massive, having wide zygomata and a shortened snout. An enormous temporal fossa accommodates the powerful temporal muscle. The conspicuous supraoccipital crest is formed by the posterior expansion of the sagittal crest. Greatest length of skull and zygomatic breadth of two males and six females (average and range) from California are, respectively: 164.8, 162.1, 145.6 (141.2–150.5); 103.6, 106.6, 94.1 (91.7–96.4).

Natural History. Generally a boreal inhabitant of tundra and forested regions, the wolverine normally stays close to the area where it was born, occasionally traveling into open country in search of food. Such wanderings probably are the basis for many of the past records on the Northern Plains.

G. gulo is a solitary creature with no enemies other than man. Anything that affords shelter, such as windfalls, rock slides, small caves, and abandoned buildings, is sufficient for a den. Often a slight depression is dug and lined with leaves or grass. Normally nocturnal, the wolverine may be active at any time of day. Individuals remain active throughout the win-

175. Wolverine, Gulo gulo *(courtesy San Diego Zoological Society).*

ter and continue to hunt for food during cold periods and in deep snow.

Though not a fossorial animal, the wolverine will dig out burrowing rodents. Small and medium-sized rodents and carrion compose most of the diet, with birds and eggs eaten when available. The wolverine has been known to attack moose, caribou, mountain goats, and deer when such animals are disabled or exhausted from walking in deep snow.

This animal normally moves at a slow walk, but it frequently will break into a bounding lope with back arched high and head and tail hung low. Though primarily terrestrial, the wolverine can swim streams and ponds and is a good climber.

Breeding takes place between April and October but is most likely to occur during midsummer. The blastocyst—a mass of about 100 to 1,500 cells resulting from a dozen or fewer generations of cell division—remains unimplanted until January; two to four young are born in late March or early April. Subsequent breeding is delayed until lactation ends. The blind, scantily haired newborns are tawny whitish at birth and are about 150 mm long, with tails about 30 mm in length. Birth weights average about 85 grams. The young are cared for solely by the female. She is extremely protective of them and will not hesitate to attack any animal in their defense. The young mature rapidly, achieving half of their growth by autumn; they leave the female when five or six months old. Sexual maturity is attained by the second summer,

when mating occurs for the first time in both sexes.

Because few people ever have seen the wolverine in its natural habitat, the animals have been shrouded in mystery and legend. Numerous stories recount their great strength, cunning, savagery, and gluttony. It is true that they raid traplines, ransack camps, and steal supplies, but the frequency of such events is low and they probably are caused by a small portion of the population. The strength of the wolverine is great, but in comparison to size is probably no greater than that of any other mustelid. Trap-wary habits are most likely a learned response, because wolverines have been captured in traps that they made no attempt to avoid, and on occasion individuals have died in traps that held them only by a toe.

When cornered, the wolverine, like many another animal, may become vicious and—making low growling or coughing sounds—will attack. Wolverines have been tamed and are reported to be faithful and gentle pets.

Selected References. General (Halfpenny et al. 1979); natural history (Rausch and Pearson 1972; Wright and Rausch 1955); systematics (Kurtén and Rausch 1959).

Genus *Taxidea*

The genus *Taxidea* is monotypic. It is one of several genera of badgerlike burrowing mustelids that include the Old World badger, *Meles,* ferret-badgers, hog-badgers, and stink-badgers. *Taxidea* is the only New World badger and is known from as early as Pliocene deposits. The dental formula is 3/3, 1/1, 3/3, 1/2, total 34, as in weasels and skunks.

Taxidea taxus: Badger

Name. The generic and specific names are of Latin derivation and mean "badge" or "badgelike." The vernacular name also refers to the white badgelike markings on the face.

Distribution. This species occurs throughout the Northern Great Plains. It is found from eastern Washington and in central Alberta and Saskatchewan, east to Michigan and Ohio, and south to Missouri, Texas, and central Mexico. Although several subspecies have been named from localities on the Northern Great Plains, probably a single race, *Taxidea taxus taxus,* should be recognized in the tristate region.

Description. The badger, a stocky member of the weasel family, is characterized by a white patch on each cheek and a white stripe extending dorsally from nose to shoulders or beyond. The neck and legs are short and muscular. The small black ears are rounded and upright. The bushy tail is only slightly longer than the hind feet. The broad feet have five clawed digits, and the forefeet are markedly larger than the hind feet. Hairs are clay-colored basally and have dark brown or black median bands and gray tips. The pelage is longer on the sides than on the back, accentuating the animal's broad appearance. Coloration becomes paler with age and varies with season and geographic location. Badgers from the Northern Great Plains are paler than those to the east; they have less black on the back and are paler below. Average and extreme external measurements of nine males and three females from Nebraska are, respectively: total length, 732.7 (692–787), 716.3 (662–762); length of tail, 130.9 (109–150), 125.5 (104–147); length of hind foot, 109.5 (98–116), 113.7 (98–127); length of ear (both sexes), 53.3 (47–64). Adults weigh 12 to 25 pounds.

The skull has a broad occiput, almost as wide as the zygomatic breadth. The tympanic bullae are highly inflated but may shrink as the animal reaches old age. The palate extends posteriorly beyond the last molar. Average and extreme cranial measurements of nine males and three females from Nebraska are: condylobasal length, 125.9 (123.6–129.5), 120.9 (119.6–121.6); zygomatic breadth, 81.1 (75.9–87.7), 74.4 (71.5–77.8).

Natural History. Badgers inhabit grasslands and forest edge communities, where their presence is indicated by characteristic large, elliptical burrows. Primarily fossorial, the badger spends much of its life underground. At the entrance, most burrows are 12 to 18 inches wide and six to 12 inches high; there are several enlarged chambers six to eight feet beneath the surface. Often abandoned, burrows may be utilized by skunks, rabbits, or snakes. In areas of dense grass, ground squirrels frequently dig their burrows in vacant badger dens below the depth of the sod.

Aboveground the badger walks in a slow gait, breaking into a lope only for short distances. The feet are plantigrade, both front and hind feet leaving a characteristic heel imprint, and the front feet leaving claw marks. The gait appears clumsy, with a pronounced swagger accentuated by the long hairs of the sides. The badger is a good swimmer and has been observed crossing lakes and streams. Although primarily cre-

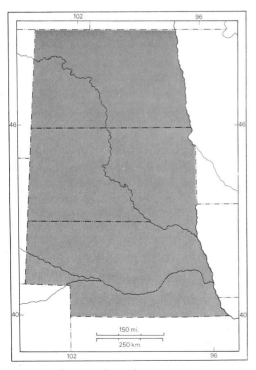

176. Distribution of Taxidea taxus.

puscular, these animals occasionally are active by day. They have been known to be active at temperatures well below freezing, but they normally sleep during cold winter periods.

Inasmuch as badgers are almost totally solitary, vocalization is not well developed, but when cornered or attacked they will hiss, growl, and gnash their teeth. Owing to their extremely muscular forelimbs and low center of gravity, badgers are efficient fighters, the long pelage and loose hide providing effective protection. The animals have a disconcerting habit of standing their ground for a moment when confronted by humans.

These carnivores have no habitual enemies other than man, who kills them both purposefully (sometimes for fur, more often because of alleged depredation on poultry and because the foraging and denning burrows can be a menace to unwary livestock), and inadvertently with vehicles. Coyotes and golden eagles have been reported to prey occasionally on badgers. Parasites include ticks, lice, fleas, and a variety of cestodes, nematodes, and trematodes. Badgers are susceptible to tularemia and rabies.

Mating probably occurs in August or September, but implantation is delayed. Specimens taken in Kansas in December and January had unimplanted blastocysts, but implantation had occurred in specimens taken in February and March. Two abdominal and two inguinal mammae are present. A litter consisting of one to five young is born in April, May, or June. The young are born nearly hairless and with closed eyes, which open at the age of one month. The litter, cared for solely by the female, is weaned at about two months. The three-quarter-grown young permanently leave the natal burrow in autumn. An occasional female mates in her first autumn, but apparently males all achieve reproductive maturity as yearlings.

Badgers are solitary animals, associating only briefly in summer and autumn to breed. Densities vary widely, but the range of movements for individuals usually is between 0.4 and 2.2 square miles. Move-

177. Badger, Taxidea taxus *(courtesy D. Randall).*

ment patterns vary widely with season, distances traveled and home ranges occupied being 10 to 100 times as great in warm weather as when it is cold. During severe weather they may not come aboveground at all. The animals are largely nocturnal. In warm weather the same den rarely is used on two consecutive days, but in winter a badger may stay in the same burrow for several days. A female may move nestling young among a series of several burrows.

During the period 1920 to 1935, the abundance of badgers was greatly reduced throughout the Great Plains and southwestern United States. In those times badger pelts brought high prices on the fur market, and overtrapping seriously reduced the population. The value of badger fur declined during the 1940s (although pelts averaged about twenty-four dollars each in Nebraska in the 1979–80 season), and since then populations have varied geographically according to food supply and suitable habitat.

Food of the badger consists primarily of small mammals, but insects are eaten when available. Studies of diets have been rare because badgers bury their dung; hence, scats seldom are found for analysis. Badgers usually capture prey alive, but they sometimes eat carrion. A study of badgers in Iowa showed food sources, listed in order of importance, to be ground squirrels, mice, cottontail rabbits, insects, birds, pocket gophers, and snakes. Composition of the diet varies seasonally, but ground squirrels and mice are of major importance the year around. Badgers are reported to eat rattlesnakes, but the extent of this is not known. Reexcavation of abandoned burrows is a common practice related to predation on ground squirrels, a primary inhabitant of abandoned badger burrows and a major component of the badger's diet. In east-central Minnesota, plains pocket gophers (Geomys bursarius) were the principal food of badgers. Foraging animals dig preliminary "prospect holes" in the gopher's tunnel system, then, having located the resident, proceed to dig a badger-sized hole to capture the prey. Nearly three-fourths of such predatory bouts are successful for badgers, and they involve moving an average of 182 liters of soil. The net result is a patch of prairie that looks something like a minefield, yet it takes only about 1.7 gophers per day to fuel such destruction.

Badgers molt from juvenile to adult pelage their first summer and annually (in summer or autumn) thereafter. Molt begins on the head and progresses in a middorsal band to the base of the tail. Then the molt progresses to the sides and venter.

Claims that badgers kill poultry and dig holes that may injure livestock and damage roadbeds are not sufficient basis to justify the past and present destruction of the animal. Such indictments seem insignificant compared with the beneficial activities of badgers in controlling rodent populations. One may hope that the importance of this animal's role in nature will overcome the stigma placed on it by an uninformed human population.

Selected References. General (Long 1973 and included citations); ecology (Lampe 1976; Lindzey 1978).

Genus *Spilogale*

Spotted skunks are distributed from Costa Rica northward to British Columbia, Minnesota, and Pennsylvania. There are three species in the genus. *Spilogale* has a fossil record extending back only to the early Pleistocene. The dental formula is 3/3, 1/1, 3/3, 1/2, total 34. These are the smallest and most weasel-like of skunks. The name *Spilogale* is compounded of two Greek roots, *spilo* ("spotted") and *gale* ("weasel").

One species, *Spilogale putorius*, is known from the Northern Great Plains, although a second, the western spotted skunk, *Spilogale gracilis*, may occur in the extreme western part of South Dakota. *S. gracilis* is considerably smaller than *S. putorius*; also, western spotted skunks have white-tipped tails, whereas those of eastern spotted skunks are tipped with black. The most trenchant difference between the two, however, is in the pattern of reproduction. Females of the western species first breed in late September or October, then exhibit delayed implantation, with a total gestation period of 210 to 230 days. Females of the eastern spotted skunk, on the other hand, breed in March, have a gestation period of 50 to 65 days, and do not show much delay in implantation. Inasmuch as the reproductive cycle of the males of each population coincides with that of the females, they are effectively reproductively isolated.

Spilogale putorius: Eastern Spotted Skunk

Name. The specific name *putorius* is derived from a Latin word meaning "stench," also the root of the English adjective "putrid." A common misnomer is "civet cat," because spotted skunks are neither civets (Viverridae) nor cats (Felidae).

Distribution. S. putorius ranges from northeastern Mexico through the central and southeastern United States, northward to Minnesota and Pennsylvania. It may have come to occur on the Northern Great Plains only in historic times. The first certain record for Nebraska, for example, was from Nemaha County in 1893; by the

1940s it was known from every county in that state. By the 1950s the species was known from the eastern half of South Dakota, and by the 1960s it was reported from southeastern North Dakota. Thus the spotted skunk seems to be well on its way to occupying the entire Northern Plains. This recorded extension of range almost certainly is real. Whether it is due to climatic amelioration or to changes in habitat owing to land-use practices is not known. Although less conspicuous than the larger striped skunk (*Mephitis mephitis*), this species readily dens about farmsteads. Where present, it cannot be ignored for long. A single subspecies, *Spilogale putorius interrupta*, occurs in the three-state region.

Information on spotted skunks from the Black Hills and adjacent Badlands is sorely needed. A report from northwestern South Dakota in 1914 held that the spotted skunk was more common than the striped skunk, but there have been no other reports of their being common in this area.

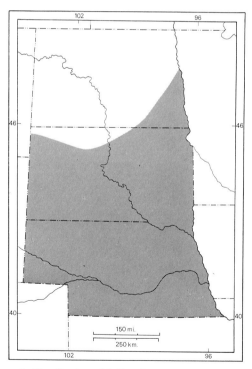

178. Distribution of Spilogale putorius.

Description. Ernest Thompson Seton—a turn-of-the-century naturalist, writer, and illustrator and a founder of the Boy Scouts of America—described it best as a "little animated checkerboard." Spotted skunks are jet black, with four or six white to yellowish white dorsal and lateral stripes. Posteriorly, the stripes are increasingly broken into spots.

These are diminutive, weasel-like skunks. As in many other mustelids, males are significantly larger than females. Average and extreme external measurements of five males, followed by measurements of two females, all from eastern Nebraska, are: total length, 516.7 (459–539), 508, 470; length of tail, 205.2 (190–228), 203, 175; length of hind foot, 48.0 (44–51), 45, 48. Length of the ear generally is between 25 and 30. Mean and extreme cranial measurements of nine males from eastern Nebraska, followed by measurements of two females from north-central Nebraska, include: condylobasal length, 58.7 (57.2–61.4), 54.2, 54.0; zygomatic breadth, 35.9 (35.1–37.0), 33.0, 31.4. Males weigh about 600 grams, females about 425 grams.

Natural History. Given their adaptable nature—with broad habitat tolerances and a catholic diet—it is hard to guess why spotted skunks are not more generally distributed. When abundant at one locality, they tend to be rare or absent at another nearby that to humans appears identical. Spotted skunks occur in brushy country, riparian woodland, and along rock rims, in thickets, fencerows, and shelterbelts, and in outbuildings or beneath dwellings. Signs consist of weasel-like tracks and cylindrical scats left conspicuously along trails, in barns, in corncribs, and elsewhere in the hunting range of an individual.

The diet is omnivorous but mostly beneficial to the economic interests of man. In winter in a study made in Iowa, these animals depended principally on mammalian flesh, primarily carrion of rabbits; some corn also was eaten. In spring the diet shifted to rodents: meadow voles, deer mice, harvest mice, and house mice, as well as insects, became increasingly important. The summer diet was composed mostly of insects, especially ground beetles, grasshoppers, and crickets. Fruit and rodents increased in importance in autumn, although arthropods continued to predominate in the menu. Spotted skunks are far more agile than striped skunks and can climb well; hence they are able to take nesting and roosting birds, preying on some passerines as well as on poultry. Enlightened farmers are inclined to pen their chickens tightly and give spotted skunks the run of the barn and corncribs, because they are excellent mousers.

S. putorius dens almost anywhere there is protection from weather, daylight, and dogs. Sheds, haylofts, outhouses, pump houses, corncribs, hogpens, drainage tiles, culverts, woodpiles, scrap heaps, abandoned vehicles—all provide cover on the prairie where little was present in prehistoric times. Farther from permanent settlements, these animals utilize burrows abandoned by ground squirrels, pocket gophers, or woodchucks, rock piles and riparian cliffs, or even hollow trees.

The young are brought forth on a nest of grass or leaves, or even on a bare attic floor. Four to nine (average about five) young compose the litter, which is born in late May or June. Females are monestrous and are spontaneous ovulators. Implantation occurs about two weeks after conception. The newborn young are highly altricial and mouse-sized (about 100 mm long and 9.5

179. *Eastern spotted skunk.* Spilogale putorius *(courtesy New York Zoological Society).*

grams in weight), but they obviously are skunks because the color pattern already is distinct. Eyes and ears are closed, and locomotion is at best a feeble squirm. The mother carries the young about by the scruff of the neck. At three weeks they are well furred. Soon thereafter the tail is lifted in the typical warning signal, but the ability to emit musk does not develop until about six weeks of age. Eyes open between weeks four and six. At week eight the young are weaned, and they are fully grown by 12 to 13 weeks of age.

In good habitat, densities of spotted skunks have been estimated at about 12 per square mile. The home range is about a fourth of a square mile in area and contains two or three dens; that of promiscuous males increases to two to four square miles in spring. These animals generally are solitary except briefly during the breeding season and during the coldest days of winter, when a communal den may hold a half dozen or more individuals of both sexes. The sex ratio in adult populations invariably favors males, but there seems to be no such bias in litters. Whether this results from differential mortality or from differential "trappability" is not known.

Like their larger striped relatives, spotted skunks are powerfully equipped to defend themselves. When approached away from easily accessible cover, the animals raise the tail, with hairs erect, arch the back, and stamp the front feet. Often, they raise the hind limbs off the ground and walk on their "hands." Only the dullest, most naive, or hungriest among potential predators fails to retreat at this point. The next line of defense is a double spray of musk from the anal glands, which seems invariably to find its way into the eyes of the adversary. This is a profoundly educational experience for most potential predators. No other mammal as small as the spotted skunk has so little to fear in its native habitat.

Spotted skunks are parasitized by roundworms, lice, fleas, mites, and ticks. They are susceptible to rabies, although they probably do not contract it to a greater degree than other wild carnivores. Certainly they do not deserve the name "hydrophobia skunk" (or, even worse, "hydrophobic cat"). More than 90 percent of the mortality of spotted skunks in a study in southeastern Iowa was due to humans: poisoning, shooting, domestic dogs, and vehicles. The pelts have been worth about three dollars each on the market in recent years, but most trappers take only a few dozen "civets" annually. Great horned owls are the only predators that seem to take these animals regularly. Changing farming practices—larger farms, removal of fences, fall plowing, continuous corn planting with no rotation to hay, and use of herbicides and pesticides—have been implicated in the decline in spotted skunks, owing to depletion of habitat for the skunks and their rodent and insect prey.

Selected References. General (Mead 1968*a*, 1968*b*); natural history (Crabb 1941, 1944, 1948; Polder 1968); systematics (Van Gelder 1959).

Genus *Mephitis*

The genus *Mephitis* includes two species, with a combined range that extends from Nicaragua to central Canada and from coast to coast in the United States. Only one of the species occurs extensively north of Mexico. The dental formula is 3/3, 1/1, 3/3, 1/2, total 34. Fossil skunks are known no earlier than Pleistocene times.

Mephitis mephitis: Striped Skunk

Name. Mephitis is a Latin word meaning "a foul odor."

Distribution. Striped skunks occur from Durango, Mexico, northward to Northwest Territories, Canada. They occur throughout the United States, except for the driest deserts of the Southwest, and are abundant and conspicuous members of the fauna throughout the Northern Great Plains. A single subspecies, *M. m. hudsonica,* is recognized in the tristate region.

Description. Striped skunks need no description, because they are familiar to near-

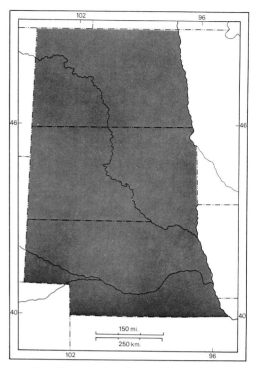

180. Distribution of Mephitis mephitis.

ly everyone, whether known in the field or only from cartoons. A jet black background and conspicuous white stripes that merge at neck and tail identify the striped skunk not only to humans, but to any other potential enemy capable of profiting from experience. The odor is somewhat less piercing than that of the spotted skunk, and those familiar with both readily can distinguish the two species by smell.

Males average considerably larger than females. Mean and extreme external measurements of seven males and six females, all from northwestern Nebraska, are: total length, 739.4 (710–795), 630.2 (560–656); length of tail, 269.7 (246–295), 259.2 (225–277); length of hind foot, 81.8 (78–85), 71.3 (68–74); length of ear, 33.0 (30–35), 28.5 (26–30). Cranial measurements of 17 males and 14 females from Nebraska include: condylobasal length, 79.6 (77.7–81.3), 70.6 (66.5–73.7); zygomatic breadth, 50.9 (48.6–57.1), 45.7 (43.4–47.7). Four males weighed 8.7 (7.6–9.2) pounds, and three females 4.1, 4.8, and 4.8.

Natural History. Striped skunks may well be the most abundant carnivore on the Northern Great Plains. If the species can be said to have a habitat preference, it is probably for woodland edge situations. These skunks will live almost anywhere they can gain adequate shelter, however, except that they avoid heavy forest and lowlands with a high water table (where dry dens are impossible). Rocky cliffs, thickets, downed timber, fencerows, shelterbelts, spaces beneath abandoned or occupied buildings, ravines, and dumps all provide shelter and places to forage. Dens usually are within a few hundred yards of water. They may be in natural cavities or in burrows made by woodchucks, badgers, red foxes, muskrats, coyotes, or other mammals.

Striped skunks are largely omnivorous, with an accent on animal food. Proportional composition of the diet varies with seasonal and local availability rather than with preference. Insects—especially beetles and grasshoppers—provide the bulk of the diet in the warm months, but mice, frogs, birds, shrews, bird and turtle eggs, lizards, berries, fruit, worms, spiders, moles, and garbage are eaten. Carrion may be important in the diet, especially in late autumn. Several descriptions of the ingenuity of skunks in procuring certain kinds of food can be found in the literature. For example, they have been known to scratch on a beehive and, when the bees come out, to kill and eat them. There is more than one report of skunks cracking eggs by "centering" them with the forepaws between the hind legs to smash them against a rock or wall. The animals sometimes do damage in poultry houses, and their depredation on the eggs and young of ground-nesting birds has earned the wrath of quail and pheasant hunters. Mostly, though, thorough study shows their food habits to be of economic benefit to man, although their malodorous presence can be a genuine nuisance about the farmstead.

Striped skunks are mostly nocturnal; they are aggressive foragers, running along a trail to a likely looking spot and then clawing furiously to root out grubs or excavate mice from their runways. Usually they are silent, although they sometimes growl or make cooing sounds. The young are much more vocal than adults and hiss when disturbed. These skunks are poor climbers and thus forage almost entirely on the ground.

Enemies of striped skunks include great horned owls, coyotes, foxes, badgers, and man. Fur-trappers harvest several thousand skunks each year on the Northern Plains; pelts are worth about two dollars each at wholesale. Most predators learn from one painful encounter not to bother these animals again. The skunk has no other defense against enemies. *M. mephitis* can run only slowly (six to eight miles per hour) and its claws and dentition are poorly adapted to fighting. The typical defense of a skunk is to turn and run. If pursued, it stops and raises its tail in warning. It may stamp the ground with its front feet. Then it expels two thin streams of musk from the anal scent glands. These may meet to form an atomized spray, or they may hit the target as a stream. The skunk usually moves its hindquarters slightly while "musking," thus achieving an arc of spray that cannot fail to be effective. A skunk usually will not spray if it cannot raise its tail out of the way. Individuals almost never spray when confined in a live trap. Thus, the closest approach to a safe way to handle a skunk is to hold the tail tightly over the anus. Even this is not guaranteed, however. These animals are more than capable of defending themselves. Still, some dogs seem never to learn. After a bath in tomato juice and gasoline and a few days in solitary confinement behind the barn, they will take on the next skunk that comes along. Highway traffic and hay mowers are even less educable. Skunks simply are poorly adapted to "predators" with no noses and no brains.

An omnivorous animal that feeds on carrion and dens in burrows is a prime candidate for a heavy parasite load, and striped skunks comply with the rule. They carry lice, fleas, mites, and ticks and harbor roundworms, flatworms, tapeworms, and botfly larvae. Among other diseases, skunks are susceptible to distemper, pneumonia,

181. Striped skunk, Mephitis mephitis *(courtesy J. L. Tveten).*

her mouth and keeps them scrupulously clean by frequent grooming. A single litter of four to 11 (average about seven) young is produced each year, in April or May. The young are blind and helpless, but they are obviously skunks from the beginning. Fine hair is visible, and the adult pattern of white on black is established in a color scheme of pink on blue. The eyes open at about four weeks, and the young are weaned at about six to eight weeks of age. The teeth erupt at about 40 days; seemingly only the permanent dentition ever emerges through the gums. At weaning, the young begin to follow the female on foraging trips. There is no postjuvenile molt. Striped skunks molt first as adults at about 11 months of age. The underfur is shed first, coming off in wads. In July the guard hairs are replaced, beginning on the head and shoulders and progressing posteriorly. The underfur also is replaced at this time. Molting is completed by September.

In the southern part of their range striped skunks are active year round, but in the north dormancy is the rule in severe winter weather. The animals do not hibernate; the body temperature stays high (about 37° C, or 99° F) through the winter. Rather, they sleep, often in a communal den, for a month or more at a time, living on a savings account of fat (which in a productive year is 20 percent of the body weight). Dormancy seems to be the rule when mean temperatures are 20° F (−4° C) or below and daily maximum temperatures are below freezing. Over most of the Northern Great Plains, such conditions can prevail from late November until mid-March.

Densities of striped skunks vary from about two to 50 animals per square mile. Adults are solitary except during the breeding season and when they enter communal winter dens. Movements range to about half a mile for females and one to one and a half miles for males. Because males wander more widely, they are captured more often than females. This difference in behavior may be why reported sex ratios tend to be biased in favor of males.

tularemia, leptospirosis, histoplasmosis, and rabies. The last is particularly important to human health. A rabid skunk is a dangerous animal. Its behavior is profoundly different from that of a healthy skunk; it is aggressive, more diurnal, and unlikely to scent. In any outbreak of rabies among wild carnivores, skunks are likely to be most affected, at least in part because they are so abundant.

Mating occurs from mid-February to mid-March. The promiscuous male approaches the estrous female and sniffs and licks her genitals. He then grasps her by the neck with his teeth and mounts her. There is no vocalization during mating. Once bred, a female chases other males away.

Gestation takes about 63 days. During the last week of pregnancy, the female is restless and defensive. The young are born on the bare floor of the nursery burrow. They are about 130 mm long and weigh 35 grams. The mother carries them about in

Selected References. Natural history (Storm 1972; Verts 1967).

Genus *Lutra*

The genus *Lutra* is the most diverse and widespread of several closely related genera of otters. Some eight species are recognized, ranging through North and South America, Eurasia, and much of Africa. A single species occurs in temperate North America. The oldest known otters are of Pliocene age. The dental formula is 3/3, 1/1, 4/3, 1/2, total 36.

Lutra canadensis: River Otter

Name. The generic name *Lutra* is a Latin word meaning "otter." The specific name *canadensis* probably refers to the source of the first records of the species.

Distribution. Riparian in habitat, the otter once was found throughout all of Alaska, Canada (except for the extreme northern portion), and the United States (with the exception of the arid Southwest). During the late 1800s, the otter was common along many of the rivers of the Northern Great Plains. With an increase of trapping and a decrease in suitable habitat, the otter gradually receded northward. Since coming under protection in many areas, otters have become more abundant and have begun to reoccupy parts of their former distribution. The presence of otters in Iowa and Minnesota, along with recent reports from southern Nebraska, suggests that the species may be reestablishing itself at least in parts of its former range on the Northern Plains.

Two subspecies, *Lutra canadensis canadensis* and *Lutra canadensis interior*, have been recognized in the tristate region. *L. c. interior* supposedly occupied all of Nebraska and a portion of southeastern South Dakota. Paler in color and larger overall than *L. c. canadensis*, *L. c. interior* was reported to have a smaller hind foot, smaller skull, and less crowded toothrow. A thorough study of geographic variation in the river otter never has been made, although only such a study would permit appraisal of the validity of these nominal subspecies.

Description. The otter is a slender, cylindrical mustelid with an elongate, rounded tail that makes up almost a third of its total length. The head is broadly flattened, with small eyes and short, round ears. The naked nose pad closely resembles the shape of a rounded ace of spades. Short, stiff bristles are present on the snout and chin. The neck is without constriction, emphasizing the pointed body outline. Forelimbs and hind limbs are short but muscular, the broad feet having five almost completely webbed digits. All digits have short claws, and there are naked pads on the soles and palms. Three or four small circular papillae are present on the posterior border of the pad on the hind feet.

A dense underfur intermixed with longer bristly guard hairs forms an effective water-repellent insulation. The color is generally a deep lustrous brown, with the throat, sides of the head, chest, and venter varying from pale brown to brownish gray. The proximal half of each hair is pale brown, the distal half dark brown. The fur is so

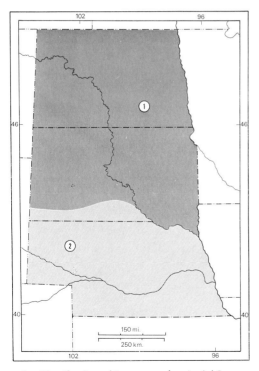

182. Distribution of Lutra canadensis: (1) L. c. canadensis; (2) L. c. interior.

dense that the pale brownish basal color is not evident. Molting occurs in spring and autumn, but the color of the pelage remains the same year round.

Measurements of a male from Nebraska followed by mean and extreme measurements of eight males and six females from northern California are: total length, 1,346, 1,126 (1,020–1,246), 1,094 (1,000–1,158); length of tail, 457, 423 (353–507), 414 (385–447); length of hind foot, 120, 127 (110–138), 122 (120–125); length of ear, 16.5 (Nebraska male only). The California males weighed 16.9 (11–23) pounds and the females 15.3 (11–17).

The skull is broad, flat, and short. The brevity of the skull is emphasized by the broadly opened zygoma. The orbit is relatively small. The occipital crest is only slightly raised, and there is no sagittal crest. Bullae are extremely flattened and small. The skull of a male from Nebraska has a condylobasal length of 112 and a zygomatic breadth of 103.

Natural History. One of the best adapted of all American mustelids to an aquatic life, the otter has a tapered body, webbed digits, short limbs, rudderlike tail, and dense pelage. With head held high out of the water and back partially exposed, the otter swims mainly by sinuous movements of the body and tail, but occasionally paddles with the hind feet. With a maximum speed of about six miles per hour, this animal can swim under the water just as well as above. It can remain submerged for up to two minutes, and otters are known to dive to depths of 30 to 45 feet. They often swim beneath ice, surfacing for air in pockets in the ice. Swimming and diving are graceful, with little noise or splashing.

L. canadensis is active the year around and, although generally crepuscular, may be observed at any time of day. Otters often emigrate when under pressure of food shortage or unfavorable environmental conditions. Individuals may range 35 to 45 miles per year, but families generally cover only two to seven miles. During such wanderings, they follow streams and rivers, but occasionally an otter will strike out across

183. River otter, Lutra canadensis *(courtesy Michigan Department of Natural Resources).*

open land. On land, the otter lopes slowly, hindered by its short legs. However, when traveling on ice or snow it increases speed by springing into the air, placing its limbs against its body, and sliding on its chest and belly for 25 feet or more.

Otters are almost exclusively carnivorous, but the composition of the diet varies from place to place, apparently with availability. Where present, crayfish are a mainstay; crayfish, turtles, and frogs are excavated from bottom mud in winter. Fish also are a staple; most studies show a prevalence of rough fish, such as suckers, in the diet. Panfish, such as sunfish, are taken in smaller amounts, and game fish (trout, for example) are rarely captured. Insects and earthworms appear in the diet in small amounts, and young muskrats and beaver are eaten in spring and early summer. Digestion is rapid; crayfish remains have appeared in feces within an hour after ingestion.

In New York, otters breed in March and April, the blastocysts remaining unimplanted until January or early February. Captive female otters in Minnesota came into heat between December and early April and were receptive to males at intervals of six days for 42 to 46 days. Copulation, which occurs during lactation, lasts 16 to 24 minutes and takes place either on land or in water but is more likely to occur in water. Young males are often unsuccessful at mating, being rejected by females. Males are sexually mature at two years of age, but females are known to breed when one year old. Parturition lasts from three to eight hours, the female giving birth while standing on all four feet. Newborn otters are about 275 mm long and weigh some 130 grams.

The one to four (usually two) young, born in March or April, are toothless, blind, and helpless at birth, although fully furred. The eyes open at about 35 days. The young remain in the den until about six weeks old, when they emerge to play about the entrance. The female teaches the young to swim and sometimes must drag them into the water. The male otter normally avoids his family until the young are six months old, when, if allowed by the female, he assumes a paternal role. Both parents are extremely patient with the young, but occasionally they punish them by a nip on the nose.

Both young and old otters spend consid-

erable time romping in grass or sliding on snow, ice, or muddy riverbanks. Vocalization includes chirps and "chuckles," and otters scream when frightened.

The otter does not dig burrows but will enlarge dens abandoned by other animals. Maternal dens vary in location and may be 150 yards or more from the nearest stream. Abandoned lodges of beaver and muskrat sometimes are used. The entrances to such dens are below water and usually below the level of the ice. Otters in the southern United States often build nests of rushes and grass in marshland.

L. canadensis usually is sociable and docile, but it can become a formidable enemy during the breeding season or when cornered. Only when frightened will the otter exude the musky secretion of its anal scent glands.

Selected References. Natural history (Greer 1955; Hamilton and Eadie 1964; Knudson and Hale 1968; Liers 1951); systematics (van Zyll de Jong 1972).

Family Felidae: Cats

Felids are highly evolved predators with sharp retractile claws, sharply pointed canine teeth, and well-developed carnassials. The fossil record of the family began in the late Eocene of North America and Europe; wild felids now occur on all continents except Australia. There is little agreement on classification of this group. House cats and tigers may differ in size by a factor of 200, but they share minute details of anatomy and behavior. Specialists argue whether two, three, four, or five Recent genera should be recognized. There are 35 to 40 modern species in the family.

Key to Felids

1. Tail long, more than one-third total length; condylobasal length greater than 150 _____*Felis concolor*

1.' Tail short, less than one-fifth total length; condylobasal length less than 125 _____2
2. Tail less than one-half length of hind foot, tip black above and below (fig. 189); jugular foramen separate from anterior condyloid foramen (fig. 186) _____ *Felis canadensis*
2.' Tail more than one-half length of hind foot, tip with black blotch above, white below (fig. 189); jugular foramen and anterior condyloid foramen confluent (fig. 186) _____*Felis rufus*

Genus *Felis*

We consider all three felids of the Northern Great Plains to be members of a single genus, *Felis*. However, some authors continue to regard the lynx and bobcat as belonging to a separate genus, *Lynx*, and some recent evidence suggests that the mountain lion ought to be classified with the cheetah in the genus *Acinonyx*. Currently, about 30 extant species of *Felis* are recognized; the geologic range of the genus is early Pliocene to Recent. The dental formula is 3/3, 1/1, 2–3/2, 1/1, total 28 or 30. *Felis* is the classical Latin word for "cat."

Felis concolor: Mountain Lion

Name. The name *concolor* is a compound of two Latin roots—*con*, "together," and *color*, "color"—a reference to the fact that the pelage is of a single shade over all the body. Frequently applied vernacular names include puma, cougar, and—especially in the eastern United States—panther.

Distribution. Historically, the mountain lion had the widest distribution of any native species of American mammal other than man, from British Columbia southward to Argentina. Over most of its range it now is limited to areas well away from human settlement. Three subspecies have been ascribed ranges on the Northern Great Plains, *Felis concolor hippolestes* in western Nebraska and South Dakota, *Felis concolor missoulensis* in western North Dakota, and *Felis concolor schorgeri* in the eastern parts of the tristate region; the last-mentioned race is extinct. In historic times these animals occurred throughout most of the Northern Plains, except perhaps in northeastern North Dakota east of the Missouri River (although at least one occurrence has been documented on the Pembina Hills of Manitoba). In Nebraska, *F. concolor* was extirpated from the eastern and central parts of the state by the 1890s, but there still are reports of mountain lions in extreme western Nebraska almost every year. In South Dakota the species occurs only on the Black Hills (and there sparingly), and a few now may be resident in North Dakota. Once they were common in the Badlands of the southwestern part of that state. The distribution is not mapped.

Description. Adult mountain lions are buffy to pale brown above, somewhat paler,

184. Mountain lion, Felis concolor *(courtesy J. L. Tveten).*

occasionally nearly white, below. The chin and chest are pale, and the tail is tipped with black. Measurements of animals from the Northern Great Plains are not available. External measurements of a male from Colorado are: total length, 2,105; length of tail, 803; length of hind foot, 292; length of ear, 105. Weights of adults range from 80 to 200 pounds or more. Some cranial measurements of three males and two females from Colorado are, respectively: condylobasal length, 202, 187, 189, 168, 164; zygomatic breadth, 148.2, 147.6, 138.8, 127.2, 128.0.

Natural History. Mountain lions are truly magnificent mammals, beautifully adapted to their role as the "top carnivore" in temperate North America. Their principal food is deer, which they stalk with astonishing grace. Imagine a house cat stalking a bird: crawling in slow motion, crouched on its belly, tail twitching metronomically. With a single graceful leap it grasps the startled prey in its paws. Now scale the image to five times the length and 30 or 40 times the weight and you have a picture of a mountain lion pursuing its prey. Our sympathies may lie with the hapless hunted, but the hunter commands deep respect.

Mountain lions occur mostly in wooded terrain. It is in the rocky, broken woodlands of the West that they are most successful today, but once they occurred in riparian woodland and forest edge communities all across the continent. Wherever there were deer, there also went *F. concolor.* In Manitoba, mountain lions seem to be making a comeback or even extending their historic range. Perhaps the same may one day be said of the Northern Plains states. Even in the early days, mountain lions seldom preyed on grazing animals of the open prairie, such as bison and pronghorns. Their hunting tactics demand good cover. Sometimes they attack from a rocky rim or a limb overhanging a deer trail. Unlike other large carnivores, they seldom eat carrion, but they do prey on rabbits, porcupines, beaver, and smaller rodents. Livestock, especially colts, are taken as available.

Most hunting is done at dawn or dusk,

although the animals may be seen abroad at any time of day. It takes about one deer a week to sustain an adult mountain lion. Having killed a deer, a lion will gorge itself on the shoulders and hams, then cover the leftovers with leaves or brush, returning within two or three days to finish the feast. After a full meal, a mountain lion will rest and digest the food for a few days before seeking another victim.

Adult mountain lions are solitary and wander widely. There is no permanent den; the animals simply camp along the trail in some unimproved shelter. Like other cats, mountain lions are active throughout the year.

This species is a promiscuous breeder, first mating at the age of two or three years. Breeding may take place at any time of the year, but there is a distinct peak of births in summer. Males may fight over the right to mate with a female, but mating is the sole end and service of paternity. The female gives birth to one to six (usually two or three) blind but well-furred spotted kittens after a gestation period of approximately 96 days. Neonates weigh about 450 grams and are about 250 mm in total length. The eyes open by about two weeks of age. Young are weaned at six weeks or so, at a weight of about 8 pounds, and at that time begin to feed on meat killed by the mother. As they grow, they soon learn to kill their own prey, such as ground squirrels and rabbits. The young weigh about 30 pounds at six months and 60 to 80 pounds as yearlings. They may stay with the female as long as two years.

Mountain lions have few enemies other than armed men. Even parasites are few, because they have no fixed den and rarely feed on carrion. Some are fatally injured in encounters with prey—a well-antlered deer or a porcupine, for example. Others incur injuries to teeth or limbs and die of starvation. Adult males may eat cubs. Because of their reliance on live meat, mountain lions cannot be drawn to poisoned baits or trapped as readily as some other carnivores. They have been pursued with hounds, however, and then shot from trees. Much of man's negative impact on populations of mountain lions must have been indirect. *F. concolor* avoids civilization, and the westward press of man forced the species into relict patches of wild land. It is also noteworthy that man hunted deer not only for his own sustenance but as a marketable commodity. This forced lions to move on or to turn to domestic stock for food; the latter circumstance was an almost certain invitation for the offender to be hunted down and killed.

Selected References. General (Young and Goldman 1946); natural history (Hornocker 1970; Robinette, Gashwiler, and Morris 1959, 1961; Seidensticker et al. 1973).

Felis lynx: Lynx

Name. The name *lynx* is a Greek word for "cat."

Distribution. The lynx is a Holarctic species, occurring in both northern Eurasia and North America. In the New World, it occurs across the taiga of Alaska and Canada, in the Central Rockies, and southward in small numbers to Colorado, Nebraska, the Great Lakes region, and New England. A single subspecies (*Felis lynx canadensis*) is recognized across this vast area. This animal rarely has been taken or observed on the Northern Great Plains in recent years. Fewer than a dozen records are available from South Dakota, scattered across the state. In North Dakota there are numerous records, although the population is highly erratic, seeming to depend on eruptive migrations from Canada in years of low populations of snowshoe hares. There are fewer than half a dozen records from Nebraska, the most recent being in 1974, at a place along the Missouri River within 40 miles of Omaha. The Northern Great Plains provide marginal habitat for the lynx at the southern edge of the range of the species. Consistently high populations would not be expected even without the negative impact of modern man.

For many years, Eurasian and North American lynx were thought to represent separate species, but for the past several

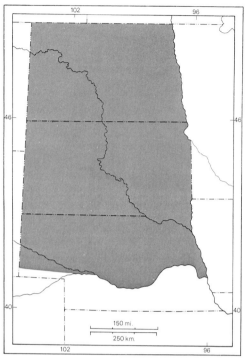

185. Distribution of Felis lynx.

decades they have been regarded as conspecific. Recent studies suggest, however, that the original arrangement is the correct one, in which case the North American population would bear the specific name *canadensis*.

Description. For comparison with the bobcat, *F. rufus*, see the account of that species. The lynx has a beautiful shadowy gray coat, with pronounced ear tufts and a short, black-tipped tail. Ranges of external measurements for three males and four females from British Columbia are, respectively: total length, 864–927, 876–902; length of tail, 102–114, 97–127; length of hind foot, 235–267, 222–242; adults weigh 25 to 35 pounds. Length of ear usually ranges between 70 and 80. Cranial measurements of males from Colorado and British Columbia include: condylobasal length, 117.2, 122.2, zygomatic breadth, 91.2, 94.2.

Natural History. The lynx typically is a denizen of the deep forest, whether in the

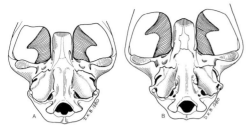

186. Ventral view of occipital region of skulls of (A) Felis rufus *and* (B) F. lynx.

North or in the mountains of the West. The animals are solitary, hunting by night and denning by day in any convenient shelter—beneath a wind-thrown tree, under a hummock of mosses or overhanging rock, or in rocky hollows or small caves. Home ranges overlap, but there is mutual avoidance, as in many other asocial carnivores. Densities vary widely, seemingly dependent upon the supply of food, especially snowshoe hares. Indeed, the tightly linked cycles of abundance of lynx and snowshoe hares have fascinated naturalists for years. The cycle lasts nine to 10 years from peak to peak, based on more than a century of fur reports from the Hudson's Bay Company. The cycle of lynx abundance tracks the cycle of the snowshoe hare, with a time lag of a year or two. Intensive study in central Alberta in recent years has confirmed the link between these cycles. In years of hare scarcity there are virtually no lynx kittens recruited to the population. Neonates starve to death, and yearling females do not breed (also a correlate of starvation stress). In Alberta, a low population was three lynx per 130 square kilometers (about 50 square miles); a high population was 13 lynx in the same area. A similar fourfold change in numbers is seen in fur company records.

Snowshoe hares are not the only food of the lynx, but they are the staple. Hare numbers seem to cycle because the denser the population the more likely the hares are to interact with one another, thus contracting "shock disease," a syndrome involving endocrine imbalance. Also, high populations of *Lepus* may be especially prone to parasitic diseases and suffer from a shortage of food. In years of abundant snowshoe hares, food consumption of an individual lynx may be a third greater than when hares are uncommon. In lean years the lynx relies increasingly on ruffed grouse—the alternative mainstay—and turns more to eating mice, voles, and carrion. Overall, the diet consists of about 70 percent hares, 15 percent carrion, and 10 percent grouse, with the last five percent comprising several kinds of birds and small mammals. An adult needs 600 grams of food per day, a juvenile about 400 grams. The nightly hunting round is a circuitous trek of three to six miles. The animals stalk their prey or lie in wait along a game trail. With a single, silent leap of six to eight feet from the crouched position, the lynx usually takes the prey before it even senses the predator's presence. Sometimes a short chase is involved, but this is not the lynx's forte.

These animals are active throughout the year, padding easily over crusty snow on their furry, oversized feet. They mate in

187. Lynx, Felis lynx *(courtesy L. E. Bingaman).*

February or March, and the female bears a litter of about four kittens in April or May, after a gestation period of about 63 days. The kittens are blind until about day nine, but they are well furred and are darker and more distinctly patterned than their mother. The adult pelage is attained at about nine months of age. Thereafter there is a single annual molt, in spring. Sexual maturity is attained (at least in females) when lynx are yearlings.

Lynx have no consistent natural enemies other than man and harbor only a few parasites. In the North they are pursued for their fur, and, in the southern parts of their range, where they are rare or only occasional, they usually are taken inadvertently by trappers or predator-hunters intent on the bobcat.

Selected References. Natural history (Brand, Keith, and Fischer 1976; Gunderson 1978; Nellis and Keith 1968; Nellis, Wetmore, and Keith 1972).

Felis rufus: Bobcat

Name. The specific epithet *rufus* is Latin for "red." Other vernacular names include wildcat and bay lynx.

Distribution. The bobcat occurs from southern Canada to the Isthmus of Tehuantepec and from coast to coast in the United States. Over most of the Northern Great Plains, bobcats occur sparingly, becoming common in the rougher country in the western third of the three-state region. Two subspecies have been ascribed ranges in our area of interest, *Felis rufus rufus* in the east and *Felis rufus pallescens* in the west. However, bobcats of the western part of the region probably represent intergrades between the two races. *F. r. pallescens* is paler and is larger externally and cranially than *F. r. rufus.*

Description. External measurements of two males and three females from western Nebraska are: total length, 940, 996, 824, 813, 848; length of tail, 152, 160, 162, 152, 134; length of hind foot, 190, 178, 165, 178,

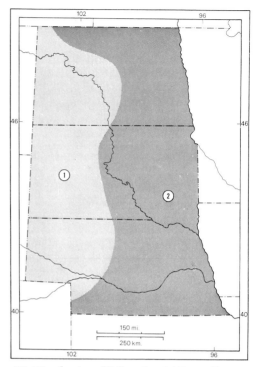

188. Distribution of Felis rufus: (1) F. r. pallescens; (2) F. r. rufus.

178; length of ear, 80, 78 (two females only). Three males and two females weighed 29, 26, 23, 18, and 15 pounds. Mean and extreme cranial measurements of six males from the same area, followed by those of two females, include: condylobasal length, 118.7 (115.9–120.6), 107.0, 109.7; zygomatic breadth, 91.2 (87.7–93.4), 80.4, 85.7.

Of mammals on the Northern Great Plains, the bobcat can be confused only with the lynx, *F. lynx.* Bobcats inhabit more open, brushy country, are smaller than lynxes, have more reddish (rather than grayish) pelage, have well-defined spots, and have bands of black hairs on the front legs. The hind foot is markedly shorter, and the black tip on the tail is incomplete, being broken below with reddish buff hairs. The pelage is shorter and the ear tufts are smaller (less than 25 mm) and sometimes apparently lacking. Mountain lion cubs, although spotted, have no "ruff" on the neck and have long tails.

Natural History. Bobcats occur in a wide range of habitats. In the western part of the tristate region, they are most typical of rugged, brushy country. They also occur in woodlands along streams and in woodlots. In the eastern portions of the Northern Great Plains, they range through lowland forests. In short, bobcats occur almost everywhere in the three states except on the featureless plains.

Open country is poorly suited to the hunting style of the bobcat. Like mountain lions, these animals are less well adapted to chase than to surprise. Often they will sit motionless next to a game trail awaiting an unsuspecting rabbit or hare. When the rabbit hops alongside, intent on its own unassuming dinner of forbs or browse, the cat dispatches it in one decisive bound, using its retractile claws and rapierlike canines. Stealth, not speed, is the bobcat's hallmark. The diet consists mostly of rabbits. Smaller mammals—such as woodrats, mice, and voles—also are eaten, as are porcupines, beavers, woodchucks, and birds, especially ground-nesting species such as grouse. A night's hunting may take the bobcat over a

190. Bobcat, Felis rufus *(courtesy D. C. Lovell and R. S. Mellott).*

circuitous route three to seven (average five and a half) miles long, although the home range is only about two to three square miles. A rabbit every other day will keep a bobcat in good condition. Some carrion is eaten, especially in winter. Bobcats have been known to kill herbivores as large as deer and pronghorns, but these species are hardly the typical fare. Although *F. rufus* climbs trees handily, individual animals seem to do little arboreal foraging. Like domestic cats, bobcats bury their dung; they usually defecate on some prominence at a spot off the hunting trail.

Bobcats hunt mostly at night. By day they rest in some unimproved shelter—under an overhanging rock or a wind-thrown tree, in a hollow log, or in a brushy thicket. Such rest-shelters seldom are used two days in succession.

Bobcats seem to be seasonally polyestrous. Females have a single litter each year but will come into estrus several times if not successfully mated. At our latitude the breeding season is February to June. The gestation period is about 10 weeks, with the bulk of births from mid-May to mid-June. Bobcats have been considered induced ovulators, but there is some evidence that unmated females ovulate spontaneously. A few females may

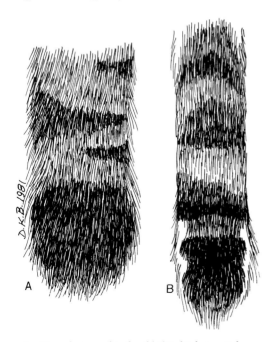

189. Dorsal view of tails of (A) Felis lynx *and* (B) F. rufus.

breed as yearlings. Males, which are seasonally fertile, reach puberty in their second spring.

Mean litter size is about 2.8, with a range of one to seven and a mode of three. The neonates are blind but covered with hair. They are brought forth in a nursery den in some out-of-the-way natural shelter: a rocky cleft among boulders, a cavern, a hollow tree, or a thicket. The eyes open at about nine days, and littermates soon engage in animated play. The young are weaned at approximately two months of age. They stay with the female into the autumn. The male usually does not participate in rearing the young.

Bobcats appear to have distinct summer and winter pelages, but this is not quite true. There is a single molt annually, in autumn. By late autumn and early winter the pelage is long and full. As it wears it becomes coarser, more brittle, and more distinctly reddish and less grayish. Thus summer pelage actually is the worn winter coat.

Bobcats have no habitual enemies other than humans. Before the advent on the Northern Plains of European man with his traps, guns, and hounds, starvation may well have been the principal regulator of populations. In years of low prey numbers, few or no young are recruited into the population. Malnourished females do not breed, and malnourished young do not survive. Bobcats have relatively few parasites because they are solitary, have no permanent dens, and depend on carrion only in restricted times of the year. Maximum longevity in the wild is about 12 years. Bobcats of that age still can reproduce, but they have worn or broken teeth and must be less effective hunters than younger animals. Given federal prohibitions on importation of cat furs from abroad, trapping pressure on bobcats has become intense in recent years. The several hundred animals trapped annually on the Northern Great Plains now sell for an average price of about one hundred fifty dollars, with select individual pelts commanding two hundred dollars or more.

Selected References. General (Young 1958); natural history (Crowe 1975; Mahan 1980; Van Wormer 1963).

Order Artiodactyla

The order Artiodactyla, the so-called even-toed ungulates, comprises nine living families. There are about 75 Recent genera and 185 Recent species. The fossil record is traceable back to the early Eocene (at least 50 million years ago) of Europe and North America, and the evolutionary history represents an extensive radiation. Members of the order are present throughout Europe, Asia, Africa, South America, North America, and the coastal regions of Greenland; additionally, there have been many introductions throughout the world. Artiodactyls were an important source of food for early explorers, travelers, and settlers in the tristate region and are valued for aesthetic and recreational purposes today. These conspicuous mammals are important elements of the ecosystems in which they occur.

Eight species of artiodactyls, representing three families, are native to the Northern Great Plains. Three of these species (bison, wapiti, mountain sheep) were extirpated by man and subsequently reintroduced. In addition, two species (mountain goat, fallow deer) have been introduced from elsewhere in North America and the Old World, respectively.

Key to Artiodactyls

1. Animals horned (both sexes); if horns shed, horn-cores permanent _____2
1.' Animals antlered (males only, except in *Rangifer*); antlers shed annually (Cervidae) _____5
2. Horns with one prong (prong lacking in many females), tip somewhat recurved, sheath shed annually; striped brown or black markings on neck (Antilocapridae) _____ *Antilocapra americana*
2.' Horns unbranched, not shed; neck lacks distinctive markings (Bovidae) _____3
3. Horns brown, heavily ridged and curving backward, forming a massive spiral in adult males; color grayish brown; rump, backs of legs, and muzzle white _____ *Ovis canadensis*

3.' Horns black, not forming a spiral; color uniformly whitish or brown _____4
4. Horns curving backward, quite slender and slightly ridged; color uniformly white or white tinged with yellowish _____*Oreamnos americanus**
4.' Horns completely smooth, curving upward and toward each other; color uniformly dark brown _____*Bison bison*
5. Antlers palmate, at least in part _____6
5.' Antlers not palmate _____8
6. Antlers palmate on most of the main beam; greatest length of skull more than 500 _____ *Alces alces*
6.' Antlers palmate distally; greatest length of skull less than 500 _____7
7. Antlers palmate distally on both the

main beam and one or more of the brow tines; canines 1/1 ___*Rangifer tarandus*

7! Antlers palmate distally only on main beam; brow tines not palmate; canines 0/1 _____*Cervus dama**

8. Underparts brownish, similar in color to back; upper canine teeth present; larger than typical deer _____*Cervus elaphus*

8! Underparts paler than back; no upper canines; smaller than wapiti _____9

9. Hindquarters mostly white, tail large and white below, often held high when running; antlers with single main beam that gives rise to a series of simple tines (fig. 193) _____*Odocoileus virginianus*

9! Hindquarters mostly grayish, tail smaller and tipped with black, held down while running; antlers with main beam forked, forming two secondary branches, each of which bears tines (fig. 193) _____ *Odocoileus hemionus*

Family Cervidae: Deer and Allies

North American representatives of this family include the wapiti, deer, moose, and caribou. The fossil record dates back to the Oligocene in the Old World and the early Miocene in North America. There are about 15 Recent genera, with approximately 38 species, generally arranged taxonomically within four subfamilies. Cervids have been introduced in many parts of the world, especially on Australia, New Zealand, Hawaii, and other islands in the Pacific Ocean, as well as Cuba and islands in the Indian Ocean.

Members of this family occur in arctic tundra, dense woodlands, grassy plains, brushy country, deserts, swamps, and tropical rain forests. All are herbivorous and feed on grass and shrubby vegetation, as well as bark, twigs, and leaves of trees.

Six species of cervids occur on the Northern Great Plains. Of these, one (caribou) apparently has been extirpated, one (wapiti) was driven to extinction and subsequently reintroduced, and one (fallow deer) was introduced from the Old World.

Genus *Cervus*

The genus *Cervus* includes the European red deer and the wapiti of North America as well as several species in Asia. The dental formula is 0/3, 1/1, 3/3, 3/3, total 34. The generic name is derived from the Latin word *cerva,* meaning "deer."

Cervus elaphus: Wapiti

Name. The species name *elaphus* comes from the Greek and means "stag" or "deer." Vernacular names include American elk, elk, and wapiti. Because the name "elk" is used in reference to the European moose, as well as a number of other artiodactyls in the Old World, the most appropriate name is wapiti, a Shawnee word meaning "white rump."

Distribution. Cervus elephus once ranged throughout much of the Northern Hemisphere and historically was one of the most widely distributed members of the family Cervidae in North America. The animals were extirpated from the eastern United States in the previous century, and the total population shrank to an all-time low about 1900, owing mostly to the pressures of market hunting and agricultural development. The last wild-living individuals in Nebraska were shot in the early 1880s, and the species was extirpated on the Black Hills by 1888.

In 1900 wapiti were reintroduced from Wyoming into Custer State Park in South Dakota. An additional 25 animals were added in 1913, and 50 more were obtained from Montana in 1916. In 1914 21 *C. elaphus* from Wyoming and Montana were released in Wind Cave National Park. At present, two herds, totaling several thousand animals, range on the Black Hills. A few hundred animals exist in Nebraska on public and private reservations, and individuals now are occasionally seen in the wild. Free-living wapiti are considered rare in North Dakota.

During late Pleistocene and early Recent times, 10 subspecies of this species inhabited North America. Of these, six are extinct, including *Cervus elaphus canadensis,* which was native to the Northern Great Plains. Wapiti that have been restocked in the tristate region are of the subspecies *Cervus elaphus nelsoni,* native to the Rocky Mountains. The former distribution on the Northern Great Plains is not mapped.

Description. Both sexes of wapiti have a heavy, dark mane that extends down to the brisket area. There is a distinct yellowish rump patch bordered with a band of dark brown or black hair. Seasonal molt occurs in May and June, resulting in a summer pelage that is deep reddish brown with little or no undercoat, which makes the animals appear sleek. In winter, after molt in September or October, the head, neck, and legs are dark brown, whereas the sides and back are grayish brown and much paler than the rest of the body.

Antlers of adult bulls consist of long, rounded beams that sweep upward and backward and usually bear six tines. The first antlers usually begin growth in May when young bulls are a year old. Antlers of yearling bulls ("spikes") are unforked, although sometimes small tines appear at the tips. The second set of antlers usually consists of four to five points on a side. Large antlers with the typical six tines are normally grown on bulls that are at least three years old.

The wapiti is the second largest cervid in North America, surpassed in size only by the moose. There is some sexual dimor-. phism, cows being smaller than bulls. Measurements for males are: total length, 2,285–2,745; length of tail, 140–157; length of hind foot, 610–710; weight, 295–375 kilograms (up to about 825 pounds). Measurements for females are: total length, 2,135–2,480; length of tail, 114–125; length of hind foot, 580–655; weight, 227–295 kilograms (up to about 650 pounds). Cranial measurements of males from the tristate area are: greatest length of skull, 436–485; zygomatic breadth, 166–190.

There is a heavy dental pad in the upper jaw to oppose the lower incisors. Unlike other North American cervids, both sexes of wapiti have a functional pair of incisiform canine teeth in the upper jaw, known as "elk teeth."

191. *Wapiti,* Cervus elaphus *(courtesy U.S. Fish and Wildlife Service).*

Natural History. The wapiti is an herbivore, consuming grasses, forbs, and browse. Analyses of rumen contents of animals taken in May, June, and July in Colorado revealed that grasses made up from 83 to 92 percent of the diet. From December through April, browse represented about 57 percent of the plant materials consumed. Although wapiti get moisture from the vegetation they eat, they regularly drink water. In winter they meet the need for water by consuming snow.

This is a gregarious species. In summer, wapiti often gather in herds of 400 to 500 animals, and natural concentrations of more than 1,000 individuals have been observed on winter ranges. Although it is not a normal situation, herds of more than 15,000 animals have been reported from the Jackson Hole National Elk Refuge during severe winters. On the Black Hills, cow-calf groups are largest during the winter months, ranging from about 40 to more than 160 animals. Young bulls usually mix with the cow-calf herds until late summer. In early September bulls begin to rut, and bugling and harem formation begin. Harems range from 15 to 30 cows. Young bulls hang around the edges of harems and opportunistically breed with the cows. By the middle of October most breeding has been accomplished, and bulls resume their solitary ways. When snow accumulates, cows, calves, and most bulls migrate to winter ranges, where they remain from about December through March.

The gestation period is about eight and a half months, most calves being born in early June. The juvenile coat is spotted dorsally and on the flanks, although not so conspicuously as in young deer. Twins are uncommon. Cows stay away from the herd for two to three weeks until the calves are capable of travel. By mid-July, cows and calves are concentrated in herds on summer ranges. During summer, bulls occur either alone or in small groups of up to five or six animals. Cows become sexually mature at three years of age. Bulls may mature at two years but probably seldom breed before the age of four.

These seem to be restless animals. Daily movements are related to the availability of food, water, and cover and to harassment, and the animals tend to follow a pattern of feeding, watering, and resting at various intervals of the day and night. In general, wapiti feed moving uphill during the morning, bed down and rest at midday, and feed moving downhill in late afternoon.

These large ungulates seem to avoid excessive heat. They frequently bed down in snowbanks in late spring and early summer as long as snow remains. On hot days the animals use north slopes and dense timber, if available. Sometimes individuals, especially bulls with growing antlers, frequent high, open, breezy areas, apparently to avoid heavy concentrations of flies.

Newborn calves make loud, high-pitched squeaks and screams. Cows bark and grunt and make sounds similar to those of calves.

Animals older than calves can give a sharp bark, and they sometimes grind their teeth. The most striking call of the wapiti is the bugle of a bull during the rut. Cows bugle also, especially during parturition.

Predators on wapiti include grizzly bears, black bears, wolves, mountain lions, and man. The species is subject to brucellosis, especially in areas where large concentrations of animals occur. Accidental mortality by drowning in rivers, especially under thin ice, or by being trapped in deep snow, and starvation because of food scarcity all are factors in population regulation. Wapiti suffer sizable prenatal and postnatal losses, apparently because of restrictive forage conditions on the winter ranges.

Cervus elaphus is an important game animal in North America and thus of considerable economic value. It is hunted in the vicinity of the Black Hills, partly to remove excess animals from the herds, and relatively small numbers are taken each year.

Selected References. General (Boyd 1978; Varland, Lovaas, and Dahlgren 1978); distribution (Bailey 1927; Jones 1964; Turner 1974).

Genus *Odocoileus*

This genus includes only the mule deer and the white-tailed deer of North America. The dental formula is 0/3, 0/1, 3/3, 3/3, total 32. The generic name is derived from three Greek roots and refers to the large pale patch on the rump.

Odocoileus hemionus: Mule Deer

Name. The species name is formed from the Greek *hemi,* "half," and *onos,* "an ass," perhaps an allusion to the large ears of this animal and thus its mulelike appearance. There is some confusion about the vernacular name for the species, because two subspecies from the Pacific Northwest are referred to as black-tailed deer. Mule deer, however, is the appropriate name for this species on the Northern Great Plains, because of the prominent ears.

Distribution. The species ranges from the southern end of the Mexican Plateau to northwestern Mexico and Baja California (including some of the islands in the Sea of Cortez), throughout most of the western United States and Great Plains, northward to southeastern Alaska. Because of their importance as food for early settlers, as well as other factors, mule deer were nearly eliminated from the Northern Great Plains in the last part of the previous century. For example, in 1900 there probably were fewer than 100 mule deer in Nebraska, and the species was relatively uncommon throughout the area until about 1950. At present there are some problems with overpopulation and overutilization of available range by mule deer on the Northern Plains.

Odocoileus hemionus hemionus is the subspecies found in the tristate region, as well as throughout the Rocky Mountain area, the Great Basin, and western Canada, except for the coastal region.

Description. The ears of the mule deer are large, the forehead is dark, and there is a brown spot on each side of the nose. The tail is small and white and has a black tip. The brisket area is brown. Pelage on the dorsum and sides is either reddish brown (in summer) or grayish brown (in winter), there being two seasonal molts annually. There is a large grayish white rump patch. The venter is generally white. Mule deer usually run with a stiff-legged bounce, with the tail held down.

Males are slightly larger than females. Representative ranges in external measurements for males include: total length, 1,370–1,800; length of tail, 150–230; length of hind foot, 410–590; length of ear, 120–250; weight, 50–215 kilograms (up to about 470 pounds). Measurements for females are: total length, 1,160–1,800; length of tail, 110–200; length of hind foot, 325–510; length of ear, 120–240; weight, 31.5–72.0 kilograms (up to about 160 pounds). Ranges in cranial dimensions for four males and four females from Nebraska are, respectively: greatest length of skull, 281–309, 253.5–283.5; zygomatic breadth, 134.6–141.5, 114.4–128.7. Mule deer have antlers with a main beam curving upward and outward. The forks are dichotomous, and the prongs are mostly equal in size (see fig. 193).

Natural History. Mule deer occur in all of the major vegetation zones in western North America. The animals inhabit brushy and wooded areas as well as open plains and broken country. In many areas, successional vegetation seems to be the preferred habitat, especially near agricultural lands.

Mule deer browse on a wide variety of woody plants, but they also graze on grasses and forbs. A total of 673 species of plants have been identified as food items for the subspecies that occurs on the Northern Great Plains. In Nebraska, studies revealed that the diet of this species was made up of about 41 percent agricultural crops, 13 percent brushy vegetation, and 18 percent leaves and young branches of trees; the rest of the diet included sunflowers, grasses, and sedges. Heavy snow can inhibit these deer by reducing the availability, and

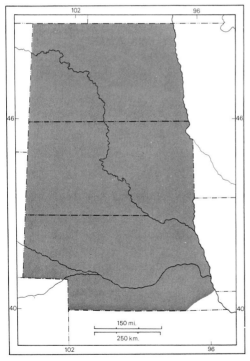

192. Distribution of Odocoileus hemionus.

193. *White-tailed deer (left) and mule deer, illustrating differences between antlers, ears, and tails (reprinted from* Nebraskaland Magazine, *published by Nebraska Game and Parks Commission).*

thereby the nutritive content, of food.

Odocoileus hemionus generally is a crepuscular species, with most activity occurring at dawn and dusk. Home ranges on the Northern Great Plains are larger than those of mule deer in other regions, size of home range being directly correlated with the availability of food, water, and cover. In general, mule deer in the tristate region are not migratory, though some herds move about for up to six or seven miles during both winter and summer. Individual animals occasionally move considerably greater distances from their home ranges,

however; in South Dakota, for example, a female was tracked for 180.2 kilometers (about 112 miles) and a male for 149.6 kilometers (about 93 miles).

The breeding season is mostly from late October through December. The gestation period is about 203 days, but variations of up to 30 days are known. The number of fawns is variable, as data for adult females from Utah indicate: nonpregnant, two percent; one fetus, 37 percent; two fetuses, 60 percent; three fetuses, one percent. In Nebraska, sex ratios of newborn mule deer are about 111 males to 100 females. Little

194. *Mule deer,* Odocoileus hemionus *(courtesy U.S. Fish and Wildlife Service, photograph by E. P. Haddon).*

is known about the survival of newborn fawns in the wild. In Utah, average weight of captive-born fawns of both sexes was about 8 pounds, males averaging slightly the larger. At 135 days of age the average weight was 27 to 36 kilograms (60 to 80 pounds); sexual maturity is reached at this time. Thereafter, these deer exhibit an annual growth cycle; they increase in weight from spring to autumn and decrease in weight from autumn to spring. Females reach maximum size at about two years of age. Males may continue to grow until they are nine to 10 years old. The antlers begin to grow in April and May; the first forks are evident in June. New antler growth is covered with a soft, highly vascularized layer of skin, the "velvet." Bucks rub off the velvet in late August and early September, and antlers are lost in late December and early January.

Accidental mortality is by starvation in areas that are either overpopulated or subject to adverse climatic conditions. In addi-

Table 3: Mule Deer Harvested on the Northern Great Plains, 1970–79

Year	North Dakota	South Dakota	Nebraska
1970	3,901	7,075	8,469
1971	4,448	7,510	7,287
1972	4,289	9,930	6,523
1973	2,222	12,100	7,936
1974	2,524	12,610	8,539
1975	3,961	11,040	8,074
1976	3,439	7,710	6,889
1977	2,978	5,165	5,012
1978	2,183	4,550	5,246
1979	2,834	5,580	5,328

tion, highway accidents are a frequent cause of deer losses on the Northern Plains. In a five-year period in Nebraska, for example, 1,300 kills of deer by motor vehicles were reported. Mountain lions, coyotes, eagles, and feral dogs prey on mule deer, uncontrolled dogs being important predators in some areas. Magpies have been known to pick out the eyes of newborn fawns. Mule deer are subject to epizootic hemorrhagic disease, brucellosis, and numerous fibromas and papillomas. These viral warts, although common, are probably not important causes of mortality. Deer lice (*Lepoptena depressa*) are commonly found on mule deer on the Northern Great Plains.

This species is an important game animal wherever it occurs in North America. After a period of almost complete protection in the early 1900s, populations of mule deer on the Northern Plains have recovered enough to support considerable hunting pressure (table 3).

Selected References. General (Menzel and Bouc 1975; Severson and Carter 1978; Taylor 1956; Wallmo 1978); distribution (Jones 1964; Turner 1974).

Odocoileus virginianus: White-tailed Deer

Name. The specific name refers to the place where this deer was first discovered, Virginia. White-tailed deer is the vernacular name used for this species throughout most of its range.

Distribution. Odocoileus virginianus is the most widely distributed artiodactyl in North America. There are records from all of the contiguous 48 states of the United States. The northern part of the range extends into all the southern provinces of Canada; to the south, the species occurs throughout Mexico and Central America and into South America about as far as 15° S latitude. As in the mule deer, numbers of this species were greatly reduced on the Northern Great Plains by the turn of the previous century; *O. virginianus* probably

persisted mostly in the Pine Ridge country, parts of the Sand Hills, along the Niobrara River, in the roughest parts of the Missouri Breaks, and on the Black Hills.

Since 1900, white-tailed deer have increased in numbers somewhat more slowly than mule deer. At present, the herd in Nebraska includes an estimated 100,000 animals. In some parts of the Northern Great Plains, white-tailed deer apparently have overpopulated the suitable range. They are more abundant than mule deer in about the eastern half of the Northern Plains; in the western part of the region mule deer outnumber whitetails. Occasional hybrids between the two species have been reported.

Three nominal subspecies are thought to occur on the Northern Great Plains: *Odocoileus virginianus dacotensis* is found throughout North Dakota and most of South Dakota; *Odocoileus virginianus macrourus* occurs in eastern Nebraska and

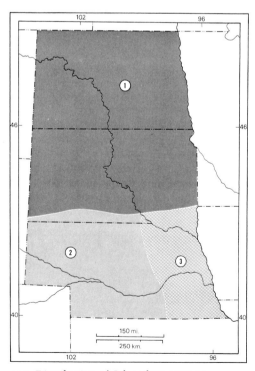

195. Distribution of Odocoileus virginianus: (1) O. v. dacotensis; (2) O. v. texanus; (3) O. v. macrourus.

extreme southeastern South Dakota; *Odocoileus virginianus texanus* ranges throughout central and western Nebraska and into the southernmost part of South Dakota. There are some real problems with the delineation of subspecies of white-tailed deer and their distributions on the Northern Great Plains, as well as elsewhere. In the absence of clearly known characteristics for distinguishing taxa and the relative lack of specimens with appropriate data from the tristate region, the aforementioned determinations should be considered only tentative.

Description. The ears of this species are smaller than those of mule deer. There is a white band around the muzzle and a white ring around the eye. The tail is broad, especially at the base, white beneath, and brown on top, except for a fringe of white on the distal half. Pelage of the head, neck, dorsum, and sides ranges from reddish brown (in summer) to grayish brown (in winter); white-tailed deer appear less yellowish than mule deer. There is a white rump patch, which is mostly covered when the tail is down. The venter is white. These deer generally run with a graceful loping action; the tail is held up and frequently waves from side to side. The young, born with a spotted dorsal coat, molt to adultlike pelage in two to three months. The pelage is replaced twice annually thereafter, as in *O. hemionus*.

Representative external measurements of males, which average larger than females, are: total length, 1,700–2,150; length of tail, 260–360; length of hind foot, 510–538; weight, 68–223 kilograms (up to about 490 pounds). Measurements for females are: total length, 1,340–2,062; length of tail, 152–325; length of hind foot, 480–520; weight, 41–110 kilograms (up to about 240 pounds). Cranial measurements for males from the tristate area are: greatest length of skull; 281–308; zygomatic breadth, 108–117. The same measurements for two females from Nebraska are: 272.5, 278; 109.3, 111.9. White-tailed deer have antlers with a main beam that grows first upward and slightly backward, then outward and forward (see fig. 193). The unbranched tines along the beam usually are upright.

Natural History. Like the mule deer, this species occurs in a wide array of habitat types. However, woodlands, forest edges, riparian vegetation, and vegetation adjacent to croplands seem to be preferred habitats. Neither dense forest nor expanses of open country are favored. Shelterbelts, especially in the eastern parts of the Northern Great Plains, provide good cover and are used heavily by these animals. Whenever adequate cover is available, white-tailed deer thrive in intensively farmed areas.

The diet of *O. virginianus* depends mostly on the food materials that are seasonally available. These animals browse on leaves, stems, buds, and bark of woody plants throughout the year. Browse is usually used most in autumn and winter. Acorns, corn, and other domestic grains are major foods of these deer. Agricultural crops constitute more than 50 percent of

196. White-tailed deer, Odocoileus virginianus *(courtesy U.S. Fish and Wildlife Service).*

their total food intake in some places, corn and soybeans being particularly important. Numerous forbs are eaten, especially in spring and early summer. In some areas, grasses and legumes are major components of the diet; fleshy fungi also are consumed whenever available. White-tailed deer drink free water, but they can go for long periods without it. They frequently eat snow in winter. These deer concentrate around salt licks, especially in spring. In the north, cover and adjacent adequate food supplies are critical factors for survival in winter.

Odocoileus virginianus is mostly active early in the morning and during the evening. Its activities are closely associated with feeding routines. Animals occasionally move about and feed during the night, and sometimes these deer feed for a brief time in the middle of the day. They may travel considerable distances in search of food within their home range and also may move outside the normal home range when food supplies are limited. In South Dakota, the home range for white-tailed deer reportedly averages about a square mile, but home ranges are expanded during times of food shortage. In some northern areas, these deer move in the early autumn to concentrate in areas with adequate shelter and food. Such concentrations of deer are called "yards" and may consist of from a few individuals up to hundreds of animals. The yarding period is generally from December to April; deer return to summer ranges when the weather is suitable and green plants become available.

In North America the breeding season for white-tailed deer is from October to January, with the major peak of activity in November. Does are in estrus for approximately 24 hours and, if not bred, come into estrus again one or two times about every 28 days. The gestation period ranges from 195 to 212 days, with an average of about 202. Most fawns are born in late May and June. The number of young is variable; single births, twins, and triplets are all common. In Nebraska, white-tailed deer have been reported to average 1.95 fetuses per doe. At birth, numbers of males and females are about equal (about 102 males to 100 females reported from Nebraska). The weight of fawns (average about 5.5 pounds at birth) may double during the first 15 days and quadruple in 30 to 40 days; growth continues at a rapid rate for the first seven months. White-tailed deer normally reach full size at five and a half to six and a half years of age. There is wide variation in body size and weight of adults, mostly a reflection of nutrition. Does can breed when they are six or seven months old, but most initial breeding occurs at a year and a half. Most bucks are a year and a half old when they first participate in the rut. Antlers are carried by males from April through February, full growth being reached in August and September, when the velvet is shed. Growth of antlers is correlated with nutrition as well as with the age of the animal; poor nutrition and old age inhibit antler development.

Major causes of mortality include starvation (due mostly to severe weather and shortages of food), entanglement in fences, and collisions with motor vehicles. Most deer are killed on highways during the breeding season and in spring, usually either at sunrise or within two hours after sunset.

Most of the predation on white-tailed deer is carried out by feral dogs and coyotes. Bobcats, mountain lions, bears, and eagles occasionally kill them, and wolves are important predators in a few

Table 4: White-tailed Deer Harvested on the Northern Great Plains, 1970–79

Year	North Dakota	South Dakota	Nebraska
1970	22,882	17,260	7,125
1971	28,673	17,760	7,029
1972	25,424	19,000	6,281
1973	27,780	23,145	7,979
1974	23,445	22,900	8,947
1975	20,666	23,630	9,583
1976	19,969	21,670	8,512
1977	19,281	17,335	6,754
1978	19,680	14,770	7,225
1979	17,912	14,290	9,158

places. Foxes, ravens, crows, and magpies may harass animals, especially fawns, but they rarely kill them. Whitetails have lived as long as 20 years in captivity but rarely live more than 10 in the wild.

White-tailed deer are heavily subject to the meningeal worm, *Parelaphostrongylus tenuis*; infestations sometimes reach 100 percent for animals in dense populations. Other parasites include liver flukes, tapeworms, lungworms, muscle worms, and nasal bots. The epizootic hemorrhagic disease is particularly common in this species. In addition, these deer are vulnerable to anthrax, brucellosis, leptospirosis, and salmonellosis. Common external parasites include ticks, mites, lice, and flies. Although nuisances, these organisms are not important causes of mortality. The cattle fever tick, *Boophilus annulatus*, also can live on white-tailed deer.

Odocoileus virginianus is certainly the most popular, and probably the most important, big game animal in North America. It long has played a significant role in the well-being of man. These animals were major sources of food and clothing both for Indians and for the early settlers in the United States. At present white-tailed deer are primarily objects of outdoor recreation; in 1974 more than eight million hunters harvested two million animals in the United States (see table 4 for statistics relating to the Northern Plains). The species is still an important source of food; in 1975, harvests of these deer produced 70.9 million pounds of boned, edible meat in this country alone.

Selected References. General (Dasmann 1971; Halls 1978; Menzel and Bouc 1975; Taylor 1956); distribution (Jones 1964; Turner 1974).

Genus *Alces*

The genus *Alces* includes a single species, which occurs both in North America and in the Old World. The dental formula is 0/3, 0/1, 3/3, 3/3, total 32. *Alces* is a Latin word meaning "elk."

Alces alces: Moose

Name. In the Old World this species is referred to as "elk" (see the discussion of names of *Cervus elaphus*). In all of North America, however, the name moose is used.

Distribution. The species is circumboreal in distribution, except for extreme northern Canada, Greenland, and Iceland. The distribution of the moose generally is associated with the presence of boreal forests. *Alces alces andersoni* is the subspecies found on the Northern Great Plains, where it occurs only in northeastern North Dakota and is considered rare (fig. 199). However, wanderers have been reported in recent years in southern Minnesota and even as far south as Iowa and northern Missouri.

Description. The moose is the largest member of the family Cervidae and is one of the largest terrestrial mammals in North America. The range of measurements for males includes: total length, 2,440–2,900; length of tail, 76–110; length of hind foot, 760–840; weight, 385–544 kilograms (up to about 1,200 pounds). Representative measurements for females are: total length, 2,040–2,590; length of tail, 90–122; length of hind foot, 725–810; weight, 329–385 kilograms (up to about 850 pounds). Ranges in cranial measurements are: greatest length of skull, 515–686; zygomatic breadth, 130–183. In addition to females averaging smaller than males, another sexually dimorphic characteristic is the extremely large antlers of males. The largest recorded measurements for moose antlers are: spread, 2,048; palm length, 1,422; palm width, 705; beam circumference, 635; greatest number of points on a single antler, 24.

The overhanging upper lip and the bell, or dewlap, are distinguishing characteristics of the moose. The pendulous dewlap of skin and hair, hanging from the throat area, may be as large as 15 inches in diameter and about 10 inches long on mature males.

Moose are generally blackish brown, but color variations range from black (especially some adult males in prime pelage) to pale brown and gray. Pelage of females is similar to that of males, but there is usually a patch of white hairs around the genital opening of females. Pelage of newborn animals is pale reddish to reddish brown, with gray to black on the chest, legs, muzzle, hooves, and ears and around the eyes. Guard hairs may grow to as much as 10 inches on the shoulder hump of Alaskan animals. Moose have an undercoat of fine, woolly grayish hairs.

Natural History. Moose are primarily browsers that feed on stems, bark, and leaves of many coniferous and deciduous trees as well as shrubs. Grasses, sedges, and aquatic plants are also eaten, especially during summer. These animals prefer early successional vegetation and therefore commonly are found in areas that have been recently cut over, burned, or otherwise manipulated. Food and climate apparently are the major influences on their choice of habitat. These animals winter successfully in areas of extreme cold and can travel and forage in deep snow. Areas with prolonged temperatures above 27° C (75° F) and lacking either shade or access to lakes and rivers do not harbor this species.

Moose are predominately solitary animals, with strong attachments to seasonally distinct home ranges that usually do not exceed eight to 15 square miles. Cows and calves are associated strongly for the first few weeks after calves are born and usually remain within a restricted area. This close association to a particular place probably enhances calf survival.

Visual signals of *Alces* are complex. Rutting bulls exhibit a repertoire of actions including swaying the head and body, hit-

197. *Moose,* Alces alces *(courtesy Montana Fish and Game Department).*

ting trees and shrubs, feeding while watching other bulls, fighting with other bulls, and driving cows and mating with them. During the rutting period, cows are aggressive toward each other and sometimes actively pursue rutting males. Communication between cows and calves usually consists of grunts. Bulls make croaking sounds during the rut; cows produce moaning sounds. Males emit a barklike sound during fights and chases.

The breeding season is from early September through November. Cows come into estrus about every 20 to 22 days; estrus lasts about 24 hours, although cows in heat are receptive for seven to 12 days. The gestation period for moose is approximately eight months, and most calving takes place in May and June. Pregnant females usually seek seclusion from other animals, and they give birth while lying down. Most cows have a single calf, but twins are common and triplets are born on occasion. The young weigh about 25 pounds at birth.

Young moose forage regularly by two weeks of age; they grow rapidly during the first five months. Sexual dimorphism in size does not become readily apparent until the animals are about three years old. Females can breed at approximately one and a half years of age and reproduce annually until they are about 18 years old. Most reproduction, however, is by animals that are from four to 12 years old. Bears, wolves, and man prey on the moose. Wolf predation is enhanced by heavy, crusted snow, and moose are also more vulnerable in winter because of lowered nutrition and concentration of individuals. Bears may take moose calves in spring.

Mortality by starvation is related mostly to depth of snow and persistence of adverse snow conditions over time. Deep snow affects the animals' mobility, the availability of food, and the related energy balance. Motor vehicles and trains cause numerous accidental deaths, because *Alces* is inclined to seek roads and similar flat surfaces for travel, especially in winter.

Moose are subject to a neurological disease caused by a roundworm, *Parelaphostrongylus tenuis*, which results in loss of coordination, weakness, and blindness and culminates in death. The white-tailed deer is the primary host for this parasite. Also, moose become infected with the liver fluke, *Fascioloides magna*, which is common in white-tailed deer, and ticks, *Dermacenter albipictus*, sometimes reach great densities on these animals. *Alces* is susceptible to brucellosis, especially in areas where animals are concentrated and in close contact with domestic livestock.

At present the total moose population in North America is estimated to be about one million. The annual harvest by hunting is about 90,000 individuals. In 1977 hunters took a total of 10 moose in North Dakota; 15 were taken each year in 1978 and 1979.

Selected References. General (Bedard 1975 and included citations); distribution (Bowles and Gladfelter 1980; Franzmann 1978); systematics (Peterson 1955).

Genus *Rangifer*

The genus *Rangifer* includes a single species, the caribou of North America and the reindeer of the Old World. The dental formula is 0/3, 1/1, 3/3, 3/3, total 34, although the upper canines are vestigial. *Rangifer* is a coined word meaning "reindeer" and refers to the wide-ranging nature of these animals.

Rangifer tarandus: Caribou

Name. The species name is from the Latin word *tarandus,* meaning "reindeer." In the Old World reindeer is the most widely used vernacular name for this species, but in North America the appropriate name is caribou, which is a French Canadian word of Algonkian origin.

Distribution. The species is circumpolar in distribution, except for the inland areas of Greenland. Formerly, caribou ranged as far south as central Idaho, the Great Lakes area, and northern New England in the United States, and into central Germany in Europe. The distribution of caribou is associated with tundra and taiga or other kinds of mature evergreen forest. Caribou were present in the vicinity of the Great Lakes until about 1935, but these animals probably were extirpated on the Northern Great Plains before then. At present the only *Rangifer* in the wild in the contiguous United States may be the small herd of only 20 to 30 animals still occupying the Selkirk Mountains in Idaho and adjacent British Columbia, although at least two individuals are known to have spent the winter of 1980–81 in northeastern Minnesota.

Rangifer tarandus caribou is the subspecies that once existed on the Northern Great Plains; it ranged across Canada from Newfoundland south of Hudson Bay and into the Yukon Territory, as well as in the Northern Rocky Mountains into the United States, and it barely reached North Dakota. The former distribution is not mapped.

Description. Caribou are unique among members of the family Cervidae in that both sexes commonly have antlers, although those of bulls generally are the larger. Antlers usually are not symmetrical; a brow tine forms a vertical, shovel-like structure over the face, and a bez tine forms a handlike structure above the brow tine.

Pelage varies from almost white to dark brown, with a white rump patch. In dark animals the venter usually is paler than the dorsum. Frequently there are patches of white on the neck, shoulders, lower legs, and flanks. Most animals have a fringe of white hairs on the ventral side of the neck.

Males are larger than females. Measurements for males from throughout the North American range of the species are: total length, 1,840–2,536; length of tail, 110–218; length of hind foot, 550–700; weight, 180–272 kilograms (up to about 600 pounds). Measurements for females are: total length, 1,369–2,030; length of tail, 100–180; length of hind foot, 381–600; weight, 91–136 kilograms (up to about 300 pounds). Ranges in cranial measurements of *Rangifer* are: greatest length of skull, 295–435; zygomatic breadth, 110–149.

198. Caribou, Rangifer tarandus *(courtesy U.S. Fish and Wildlife Service).*

Natural History. Caribou are rather opportunistic feeders; they consume a wide variety of plants, including leaves, buds, and bark of deciduous and evergreen shrubs and trees, grasses, sedges, horsetails, and a wide array of green vascular plants, as well as mushrooms and both terrestrial and arboreal lichens. In summer the animals shift their ranges to find new plant growth, and they seem to follow receding snowfields in search of newly sprouting plants. In winter caribou dig holes in the snow with their hooves to uncover food. To find food, they congregate in places with little snow cover, south-facing exposures, and windswept areas. However, individuals frequently walk on top of crusted snow to browse on woody plants and lichens. The major habitat requirements for *Rangifer* include sources of food during periods of deep snow in winter and escape from flying insects in summer; these needs generally influence the locations in which these animals are found.

Caribou are highly gregarious and nomadic. They move almost continuously, usually in large groups. There is some seasonal segregation of herds by sex. During winter, herds consist of separate male and female-young groups. In early spring these units include pregnant females, a few males, and young animals, but groups of adult bulls are segregated. In late spring females become scattered during calving, but afterward they congregate in huge groups of females, calves, and yearlings. In late summer herds are made up of scattered small groups, but in autumn males and females form large mixed herds.

The breeding season is mostly during October. Females come into heat at intervals of 10 to 12 days. Calves are born in May and June after a gestation period of about 227 to 230 days. Cows usually bear single calves, though twins are produced infrequently. Calves are able to follow their mothers within an hour after birth. In Alaska, about 51 to 55 percent of the calves are males. In some herds, calf mortality is as high as 25 percent before the animals attain one month of age. Among adults, males make up about 40 percent of most populations.

In areas where caribou still range freely, calves may die from an array of accidental causes, such as trampling, wind chill and exposure, and desertion. Predation by

wolves, lynx, and grizzly bears also is an important cause of death in calves. Studies in Alaska show that *Rangifer* in small groups are much more susceptible to predation than animals in large aggregations. Accidental mortality in adults may be caused by drowning and by starvation, especially during severe winters.

Like moose, caribou are subject to the neurological disease caused by *Parelaphostrongylus tenuis*. Infestation by this parasite usually results in death. Some authors believe that the northward spread of white-tailed deer (the primary host of this parasite), in response to clearing of forests, may have been the major cause of the decline of caribou on the Northern Great Plains and areas adjacent to the Great Lakes.

Wherever they occur, caribou are hunted heavily by man. Most caribou populations probably are below desirable densities, mostly because of overharvesting.

Selected References. General (Bergerud 1978; Skoog 1968); systematics (Banfield 1962).

Family Antilocapridae: Antilocaprids

The pronghorn is the sole living survivor of the New World family Antilocapridae. The fossil record of this family dates back to the Miocene, and many genera were abundant during the Pliocene. The distribution of the family appears to have been greater in the Pliocene and Pleistocene than in Recent times; fossil records are known from Wisconsin, Illinois, Nebraska, and Florida and from California, Oregon, and other western states. Once numbering an estimated 30 to 40 million head in North America, pronghorns were driven to near extinction at the turn of the previous century. Mostly because of conservation efforts and reintroductions, the total population of pronghorns in the United States now probably numbers about 400,000 animals. This

animal inhabits the dry grassland areas of the western United States, southern Canada, and northern Mexico. It is well adapted for rapid travel on hard ground, being one of the fastest of land mammals. Recent biochemical evidence suggests that pronghorns are closely related to bovids, and some authorities regard them as a subfamily of that group.

Genus *Antilocapra*

The name *Antilocapra* is derived from the Greek word *antholops* (from which the English word antelope derives), meaning "horned animal," and the Latin *capri*, meaning "goat." The name is a misnomer,

however, because both true antelopes and
goats represent distinct subfamilies in the
family Bovidae. The dental formula is 0/3,
0/1, 3/3, 3/3, total 32. *Antilocapra* is a
monotypic genus, represented only by the
pronghorn, *Antilocapra americana.*

Antilocapra americana: Pronghorn

Name. The specific name *americana* is in
reference to the North American distribution of the pronghorn. In the early literature this animal often was referred to in
the vernacular as "antelope," a term hand-
ed down through many generations, resulting in its common usage today (sometimes
as "prong-horned antelope"). However,
pronghorn is the preferred name, the antelopes proper being an Old World group.

Distribution. This ungulate once was found
in much of the dry prairie country west of
the Mississippi River. Its range formerly
extended northward into the central prairie
provinces of Canada, south onto the Mexican Plateau, and into western Mexico and
Baja California. Before its decimation by
man, the pronghorn was found throughout
most of the Northern Great Plains, but it
now has a limited range resulting from
protection and from reintroductions since
about 1920.

Populations of pronghorns on the Northern Plains were extremely low around
1900, at which time there were thought to
be only 17,000 remaining in all the United
States. Only 680 of these animals remained
in South Dakota in 1911, when they were
declared illegal game; North Dakota out-
lawed hunting of them in 1899. In 1908 the
pronghorn was nearly extinct in Nebraska,
and only 187 individuals were known to
exist there in 1925. Recovery has been
good, however, and all three states have
had hunting seasons on pronghorns since
the 1950s (table 5), albeit restricted in the
past several years.

Pronghorns of the Great Plains are classified in the subspecies *Antilocapra americana americana,* the range of which
extended from Manitoba and Alberta south-

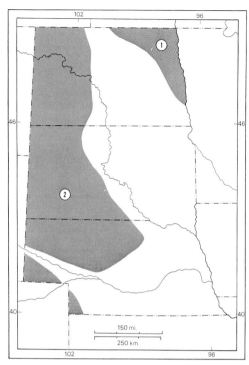

199. Distribution of (1) Alces alces *and*
(2) Antilocapra americana *(present range).*

ward to California and the panhandle of
Texas.

Description. This animal is one of the
smaller artiodactyls in North America,
standing about three feet at the shoulder.
Males average about 115 pounds and
females average about 90 pounds. Weights
vary seasonally. Measurements of males
are: total length, 1,245–1,472; height at
shoulder, 860–875; length of tail, 89–178;
length of ear, 142–149. Representative cranial measurements of animals from the tristate region include: greatest length of
skull, 270–289; zygomatic breadth,
127–143.

The pelage is generally tan, the neck has
a black mane, and the underparts and rump
are white. The hairs on the back and rump
can be erected, serving as a warning signal
to other members of a herd in case of
danger. Two white bands across the throat
alternate with black or tan markings, and

males have black markings on the snout. The feet lack the lateral digits present in other Northern Plains artiodactyls.

Horns are present in both sexes and are composed of fused hairs. They surround laterally flattened horn-cores that are unbranched and permanent. The branched horn-sheaths of males are shed after breeding, which seems to indicate a relationship between the sexual cycle and horn development. Shedding of branched or simple horn-sheaths of females is far less regular, however.

Natural History. Pronghorns are probably best known for their great speed, which has been clocked at 60 miles an hour on flat ground. Although they can maintain highest speeds only for a half mile or less, they can easily cruise at 30 miles an hour for as much as seven miles. Their legs are slender; cartilaginous pads protect their feet from rock abrasion and reduce noise.

Pronghorns are mainly browsers, sagebrush and various forbs making up the bulk of their diet. Sagebrush is the most important food item, particularly in winter. Although unpalatable to cattle, it has a protein content as high as that of alfalfa meal, and more carbohydrate and fat. In a study in South Dakota, stomach analyses indicated that ingested food was composed of two-thirds browse species, one-sixth

forbs, one-eighth crops, and less than one percent grasses. Because grass composes so little of the diet, removing livestock from overgrazed areas results in recovery of grasses despite the presence of large herds of pronghorns.

The principal predators are coyotes and bobcats, which prey primarily on the young. Golden eagles also prey on kids. Ravens sometimes pluck out the eyes of young pronghorns, causing blindness or death, and magpies bother the animals by pecking at them. Groups of pronghorns and sometimes individuals (usually females) have been reported to chase off potential predators.

The primary losses to populations on the Great Plains at present are due to illegal kills by man, especially during the tourist season. As access to the herds improves and the number of humans increases, more illegal kills can be expected. Accidental deaths occur in pronghorn populations; herds are known to stampede off cliffs for unknown reasons, and casualties also are related to deep snowfalls and to miring in mud.

Epizootics strike pronghorn herds in mild winter weather on occasion, and extremely serious instances of this were reported in 1873 and around 1900. As a result of the 1873 epizootic, three-fourths to nine-tenths of the animals between the Yellowstone and Missouri rivers are thought to have perished.

Pronghorns usually are involved in either resting or feeding. Feeding most commonly occupies the morning and evening hours, and resting tends to be more common near midday. These animals are gregarious, especially in winter, when all age classes band together; such groups tend to be composed of fewer than 100 individuals, but before numbers were reduced by man congregations of 1,000 or more were known.

In spring and summer, pronghorns separate into bachelor herds and female-kid groups. In late summer and early autumn, the orderly hierarchy that exists in bachelor herds breaks down, and males associate with the female-kid groups during the rut-

Table 5: Pronghorns Harvested on the Northern Great Plains, 1970–79

Year	North Dakota	South Dakota	Nebraska
1970	1,807	3,810	1,329
1971	1,486	5,450	972
1972	1,132	6,370	857
1973	1,378	6,830	1,145
1974	1,816	8,540	1,317
1975	1,168	10,330	1,470
1976	1,225	6,720	1,366
1977	1,194	7,590	1,335
1978	No season	4,820	1,137
1979	No season	2,470	554

ting season. Male pronghorns studied in Montana were found to be territorial during the rut (September and the first week of October). They defended discernible areas against other males by ritualized displays and occasional fights. These areas were almost always defended successfully, intruding males being driven out in 202 of 204 encounters. The boundaries of the territories appeared to be determined by the presence of a standing male, by marking of vegetation with the postmandibular scent glands, and possibly by the "laugh" of the pronghorn, which consists of descending sneezelike sounds.

Studies have shown that undisturbed pronghorns stay within a single square-mile area. If disturbed, however, they will move as much as a mile away. These animals are slow to invade new areas, and when forced from their native range they are slow to return. This pattern tends to be true even when the new habitat is atypical, such as areas occupied by pronghorns in Yellowstone Park. Because of this behavior, restocking must be used to enlarge the present distribution of the species.

Pronghorns have keen eyesight, but they approach strange objects with curiosity, and individuals often will not take flight until they obtain a scent. Flight distances decrease in heavy rains but increase in strong winds, light rain, or fog. For females with kids, typical flight distances are 300 to 500 yards, whereas only 200 to 300 yards is typical for bachelor herds.

Pronghorns carry out comfort movements of several types. They lick themselves and their offspring, rub their horns against fence posts, and often use a hind foot to scratch the head, neck, or shoulders. To ward off insects they wiggle their ears, shake their heads, flip their tails, stamp their feet, and flex their skins.

200. *Pronghorn,* Antilocapra americana *(courtesy D. Randall).*

The pelage of the pronghorn is shed continuously, but the most rapid change of coat is between March and May. The coat is in prime condition by the middle of July. The hairs are brittle and often are torn from the hide when an animal is shot while running or when, as is their habit, individuals negotiate barbed-wire fences by diving under the lowest strand.

The gestation period is 230 to 250 days, usually from early October until mid-June. In Oregon the sex ratio of 300 young in one study was 45 males to 55 females, which is similar to sex ratios in adult herds in that area. The young are usually twins, but occasionally single births occur. They average about 7 pounds when born. In giving birth, the female pronghorn actually "drops" the kids, which is also common in sheep and goats, in that she stands during expulsion of the young, allowing them to fall to the ground. It is thought that the shock of

falling stimulates breathing in the newborn. The female feeds her kids every 90 minutes or so at first, but time between nursing periods gradually increases. The kids are well camouflaged and often are left in a hiding place while the mother feeds. By the age of 10 days the kids can outrun an average dog and often play with other kids in the female-kid bands. They still rest in concealment at this stage, however.

The pronghorn is of considerable economic importance in that it is a popular game animal and tends to improve the range by, among other things, reducing sagebrush growth. Its continued well-being depends on the condition of preferred habitat, wise management policies, and tolerance on the part of private landowners.

Selected References. General (O'Gara 1978 and included citations; Suetsugu 1977; Yoakum 1978 and included citations).

Family Bovidae: Bovids

This family includes the bison, muskox, mountain goat, and sheep native to North America, as well as the buffaloes, antelopes, gazelles, goats, sheep, and cattle of the Old World. The fossil record dates back to the early Miocene in Europe, with most of the radiation in North America taking place during the Pleistocene. There are about 44 Recent genera, with approximately 125 species; these generally are arranged taxonomically within five subfamilies. The group has been introduced in many parts of the world, especially Australia and New Zealand, where there were no native bovids. Members of this family occur virtually worldwide in domestication, which first began in southwestern Asia about 8,000 years ago.

Some species prefer rocky, mountainous areas, and some occur in forests and swamps, but most bovids inhabit grassland, scrub, savanna, desert, and tundra. All members of this family are primarily grazers that feed by twisting grasses around

the tongue and breaking them off against the lower incisors.

Three species of bovids occur on the Northern Great Plains. Of these, two (bison and mountain sheep) were extirpated and subsequently reintroduced, and one (mountain goat) was introduced from the Rocky Mountains.

Genus *Bison*

The genus *Bison* includes the bison of North America, as well as the woodland bison, or wisent, of northern Europe. The dental formula is 0/3, 0/1, 3/3, 3/3, total 32. *Bison* is a Greek word referring to an oxlike animal.

Bison bison: Bison

Name. Vernacular names applied to this species include bison, American bison, and buffalo. *Bison bison bison* (plains bison) is the subspecies that once inhabited all of the Northern Great Plains.

Distribution. The former range of the bison extended from Alaska and the Northwest Territories along the Rocky Mountains, including a westward extension into the prairie region of eastern Oregon, to the Gulf of Mexico and eastward to the Appalachian Mountains and, in places, the Atlantic Coast. These animals occurred mostly on plains but ranged into grasslands of the foothills and even mountainous country.

When European man arrived in North America, large herds of bison numbered in the hundreds of thousands, and the total population on the continent may have been about 70 million. The bison had been completely extirpated east of the Missouri River before 1870. Large herds were known from west of the James River in South Dakota as late as 1866, and the species occurred in the area between the Grand and Cheyenne rivers until about 1869. Occasional animals were reported on the Plains in the vicinity of the Black Hills until 1877. The latest known kill of a wild animal in Nebraska was in 1878, and the species certainly was gone from North Dakota by 1884. By 1889 the total population in North America had been reduced to an estimated 1,000 or fewer individuals. Fortunately, a few semicaptive herds were preserved, and bison remain today in Canada and the United States in herds maintained by both public and private organizations, special interest groups, and individual ranchers. The distribution is not mapped.

A nucleus herd of seven bulls and seven cows was reintroduced on the Black Hills in 1913 with funds provided by the National Bison Society. Two bulls and four cows from Yellowstone National Park were added to the herd in 1916. At present, about 350 animals occur in Wind Cave National Park. The Custer State Park bison herd, which now numbers about 1,600, was reintroduced in 1914. With funds provided by the South Dakota State Legislature, 36 individuals (six bulls, 18 cows, 12 calves) were used to restock the park. An additional 60 bison from the Pine Ridge Indian Reservation, Rosebud, South Dakota, were obtained by the state park in 1951. The largest herd of bison in Nebraska occurs in the Fort Niobrara National Wildlife Refuge.

Of those wild bison obtained from the Northern Great Plains, it seems that few, if any, museum specimens were preserved. This is of considerable interest in view of the hundreds of thousands of animals that were slaughtered by man in the region.

Description. The bison is generally dark brown, with massive forequarters. The head, forelegs, shoulders, and front part of the back are covered with hairs longer than those on the rest of the body. Hairs on the back are short and woolly. As the pelage ages, the long hairs on the forequarters become pale, especially in old bulls. Calves are reddish tan at birth, changing to the brownish black adult coat at about three months of age. Adults molt once a year, in spring and early summer.

There is considerable sexual dimorphism in size. Representative measurements for males are: total length, 3,050–3,800; length of tail, 550–815; length of hind foot, 585–660; weight, 725–910 kilograms (up to about 2,000 pounds). Measurements for females are: total length, 1,980–2,280; length of tail, 430–535; length of hind foot, 460–555; weight, 410–500 kilograms (up to about 1,100 pounds). Cranial measurements include: greatest length of skull, 491–570; zygomatic breadth, 242–290. Both sexes have permanent horns. Bulls generally have a high and sharply angled shoulder hump that is larger than that in females. Length of the ear has been recorded as varying from about 120 in females up to about 150 in males.

Natural History. Bison are predominantly grazers, with a special preference for warm-season grasses. They will consume sedges in some areas and, when food is scarce, browse on plants such as sagebrush. It seems that these herbivores have considerable ability to digest low-protein, poor-quality forage; thus they are better able than cattle to exploit the short-grass plains. Bison use forested areas when available,

201. Bison, Bison bison *(courtesy J. L. Tveten).*

especially for shade and to escape from insects. Also, they forage successfully in forested areas during periods of heavy, deep snow. These animals can reach food beneath several feet of loose snow by a side-to-side head-swinging movement. They apparently require daily access to either free water or snow.

Bison are remarkably gregarious. Cows, calves, and other young animals (yearlings and two- to three-year-old bulls) remain in herd groups throughout the year. A basic herd unit of 11 to 20 animals is common, with larger herds formed by temporary bunching of such smaller units. Throughout most of the year, bulls remain either solitary or in small groups. In mid- to late July, the bulls begin to mix with the cow and calf herds, and mating is completed by late August and early September. Bulls do not maintain harems but are polygynous in that they pair temporarily and breed a number of cows sequentially.

Most cows breed at three years of age.

Males also normally are mature at three years old, but most breeding is done by fully mature males (seven or eight years old). The gestation period is about nine and a half months. Cows usually separate briefly from the herd to calve. In Yellowstone National Park, most calves are born in May, with occasional births later in the season. One calf is usual, but twins are not uncommon. Newborn young may weigh 20 pounds or more and are able to follow their mothers within several hours after birth.

Studies of fetal sex ratios reveal that males slightly outnumber females. In Yellowstone National Park, females are more numerous than males in the age range of 10 months to three and a half years. Thereafter, male survival predominates until the animals are fully mature, when females again outnumber males. Longevity varies considerably, but apparently more cows than bulls live to old age. Cows occasionally produce young until about 28 years old. Bulls rarely live more than 20 years.

Communication between cows, calves, and other members of the herd consists mostly of nasal grunts and snorts. Roars and bellows of rutting bulls are the most distinctive sounds. Bulls also snort and stamp. Visual signals of bison include head movements, broadside displays, turning and moving away, and grazing.

Predators on wild bison included grizzly bears, wolves, and man. Instances of predation by bears and wolves must have been opportunistic and infrequent. The physical structure of the bison, with massive forequarters, and its stolid temperament that led it to face danger rather than flee, made the animals vulnerable to man, especially when he was armed with a modern rifle.

Bison are subject to tuberculosis, brucellosis, and anthrax, and these diseases act to regulate most herds. External parasites, especially biting flies, influence movements and distribution of bison in some seasons of the year. Accidental mortality by drowning in rivers and starving when there is extensive crusting of snow is apparently a minor factor in population regulation. Severe weather in late spring can cause death among newborn calves.

Modern bison herds are managed by culling and selective harvesting. Excess animals are removed for slaughter and live sales. Occasionally, hunting permits are issued to remove bison from the semicaptive herds on the Northern Great Plains.

Selected References. General (Meagher 1973, 1978); distribution (Bailey 1927; Jones 1964; Turner 1974); systematics (Skinner and Kaisen 1947).

Genus *Oreamnos*

The monotypic genus *Oreamnos* includes only the mountain goat of North America. The dental formula is 0/3, 0/1, 3/3, 3/3, total 32. The generic name is from the Greek words *oros*, "mountain," and *amnos*, "lamb."

Oreamnos americanus: Mountain Goat

Name. The specific name alludes to the American distribution. Mountain goat is the vernacular name that is uniformly applied to this species. These are not true goats, however, but belong to a group that is better thought of as "mountain antelope."

Distribution. O. americanus ranges from the Rocky Mountains and the Cascades of the Pacific Northwest across western Canada, and into the Yukon Territory and southeastern Alaska. The species is not native to the Northern Great Plains.

In February 1924 six mountain goats (four females, two males) that were trapped near Banff, Alberta, were placed in an enclosure in Custer State Park, in the Black Hills of South Dakota. During the first night, a female and a male escaped. In 1929 the remaining captive animals (which then numbered eight) escaped when a tree fell across the fence of the enclosure. All the animals took up residence in the rough country around Harney Peak. This area and the adjacent Needles (a total of about 2,080 acres) is the current primary range of mountain goats in the area, although the total range on the Black Hills includes approximately 32,000 acres. By 1949 the herd on the Black Hills had increased to about 300 animals, and at present it is estimated at 300 to 400 head. Seemingly, the number of individuals and the extent of their range has been static for the past 30 years, which probably indicates that all suitable habitat has been occupied. Some mountain goats were transplanted to Spearfish Canyon, north of Savoy, about 25 years ago, but none has been seen there since, and it is assumed that this translocation failed.

Oreamnos americanus missoulae, which occurs in the Rocky Mountains of southern Canada and the adjacent United States, is the subspecies that was introduced on the Black Hills. This distribution is not mapped.

Description. The mountain goat is the only artiodactyl in North America that has scent glands behind the bases of the horns, which is true of both sexes, and is the only one in the 48 contiguous states that is whitish in color. The pelage is made up mostly of all white wool and guard hairs, although some

202. *Mountain goat,* Oreamnos americanus
(courtesy C. B. Rideout).

dark brown hairs may be found on the back and rump. There are shorter hairs on the lower legs and face than on the rest of the body. Adults have a tuft of hairs beneath the throat that is about 130 mm long in winter. Both sexes have black horns from 200 to 305 mm in length. The horns of adult males are thicker at the bases and taper more from base to tip than those of adult females. External measurements of males are: total length, 1,245–1,787; length of tail, 84–203; length of hind foot, 300–368; weight, 46–136 kilograms (up to about 300 pounds). Cranial measurements of males include: greatest length of skull, 299–336; zygomatic breadth, 100–113. Females generally are 10 to 30 percent smaller than males in linear measurements and weight.

Natural History. Mountain goats inhabit high, rocky areas with steep slopes and cliffs. In the Northern Rocky Mountains, the range of the species includes alpine tundra and subalpine areas. These habitats are characterized by low temperatures, heavy snowfalls, and high winds. On the Black Hills, *Oreamnos* inhabits areas of granite ridges with abundant growths of mosses and lichens, the most rugged terrain

on the Northern Plains. Mountain goats are excellent climbers, and they frequent places that are not accessible to other animals of the same size. This habitat preference may result in protection from predators and lack of competition for food.

O. americanus consumes a broad variety of foods. Studies in Alberta showed that 77 percent grasses and 23 percent willows are consumed in summer months. In Montana, grazing provides four percent of the summer diet, and browse makes up the other 96 percent. On the Black Hills, diets in winter are made up of 60 percent mosses and lichens, 20 percent bearberry, 10 percent pine twigs and needles, and 10 percent miscellaneous browse, grasses, and weeds. In the habitats frequented by mountain goats, snow accumulations are great on north- and east-facing slopes, where snow, water, and succulent forage occur during the summer. As a result of limitations due to wind and sun, food mostly is available on steep south- and west-facing slopes in the winter months. During severe weather, *Oreamnos* seeks the protection of caves and overhanging ledges; in the Black Hills region the animals also use abandoned mines. Mountain goats usually occur in groups of two to four in summer. Larger groups form in winter, especially on bedding grounds during severe storms, and at sources of salt in the summer. Females move to isolated places in areas of steep cliffs to bear young.

Breeding can begin when animals are two years old. Mating occurs from early November through December. The gestation period varies from 147 to 178 days. Kids are born in late May and early June. Single births are the rule, twins occur frequently, and triplets are rare. Kids are precocial and follow their mothers from shortly after birth until the next year's parturition. Kid-to-female ratios reported for mountain goats include: Montana, 77:100; Idaho, 22:100 to 79:100; Alberta, 45:100; South Dakota, 86:100. Accurate data on sex ratios are lacking because of difficulties of access to the goats and of determining sex in the field. Longevity records include a female 18 years old and a male of 14 years, but researchers seldom have reported mountain goats more than 11 years old in the wild.

Daily movements range from a few hundred yards up to several miles. In August animals move about 20 times farther than in January. Peaks of activity usually occur at dawn and dusk, but some activities take place at all times of the day and night. These goats migrate to wintering areas with the first heavy snows. Winter ranges are smaller and separate from the areas usually occupied in summer.

Communication between females and kids consists of bleats, especially when they become separated from each other. Males emit warning snorts and bleats, as well as low-pitched grunts and buzzing or humming sounds. These goats rarely butt heads when fighting. Common visual signals include displaying the horns, rushing toward another animal, making swipes with the horns, staring intently at one another, and broadside displays termed "present threats." From the position of the broadside displays, blows with the horns are made to the flanks and rump; these may result in injury or death, especially in females during the mating season.

Predators on mountain goats include mountain lions, bobcats, coyotes, grizzly bears, black bears, and eagles. Mountain lions and golden eagles are the most significant predators; the former can kill goats, and the latter can carry off small kids or knock animals from cliffs. Accidental mortality by starvation is related mostly to the severity and length of winter conditions. Animals sometimes slip from snow- and ice-covered ledges and fall to their deaths, and they occasionally are trapped by getting their legs caught between rocks or logs.

Mountain goats are subject to parasitism by helminths and nematodes, especially *Protostrongylus stilesi*, *P. rushi*, and *Nematodirus maculosus*. These parasites are extremely common in goats on the Black Hills and may be a major limiting factor to that introduced population. Infestations of wood ticks and biting lice also are common in *Oreamnos*.

The first hunting season on the Black Hills was held in 1967, and 25 mountain goats were harvested. During the next four seasons, a total of 105 permits was issued, and 94 animals were taken. In 1979 hunters obtained 20 mountain goats (10 females, 10 males) from the Black Hills. The meat is generally unpalatable, and hunting therefore is for sport and trophies.

Selected References. General (Hanson 1950; Richardson 1971; Rideout 1978; Rideout and Hoffmann 1975 and included citations); distribution (Turner 1974); systematics (Cowan and McCrory 1970).

Genus *Ovis*

The genus *Ovis* includes all the wild and domestic sheep. The dental formula, as in all bovids, is 0/3, 0/1, 3/3, 3/3, total 32. *Ovis* is a Latin word meaning "sheep."

Ovis canadensis: Mountain Sheep

Name. The species name refers to Canada. Other vernacular names include bighorn sheep, bighorn, and Rocky Mountain bighorn sheep. Names occasionally assigned to the subspecies that formerly occurred in the tristate region include Audubon's bighorn and badlands bighorn.

Distribution. The former range of the mountain sheep extended from the Canadian Rockies and Montana south to Mexico and Baja California, including mountainous country, as well as foothills and even plains. Populations on the Northern Great Plains were entirely eliminated by man by the early 1920s, earlier in many areas. The former distribution is not mapped.

Early reports indicate that the mountain sheep originally was found more commonly in foothill regions and even river valleys, far from mountain ranges, than in the high mountain country that typifies its range today. In some areas, populations migrated seasonally from the mountains to winter in the lowlands. It appears that lowland bands were unable to compete with domestic livestock or to escape hunting pressure. Foothills herds appear to have been forced back into areas of mountain wilderness much as in the case of the wapiti, the plains populations being extirpated for the most part.

The subspecies *Ovis canadensis auduboni* originally occupied eastern Montana and Wyoming and western parts of North Dakota, South Dakota, and Nebraska, but it is now extinct. Mountain sheep have been reintroduced into several areas in the western part of the tristate region, and the introductions appear to have been successful. In 1923 mountain sheep from the Rocky Mountains (subspecies *Ovis canadensis canadensis*) were moved into an enclosure in Custer State Park, South Dakota, and others were added in 1965; this herd now numbers from 50 to 70 animals. In 1961 12 mountain sheep from Alberta escaped from a pen in the Slim Buttes area and fostered a free-living herd in Harding County, South Dakota. Twenty-two more sheep were released in Badlands National Park in 1963.

The first reintroduction in North Dakota took place in 1956 when 18 animals from British Columbia (subspecies *Ovis canadensis californiana*) were moved into an enclosure on Magpie Creek in McKenzie County. This and 14 subsequent transplants into four other badland areas have resulted in a total population of about 250 animals in the state at present. No sheep have yet been released in Nebraska, but a few free-living individuals have been observed in the far western part in recent years, presumably immigrants from the north or west, and a small herd is scheduled to be moved to Fort Robinson on the Pine Ridge within the next few years.

Few museum specimens of *O. c. auduboni* exist, and the validity of its classification has been questioned. Some investigators have commented that this subspecies, because it is based on slight differences in cranial characters in relatively few specimens, should not be recognized.

Description. The mountain sheep is grayish brown to dark brown, somewhat dependent on season, with a grayish white muzzle and

203. *Mountain sheep,* Ovis canadensis *(courtesy E. P. Haddon).*

rump. Brown ridged horns curve backward in both sexes, but they form a massive spiral in adult males. Adults molt only once a year, in summer, when the old, matted, and faded pelage is replaced by a glossy new coat.

Representative ranges of external measurements of males and females of the subspecies *O. c. canadensis* are, respectively: total length, 1,600–1,953; 1,369–1,490; length of tail, 95–150, 80–121; length of hind foot, 394–482, 370–420; length of ear (females only), 112–127. A recent study of mountain sheep four years of age and older in the Canadian Rockies showed weights of 160 to 265 pounds for males and 117 to 200 pounds for females. The mountain sheep from the badlands is recorded as having been equal in size to, or larger than, the other subspecies, males reaching 344 pounds and females weighing as much as 240 pounds. Cranial measurements for males from North Dakota are: greatest length of skull, 284–318; zygomatic breadth, 151–181. Adult males have horns that may measure as much as 18.5 inches in circumference at the base, 44 inches in length, and 26 inches from tip to tip.

Natural History. The extirpation of the mountain sheep on the Northern Plains can be attributed to extensive hunting with rifles, both by white men and by Indians. Wolves and coyotes were still abundant at the time and perhaps hastened the rapid decline in numbers. Last known records from the three states include: 1905, on the hills along Magpie Creek in North Dakota; 1918, near Scottsbluff in Nebraska; and 1924, in eastern Pennington County, South Dakota.

In Idaho the diet of mountain sheep is made up primarily of grasses and forbs, with the major emphasis on grasses and grasslike plants. It was reported that the December diet is 83 percent grasses, 10 percent forbs, and seven percent shrubs. Browse may make up 40 percent or more of the diet in winter but frequently constitutes a smaller portion of the summer intake. However, a study in North Dakota revealed approximately 90 percent browse in the diet in both summer and winter, with buffalo berry the dominant browse species in summer and winter fat in other seasons. These animals use their hooves to dig down through the snow in search of food. They seldom have been known to drink free water, but in winter and spring they obtain moisture by eating snow.

The movements of mountain sheep seldom adhere to a stereotyped pattern, being erratic for the most part. In their feeding and directional movements, the sheep rarely follow a particular leader. Adult males and females, however, are more independent in their actions than are younger sheep and therefore determine the direction of travel. Herds lack cohesiveness and often break up when two adults range in different directions for food. Ewes generally tend to remain apart from rams in summer, occupying somewhat steeper terrain. This separation is never great, but there do seem to be bachelor herds and female-kid groups.

The large spiral horns of the mountain sheep serve as broad areas of contact in the jarring encounters during the rutting season. Some investigators have proposed that these horns serve another function—as display organs that advertise social rank. This conclusion was supported by data indicating that large-horned rams dominated those with small horns, that large horns conveyed dominance despite age, that strange rams with large horns entering a new herd

were dominant, and that large-horned rams did virtually all the breeding. The advantages of this outward rank symbol are that it allows predictability of social relationships and hence the smallest possible waste of energy in rank orientation. But it may be that these huge horns, which make up as much as 10 to 13 percent of a male's total weight, shorten life expectancy.

In their rutting behavior, males differ from those of other ungulates in that they do not establish or defend territories and guard no harems. When a female is in estrus a large ram guards her, chasing smaller rams away. Subordinate males are treated like females, and a male, upon establishing dominance, often mounts the other male.

Puberty in the mountain sheep is reached in two and a half to three years, but owing to social structure males may not breed until they are considerably older. The breeding season normally extends from mid-November to mid-December. The gestation period is about 180 days, lambs being born in late May and June. Twins are rare, single births being the rule.

Young lambs feed at 30-minute intervals, but feedings become less frequent as they grow older. They generally are weaned by the age of six months, although this varies considerably. During the first few days of life, a lamb follows its mother closely in what appears to be a period of imprinting. Females with young lambs join together into nursery bands in which the lambs are free to play and interact with their equals. Occasionally a few mothers leave their lambs in these groups while they go off by themselves, but they never hide their kids as pronghorns do.

Predators on mountain sheep include bobcats, coyotes, wolves, cougars, and eagles, all of which typically prey on the young. The present distributions of all these predators are much reduced, however, and the incidence of predation on present-day herds depends on the types and abundance of predators found locally. Mountain sheep utilize a group formation on occasion when opposing a predator. Ten rams banding closely together to charge a coyote has been reported; seven sheep were observed to run downhill from a bobcat, then turn and chase it back up the hill. When confronted alone, these animals will lash out at a predator with their front hooves.

Ovis canadensis is considered a prize big-game animal, but populations are so low in the United States that at present the species is protected in most areas. The animals have been displaced from many parts of their former range and destroyed in others. Success of the species ultimately will depend on the preservation of remaining wilderness areas, as well as on intelligent

programs of trapping individuals and transplanting them into areas of suitable habitat.

Mountain sheep are hunted to some extent on the Northern Great Plains. In North Dakota, for example, 12 were harvested in 1975, 12 in 1976, and nine in 1977. A few also have been taken in Custer State Park, in the Black Hills of South Dakota, in recent years.

Selected References. General (Geist 1971; Wishart 1978); natural history (Welles and Welles 1961); distribution (Jones 1964; Turner 1974); systematics (Cowan 1940).

Introduced Species

Five species of mammals not native to North America have been taken in the wild on the Northern Plains and are treated in this section. One, the coypu, is a native of South America; the others were introduced to this continent from Eurasia. Only two of the five species, the two murid rodents, are widely distributed in the tristate region and thus likely to be encountered by most users of this book. Domesticated mammals imported by man are not treated here, although occasional individuals of some taxa (dogs and house cats, for example) may exist in the feral state.

Insofar as we are aware, only one North American species not indigenous to the Northern Great Plains has become established there—the mountain goat (*Oreamnos americanus*), which was introduced onto the Black Hills some years ago. Because it is of North American origin, it is treated along with other native Artiodactyla. However, stocks from elsewhere of some species native to the Dakotas and Nebraska have been released in the region, frequently by conservation officials in an attempt to reestablish them where they once occurred.

All five species listed below are treated in the appropriate keys in the foregoing text.

Family Leporidae

Oryctolagus cuniculus: European Rabbit

A specimen of *O. cuniculus* in wild-type pelage was taken in northeastern Nebraska in 1958. We have no other information concerning this species on the Northern Plains, and it is extremely doubtful that it has become established there. However, individuals of domesticated strains, the so-called Belgian hare, probably are released or escape from captivity from time to time.

The European rabbit is smaller than native species of *Lepus* but somewhat larger than *Sylvilagus*. The skull resembles those of cottontails in that the interparietal bone is fused with the parietals, but it is larger. External and cranial measurements of the specimen from Nebraska, a male, are: total length, 514; length of tail, 52; length of hind foot, 110; length of ear, 79; greatest length of skull, 92.5; zygomatic breadth, 43.5; weight, 5 pounds.

Family Muridae

Mus musculus: House Mouse

The house mouse is a common, often abundant resident in and near the habitations of man throughout the Northern Plains. Feral populations occasionally are encountered, especially in the eastern part of the region. This species breeds throughout much of the year, and females in laboratory strains are capable of producing as many as 14 litters annually, although the usual number is considerably less in the wild. Litters contain three to 12 young, but most commonly there are five to eight.

Because of the cold winters on the Northern Great Plains, house mice living in the wild adjacent to human habitations have a tendency to move inside with the advent of cold weather and to move out again in spring. Unwanted guests easily can be trapped using the familiar break-back trap baited with cereals or grains, cheese, nuts, or meat. Concentrations in outbuildings, seasonally used facilities, and the like best can be controlled by the careful use of poisoned baits, which are available commercially. A pet cat or two on the premises also aids in control of these mice.

M. musculus sometimes is confused with species of the native cricetid genera *Reithrodontomys* and *Peromyscus*, especially when immature specimens are at hand. From these species, the house mouse differs in having nearly naked ears, a scantily haired tail bearing conspicuous scaly annulations, a grayish or dark buffy (rather than whitish) venter, three longitudinal rows of tubercles on the molars, and a distinct notch on the occlusal surface of each upper incisor (seen best in lateral view—fig. 108). It also differs from *Reithrodontomys* in lacking a groove on the face of each upper incisor.

Average and extreme external measurements of a series of 10 adults from western Nebraska are: total length, 173.4 (166–178); length of tail, 83.7 (80–87); length of hind foot, 19.2 (18–20); length of ear, 13.8 (13–15). The greatest length of skull generally ranges between 21 and 23, and the zygomatic breadth ranges between 10.5 and 12. Weights of adults average between 20 and 25 grams.

Rattus norvegicus: Norway Rat

This rat was introduced into eastern North America about the time of the American Revolution and subsequently came to occupy most areas inhabited by humans, through repeated introductions at new ports of entry and by following the spread of European man across the continent. The species now occurs throughout the Northern Great Plains, but it is more likely to be found only in urban areas or strictly associated with other human habitations than is the house mouse. Feral populations of Norway rats are therefore rare, at least in the tristate region.

R. norvegicus is likely to be confused, among native rodents of the Dakotas and Nebraska, only with the two species of woodrats, *Neotoma*. From them, the Norway rat can be distinguished readily as follows: tail scantily haired, scaly annulations on it clearly evident; crowns of molar teeth cuspidate and more or less cuboid rather than elongate and semiprismatic; skull with well-developed temporal ridges; pelage of venter in adults tending to be grayish, sometimes washed with yellowish, as opposed to whitish. Another Old World rat, *Rattus rattus*, also has been introduced into the Americas and is widely distributed in the warmer parts of the two continents. It has not been reported from the Northern Great Plains, however, and probably does not occur there. *R. rattus* differs from species of *Neotoma* in the same way as does *R. norvegicus*. From the latter, it can be identified as follows: tail longer, rather than shorter, than head and body; ears larger; females normally with 10 mammae instead of 12; temporal ridges on skull bowed outward posteriorly rather than essentially parallel.

In urban areas the Norway rat occurs primarily in and around older, run-down buildings, in trash-filled alleys, at municipal garbage dumps, and in similar surroundings. In rural environs it is to be looked for in and under farm outbuildings, grain bins, stock feeding platforms, and such. Like the house mouse, this species breeds year round. The number of young per litter ranges between two and 14, although litters of more than 14 occasionally are reported. Where it is common, this rat can destroy grain and other foods and also is a notorious carrier of diseases that affect man.

An environment clean of trash and solidly built ("rat proof") buildings guard against the invasion of a neighborhood or farmstead by this pest. Where necessary, cautious use of commercially available poisoned baits is recommended for its control.

Three adult males from Nebraska had the following ranges in external measurements: total length, 372–440; length of tail, 164–199; length of hind foot, 41–43.5; length of ear, 21.5–24. The greatest length of skull in one of those specimens was

51.9, and the zygomatic breadth measured 24.3. Weights in adults normally range from about 250 up to 400 grams.

Family Myocastoridae

Myocastor coypus: Coypu or Nutria

The coypu, a native of temperate South America, first was introduced into the southern United States in the early 1930s. It currently is well established in the wild in several Gulf Coast states, especially Louisiana. Because of anticipated large returns for the fur of this animal, many were transported to enclosures elsewhere in the country, especially in the years following the Second World War. The expected value of pelts never was realized, however, and many captives subsequently were released or escaped, probably accounting for the few recorded occurrences on the Northern Great Plains. We are not aware of any currently viable populations in the Dakotas or Nebraska. If the coypu does occur in the wild in this region, it is to be looked for along major river systems and around large, permanent lakes, especially those with extensive marshes.

The semiaquatic *M. coypus* is likely to be confused only with the beaver, from which it easily can be distinguished by the cylindrical (as opposed to broad and flattened) tail and the enormously enlarged infraorbital foramen in the skull. Also, the nutria tends to be more yellowish than the beaver, with a whitish patch around the nose and mouth, and its pelage is coarser. Young animals possibly could be confused with muskrats but can be immediately identified by the tail (laterally compressed in the muskrat) and by the characters of the skull.

External and cranial measurements of a series of *M. coypus* from Louisiana are: total length, 940 (837–1,010); length of tail, 344 (300–450); length of hind foot, 131 (100–150); length of ear, 27 (21–30); greatest length of skull, 116.3 (109.6–125.6); zygomatic breadth, 71.9 (68.4–79.6). Adults generally weigh between 10 and 20 pounds.

Family Cervidae

Cervus dama: Fallow Deer

The European fallow deer was introduced in central Nebraska just before the Second World War, and by the late 1950s and early 1960s wild populations had become rather well established in the valleys of Beaver Creek and adjacent Cedar Creek, and in the lower valley of the Loup River. One estimate in 1960 put the total in the wild in Nebraska at approximately 500. Since that time the fallow deer has declined in numbers to an estimated 35 to 40 in 1978, and it is uncertain at this juncture whether it will survive in the wild state.

C. dama differs most conspicuously from our native deer (*Odocoileus*) in that the antlers are palmate, or flattened, distally on the main beam, the internal nares are not completely divided by the vomer bone, and the premaxillary bones are distinctly separate (as viewed from below) for a third or more of their length anteriorly. Additionally, the fallow deer is spotted in summer pelage and lacks tarsal glands. In general its size approximates that of *Odocoileus*.

Glossary

The following terms are used frequently in descriptions or discussions of mammals. Many, but not all, have been used in the foregoing text. Some that appear on one or more of the three figures in the Glossary are not otherwise defined. For supplemental information the reader is referred to glossaries in Hall (1955), DeBlase and Martin (1974, 1980), and similar sources. Where a term listed has two or more meanings in the English language, only the one applying to mammalian biology is given.

abdomen. Ventral part of body, lying between thorax (rib cage) and pelvis.

aestivation. Torpidity in summer.

agouti hair. Hair with alternate pale and dark bands of color.

albinism. Lacking external pigmentation; albino (may be only partial).

allopatric. Pertaining to two or more populations that occupy disjunct or non-overlapping geographic areas.

altricial. Pertaining to young that are blind, frequently naked, and entirely dependent on parental care at birth.

alveolus. Socket in jawbone that receives root(s) of tooth.

angular process. Posterior projection of dentary ventral to condyloid (articular) process; evident but not labeled on figure 205.

annulation. Circular or ringlike formation as in dermal scales on tail of a mammal or in dentine of a tooth.

anterior. Pertaining to or toward front end.

antler. Branched (usually), bony head orna-ment found on cervids, covered with skin (velvet) during growth; shed annually.

apposable. Capable of being brought together (as thumb with other digits of hand).

arboreal. Pertaining to activity in trees.

articular condyle. Surface of condyloid (articular) process (fig. 205), articulating lower jaw with skull.

auditory bulla. Bony capsule enclosing middle ear; when formed by tympanic bone, termed tympanic bulla (see fig. 206).

auditory meatus. Opening leading from external ear to eardrum (see fig. 205).

baculum. Sesamoid bone (os penis) in penis of males of certain mammalian groups.

basal. Pertaining to base.

baubellum. See os clitoridis.

beam. Main trunk of antler.

bez tine. First tine above brow tine of antler.

bifid. Divided into two more-or-less equal lobes.

bifurcate. Divided into two branches.

bipedal. Pertaining to locomotion on two legs.

blastocyst. In embryonic development of mammals, a ball of cells produced by repeated division (cleavage) of fertilized egg; stage of implantation in uterine wall.

braincase. Posterior portion of skull; part that encloses and protects brain.

breech birth. Birth in which posterior part of body emerges first.

brisket. Breast or lower part of chest.

brow tine. First tine above base of antler.

buccal. Pertaining to cheek.

bunodont. Low-crowned, rectangular grinding teeth, typical of omnivores.

calcar. Spur of cartilage or bone that projects medially from ankle of many species of bats and helps support uropatagium.

canine. One of four basic kinds of mammalian teeth; anteriormost tooth in maxilla (and counterpart in dentary), frequently elongate, unicuspid, and single-rooted; never more than one per quadrant (see figs. 204–205). Also a term pertaining to dogs or to Canidae.

carnassials. Pair of large, bladelike teeth (last upper premolar and first lower molar) that occlude with scissorlike action; possessed by most modern members of order Carnivora.

carnivore. Animal that consumes meat as primary component of diet.

carpal. Any one of group of bones in wrist region, distal to radius and ulna and proximal to metacarpals.

caudal. Pertaining to tail or toward tail (caudad).

centimeter (cm). Unit of linear measure in metric system equal to 10 millimeters; 2.5 cm equal one inch.

cheekteeth. Collectively, postcanine teeth (premolars and molars).

cingulum. Enamel shelf bordering margin(s) of a tooth (cingulid used for those of teeth in lower jaw).

claw. Sheath of keratin on digits; usually long, curved, and sharply pointed.

cline. Gradual change in morphological character through a series of interbreeding populations; character gradient.

cloaca. A common chamber into which digestive, reproductive, and urinary systems empty and from which products of these systems leave the body.

condylobasal length. See figure 204.

coprophagy. Feeding upon feces.

coronoid process. Projection of posterior portion of dentary dorsal to mandibular condyle (see fig. 205).

cosmopolitan. Common to all the world; not local or limited, but widely distributed.

cranial breadth. Measurement of cranium taken across its broadest point perpendicular to long axis of skull; frequently used for insectivores.

crepuscular. Pertaining to periods of dusk and dawn (twilight); active by twilight.

Cretaceous. See geologic time.

cursorial. Pertaining to running; running locomotion.

cusp. Point, projection, or bump on crown of a tooth.

cuspidate. Presence of cusp or cusps on a tooth.

deciduous dentition. Juvenile or milk teeth, those that appear first in lifetime of a mammal, consisting (if complete) of incisors, canines, and premolars; generally replaced by adult (premanent) dentition.

delayed implantation. Postponement of embedding of blastocyst (embryo) in uterine epithelium for several days, weeks, or months; typical of some carnivores.

dental formula. Convenient way of designating number and arrangement of mammalian teeth (for example, i 3/3, c 1/1, p 4/4, m 3/3); letters indicate incisors, canines, premolars, and molars, respectively; numbers before slashes (or above line) indicate number of teeth on one side of upper jaw, whereas those following (or below) indicate number on one side of lower jaw.

dentary. One of pair of bones that constitute entire lower jaw (mandible) of mammals.

dentine. Hard, generally acellular material between pulp and enamel of tooth; sometimes exposed on surface of crown.

dentition. Teeth of mammal considered collectively.

dewclaw. Vestigial digit on foot.

dewlap. Pendulous fold of skin under neck.

diastema. Space between adjacent teeth; for example, space between incisors and cheekteeth in species lacking canines.

dichromatism. Having two distinct color phases.

digit. Finger or toe.

digitigrade. Pertaining to walking on digits, with wrist and heel bones held off ground.

distal. Away from base or point of attachment or from any named reference point (as opposed to proximal).

distichous. Arranged alternately in two vertical rows on opposite sides of an axis, as in hairs on tails of some rodents.

diurnal. Pertaining to daylight hours; active by day.

dorsal. Pertaining to back or upper surface (dorsum).

ectoparasite. Parasite living on, and feeding from, external surface of an animal (for example, fleas, lice, ticks, mites).

enamel. Hard outer layer of tooth, usually white, consisting of calcareous compounds and small amount of organic matrix.

endoparasite. Parasite living within host (for example, flukes, tapeworms).

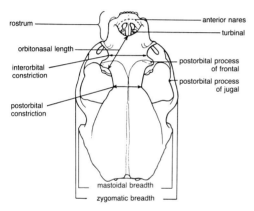

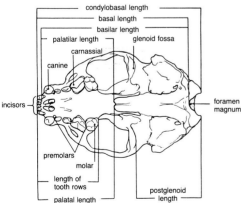

204. Dorsal and ventral views of skull of a river otter, showing cranial features and measurements (after Hall 1955).

Eocene. See geologic time.

epiphysis. Secondary growth center near end of long bone.

estrous cycle. Recurring growth and development of uterine endometrium, culminating in time when female is receptive to male.

estrus. Stage of reproductive receptivity of female for male; "heat."

excrescence. Dermal projection, as on face and ears of bats.

familial name. Name applying to a group of organisms of family rank among animals, ending in -idae (names of subfamilies end in -inae).

feces. Excrement.

fecundity. Rate of producing offspring; fertility.

femur. Single bone of upper (proximal) part of each hind (pelvic) limb.

fenestrate. Having openings.

feral. Having reverted from domestic to wild state.

fetal membranes. Tissue layers that surround, or attach to, the growing mammalian embryo (chorion, amnion, allantois).

fetus. Embryo in later stages of development (still in uterus).

fibula. Smaller of two bones in lower part of hind (pelvic) limb.

flank. Side of animal between ribs and hip.

foramen. Any opening, orifice, or perforation, especially through bone.

foramen magnum. Large opening at posterior of skull through which spinal cord emerges from braincase (see figs. 204, 206).

fossa. Pit or depression in bone; frequently site of bone articulation or muscle attachment.

fossorial. Pertaining to life under surface of ground.

frontal bone. See figure 205.

fusiform. Compact, tapered; pertaining to body form with shortened projections and no abrupt constrictions.

geologic time. Mammals arose in the *Mesozoic* era, which began some 230 million years ago. Periods of the Mesozoic (from oldest to youngest) are: *Triassic; Jurassic,* which began about 180 million years ago; and *Cretaceous*, which began about 135 million years ago. The following *Cenozoic* era, the "Age of Mammals," was the time of evolution and radiation of major modern groups. The Cenozoic is divided into two periods, *Tertiary* (beginning about 63 million years ago and continuing until two million years ago) and *Quaternary* (two million years ago to present). Subdivisions of the Tertiary (termed epochs) are (oldest to youngest): *Paleocene*; *Eocene*, which began about 58 million years ago; *Oligocene*, which began about 36 million years ago; *Miocene*, which began about 25 million years ago; and *Pliocene*, which began about 12 million years ago. The Quarternary has only two epochs, *Pleistocene* (two million years ago to 10,000 years ago) and *Holocene* or *Recent* (10,000 years before the present until now).

gestation period. Period of embryonic development during which developing zygote is in uterus; period between fertilization and parturition.

gram (g). Unit of weight in metric system; there are 28.5 grams in an ounce.

granivorous. Subsisting on diet of grains, seeds from cereal grasses.

gravid. Pregnant.

greatest length of skull. Measurement encompassing overall length of skull, including teeth that project anterior to premaxilla; frequently recorded instead of condylobasal length in some kinds of mammals.

guano. Excrement of bats or birds; sometimes sold commercially as fertilizer.

guard hairs. Outer coat of coarse protective hairs found in most mammals.

hallux. First (most medial) digit of hind foot (pes).

hamular process. Hooklike projection, such as hamular process of pterygoid bone.

hectare. Unit of land area in metric system equal to 10,000 square meters or 2.47 acres.

herbivore. Animal that consumes plant material as primary component of diet.

hibernaculum. Shelter in which animal hibernates.

hibernation. Torpidity in winter.

home range. Area in which an animal lives, which contains all necessities of life; generally is not entirely defended and therefore can overlap with those of other individuals.

horn. Structure projecting from head of mammal and generally used for offense, defense, or social interaction; members of family Bovidae have horns formed by permanent hollow keratin sheaths growing over bony cores (see also pronghorn).

humerus. Single bone in upper (proximal) portion of each front (pectoral) limb.

hypsodont. Pertaining to a particularly high-crowned tooth; such teeth have shallow roots.

imbricate. Overlapping, like shingles of a roof.

implantation. Process by which blastocyst (embryo) embeds in uterine lining.

incisive foramen. See figure 206.

incisor. One of four basic kinds of teeth in mammals, usually chisel-shaped; anteriormost of teeth; always rooted in premaxilla in upper jaw (see figs. 204–205).

infraorbital canal. Canal through zygomatic process of maxilla, from anterior wall of orbit to side of rostrum (see fig. 205).

infraorbital foramen. Foramen through zygomatic process of maxilla (see fig. 205 and above).

inguinal. Pertaining to region of groin.

insectivorous. Preying on insects.

interfemoral membrane. See uropatagium.

interorbital. Referring to region of skull between eye sockets.

interparietal. Unpaired bone on dorsal part of braincase between parietals and just anterior to supraoccipital (see fig. 206).

interspecific. Between or among species (interspecific competition, for example).

intraspecific. Within a species (intraspecific variation, for example).

jugal bone. Bone connecting maxillary and squamosal bones to form midpart of zygomatic arch (see fig. 206).

juvenile pelage. Pelage characteristic of juvenile (young) mammals.

karyotype. Morphological description of chromosomes of cell, including size, shape, position of centromere, and number.

keel. Ridge that provides expanded surface for attachment.

keratin. Tough, fibrous protein especially abundant in epidermis and epidermal derivatives.

kilogram (kg). Unit of weight in metric system (frequently shortened to "kilo") equal to 1,000 grams or 2.2 pounds.

kilometer (km). Unit of linear measure in metric system equal to 1,000 meters, slightly more than six-tenths of a mile.

labial. Pertaining to lips; for example, labial side of a tooth is side nearer lips (as opposed to lingual side, which is nearer tongue).

lactating. Secreting milk.

lateral. Situated away from midline; at or near sides.

lingual. Pertaining to tongue; lingual side of a tooth is side nearer tongue.

live trap. Any one of several kinds of traps designed to catch mammals alive.

loph. Ridge on occlusal surface of tooth formed by elongation and fusion of cusps.

malar process. Projection of maxillary that makes contact with jugal bone and forms base of zygomatic arch.

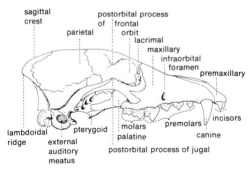

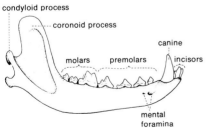

205. Lateral views of skull and lower jaw of a coyote, showing cranial and mandibular features (after Bekoff 1977).

mammae. Milk-producing glands unique to mammals; growth and activity governed by hormones of ovary, uterus, and pituitary; developed in both sexes but degenerate in males.

mandible. Lower jaw; in mammals composed of single pair of bones, the dentaries.

marsupium. External pouch formed by fold of skin in abdominal wall and supported by epipubic bones; found in most marsupials and some monotremes; encloses mammary glands and serves as incubation chamber.

masticate. To chew.

mastoid. Bone bounded by squamosal, exoccipital, and tympanic bones.

mastoid process. Exposed portion of petromastoid bone; situated anterior to auditory bulla.

maturational molt. Molt from juvenile or subadult pelage.

maxilla. Either of pair of relatively large bones that form major portion of side and ventral part of rostrum; contributes to hard palate, forms anterior root of zygomatic arch, and bears all upper teeth except incisors; also termed maxillary (see figs. 205–206).

maxillary toothrow. That part of toothrow in cranium of a mammal seated in maxilla; length includes all postincisor teeth and generally is taken parallel to long axis of skull (measurement shown in fig. 204 is of total upper toothrow).

medial. Pertaining to middle, as of a bone or other structure.

melanism. Unusual darkening of coloration owing to deposition of abnormally large amounts of melanins in integument.

Mesozoic. See geologic time.

metabolic water. Water formed biochemically as end product of fat and carbohydrate metabolism in body of animal.

meter (m). Unit of linear measure in metric system equal to 100 centimeters or 39.37 inches.

milk dentition. See deciduous dentition.

millimeter (mm). Unit of linear measure in metric system; 25.4 mm equal one inch.

Miocene. See geologic time.

mist net. Net of fine mesh used to capture

birds and bats; usually two meters high and ranging from six to 30 meters wide.

molar. One of four basic kinds of mammalian teeth; any cheektooth situated posterior to premolars and having no deciduous precursor; normally not exceeding three per quadrant (see figs. 204–205).

molariform. Pertaining to teeth with a form that is molarlike.

molt. Process by which hair is shed and replaced.

monestrous. Having a single estrous cycle each year.

monotypic. Pertaining to taxon that contains only one immediately subordinate taxon; for example, genus that contains only one species.

musk gland. One of several kinds of glands in mammals with secretions that have a musky odor.

muzzle. Projecting snout.

nail. Flat, keratinized, translucent, epidermal growth protecting upper portion of tips of digits in some mammals; a modified claw.

nape. Back of neck.

nares. Openings of nose.

natal. Pertaining to birth.

neonate. Newborn.

nictitating membrane. Thin membrane at inner angle of eye in some species, which can be drawn over surface of eyeball; a "third" eyelid.

nocturnal. Pertaining to night (hours without daylight); active by night.

nomadic. Wandering.

nominal species. Named species (sometimes implies species in name only); also, in a more restrictive sense, species that contains type of a genus and bears same name.

occipital bone. Bone surrounding foramen magnum and bearing occipital condyles (see occiput on fig. 206); formed from four embryonic elements, a ventral basioccipital, a dorsal supraoccipital, and two lateral exoccipitals.

occiput. General term for posterior portion of skull (see fig. 206).

occlusal. Pertaining to contact surfaces of upper (cranial) and lower (mandibular) teeth.

Oligocene. See geologic time.

omnivorous. Pertaining to animals that eat both animal and vegetable food.

orbit. Bony socket in skull in which eyeball is situated (see fig. 205).

ordinal name. Name applying to an order of organisms.

os clitoridis. Small sesamoid bone present in clitoris of females of some mammalian species; homologous to baculum in males.

ossify. To become bony or hardened and bonelike.

palate. Bony plate formed by palatine bones and palatal branches of maxillae and premaxillae (see fig. 206).

Paleocene. See geologic time.

palmate. Pertaining to presence of webbing between digits or to flattening of tines of antlers.

papilla. Any blunt, rounded, or nipple-shaped projection.

parapatric. Pertaining to two or more populations that occupy locally contiguous, but not overlapping, geographic areas.

parietal. Either of pair of bones contributing to roof of cranium posterior to frontals and anterior to occipital (see figs. 205–206).

parturition. Process by which fetus of therian mammals separates from uterine wall of mother and is born; birth.

patagium. Web of skin; in bats, the wing membrane.

patronym. Scientific name based on name of a person or persons.

pectoral. Pertaining to chest.

pectoral girdle. Shoulder girdle, composed in most mammals of clavicle and scapula, or scapula alone.

pelage. Collectively, all the hairs on a mammal.

pelvic girdle. Hip girdle, composed of ischium, ilium, and pubis.

penicillate. Ending in tuft of fine hairs.

phalangeal epiphyses. Growth centers just proximal to articular surfaces of phalanges; fusion of epiphyses to shafts of phalanges used as a means of determining age in bats.

phalanges. Bones of fingers and toes, distal to metacarpals and metatarsals.

pigment. Minute granules that impart color to an organism; such granules are usually metabolic wastes and may be shades of black, brown, red, or yellow.

pinna. Externally projecting part of ear.

placenta. Composite structure formed by maternal and fetal tissues across which gases, nutrients, and wastes are exchanged.

placental scar. Scar that remains on uterine wall after deciduate placenta detaches at parturition.

plantar pad (tubercle). Cutaneous pad or tubercle on sole of foot.

plantigrade. Foot structure in which phalanges and metatarsals or metacarpals touch ground; basic structure for ambulatory (walking) locomotion.

Pleistocene. See geologic time.

Pliocene. See geologic time.

pollex. Thumb; first (most medial) digit on hand (manus).

polyestrous. Having more than one estrous cycle each year.

polygynous. Male mating with several females.

postauricular. Pertaining to behind the ear.

posterior. Pertaining to or toward rear end.

postorbital process. Projection of frontal bone that marks posterior margin of orbit (see figs. 204–205).

postpartum estrus. Ability of female to become receptive to male directly after giving birth.

precocial. Pertaining to young that are born at relatively advanced stage, capable of moving about shortly after birth and usually of some feeding without parental assistance.

prehensile. Adapted for grasping by curling or wrapping around.

premaxilla. One of paired bones at anterior end of rostrum (see figs. 205–206); also termed premaxillary.

premolar. One of four basic kinds of teeth in mammals; situated anterior to molars and posterior to canines; only cheekteeth usually present in both permanent and milk dentitions; normally not exceeding four per quadrant (see figs. 204–205).

pronghorn. Modified horn (in both sexes of Antilocapridae) that grows over perma-

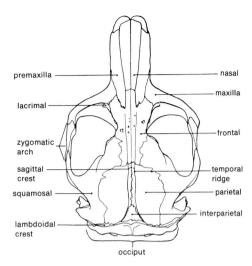

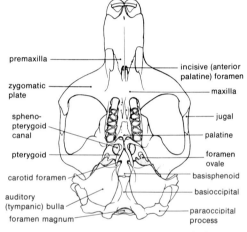

206. Dorsal and ventral views of skull of a pocket gopher (Thomomys), showing cranial features (after Hall 1955).

nent bony cores and is shed annually; each is slightly curved, with one anterolateral prong in males and some females.

proximal. Situated toward or near a point of reference or attachment (as opposed to distal).

pterygoid. Either of paired bones in ventral wall of braincase, posterior to palatines (see fig. 206).

pubic symphysis. Midventral plane of contact between two halves of pelvic girdle.

quadrupedal. Pertaining to use of all four limbs for locomotion.

race. Informal term for subspecies (which see).

radius. Medial of two bones in lower part of front (pectoral) limb.

ramus. Horizontal portion of dentary, that part in which teeth are rooted.

range. Geographic area inhabited by a particular taxon.

Recent. See geologic time.

recurved. Curved downward or backward.

reentrant angle. Inward infolding of enamel layer on side, front, or back of cheektooth.

refugium. Geographic area to which species retreats in time of stress, such as glacial episode.

retractile. Capable of being drawn back or in (as retractile claws of felids).

root. Portion of tooth that lies below gum line and fills alveolus.

rooted tooth. Tooth with definitive growth; not ever-growing.

rootless tooth. Tooth that is ever-growing, having continuously open root canal.

rostrum. Facial region of skull anterior to plane drawn through anterior margin of orbits (see fig. 204).

rugose. Wrinkled.

rumen. First "stomach" of ruminant mammals; modification of esophagus.

ruminant. Any of the Artiodactyla (including all those occurring on the Northern Plains) that possesses a rumen; cud-chewing.

runway. Worn or otherwise detectable pathway made by repeated usage.

rutting season. Season of sexual activity when mating occurs; particularly applied to deer and other artiodactyls.

sagittal crest. Medial dorsal ridge on braincase, often formed by coalescence of temporal ridges.

saltatory. Adapted for leaping; usually with elongate and unusually well-developed hind legs.

scansorial. Pertaining to arboreal animals that climb by means of sharp, curved claws.

scapula. Shoulder blade.

scent glands. Sweat or sebaceous glands, or combination of these two, modified to produce odoriferous secretions.

scrotum. Pouch of skin in which testes are contained outside the abdominal cavity; permanently present in some species, seasonally present in some, and lacking in others.

scute. Thin plate or scale.

sebaceous glands. Epidermal glands that secrete fatty substance and usually open into hair follicle.

septum. A dividing wall, such as one formed by membrane or thin bone.

sexual dimorphism. Difference in size or other features between males and females of species.

snap trap. Kill trap that usually consists of wooden base, with wire bail, spring, and trigger mechanism; designed primarily to catch rodents and insectivores.

snout. Nose, muzzle.

species. Group of naturally or potentially interbreeding populations reproductively isolated (or mostly so) from other such groups.

squamosal. Either of pair of bones contributing to side of cranium (see fig. 206) and forming posterior part of zygomatic arch.

subspecies. Relatively uniform and genetically distinctive population of a species that represents a separately or recently evolved lineage with its own evolutionary tendencies, definite geographic range, and actual or potential zone of intergradation (interbreeding) with another such group or groups.

superciliary. Pertaining to eyebrow.

supraorbital process. Projection of frontal bone on superior rim of orbit, as in hares and rabbits.

sweat gland. Tubular epidermal gland that extends into dermis and secretes sweat or perspiration and also various scents.

sympatric. Pertaining to two or more populations that occupy overlapping geographic areas.

symphysis. Relatively immovable articulation between bones.

tarsal. Any one of group of bones in ankle region, distal to fibula and proximal to metatarsals.

taxon. Any group (in this case, of mammals) distinctive from other groups at same taxonomic level.

teat. Protruberance of mammary gland in which numerous small ducts empty into common collecting structure that in turn opens to exterior through one or a few pores.

temporal ridges. Pair of ridges atop brain-case of many mammals; usually originating on frontal bones near postorbital processes and converging posteriorly to form middorsal sagittal crest (see fig. 206).

territory. Portion of home range that individual defends against members of same (and sometimes different) species.

tibia. Larger of two bones in lower part of hind (pelvic) limb.

tine. Spike on an antler.

torpor. Dormancy; body temperature approximates that of surroundings; rate of respiration and heartbeat ordinarily much slower than in active animal.

tragus. Fleshy projection from lower medial margin or ear of most microchiropteran bats.

tricolored. Having three colors.

trifid. Divided into three more-or-less equal lobes.

trifurcate. Having three branches.

tularemia. Bacterial disease contracted by man through bite of tick harbored principally by hares and rabbits, but also by rodents and birds; also may be transmitted by direct contact with infected animal.

tympanic bulla. See auditory bulla and figure 206.

tympanum. Eardrum; thin membranous structure that receives external vibrations from air and transmits them to middle ear ossicles.

type specimen. Specimen (holotype) on which a species or subspecies name is based.

ulna. Outermost of two bones in lower part of front (pectoral) limb.

underfur. Short hairs of mammal that serve primarily as insulation.

ungulate. Hoofed mammals of extant order Perissodactyla or Artiodactyla; term has no formal taxonomic status but refers to broad group of herbivorous mammals.

unguligrade. Foot structure in which only the unguis (hoof) is in contact with ground.

unicuspid. Single-cusped tooth in shrews posterior to large, anteriormost tooth (an incisor) and anterior to fourth premolar in upper jaw and first true molar in lower jaw.

uropatagium. Web of skin between hind legs of bats, frequently enclosing tail; interfemoral membrane.

valvular. Capable of being closed, like a valve.

ventral. Pertaining to under or lower surface (venter).

vernacular name. Common (as opposed to scientific) name.

vernal. Seasonal term pertaining to spring.

vibrissae. Long, stiff hairs that serve primarily as tactile receptors.

volant. Able to fly.

vomer. Unpaired bone that forms septum between nasal passages.

vulva. External genitalia of female.

wool. Underhair with angora growth; serves primarily for insulation.

zygomatic arch. Arch of bone enclosing orbit and temporal fossa formed by jugal bone and parts of maxilla (malar process) and squamosal.

zygomatic breadth. See figure 205.

zygomatic plate. Bony plate, part of zygomatic process of maxilla, forming anterior face of zygomatic arch (see fig. 206).

Addenda

Listed below are publications and information that may be of interest to readers of this book and that came to the attention of the authors after the manuscript was delivered to the University of Nebraska Press. Many of the publications would have been cited in appropriate sections of the text had they been available to us. A few changes were made in accounts at the time galleys were read.

General

Chapman, J. A., and G. E. Feldhamer (eds.). 1982. Wild mammals of North America. The John Hopkins Univ. Press, Baltimore, xiii + 1147 pp. A compilation of information on the biology, status, and management of many economically important species of mammals found in North America to the north of Mexico.

Honacki, J. H., K. E. Kinman, and J. W. Koeppl (eds.). 1982. Mammal species of the world. Allen Press, Inc., and Assoc. Syst. Collections, Lawrence, Kansas, ix + 694 pp. A briefly annotated checklist of the 4,170 species of Recent mammals of the world; this source was used by us to verify or correct in galley stage the stated numbers of included taxa in genera, families, and orders.

Jones, J. K., Jr., D. C. Carter, H. H. Genoways, R. S. Hoffmann, and D. W. Rice. 1982. Revised checklist of North American mammals north of Mexico, 1982. Occas. Papers Mus., Texas Tech Univ., 80:1–22. A list of scientific and vernacular names of all species of North American mammals occurring north of Mexico recognized in the published literature as of March 1982.

Mumford, R. E., and J. O. Whitaker, Jr. 1982. Mammals of Indiana. Indiana Univ. Press, Bloomington, xvii + 537 pp. A comprehensive coverage of the 54 species of mammals known from Indiana, a number of which also occur on the Northern Great Plains.

Risser, P. G., E. C. Birney, H. D. Blocker, S. W. May, W. J. Parton, and J. A. Wiens. 1982. The true prairie ecosystem. Hutchinson Ross Publ. Co., Stroudsburg, Pennsylvania, xiv + 557 pp. A volume containing much useful information on the environments of the Northern Great Plains.

Mammalian Species

This series, published by the American Society of Mammalogists, is intended to provide a concise, yet complete, account of the biology of individual species of mammals. As of November 1982, 194 accounts had been issued (available from G. L. Kirkland, Jr., The Vertebrate Museum, Shippensburg State College, Shippensburg, Pennsylvania 17257). We routinely cited available Mammalian Species under "Selected References" in text and thus would have included all the following had they been published, or had we had knowledge they were already in press, at the time our manuscript was completed.

Lasionycteris noctivagans. T. H. Kunz. Mamm. Species, 172:1–5, 1982.

Lasiurus borealis. K. A. Shump, Jr., and A. U. Shump. Mamm. Species, 183:1–6, 1982.

Lasiurus cinereus. K. A. Shump, Jr., and A. U. Shump. Mamm. Species, 185:1–5, 1982.

Plecotus townsendii. T. H. Kunz and R. A. Martin. Mamm. Species, 175:1–6, 1982.

Tamias striatus. D. P. Snyder. Mamm. Species, 168:1–8, 1982.

Reithrodontomys megalotis. W. D. Webster and J. K. Jones, Jr. Mamm. Species, 167:1–5, 1982.

Sigmodon hispidus. G. N. Cameron and S. R. Spencer. Mamm. Species, 158:1–9, 1981.

Microtus pennsylvanicus. L. M. Reich. Mamm. Species, 159:1–8, 1981.

Microtus pinetorum. M. J. Smolen. Mamm. Species, 147:1–7, 1981.

Urocyon cinereoargenteus. E. K. Fritzell and K. J. Haroldson. Mamm. Species, 189:1–8, 1982.

Martes pennanti. R. A. Powell. Mamm. Species, 156:1–6, 1981.

Mephitis mephitis. J. Wade Smith and B. J. Verts. Mamm. Species, 173:1–7, 1982.

Alces alces. A. W. Franzmann. Mamm. Species, 154:1–7, 1981.

Additional Information

Listed here are addenda items concerning individual species treated in this book that may prove useful to readers.

Blarina carolinensis.—George, Choate, and Genoways (Ann. Carnegie Mus., 50:493–513, 1981) and George, Genoways, Choate, and Baker (J. Mamm., 63:639–645, 1982) have demonstrated that short-tailed shrews from the southwestern part of the range of the genus in the United States differ morphometrically and chromosomally from those occurring in the Southeast. The specific name *carolinensis* applies to the latter. The correct name for the shrews here treated as *Blarina carolinensis* thus becomes *Blarina hylophaga* (Elliot's short-tailed shrew).

Tadarida brasiliensis.—Glass (Southwestern Nat., 27:127–133, 1982) reported a banded individual of this bat recovered at Menno, Hutchinson Co., South Dakota, the first record from the state and the northernmost in the United States. Glass also reported a banded *T. brasiliensis* taken at Oxford, Furnas Co., Nebraska, which constitutes the fourth locality of record in that state. Both records evidently resulted from animals that lost their way during northward migration in spring and thus moved far beyond normal summer roosting areas.

Sylvilagus audubonii and *Sylvilagus floridanus.*—Bergeron and Seabloom (Prairie Nat., 13:105–110, 1982) reported on habitat partitioning by these two species in southwestern North Dakota. Useful information on these and other Plains lagomorphs can be found in Myers and MacInnes (eds.), *Proceedings of the World Lagomorph Conference . . .* (Univ. Guelph, Canada, xix + 983 pp., 1982).

Spermophilus richardsonii.—Long (Proc. South Dakota Acad. Sci., 43:207, 1964) reported this species from 7 mi. S Artesian, Sanborn Co., South Dakota, a locality to the southwest of the limits of distribution as shown on Fig. 69. This ground squirrel also has recently been reported from northwestern Iowa by Lampe, Bowles, and Spengler (Prairie Nat., 13:94–96, 1982).

Neotoma cinerea.—Escherich (Univ. California Publ. Zool., 110:xiii + 1–132, 1982) has provided a good laboratory- and field-based study of the social biology of this rodent.

Martes pennanti.—The first thorough study of the biology of the fisher can be found in Powell, *The fisher: life history, ecology, and behaviour* (Univ. Minnesota Press, xvi + 219 pp., 1982).

Felis concolor.—A study of the mountain lion in Manitoba by Wrigley and Nero, which includes mention of occurrences in North Dakota, recently has been published (*Manitoba's Big Cat*, Manitoba Museum of Man and Nature, 68 pp., 1982).

Odocoileus hemionus.—An excellent summary of the biology of this species can be found in Walmo (ed.), *Mule and black-tailed deer of North America* (Wildlife Manag. Inst. and Univ. Nebraska Press, xvii + 605 pp., 1981).

Alces alces.—Tillman and Edens (Prairie Nat., 14:99, 1982) reported the first known occurrence in South Dakota of the moose. In the summer of 1980, an adult cow and young bull were observed on the Sand Lake National Wildlife Refuge in Brown County.

Bison bison.—A systematic study of the bison by McDonald, *North American bison: their classification and evolution* (Univ. California Press, Berkeley, 316 pp., 1981) is a noteworthy addition to the literature on this bovid.

Literature Cited

Adams, C. E. 1976. Measurement and characteristics of fox squirrel, Sciurus niger rufiventer, home ranges. Amer. Midland Nat., 95:211–215.

Alcoze, T. M., and E. G. Zimmerman. 1973. Food habits and dietary overlap of two heteromyid rodents from the mesquite plains of Texas. J. Mamm., 54:900–908.

Aldous, C. M. 1937. Notes on the life history of the snowshoe hare. J. Mamm., 18:46–57.

Allen, H. 1864. Monograph of the bats of North America. Smithsonian Misc. Coll., 7(165):xxiii + 1–85.

———. 1894. A monograph of the bats of North America. Bull. U.S. Nat. Mus., 43:ix + 1–198.

Ambrose, H. W., III. 1973. An experimental study of some factors affecting the spatial and temporal activity of Microtus pennsylvanicus. J. Mamm., 54:79–110.

Andersen, D. C. 1978. Observations on reproduction, growth, and behavior of the northern pocket gopher (Thomomys talpoides). J. Mamm., 59:418–422.

Andersen, K. W., and J. K. Jones, Jr. 1971. Mammals of northwestern South Dakota. Univ. Kansas Publ., Mus. Nat. Hist., 19:361–393.

Anderson, E. 1970. Quaternary evolution of the genus Martes (Carnivora, Mustelidae). Acta Zool. Fennica, 130:1–132.

Anderson, S. 1956. Subspeciation in the meadow mouse, Microtus pennsylvanicus, in Wyoming, Colorado, and adjacent areas. Univ. Kansas Publ., Mus. Nat. Hist., 9:85–104.

Anderson, S., and J. K. Jones, Jr. (eds.). 1967. Recent mammals of the World . . . . Ronald Press Co., New York, viii + 453 pp.

Armstrong, D. M. 1971. Notes on variation in Spermophilus tridecemlineatus (Rodentia, Sciuridae) in Colorado and adjacent states, and description of a new subspecies. J. Mamm., 52:528–536.

———. 1972. Distribution of mammals in Colorado. Monogr. Mus. Nat. Hist., Univ. Kansas, 3:x + 1–415.

Armstrong, D. M., and J. K. Jones, Jr. 1971. Sorex merriami. Mamm. Species, 2:1–2.

Baar, S. L., and E. D. Fleharty. 1976. A model of the daily energy budget and energy flow through a population of the white-footed mouse. Acta Theriol., 12:179–193.

Bailey, V. 1927. A biological survey of North Dakota. N. Amer. Fauna, 49:vi + 1–226.

Baird, S. F. 1858. Mammals, in Reports of explorations and surveys . . . from the Mississippi River to the Pacific Ocean . . . , Washington, D.C., 8(1):xxi-xlviii + 1–757.

Banfield, A. W. F. 1962. A revision of the reindeer and caribou, genus Rangifer. Bull. Nat. Mus. Canada, 177:vi + 1–137.

———. 1974. The mammals of Canada. Univ. Toronto Press, xxv + 438 pp.

Banks, R. C. (ed.). 1979. Museum studies and wildlife management. Smithsonian Inst. Press, Washington, D.C., 295 pp.

Barbour, R. W., and W. H. Davis. 1969. Bats of America. Univ. Press Kentucky, Lexington, 286 pp.

Batzli, G. O., L. L. Getz, and S. S. Hurley. 1977. Suppression of growth and reproduction of microtine rodents by social factors. J. Mamm., 58:583–591.

Bear, G. H., and R. M. Hansen. 1966. Food habits, growth and reproduction of white-tailed jackrabbits in southern Colorado. Tech. Bull. Agric. Exper. Sta., Colorado State Univ., 90:viii + 1–59.

Beck, W. H. 1958. A guide to Saskatchewan mammals. Spec. Publ. Saskatchewan Nat. Hist. Soc., 1:1–52.

Bedard, J. (ed.). 1975. Moose ecology. Internat. Symp. Moose Ecol. (Quebec, Canada, March 26–28, 1973), Presses de L'Université Laval, Quebec, xii + 741 pp.

Bee, J. W., G. E. Glass, R. S. Hoffmann, and R. R. Patterson. 1981. Mammals in Kansas. Public Ed. Ser., Mus. Nat. Hist., Univ. Kansas, 7:ix + 1–300.

Beer, J. R. 1961. Hibernation in *Perognathus flavescens*. J. Mamm., 42:103.

Bekoff, M. 1977. Canis latrans. Mamm. Species, 79:1–9.

Benton, A. H. 1955. Observations on the life history of the northern pine mouse. J. Mamm., 36:52–62.

Bergerud, A. T. 1978. Caribou. Pp. 83–101, *in* Big game of North America (J. L. Schmidt and D. L. Gilbert, eds.), Stackpole Books, Harrisburg, Pennsylvania, xv + 494 pp.

Birney, E. C., W. E. Grant, and D. D. Baird. 1976. Importance of vegetative cover to cycles of *Microtus* populations. Ecology, 57:1043–1051.

Blair, W. F. 1937. Burrows and food of the prairie pocket mouse. J. Mamm., 18:188–191.

———. 1940. Notes on home ranges and populations of the short-tailed shrew. Ecology, 21:284–288.

Bowles, J. B. 1975. Distribution and biogeography of mammals of Iowa. Spec. Publ. Mus., Texas Tech Univ., 9:1–184.

Bowles, J. B., and H. L. Gladfelter. 1980. Movement of moose south of traditional range in the upper midwestern United States. Proc. Iowa Acad. Sci., 87:124–125.

Boyd, R. J. 1978. American elk. Pp. 11–29, *in* Big game of North America (J. L. Schmidt and D. L. Gilbert, eds.), Stackpole Books, Harrisburg, Pennsylvania, xv + 494 pp.

Brand, C. J., L. B. Keith, and C. A. Fischer. 1976. Lynx responses to changing snowshoe hare densities in central Alberta. J. Wildlife Manag., 40:416–428.

Brown, J. H., and R. C. Lasiewski. 1972. Metabolism of weasels: the cost of being long and thin. Ecology, 53:939–943.

Brown, L. N. 1967a. Seasonal activity patterns and breeding of the western jumping mouse (Zapus princeps) in Wyoming. Amer. Midland Nat., 78:460–470.

———. 1967b. Ecological distribution of six species of shrews and comparison of sampling methods in the Central Rocky Mountains. J. Mamm., 48:617–623.

Brown, L. N., and D. Metz. 1966. First record of *Perognathus flavescens* in Wyoming. J. Mamm., 47:118.

Buckner, C. H. 1964. Metabolism and feeding behavior of three species of shrews. Canadian J. Zool., 42:259–279.

———. 1966. Populations and ecological relationships of shrews in tamarack bogs of southeastern Manitoba. J. Mamm., 47:181–194.

Burk, D. (ed.). 1979. The black bear in modern North America. Boone and Crockett Club and Amwell Press, Clinton, New Jersey, 300 pp.

Carroll, L. E., and H. H. Genoways. 1980. Lagurus curtatus. Mamm. Species, 124:1–6.

Chapman, J. A. 1975. Sylvilagus nuttallii. Mamm. Species, 56:1–3.

Chapman, J. A., J. G. Hockman, and M. M. Ojeda C. 1980. Sylvilagus floridanus. Mamm. Species, 136:1–8.

Chapman, J. A., and G. R. Willner. 1978. Sylvilagus audubonii. Mamm. Species, 106:1–4.

Choate, J. R., and J. K. Jones, Jr. 1982. Provisional checklist of South Dakota mammals. Prairie Nat., 13:65–77.

Choate, J. R., and S. L. Williams. 1978. Biogeographic interpretation of variation within and among populations of the prairie vole, Microtus ochrogaster. Occas. Papers Mus., Texas Tech Univ., 49:1–25.

Clark, T. W. 1970a. Richardson's ground squirrel (*Spermophilus richardsonii*) in the Laramie Basin, Wyoming. Great Basin Nat., 30:55–70.

———. 1970b. Early growth, development, and behavior of the Richardson ground squirrel (Spermophilus richardsoni elegans). Amer. Midland Nat., 83:197–205.

———. 1971. Ecology of the western jumping mouse in Grand Teton National Park, Wyoming. Northwest Sci., 45:229–238.

Cleveland, A. G. 1979. Natural history of the hispid cotton rat. Proc. Welder Wildlife Found. Symp., 1:229–241.

Clough, G. C. 1963. Biology of the arctic shrew, Sorex arcticus. Amer. Midland Nat., 69:69–81.

Conaway, C. H. 1952. Life history of the water shrew (Sorex palustris navigator). Amer. Midland Nat., 48:219–248.

Conley, W. 1976. Competition between Microtus: a behavioral hypothesis. Ecology, 57:224–237.

Connor, P. F. 1959. The bog lemming Synaptomys cooperi in southern New Jersey. Publ. Mus. (Biol. Ser.), Michigan State Univ., 1:165–248.

Corbet, G. B., and H. N. Southern (eds.). 1977. The handbook of British mammals. Blackwell Sci. Publ., Oxford, 2nd ed., xxxii + 520 pp.

Costello, D. F. 1966. The world of the porcupine. J. B. Lippincott Co., Philadelphia, 157 pp.

Coues, E., and J. A. Allen. 1877. Monographs of North American Rodentia. Bull. U.S. Geol. Surv. Territories, 11:xii + 1–1091.

Cowan, I. McT. 1940. Distribution and variation in the native sheep of North America. Amer. Midland Nat., 14:505–580.

Cowan, I. McT., and W. McCrory. 1970. Variation in the mountain goat, Oreamnos americanus (Blainville). J. Mamm., 51:60–73.

Crabb, W. D. 1941. Food habits of the prairie spotted skunk in southeastern Iowa. J. Mamm., 22:349–364.

———. 1944. Growth, development and seasonal weight of spotted skunks. J. Mamm., 25:213–221.

———. 1948. Ecology and management of the prairie spotted skunk in Iowa. Ecol. Monogr., 18:201–232.

Cranford, J. A. 1978. Hibernation in the western jumping mouse (Zapus princeps). J. Mamm., 59:496–509.

Crowe, D. M. 1975. Aspects of aging, growth, and reproduction of bobcats from Wyoming. J. Mamm., 56:177–198.

Czaplewski, N. J. 1976. Vertebrate remains in great horned owl pellets in Nebraska. Nebraska Bird Rev., 44:12–15.

Czaplewski, N. J., J. P. Farney, J. K. Jones, Jr., and J. D. Druecker. 1979. Synopsis of bats of Nebraska. Occas. Papers Mus., Texas Tech Univ., 61:1–24.

Dapson, R. W. 1968. Reproduction and age structure in a population of short-tailed shrews, Blarina brevicauda. J. Mamm., 49:205–214.

Dasmann, W. 1971. If deer are to survive. Stackpole Books, Harrisburg, Pennsylvania, 128 pp.

Davis, W. H. 1966. Population dynamics of the bat Pipistrellus subflavus. J. Mamm., 47:383–396.

DeBlase, A. F., and R. E. Martin. 1974. A manual of mammalogy. . . . Wm. Brown Co., Dubuque, Iowa, xv + 329 pp.

———. 1980. A manual of mammalogy. . . . Wm. Brown Co., Dubuque, Iowa, 2nd ed., xii + 436 pp.

de Vos, A., and D. I. Gillespie. 1960. A study of woodchucks on an Ontario farm. Canadian Field-Nat., 74:130–145.

Diersing, V. E. 1980a. Systematics and evolution of the pygmy shrews (subgenus Microsorex) of North America. J. Mamm., 61:76–101.

———. 1980b. Systematics of flying squirrels, Glaucomys volans (Linnaeus), from Mexico, Guatemala, and Honduras. Southwestern Nat., 25:157–172.

Diersing, V. E., and D. F. Hoffmeister. 1977. Revision of the shrews Sorex merriami and a description of a new species of the subgenus Sorex. J. Mamm., 58:321–333.

Dolan, P. G., and D. C. Carter. 1977. Glaucomys volans. Mamm. Species, 78:1–6.

Downhower, J. F., and E. R. Hall. 1966. The pocket gopher in Kansas. Misc. Publ. Mus. Nat. Hist., Univ. Kansas, 44:1–32.

Egoscue, H. J. 1979. Vulpes velox. Mamm. Species, 122:1–5.

Eisenberg, J. F. 1963. The behavior of heteromyid rodents. Univ. California Publ. Zool., 69:1–114.

Elliott, L. 1978. Social behavior and foraging ecology of the eastern chipmunk (*Tamias striatus*) in the Adirondack Mountains. Smithsonian Contrib. Zool., 265:vi + 1–107.

Enders, R. K. 1952. Reproduction in the mink (*Mustela vison*). Proc. Amer. Philos. Soc., 96:691–755.

Engstrom, M. D., and J. R. Choate. 1979. Systematics of the northern grasshopper mouse (*Onychomys leucogaster*) on the Central Great Plains. J. Mamm., 60:723–739.

Errington, P. L. 1943. An analysis of mink predation upon muskrats in north-central United States. Res. Bull. Agric. Exper. Sta., Iowa State Coll., 320:798–924.

Farney, J. P. 1975. Natural history and northward dispersal of the hispid cotton rat in Nebraska. Platte Valley Rev., 3:11–16.

Farney, J. P., and J. K. Jones, Jr. 1980. Notes on the natural history of bats from Badlands National Monument, South Dakota. Prairie Nat., 12:9–12.

Fenneman, N. M. 1931. Physiography of western United States. McGraw-Hill Book Co., New York, xiii + 534 pp.

———. 1938. Physiography of eastern United States. McGraw-Hill Book Co., New York, xiii + 691 pp.

Fenton, M. B., and R. M. R. Barclay. 1980. Myotis lucifugus. Mamm. Species, 142:1–8.

Findley, J. S. 1956. Mammals of Clay County, South Dakota. Univ. South Dakota Publ. Biol., 1:1–45.

Findley, J. S., and C. Jones. 1964. Seasonal distribution of the hoary bat. J. Mamm., 45:461–470.

Finley, R. B., Jr. 1958. The wood rats of Colorado: distribution and ecology. Univ. Kansas Publ., Mus. Nat. Hist., 10:213–552.

Fitch, J. H., and K. A. Shump, Jr. 1979. Myotis keenii. Mamm. Species, 121:1–3.

Fitzgerald, B. M. 1977. Weasel predation on a cyclic population of the montane vole (*Microtus montanus*) in California. J. Anim. Ecol., 46:367–397.

Flake, L. D. 1973. Food habits of four species of rodents on a short-grass prairie in Colorado. J. Mamm., 54:636–647.

———. 1974. Reproduction of four rodent species in a shortgrass prairie of Colorado. J. Mamm., 55:213–216.

Fleharty, E. D., and J. R. Choate. 1973. Bioenergetic strategies of the cotton rat, *Sigmodon hispidus*. J. Mamm., 54:680–692.

Fleharty, E. D., M. E. Krause, and D. P. Stinnett. 1973. Body composition, energy content, and lipid cycles of four species of rodents. J. Mamm., 54:426–438.

Fleharty, E. D., and M. H. Mares. 1973. Habitat preference and spatial relations of *Sigmodon hispidus* on a remnant prairie in west-central Kansas. Southwestern Nat., 18:21–29.

Forbes, R. B. 1964. Some aspects of the life history of the silky pocket mouse, Perognathus flavus. Amer. Midland Nat., 72:438–443.

———. 1966. Studies of the biology of Minnesotan chipmunks. Amer. Midland Nat., 76:290–308.

Forsyth, D. J. 1976. A field study of growth and development of nestling masked shrews (*Sorex cinereus*). J. Mamm., 57:708–721.

Franzmann, A. W. 1978. Moose. Pp. 67–81, *in* Big game of North America (J. L. Schmidt and D. L. Gilbert, eds.), Stackpole Books, Harrisburg, Pennsylvania, xv + 494 pp.

Frase, B. A., and R. S. Hoffmann. 1980. Marmota flaviventris. Mamm. Species, 135:1–8.

Gaines, M. S., C. L. Baker, and A. M. Vivas. 1979. Demographic attributes of dispersing southern bog lemmings (*Synaptomys cooperi*) in eastern Kansas. Oecologia, 40:91–101.

Gaines, M. S., and R. K. Rose. 1976. Population dynamics of *Microtus ochrogaster* in eastern Kansas. Ecology, 57:1145–1161.

Gardner, A. L. 1973. The systematics of the genus Didelphis (Marsupialia: Didelphidae) in North and Middle America. Spec. Publ. Mus., Texas Tech Univ., 4:1–81.

Geist, V. 1971. Mountain sheep. Univ. Chicago Press, xv + 383 pp.

Genoways, H. H., and J. R. Choate. 1972. A multivariate analysis of systematic relationships among populations of the short-tailed shrew (genus *Blarina*) in Nebraska. Syst. Zool., 21:106–116.

Genoways, H. H., and J. K. Jones, Jr. 1972. Mammals from southwestern North Dakota. Occas. Papers Mus., Texas Tech Univ., 6:1–36.

Genoways, H. H., J. C. Patton, III, and J. R. Choate. 1977. Karyotypes of shrews of the genera *Cryptotis* and *Blarina* (Mammalia: Soricidae). Experientia, 33:1294–1295.

Gentry, J. B. 1968. Dynamics of an enclosed population of pine mice. *Microtus pinetorum*. Res. Pop. Ecol., 10:21–30.

Gettinger, R. D. 1975. Metabolism and thermoregulation of a fossorial rodent, the northern pocket gopher (*Thomomys talpoides*). Physiol. Zool., 48:311–322.

Getz, L. L. 1960. A population study of the vole, Microtus pennsylvanicus. Amer. Midland Nat., 64:392–405.

———. 1972. Social structure and aggressive behavior in a population of *Microtus pennsylvanicus*. J. Mamm., 53:310–317.

Gier, H. T., B. R. Mahan, and W. F. Andelt. 1977. Parasites collected from coyotes, *Canis latrans* (Say, 1823), in Nebraska. Proc. Nebraska Acad. Sci., 87:14 (abstract).

Glass, B. P. 1947. Geographic variation in *Perognathus hispidus*. J. Mamm., 28:174–179.

Goertz, J. W. 1963. Some biological notes on the plains harvest mouse. Proc. Oklahoma Acad. Sci., 43:123–125.

———. 1971. An ecological study of Microtus pinetorum in Oklahoma. Amer. Midland Nat., 86:1–12.

Grange, W. B. 1932a. Observations on the snowshoe hare, Lepus americanus phaeonotus Allen. J. Mamm., 13:1–9.

———. 1932b. The pelages and color changes of the snowshoe hare, Lepus americanus phaeonotus Allen. J. Mamm., 13:99–116.

Greer, K. R. 1955. Yearly food patterns of the river otter in the Thompson Lakes region, northwestern Montana, as indicated by scat analysis. Amer. Midland Nat., 54:299–313.

Grizzell, R. A. 1955. A study of the southern woodchuck, Marmota monax monax. Amer. Midland Nat., 53:257–293.

Gunderson, H. L. 1978. A mid-continent irruption of Canada lynx, 1962–63. Prairie Nat., 10:71–80.

Gunderson, H. L., and J. R. Beer. 1953. The mammals of Minnesota. Univ. Minnesota Press, Minneapolis, xii + 190 pp.

Halfpenny, J. C., D. Nead, S. J. Bissell, and R. J. Auerlich. 1979. Bibliography of mustelids, part IV: wolverine. Michigan Agric. Exper. Sta., East Lansing, 51 pp.

Hall, E. R. 1951. American weasels. Univ. Kansas Publ., Mus. Nat. Hist., 4:1–466.

———. 1955. Handbook of mammals of Kansas. Misc. Publ. Mus. Nat. Hist., Univ. Kansas, 7:1–303.

———. 1962. Collecting and preparing study specimens of vertebrates. Misc. Publ. Mus. Nat. Hist., Univ. Kansas, 30:1–46.

———. 1981. Mammals of North America. John Wiley and Sons, New York, 2 vols. (1:xv + 1–600 + 90, 2:vi + 601–1181 + 90).

Hall, E. R., and K. R. Kelson. 1959. The mammals of North America. Ronald Press Co., New York, 2 vols. (1:xxx + 1–546 + 79, 2:viii + 547–1083 + 79).

Halls, L. K. 1978. White-tailed deer. Pp. 43–65, *in* Big game of North America (J. L. Schmidt and D. L. Gilbert, eds.), Stackpole Books, Harrisburg, Pennsylvania, xv + 494 pp.

Hamilton, W. J., Jr. 1929. Breeding habits of the short-tailed shrew, Blarina brevicauda. J. Mamm., 10:125–134.

———. 1931. Habits of the short-tailed shrew, Blarina brevicauda (Say). Ohio J. Sci., 31:97–106.

———. 1934. The life history of the rufescent woodchuck, *Marmota monax rufescens* Howell. Ann. Carnegie Mus., 23:85–178.

_____. 1938. Life history notes on the northern pine mouse. J. Mamm., 19:163–170.

Hamilton, W. J., Jr., and A. H. Cook. 1955. The biology and management of the fisher in New York. New York Fish Game J., 2:13–35.

Hamilton, W. J., Jr., and W. R. Eadie. 1964. Reproduction in the otter, *Lutra canadensis.* J. Mamm., 45:242–252.

Handley, C. O., Jr. 1959. A revision of American bats of the genera *Euderma* and *Plecotus.* Proc. U.S. Nat. Mus., 110:95–246.

Hansen, R. M., and J. T. Flinders. 1969. Food habits of North American hares. Sci. Ser., Range Sci. Dept., Colorado State Univ., 1:ii + 1–18.

Hansen, R. M., and M. K. Johnson. 1976. Stomach content weight and food selection by Richardson ground squirrels. J. Mamm., 57:749–757.

Hanson, W. O. 1950. The mountain goat in South Dakota. Unpublished Ph.D. dissertation, Univ. Michigan, viii + 92 pp.

Hart, E. B. 1978. Karyology and evolution of the plains pocket gopher, *Geomys bursarius.* Occas. Papers Mus. Nat. Hist., Univ. Kansas, 71:1–20.

Hatfield, D. M. 1939. Winter food habits of foxes in Minnesota. J. Mamm., 20:202–206.

Hawley, V. D., and F. E. Newby. 1957. Marten home ranges and population fluctuations. J. Mamm., 38:174–184.

Hazard, E. B. 1982. The mammals of Minnesota. Univ. Minnesota Press, Minneapolis, xii + 280 pp.

Hennings, D., and R. S. Hoffmann. 1977. A review of the taxonomy of the *Sorex vagrans* species complex from western North America. Occas. Papers Mus. Nat. Hist., Univ. Kansas, 68:1–35.

Hibbard, C. W. 1938. Distribution of the genus *Reithrodontomys* in Kansas. Univ. Kansas Sci. Bull., 25:173–179.

Hill, J. E., and C. W. Hibbard. 1943. Ecological differentiation between two harvest mice (*Reithrodontomys*) in western Kansas. J. Mamm., 24:22–25.

Hillman, C. N., and T. W. Clark. 1980. *Mustela nigripes.* Mamm. Species, 126:1–3.

Hines, T. 1980. An ecological study of *Vulpes velox* in Nebraska. Unpublished M.S. thesis, Univ. Nebraska, 103 pp.

Hines, T., R. Case, and R. Lock. 1981. The swiftest fox. Nebraskaland, 59(8):20–27.

Hoffmann, R. S., and J. K. Jones, Jr. 1970. Influence of late-glacial and post-glacial events on the distribution of Recent mammals of the Northern Great Plains. Pp. 355–394, *in* Pleistocene and Recent environments of the Central Great Plains (W. Dort, Jr., and J. K. Jones, Jr., eds.), Univ. Press Kansas, (x) + 433 pp.

Hoffmann, R. S., and J. G. Owen. 1980. Sorex tenellus and Sorex nanus. Mamm. Species, 131:1–4.

Hoffmann, R. S., and D. L. Pattie. 1968. A guide to Montana mammals. . . . Univ. Montana, Missoula, x + 133 pp.

Hoogland, J. L. 1979. The effect of colony size on individual alertness of prairie dogs (Sciuridae: *Cynomys* spp.). Anim. Behav., 27:394–407.

Hornocker, M. G. 1970. An analysis of mountain lion predation upon mule deer and elk in the Idaho Primitive Area. Wildlife Monogr., 21:1–39.

Horwich, R. H. 1972. The ontogeny of social behavior in the gray squirrel (*Sciurus carolinensis*). Adv. Ethol., 8:1–103.

Howell, A. H. 1915. Revision of the American marmots. N. Amer. Fauna, 37:1–80.

_____. 1942. A new red squirrel from North Dakota. Proc. Biol. Soc. Washington, 55:13–14.

Humphrey, S. R., and T. H. Kunz. 1976. Ecology of a Pleistocene relict, the western big-eared bat (*Plecotus townsendii*), in the Southern Great Plains. J. Mamm., 57:470–494.

Iverson, S. L., and B. N. Turner. 1972. Natural history of a Manitoba population of Franklin's ground squirrels. Canadian Field-Nat., 86:145–149.

Izor, R. J. 1979. Winter range of the silver-haired bat. J. Mamm., 60:641–643.

Jackson, H. H. T. 1928. A taxonomic review of the American long-tailed shrews (genera Sorex and Microsorex). N. Amer. Fauna, 51:vi + 1–238.

_____. 1961. Mammals of Wisconsin. Univ. Wisconsin Press, Madison, xii + 504 pp.

James, T. R., and R. Seabloom. 1969a. Reproductive biology of the white-tailed jack rabbit in North Dakota. J. Wildlife Manag., 33:558–568.

_____. 1969b. Aspects of growth in the white-tailed jackrabbit. Proc. North Dakota Acad. Sci., 23:7–14.

Jameson, E. W., Jr. 1947. Natural history of the prairie vole (mammalian genus Microtus). Univ. Kansas Publ., Mus. Nat. Hist., 1:125–151.

Jenkins, S. H., and P. E. Busher. 1979. Castor canadensis. Mamm. Species, 120:1–8.

Johnson, D. R. 1961. The food habits of rodents on rangelands of southern Idaho. Ecology, 42:407–410.

Jones, C., and J. Pagels. 1968. Notes on a population of Pipistrellus subflavus in southern Louisiana. J. Mamm., 49:134–139.

Jones, J. K., Jr. 1964. Distribution and taxonomy of mammals of Nebraska. Univ. Kansas Publ., Mus. Nat. Hist., 16:1–356.

Jones, J. K., Jr., D. C. Carter, and H. H. Genoways. 1979. Revised checklist of North American mammals north of Mexico, 1979. Occas. Papers Mus., Texas Tech Univ., 62:1–17.

Jones, J. K., and J. R. Choate. 1978. Distribution of two species of long-eared bats of the genus Myotis on the Northern Great Plains. Prairie Nat., 10:49–52.

_____. 1980. Annotated checklist of mammals of Nebraska. Prairie Nat., 12:43–53.

Jones, J. K., Jr., J. R. Choate, and R. B. Wilhelm. 1978. Notes on the distribution of three species of mammals in South Dakota. Prairie Nat., 10:65–70.

Jones, J. K., Jr., and G. L. Cortner. 1960. The subspecific identity of the gray squirrel (Sciurus carolinensis) in Kansas and Nebraska. Trans. Kansas Acad. Sci., 63:285–288.

Jones, J. K., Jr., E. D. Fleharty, and P. B. Dunnigan. 1967. The distributional status of bats in Kansas. Misc. Publ. Mus. Nat. Hist., Univ. Kansas, 46:1–33.

Jones, J. K., Jr., and H. H. Genoways. 1967a. A new subspecies of the fringe-tailed bat, Myotis thysanodes, from the Black Hills of South Dakota and Wyoming. J. Mamm., 48:231–235.

_____. 1967b. Annotated checklist of bats from South Dakota. Trans. Kansas Acad. Sci., 70:184–196.

Jones, J. K., Jr., R. P. Lampe, C. A. Spenrath, and T. H. Kunz. 1973. Notes on the distribution and natural history of bats in southeastern Montana. Occas. Papers Mus., Texas Tech Univ., 15:1–12.

Jones, J. K., Jr., and B. Mursaloğlu. 1961. Geographic variation in the harvest mouse, Reithrodontomys megalotis, on the Central Great Plains and in adjacent regions. Univ. Kansas Publ., Mus. Nat. Hist., 14:9–27.

Keen, R., and H. B. Hitchcock. 1980. Survival and longevity of the little brown bat (Myotis lucifugus) in southeastern Ontario. J. Mamm., 61:1–7.

Keith, L. B., and L. A. Windberg. 1978. A demographic analysis of the snowshoe hare cycle. Wildlife Monogr., 58:1–70.

Keller, B. L., and C. J. Krebs. 1970. Microtus population biology. III. Reproductive changes in fluctuating populations of M. ochrogaster and M. pennsylvanicus in southern Indiana, 1965–67. Ecol. Monogr., 40:263–294.

Kennedy, M. L., and G. D. Schnell. 1978. Geographic variation and sexual dimorphism in Ord's kangaroo rat, Dipodomys ordii. J. Mamm., 59:45–59.

Kennerly, T. E., Jr. 1959. Contact between the ranges of two allopatric species of pocket gophers. Evolution, 13:247–263.

Kilgore, D. L., Jr. 1970. The effects of northward dispersal on growth rate of young, size of young at birth, and litter size in Sigmodon hispidus. Amer. Midland Nat., 84:510–520.

King, C. M., and P. J. Moors. 1979. On coexistence, foraging strategy and the biogeography of weasels and stoats (Mustela nivalis and M. erminea) in Britain. Oecologia, 39:129–150.

King, J. A. 1951. The subspecific identity of the Black Hills flying squirrels (Glaucomys sabrinus). J. Mamm., 32:469–470.

_____. 1955. Social behavior, social organization, and population dynamics in a black-tailed prairiedog town in the Black Hills of South Dakota. Contrib. Lab. Vert. Biol., Univ. Michigan, 67:1–123.

——— (ed.). 1968. Biology of Peromyscus (Rodentia). Spec. Publ., Amer. Soc. Mamm., 2:xiii + 1–593.

Kirkpatrick, R. L., and G. L. Valentine. 1970. Reproduction in captive pine voles, *Microtus pinetorum*. J. Mamm., 51:779–785.

Kirsch, J. A. W. 1977. The comparative serology of Marsupialia, and a classification of marsupials. Australian J. Zool., Suppl. Ser., 52:1–152.

Knudsen, G. J., and J. B. Hale. 1968. Food habits of otters in the Great Lakes Region. J. Wildlife Manag., 32:89–93.

Koeppl, J. W., and R. S. Hoffmann. 1981. Comparative postnatal growth of four ground squirrel species. J. Mamm., 62:41–57.

Koford, C. B. 1958. Prairie dogs, whitefaces, and blue grama. Wildlife Monogr., 3:1–78.

Korschgen, L. J. 1959. Food habits of the red fox in Missouri. J. Wildlife Manag., 23:168–176.

Krebs, C. J., B. L. Keller, and R. H. Tamarin. 1969. *Microtus* population biology: demographic changes in fluctuating populations of *M. ochrogaster* and *M. pennsylvanicus* in southern Indiana. Ecology, 50:587–607.

Krutzsch, P. H. 1954. North American jumping mice. Univ. Kansas Publ., Mus. Nat. Hist., 7:349–472.

Küchler, A. W. 1964. The potential natural vegetation of the conterminous United States, 1:3,168,000. Spec. Publ. Amer. Geogr. Soc., v + 1–37 pp.

Kunz, T. H. 1973. Resource utilization: temporal and spatial components of bat activity in central Iowa. J. Mamm., 54:14–32.

———. 1974. Reproduction, growth, and mortality of the vespertilionid bat, *Eptesicus fuscus*, in Kansas. J. Mamm., 55:1–13.

Kurtén, B., and E. Anderson. 1980. Pleistocene mammals of North America. Columbia Univ. Press, New York, xvii + 442 pp.

Kurtén, B., and R. Rausch. 1959. Biometric comparisons between North American and European mammals. Acta Arctica, 11:1–44.

Lampe, R. P. 1976. Aspects of the predatory strategy of the North American badger, *Taxidea taxus*. Unpublished Ph.D. dissertation, Univ. Minnesota, 106 pp.

Lampe, R. P., J. K. Jones, Jr., R. S. Hoffmann, and E. C. Birney. 1974. The mammals of Carter County, southeastern Montana. Occas. Papers Mus. Nat. Hist., Univ. Kansas, 25:1–39.

Latham, R. M. 1952. The fox as a factor in the control of weasel populations. J. Wildlife Manag., 16:516–517.

LaVal, R. K., and M. L. LaVal. 1979. Notes on reproduction, behavior, and abundance of the red bat, *Lasiurus borealis*. J. Mamm., 60:209–212.

Layne, J. N. 1954. The biology of the red squirrel, *Tamiasciurus hudsonicus loquax* (Bangs), in central New York. Ecol. Monogr., 24:227–267.

Lechleitner, R. R. 1958. Movements, density, and mortality in a black-tailed jackrabbit population. J. Wildlife Manag., 22:371–384.

Leraas, H. J. 1938. Observations on growth and behavior of harvest mice. J. Mamm., 19:441–444.

Liers, E. 1951. Notes on the river otter (*Lutra canadensis*). J. Mamm., 32:1–9.

Lindzey, F. G. 1978. Movement patterns of badgers in northwestern Utah. J. Wildlife Manag., 42:418–422.

Long, C. A. 1965. The mammals of Wyoming. Univ. Kansas Publ., Mus. Nat. Hist., 14:493–758.

———. 1973. Taxidea taxus. Mamm. Species, 26:1–4.

———. 1974. Microsorex hoyi and Microsorex thompsoni. Mamm. Species, 33:1–4.

Long, C. A., and R. G. Severson. 1969. Geographical variation in the big brown bat in the north-central United States. J. Mamm., 50:621–624.

Longley, W. H. 1963. Minnesota gray and fox squirrels. Amer. Midland Nat., 69:82–98.

Lord, R. D., Jr. 1961. A population study of

the gray fox. Amer. Midland Nat., 66:87–109.

Lotze, J.-H., and S. Anderson. 1979. Procyon lotor. Mamm. Species, 119:1–8.

Lowery, G. H., Jr. 1974. The mammals of Louisiana and its adjacent waters. Louisiana State Univ. Press, Baton Rouge, xxii + 565 pp.

McBee, K., and R. J. Baker. 1982. Dasypus novemcinctus. Mamm. Species, 162:1–9.

McCarty, R. 1978. Onychomys leucogaster. Mamm. Species, 87:1–6.

McClenaghan, L. R., Jr., and M. S. Gaines. 1978. Reproduction in marginal populations of the hispid cotton rat (*Sigmodon hispidus*) in northeastern Kansas. Occas. Papers Mus. Nat. Hist., Univ. Kansas, 74:1–16.

McManus, J. J. 1974. Didelphis virginiana. Mamm. Species, 40:1–6.

Madison, D. M. 1978. Movement indicators of reproductive events among female meadow voles as revealed by radio-telemetry. J. Mamm., 59:835–843.

Mahan, B. R. 1977. Comparison of coyote and coyote × dog hybrid food habits in southeastern Nebraska. Prairie Nat., 9:50–52.

Mahan, B. R., P. S. Gipson, and R. M. Case. 1978. Characteristics and distribution of coyote × dog hybrids collected in Nebraska. Amer. Midland Nat., 100:408–415.

Mahan, C. J. 1980. Winter food habits of Nebraska bobcats (*Felis rufus*). Prairie Nat., 12:59–63.

Mares, M. A., M. D. Watson, and T. E. Lacher, Jr. 1976. Home range perturbations in *Tamias striatus*: food supply as a determinant of home range and density. Oecologia, 25:1–12.

Martin, R. A., and B. G. Hawks. 1972. Hibernating bats of the Black Hills of South Dakota. I. Distribution and habitat selection. Bull. New Jersey Acad. Sci., 17:24–30.

Mead, R. A. 1968a. Reproduction in western forms of the spotted skunk (genus *Spilogale*). J. Mamm., 49:373–390.

———. 1968b. Reproduction in eastern forms of the spotted skunk (genus *Spilogale*). J. Zool., 156:119–136.

Meagher, M. M. 1973. The bison of Yellowstone National Park. Sci. Monogr. Nat. Park Serv., 1:xv + 1–161.

———. 1978. Bison. Pp. 123–133, *in* Big game of North America (J. L. Schmidt and D. L. Gilbert, eds.), Stackpole Books, Harrisburg, Pennsylvania, xv + 494 pp.

Mech, L. D. 1974. Canis lupus. Mamm. Species, 37:1–6.

Menzel, K., and K. Bouc. 1975. The deer of Nebraska. Nebraska Game and Parks Commission, Lincoln, 30 unnumbered pp.

Merriam, C. H. 1918. Review of the grizzly and big brown bears of North America (genus Ursus) with description of a new genus, Vetularctos. N. Amer. Fauna, 41:1–136.

Merritt, J. F. 1981. Clethrionomys gapperi. Mamm. Species, 146:1–9.

Meserve, P. L. 1971. Population ecology of the prairie vole, Microtus ochrogaster, in the western mixed prairie of Nebraska. Amer. Midland Nat., 86:417–433.

Michener, G. R. 1979. The circannual cycle of Richardson's ground squirrels in southern Alberta. J. Mamm., 60:760–768.

———. 1980. Estrous and gestation periods in Richardson's ground squirrels. J. Mamm., 61:531–534.

Michener, G. R., and D. R. Michener. 1977. Population structure and dispersal in Richardson's ground squirrels. Ecology, 58:359–368.

Miller, D. F., and L. L. Getz. 1969. Botfly infections in a population of *Peromyscus leucopus*. J. Mamm., 50:277–283.

Miller, R. S. 1964. Ecology and distribution of pocket gophers (Geomyidae) in Colorado. Ecology, 45:256–272.

Mitchell, J. L. 1961. Mink movements and populations on a Montana river. J. Wildlife Manag., 25:48–54.

Moore, J. C. 1957. The natural history of the fox squirrel, *Sciurus niger shermani*. Bull. Amer. Mus. Nat. Hist., 113:1–72.

Moore, T. M., L. M. Spence, and C. E. Dugnolle. 1974. Identification of the dorsal guard hairs of some mammals of Wyoming. Bull. Wyoming Game and Fish Dept., 14: x + 1–177.

Murie, J. O. 1973. Population characteristics and phenology of a Franklin ground squirrel (Spermophilus franklinii) colony in Alberta, Canada. Amer. Midland Nat., 90:334–340.

Murie, O. J. 1954. A field guide to animal tracks. Houghton Mifflin Co., Boston, xxii + 374 pp.

Nadler, C. F., R. S. Hoffmann, J. H. Honacki, and D. Pozin. 1977. Chromosomal evolution of chipmunks, with special emphasis on A and B karyotypes of the subgenus Neotamias. Amer. Midland Nat., 98:343–353.

Nagel, H. G., C. True, S. Longfellow, J. P. Farney, D. Lund, and O. A. Kolstad. 1974. Identification and evaluation of present zoologic resources, Nebraska mid-state division Pick-Sloan Missouri Basin Program and associated areas. U.S. Dept. Interior, Bur. Reclamation, viii + 390 pp. (processed literature).

Nellis, C. H., and L. B. Keith. 1968. Hunting activities and success of lynx in Alberta. J. Wildlife Manag. 32:718–722.

Nellis, C. H., S. P. Wetmore, and L. B. Keith. 1972. Lynx-prey interactions in central Alberta. J. Wildlife Manag., 36:320–329.

Nero, R. W., and R. E. Wrigley. 1977. Rare, endangered, and extinct wildlife in Manitoba. Manitoba Nature, 18(2):4–37.

O'Farrell, M. J., and E. H. Studier. 1980. Myotis thysanodes. Mamm. Species, 137:1–5.

O'Gara, B. W. 1978. Antilocapra americana. Mamm. Species, 90:1–7.

Over, W. H., and E. P. Churchill. 1945. Mammals of South Dakota. Univ. South Dakota Mus., (2) + 56 + 3 pp. (processed literature).

Packard, R. L. 1956. The tree squirrels of Kansas. Misc. Publ. Mus. Nat. Hist., Univ. Kansas, 11:1–67.

Pearson, O. P. 1944. Reproduction in the shrew (Blarina brevicauda Say). Amer. J. Anat., 75:39–93.

———. 1945. Longevity of the short-tailed shrew. Amer. Midland Nat., 34:531–546.

———. 1960. Habits of harvest mice revealed by automatic photographic recorders. J. Mamm., 41:58–74.

Pearson, O. P., M. R. Koford, and A. K. Pearson. 1952. Reproduction of the lump-nosed bat (Corynorhinus rafinesquei) in California. J. Mamm., 33:273–320.

Pefaur, J. E., and R. S. Hoffmann. 1971. Merriam's shrew and hispid pocket mouse in Montana. Amer. Midland Nat., 86:247–248.

———. 1974. Notes on the biology of the olive-backed pocket mouse Perognathus fasciatus on the Northern Great Plains. Prairie Nat., 6:7–15.

———. 1975. Studies of small mammal populations at three sites on the Northern Great Plains. Occas. Papers Mus. Nat. Hist., Univ. Kansas, 37:1–27.

Peterson, R. L. (ed.). 1955. North American moose. Univ. Toronto Press, Toronto, 280 pp.

Phillips, G. L. 1966. Ecology of the big brown bat (Chiroptera: Vespertilionidae) in northeastern Kansas. Amer. Midland Nat., 75:168–198.

Pils, C. M., and M. A. Martin. 1978. Population dynamics, predator-prey relationships and management of the red fox in Wisconsin. Tech. Bull., Wisconsin Dept. Nat. Res., 105:1–56.

Pizzimenti, J. J. 1975. Evolution of the prairie dog genus Cynomys. Occas. Papers Mus. Nat. Hist., Univ. Kansas, 39:1–73.

Polder, E. 1968. Spotted skunk and weasel populations, den and cover usage in northeast Iowa. Proc. Iowa Acad. Sci., 75:142–146.

Polderboer, E. B., L. W. Kuhn, and G. O. Hendrickson. 1941. Winter and spring habits of weasels in central Iowa. J. Wildlife Manag., 5:115–119.

Pound, R., and F. E. Clements. 1898. Phytogeography of Nebraska. Jacob North and Co., Lincoln, viii + 329 pp.

Quick, H. F. 1951. Notes on the ecology of weasels in Gunnison County, Colorado. J. Mamm., 32:281–290.

Randall, J. A. 1978. Behavioral mechanisms of habitat segregation between sympatric species of Microtus: habitat preference and interspecific dominance. Behav. Ecol. Sociobiol., 3:187–202.

Rausch, R. A., and A. M. Pearson. 1972. Notes on the wolverine in Alaska and the Yukon Territory. J. Wildlife Manag., 36:249–268.

Reid, V. H. 1973. Population biology of the northern pocket gopher. Bull. Agric. Exper. Sta., Colorado State Univ., 554-S:21–41.

Reith, C. C. 1980. Shifts in times of activity by *Lasionycteris noctivagans*. J. Mamm., 61:104–108.

Repenning, C. A. 1967. Subfamilies and genera of the Soricidae. U.S. Geol. Surv. Prof. Paper, 565:iv + 1–74.

Richardson, A. H. 1971. The Rocky Mountain goat in the Black Hills. Bull. South Dakota Dept. Game, Fish, and Parks, 2:1–15.

Ride, W. D. L. 1964. A review of Australian fossil marsupials. J. Proc. Royal Soc. Western Australia, 47:97–131.

Rideout, C. B. 1978. Mountain goat. Pp. 149–159, *in* Big game of North America (J. L. Schmidt and D. L. Gilbert, eds.), Stackpole Books, Harrisburg, Pennsylvania, xv + 494 pp.

Rideout, C. B., and R. S. Hoffmann. 1975. Oreamnos americanus. Mamm. Species, 63:1–6.

Robinette, W. L., J. S. Gashwiler, and O. W. Morris. 1959. Food habits of the cougar in Utah and Nevada. J. Wildlife Manag., 23:261–273.

———. 1961. Notes on cougar productivity and life history. J. Mamm., 42:204–217.

Robinson, J. W., and R. S. Hoffmann. 1975. Geographical and interspecific cranial variation in big-eared ground squirrels (*Spermophilus*): a multivariate study. Syst. Zool., 24:79–88.

Rue, L. L., III. 1969. The world of the red fox. J. B. Lippincott Co., Philadelphia, 202 pp.

Rusch, D. A., and W. G. Reeder. 1978. Population ecology of Alberta red squirrels. Ecology, 59:400–420.

Sargeant, A. B. 1972. Red fox spatial characteristics in relation to waterfowl predation. J. Wildlife Manag., 36:225–236.

Sargeant, A. B., G. A. Swanson, and H. A. Doty. 1973. Selective predation by mink, *Mustela vison*, on waterfowl. Amer. Midland Nat., 89:208–214.

Schemnitz, S. D. (ed.). 1980. Wildlife management techniques manual. Wildlife Soc., Washington, D. C., 4th ed., viii + 686 pp.

Schmidt-Nielsen, K., and B. Schmidt-Nielsen. 1952. The water metabolism of desert animals. Physiol. Rev., 32:135–166.

Schoonmaker, W. J. 1968. The world of the grizzly bear. J. B. Lippincott Co., Philadelphia, 190 pp.

Schowalter, D. B., J. R. Gunson, and L. D. Harder. 1979. Life history characteristics of little brown bats (*Myotis lucifugus*) in Alberta. Canadian Field-Nat., 93:243–251.

Schwartz, C. W., and E. R. Schwartz. 1981. The wild mammals of Missouri. Univ. Missouri Press and Missouri Dept. Conserv., revised ed., viii + 356 pp.

Scott, T. G. 1955. Dietary patterns of red and gray foxes. Ecology, 36:366–367.

Seabloom, R. W., R. D. Crawford, and M. G. McKenna. 1978. Vertebrates of southwestern North Dakota: amphibians, reptiles, birds, mammals. Res. Rept. Inst. Ecol. Studies, Univ. North Dakota, 24:xiv + 1–549 (processed literature).

Seidensticker, J. C., M. G. Hornocker, W. V. Wiles, and J. P. Messick. 1973. Mountain lion social organization in the Idaho Primitive Area. Wildlife Monogr., 35:1–60.

Seton, E. T. 1929. Lives of game animals. . . . Doubleday, Doran and Co., Garden City, New York, 4 vols.

Setzer, H. W. 1949. Subspeciation in the kangaroo rat, Dipodomys ordii. Univ. Kansas Publ., Mus. Nat. Hist., 1:473–573.

Severinghaus, W. D. 1977. Description of a new subspecies of prairie vole, *Microtus ochrogaster*. Proc. Biol. Soc. Washington, 90:49–54.

Severson, K. E., and A. V. Carter. 1978. Movements and habitat use by mule deer in the Northern Great Plains, South Dakota. Pp. 466–468, *in* Proceedings of the First International Rangeland Congress (D. N. Hyder, ed.), Soc. Range Manag., Denver, Colorado, 742 pp.

Sheldon, W. G. 1949. Reproductive behavior of foxes in New York State. J. Mamm., 30:236–246.

Sheppard, D. H. 1972. Home ranges of chipmunks (*Eutamias*) in Alberta. J. Mamm., 53:379–380.

———. 1979. Richardson's ground squirrel. Canadian Wildlife Serv., Hinterland Who's Who, 4 pp.

Skinner, M. R., and O. C. Kaisen. 1947. The fossil *Bison* of Alaska and preliminary revision of the genus. Bull. Amer. Mus. Nat. Hist., 89:123–256.

Skoog, R. O. 1968. Ecology of the caribou (*Rangifer tarandus granti*) in Alaska. Unpublished Ph.D. dissertation, Univ. California, Berkeley, 699 pp.

Skryja, D. D. 1974. Reproductive biology of the least chipmunk (*Eutamias minimus operarius*) in southeastern Wyoming. J. Mamm., 55:221–224.

Smith, C. C. 1968. The adaptive nature of social organization in the genus of tree squirrels *Tamiasciurus*. Ecol. Monogr., 38:31–63.

———. 1970. The coevolution of pine squirrels (*Tamiasciurus*) and conifers. Ecol. Monogr., 40:349–371.

Smith, R. E. 1967. Natural history of the prairie dog in Kansas. Misc. Publ. Mus. Nat. Hist., Univ. Kansas, 49:1–39.

Soil Survey Staff. 1960. Soil classification, a comprehensive system, seventh approximation. Soil Conserv. Serv., U.S. Dept. Agric., v + 265 pp.

Soper, J. D. 1961. The mammals of Manitoba. Canadian Field-Nat., 75:171–219.

Sorenson, M. W. 1962. Some aspects of water shrew behavior. Amer. Midland Nat., 68:445–462.

Sowls, L. K. 1948. The Franklin ground squirrel, *Citellus franklinii* (Sabine), and its relation to nesting ducks. J. Mamm., 29:113–137.

Stanley, W. C. 1963. Habits of the red fox in northeastern Kansas. Misc. Publ. Mus. Nat. Hist., Univ. Kansas, 34:1–31.

Storm, G. L. 1972. Daytime retreats and movements of skunks on farmlands in Illinois. J. Wildlife Manag., 36:31–45.

Storm, G. L., R. D. Andrews, R. L. Phillips, R. H. Bishop, D. B. Siniff, and J. R. Tester. 1976. Morphology, reproduction, dispersal, and mortality of midwestern red fox populations. Wildlife Monogr., 49:1–82.

Streubel, D. P., and J. P. Fitzgerald. 1978a. Spermophilus spilosoma. Mamm. Species, 101:1–4.

———. 1978b. Spermophilus tridecemlineatus. Mamm. Species, 103:1–5.

Stromberg, M. R., D. E. Biggins, and M. Bidwell. 1981. New record of the least weasel in Wyoming. Prairie Nat., 13:45–46.

Suetsugu, H. 1977. The pronghorn antelope. Nebraska Game and Parks Comm., Lincoln, 14 unnumbered pp.

Svendsen, G. 1964. Comparative reproduction and development in two species of mice in the genus *Peromyscus*. Trans. Kansas Acad. Sci., 67:527–538.

Svendsen, G. E., and R. H. Yahner. 1979. Habitat preference and utilization by the eastern chipmunk (*Tamias striatus*). Kirtlandia, 31:1–14.

Swenk, M. H. 1908. A preliminary review of the mammals of Nebraska, with synopses. Proc. Nebraska Acad. Sci., 8:61–144.

———. 1941. A study of subspecific variation in the Richardson pocket-gopher (Thomomys talpoides) in Nebraska, with descriptions of two new subspecies. Missouri Valley Fauna, 4:1–8.

Swenson, J. E., and G. F. Shanks, Jr. 1979. Noteworthy records of bats from northeastern Montana. J. Mamm., 60:650–652.

Taylor, W. P. (ed.). 1956. The deer of North America. Stackpole Books, Harrisburg, Pennsylvania, 668 pp.

Thaeler, C. S., Jr. 1972. Taxonomic status of the pocket gophers, *Thomomys idahoensis* and *Thomomys pygmaeus* (Rodentia, Geomyidae). J. Mamm., 53:417–428.

———. 1980. Chromosome numbers and systematic relations in the genus *Thomomys* (Rodentia: Geomyidae). J. Mamm., 61:414–422.

Thaeler, C. S., Jr., and L. L. Hinesley. 1979. *Thomomys clusius*, a rediscovered species of pocket gopher. J. Mamm., 60:480–488.

Tiemeier, O. W., M. F. Hansen, M. H. Bartel, E. T. Lyon, B. M. El-Rawi, K. J. McMahon, and E. H. Herrick. 1965. The black-tailed jack rabbit in Kansas. Tech. Bull. Kansas Agric. Exper. Sta., 140:1–75.

Tryon, C. A., Jr. 1947. The biology of the pocket gopher (*Thomomys talpoides*) in Montana. Tech. Bull. Agric. Exper. Sta., Montana State Coll., 448:1–30.

Tryon, C. A., and D. P. Snyder. 1973. Biology of the eastern chipmunk, *Tamias striatus:* life tables, age distributions, and trends in population numbers. J. Mamm., 54:145–168.

Turner, B. N., S. L. Iversen, and K. L. Severson. 1976. Postnatal growth and development of captive Franklin's ground squirrels (Spermophilus franklinii). Amer. Midland Nat., 95:93–102.

Turner, R. W. 1974. Mammals of the Black Hills of South Dakota and Wyoming. Misc. Publ. Mus. Nat. Hist., Univ. Kansas, 60:1–178.

Turner, R. W., and J. B. Bowles. 1967. Comments on reproduction and food habits of the olive-backed pocket mouse in western North Dakota. Trans. Kansas Acad. Sci., 70:266–267.

Tuttle, M. D., and L. R. Heaney. 1974. Maternity habits of *Myotis leibii* in South Dakota. Bull. So. California Acad. Sci., 73:80–83.

Uhlig, H. G. 1955. The gray squirrel, its life history, ecology and population characteristics in West Virginia. West Virginia Conserv. Comm., P.-R. Proj. Rep. 31-R, 175 pp.

Van Ballenberghe, V., A. W. Erickson, and D. Byman. 1975. Ecology of the timber wolf in northeastern Minnesota. Wildlife Monogr., 43:1–44.

van der Meulen, A. J. 1978. *Microtus* and *Pitymys* (Arvicolidae) from Cumberland Cave, Maryland, with a comparison of some New and Old World species. Ann. Carnegie Mus. Nat. Hist., 47:101–145.

Van Gelder, R. G. 1959. A taxonomic revision of the spotted skunks (genus *Spilogale*). Bull. Amer. Mus. Nat. Hist., 117:229–392.

Van Wormer, J. 1963. The world of the bobcat. J. B. Lippincott Co., Philadelphia, 128 pp.

———. 1966. The world of the black bear. J. B. Lippincott Co., Philadelphia, 163 pp.

van Zyll de Jong, C. G. 1972. A systematic review of the Nearctic and Neotropical river otters (genus *Lutra,* Mustelidae, Carnivora). Life Sci. Contrib., Royal Ontario Mus., 80:1–104.

———. 1976a. A comparison between woodland and tundra forms of the common shrew (*Sorex cinereus*). Canadian J. Zool., 54:963–973.

———. 1976b. Are there two species of pygmy shrews (*Microsorex*)? Canadian Field-Nat., 90:485–487.

———. 1979. Distribution and systematic relationships of long-eared *Myotis* in western Canada. Canadian J. Zool., 57:987–994.

———. 1980. Systematic relationships of woodland and prairie forms of the common shrew, *Sorex cinereus cinereus* Kerr and *S. c. haydeni* Baird, in the Canadian prairie provinces. J. Mamm., 61:66–75.

Varland, K. L., A. L. Lovaas, and R. B. Dahlgren. 1978. Herd organization and movements of elk in Wind Cave National Park, South Dakota. U.S. Dept. Interior, National Park Serv., Nat. Res. Rept., 13:1–28.

Vaughan, T. A. 1961. Vertebrates inhabiting pocket gopher burrows in Colorado. J. Mamm., 42:171–174.

———. 1962. Reproduction in the plains pocket gopher in Colorado. J. Mamm., 43:1–13.

———. 1967. Food habits of the northern pocket gopher on shortgrass prairie. Amer. Midland Nat., 77:176–189.

Verts, B. J. 1967. The biology of the striped skunk. Univ. Illinois Press, Urbana, xii + 218 pp.

Vorhies, C. T., and W. P. Taylor. 1933. The life histories and ecology of the jack rabbits, *Lepus alleni* and *Lepus californicus* ssp. in relation to grazing in Arizona. Tech. Bull., Univ. Arizona Exper. Sta., 49:471–587.

Walker, E. P., et al. 1964. Mammals of the World. Johns Hopkins Press, Baltimore, 3 vols.

Wallmo, O. C. 1978. Mule and black-tailed deer. Pp. 31–41, *in* Big game of North America (J. L. Schmidt and D. L. Gilbert, eds.), Stackpole Books, Harrisburg, Pennsylvania, xv + 494 pp.

Watkins, L. C. 1972. Nycticeius humeralis. Mamm. Species, 23:1–4.

Weaver, J. E. 1954. North American prairie. Johnsen Publ. Co., Lincoln, Nebraska, xi + 348 pp.

Weaver, J. E., and F. W. Albertson. 1956. The grasslands of the Great Plains. Johnsen Publ. Co., Lincoln, Nebraska, 395 pp.

Weckwerth, R. P., and V. D. Hawley. 1962. Marten food habits and population fluctuations in Montana. J. Wildlife Manag., 26:55–74.

Weigl, P. D. 1978. Resource overlap, interspecific interactions and the distribution of the flying squirrels, Glaucomys volans and G. sabrinus. Amer. Midland Nat., 100:83–96.

Welles, R. E., and F. B. Welles. 1961. The bighorn of Death Valley. Fauna of the National Parks of the United States, 6:xv + 1–242.

Wetzel, R. M. 1955. Speciation and dispersal of the southern bog lemming, Synaptomys cooperi (Baird). J. Mamm., 36:1–20.

Wetzel, R. M., and E. Mondolfi. 1979. The subgenera and species of long-nosed armadillos, genus Dasypus L. Pp. 43–63, in Vertebrate ecology in the northern Neotropics (J. F. Eisenberg, ed.). Smithsonian Inst. Press.

Whitaker, J. O., Jr. 1972. Zapus hudsonius. Mamm. Species, 11:1–7.

———. 1974. Cryptotis parva. Mamm. Species, 43:1–8.

Wiehe, J. M., and J. F. Cassel. 1978. Checklist of North Dakota mammals (revised). Prairie Nat., 10:81–88.

Wiley, R. W. 1980. Neotoma floridana. Mamm. Species, 139:1–7.

Wilhelm, R. B., J. R. Choate, and J. K. Jones, Jr. 1981. Mammals of Lacreek National Wildlife Refuge, South Dakota. Spec. Publ. Mus., Texas Tech Univ., 17:1–39.

Williams, D. F. 1978. Systematics and eco-geographic variation of the Apache pocket mouse (Rodentia: Heteromyidae). Bull. Carnegie Mus. Nat. Hist., 10:1–57.

Williams, D. F., and H. H. Genoways. 1979. A systematic review of the olive-backed pocket mouse, Perognathus fasciatus (Rodentia, Heteromyidae). Ann. Carnegie Mus., 48:73–102.

Willner, G. R., G. A. Feldhamer, E. E. Zucker, and J. A. Chapman. 1980. Ondatra zibethicus. Mamm. Species, 141:1–8.

Wishart, W. 1978. Bighorn sheep. Pp. 161–171, in Big game of North America (J. L. Schmidt and D. L. Gilbert, eds.), Stackpole Books, Harrisburg, Pennsylvania, xv + 494 pp.

Wood, J. E. 1958. Age structure and productivity of a gray fox population. J. Mamm., 39:74–86.

Woods, C. A. 1973. Erethizon dorsatum. Mamm. Species, 29:1–6.

Wright, P. L., and M. W. Coulter. 1967. Reproduction and growth in Maine fishers. J. Wildlife Manag., 31:70–87.

Wright, P. L., and R. Rausch. 1955. Reproduction in the wolverine, Gulo gulo. J. Mamm., 36:346–355.

Wrigley, R. E. 1971. Manitoba pocket gophers. Zoolog, 12(2):4–7.

Wrigley, R. E., J. E. Dubois, and H. W. R. Copland. 1979. Habitat, abundance, and distribution of six species of shrews in Manitoba. J. Mamm., 60:505–520.

Yahner, R. H. 1978a. Burrow system and home range use by eastern chipmunks, Tamias striatus: ecological and behavioral considerations. J. Mamm., 59:324–329.

———. 1978b. The sequential organization of behavior in Tamias striatus. Behav. Biol., 24:229–243.

———. 1978c. The adaptive nature of the social system and behavior in the eastern chipmunk, Tamias striatus. Behav. Ecol. Sociobiol., 3:397–427.

Yates, T. L., and D. J. Schmidly. 1977. Systematics of Scalopus aquaticus (Linnaeus) in Texas and adjacent states. Occas. Papers Mus., Texas Tech Univ., 45:1–36.

———. 1978. Scalopus aquaticus. Mamm. Species, 105:1–4.

Yoakum, J. D. 1978. Pronghorn. Pp. 103–121, in Big game of North America (J. L. Schmidt and D. L. Gilbert, eds.), Stackpole Books, Harrisburg, Pennsylvania, xv + 494 pp.

Young, S. P. 1958. The bobcat of North America, its history, life habits, economic status and control, with a list of currently recognized subspecies. Wildlife Manag. Inst., Washington, D.C., xi + 193 pp.

Young, S. P., and E. A. Goldman. 1944. The wolves of North America. Amer. Wildlife

Inst., Washington, D.C., xx + 636 pp.
———. 1946. The puma, mysterious American cat. Amer. Wildlife Inst., Washington, D.C., xiv + 358 pp.
Youngman, P. M. 1975. Mammals of the Yukon Territory. Nat. Mus. Nat. Sci. (Nat. Mus. Canada), Publ. Zool., 10:1–192.

Zegers, D. A., and O. Williams. 1977. Seasonal cycles of body weight and lipids in Richardson's ground squirrel, *Spermophilus richardsoni elegans.* Acta Theriol., 22:380–383.
Zimmerman, E. G. 1965. A comparison of habitat and food of two species of *Microtus.* J. Mamm., 46:605–612.

Index to Scientific and Vernacular Names